TABLE III   Student's *t*-distribution (Values of t_α)

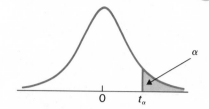

df	$t_{0.10}$	$t_{0.05}$	$t_{0.025}$	$t_{0.01}$	$t_{0.005}$	df
1	3.078	6.314	12.706	31.821	63.657	1
2	1.886	2.920	4.303	6.965	9.925	2
3	1.638	2.353	3.182	4.541	5.841	3
4	1.533	2.132	2.776	3.747	4.604	4
5	1.476	2.015	2.571	3.365	4.032	5
6	1.440	1.943	2.447	3.143	3.707	6
7	1.415	1.895	2.365	2.998	3.499	7
8	1.397	1.860	2.306	2.896	3.355	8
9	1.383	1.833	2.262	2.821	3.250	9
10	1.372	1.812	2.228	2.764	3.169	10
11	1.363	1.796	2.201	2.718	3.106	11
12	1.356	1.782	2.179	2.681	3.055	12
13	1.350	1.771	2.160	2.650	3.012	13
14	1.345	1.761	2.145	2.624	2.977	14
15	1.341	1.753	2.131	2.602	2.947	15
16	1.337	1.746	2.120	2.583	2.921	16
17	1.333	1.740	2.110	2.567	2.898	17
18	1.330	1.734	2.101	2.552	2.878	18
19	1.328	1.729	2.093	2.539	2.861	19
20	1.325	1.725	2.086	2.528	2.845	20
21	1.323	1.721	2.080	2.518	2.831	21
22	1.321	1.717	2.074	2.508	2.819	22
23	1.319	1.714	2.069	2.500	2.807	23
24	1.318	1.711	2.064	2.492	2.797	24
25	1.316	1.708	2.060	2.485	2.787	25
26	1.315	1.706	2.056	2.479	2.779	26
27	1.314	1.703	2.052	2.473	2.771	27
28	1.313	1.701	2.048	2.467	2.763	28
29	1.311	1.699	2.045	2.462	2.756	29
∞	1.282	1.645	1.960	2.326	2.576	∞

Elementary Statistics

Elementary Statistics

NEIL A. WEISS

Arizona State University

ADDISON-WESLEY PUBLISHING COMPANY

Reading, Massachusetts ♦ Menlo Park, California ♦ Don Mills, Ontario
Wokingham, England ♦ Amsterdam ♦ Bonn ♦ Sydney
Singapore ♦ Tokyo ♦ Madrid ♦ San Juan

On the cover: Blue lobsters do exist, although they are extremely rare. According to the State Lobster Hatchery of Massachusetts, approximately one in every 20,000,000 lobsters hatched is blue. How many lobsters must be hatched before we can be at least 90% certain that at least one is blue? Out of 100,000,000 lobsters hatched, what is the probability that between 10 and 20, inclusive, are blue?

Sponsoring Editor: **Charles B. Glaser**
Production Supervisor: **Marion E. Howe**
Technical Art Consultant: **Loretta Bailey**
Illustrator: **Scientific Illustrators**
Cover Design: **Marshall Henrichs**
Manufacturing Supervisor: **Roy Logan**

Library of Congress Cataloging-in-Publication Data

Weiss, N. A. (Neil A.)
    Elementary statistics / by Neil A. Weiss.
        p.   cm.
    Includes index.
    ISBN 0–201–18491–5 :
    1. Statistics.   I. Title.
    QA276.12.W445 1989
    519.5—dc19   88-22600

To My Father
and the
Memory of My Mother

Preface

Today, more than ever, statistics is a part of our everyday lives. It has become an essential tool used by government, business, industry, and virtually every academic discipline. The purpose of this book is to provide the reader with a clear understanding of basic statistical techniques and to present well-organized procedures for applying those techniques. The book is designed to be used for either a *one-semester* or *one-quarter* course in beginning statistics. Introductory high school algebra is a sufficient prerequisite.

Features

We have included the following features in our text to assist the reader in learning elementary statistics:

Emphasis on application. We have concentrated on the application of statistical techniques to the analysis of data. Although the statistical theory has been kept to a minimum, we have made every effort to give a thorough development of the rationale for using each statistical procedure.

Detailed and careful explanations. We have attempted to include every step of explanation that a typical reader might need. Our guiding principle is to avoid "cognitive jumps" and thereby make the learning process smooth and enjoyable. We feel that detailed and careful explanations will result in a better understanding.

Data sets. In most examples and exercises we have provided the raw data sets instead of or in addition to the summary statistics. There are two advantages to this. First, it tends to give the reader a more concrete and real picture of statistics.

Supplying only the summary statistics places the reader in a position of being once removed from the statistical analysis. Second, having the raw data sets allows the reader to solve the problems using a computer, if that is desired.

Chapter introductions and chapter outlines. At the beginning of each chapter we have presented a brief introduction of what is contained in the chapter and how the chapter relates to the text as a whole. As a further aid, a chapter outline follows the chapter introduction. The outline lists the sections in the chapter along with a short description of the content of each section.

Interesting and real examples. It is our feeling that the vast majority of students learn by example. Therefore, every concept discussed in the book is illustrated by at least one example. The examples are, for the most part, based on real-life situations and have been chosen for their interest as well as for their illustrative value.

Definitions, formulas, key facts. As an aid to learning and for reference, we have set off all definitions, formulas, and key facts. These items are printed in color to make them easy to locate.

Procedures. To help the reader with the application of statistical procedures, we have presented easy-to-follow, step-by-step methods for carrying out these procedures. Each procedure is displayed between two vertical color rules. A unique feature of this book is that when a procedure is illustrated by an example, each step in the procedure is presented (in italic) within the example. This serves a two-fold purpose. It shows the reader how the procedure is applied and also helps the reader master the steps in the procedure.

Procedure index. With the vast number of statistical procedures, it is sometimes difficult to find a particular procedure, especially when the book is being used for reference purposes. Thus we have included a *procedure index,* which is located on the back inside cover of the book. This provides a quick and easy way to find the appropriate procedure for performing a specific statistical analysis.

Extensive and diverse exercise sets. Since most students master statistics by practice, we have presented both extensive and diverse exercise sets. The majority of exercises are based on real applications found in newspapers, magazines, statistical abstracts, and so forth. The exercises are not only used to help the reader master the material, but they are also designed to show that statistics is a lively discipline.

Because the students in an elementary statistics course often have different mathematical backgrounds, we have included a wide variety of exercises and have labeled them by difficulty. Every exercise set contains a number of basic exercises. The *basic exercises* are routine applications of material presented in the text. We have organized the basic exercises so that each concept is covered by at least two problems. For each odd-numbered basic exercise that involves a given concept, there is also an even-numbered basic exercise that involves that same concept. The answers to the odd-numbered basic exercises are given in the appendix and the answers to the even-numbered basic exercises can be found in the *Instructor's Manual.*

In addition to the basic exercises, most of the exercise sets contain intermediate and advanced problems. The *intermediate exercises* are underscored and contain supplementary material that is not necessarily covered in the text, but that may be of interest to some of the more highly motivated students. The *advanced exercises* are double underscored and cover abstract concepts, theory, and algebraic derivations. These exercises are intended for the students with special mathematical background and aptitude. The solutions to all the intermediate and advanced exercises can be found in the *Instructor's Manual.*

Computer packages. The computer plays a fundamental role in the application of statistics. It is therefore important for every student of statistics to have some familiarity with statistical computer packages such as Minitab, SPSS, SAS, or BMDP. In most chapters, we have included an optional section on computer packages. We have chosen Minitab to illustrate the basic ideas of computer packages. In each computer section, we discuss briefly how Minitab can be used to solve problems that were solved by hand earlier in the chapter. Each solution consists of explaining the appropriate command, displaying the computer output, and interpreting the results. Additionally, we have prepared a separate supplement that provides a detailed explanation of the computer package Minitab. This is cross-referenced to the book.

Chapter reviews. Frequently students in elementary statistics courses feel a certain amount of anxiety and confusion about how they should study and review. To help the reader, we have written a chapter-review section at the end of every chapter. The chapter reviews include (1) a list of key terms with page references, (2) formulas, (3) chapter objectives, and (4) a review test. These pedagogical aids provide the reader with an organized method for reviewing and studying the material discussed in each chapter. The answers to the review tests are given in the appendix.

Organization

We have attempted to write a text that offers a great deal of flexibility in the choice of material to be covered. Chapters 1 through 3 introduce the nature of statistics and the fundamentals of descriptive statistics. In Chapters 4 through 6, we cover probability, discrete random variables, and the normal distribution. Chapter 7 discusses sampling and the sampling distribution of the mean. Following that, in Chapters 8 and 9, we examine confidence intervals and hypothesis tests for one population mean or one population proportion. We consider Chapters 1 through 9 the core of an elementary statistics course.

Chapter 10 presents inferences concerning two population means or two population proportions. It also contains an optional section on one-way analysis of variance. In Chapter 11, we discuss chi-square procedures (goodness-of-fit test, independence test, and inferences for a population standard deviation).

We have divided the traditional material on regression and correlation into two chapters. Chapter 12 discusses *descriptive* methods in regression and correlation. This chapter has been written so that it can be covered at any time after Chapter 3. Chapter 13 presents *inferential* methods in regression and correlation and can be covered once Chapters 9 and 12 have been completed.

The following flowchart summarizes the preceding discussion. The prerequisite for a given chapter consists of all chapters that have a path leading to the given chapter.

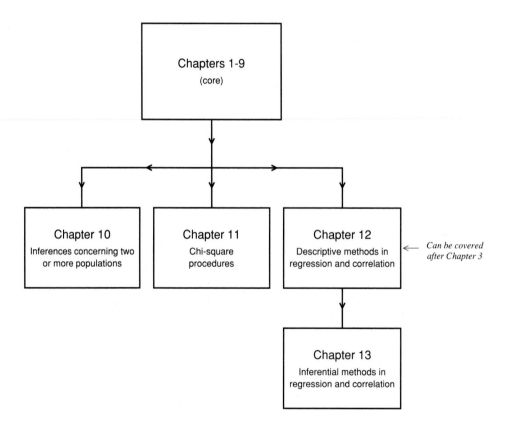

Supplements

The following supplements have been prepared to accompany the book *Elementary Statistics.*

Minitab. This supplement teaches the reader how to use the statistical computer package Minitab. It is designed specifically to be used in conjunction with *Elementary Statistics* and is keyed to the book. No prerequisite knowledge of computers or statistical computer packages is presumed.

Instructor's Manual. This manual contains the solutions to all exercises whose answers are not given in the appendix of *Elementary Statistics*. Additionally, several course syllabi are included.

AWTest. This is a computerized testing system that generates from algorithms a limitless number of quizzes, tests, midterms, final exams, etc. It is available free to adopters for use on an IBM PC or compatibles.

Printed Test Bank. This supplement provides four test forms for each chapter.

We should also point out that Addison-Wesley Publishing Company and Minitab, Inc., have recently published *The Student Edition of MINITAB*. This makes it possible for students with access to an IBM PC or compatible to purchase their own Minitab software. The package includes three floppy discs, a quick-reference card, and *The Student Edition of MINITAB* manual.

Acknowledgments

It is our pleasure to thank the following reviewers whose comments and suggestions were invaluable in writing *Elementary Statistics:*

Jasper Adams
Stephen F. Austin State University

Nancy Carter
California State University at Chico

Timothy Green
University of Georgia

Shu-ping Hodgson
Central Michigan University

James Lang
Valencia Community College

LeRoy Sathre
Valencia Community College

We also thank Professor Michael Driscoll, Professor Dennis Young, Dr. Michael Aicken, and Dr. Terry Woodfield for their assistance. Our special thanks go to Professor Ronald Jacobowitz for his many suggestions and helpful comments. We are grateful to Carol Weiss for reading and critiquing the manuscript, for participating in the production aspects of the book, for working all the exercises, and for providing support and encouragement. Finally, we would like to thank Mr. Tom Taylor, Ms. Marion Howe, Ms. Stephanie Botvin, and everyone else at Addison-Wesley Publishing Company who participated in the development, production, and publication of the book.

Tempe, Arizona

N.A.W.

Contents

* Denotes an optional section.

What does the word "statistics" bring to mind? Most people immediately think of numerical facts or data, such as unemployment figures, farm prices, or the number of marriages and divorces. *Webster's New World Dictionary* gives two definitions of the word "statistics":

> 1. facts or data of a numerical kind, assembled, classified, and tabulated so as to present significant information about a given subject. 2. [construed as sing.], the science of assembling, classifying, and tabulating such facts or data.

Technically, however, statistics means more than this. Not only do statisticians assemble, classify, and tabulate data, but they also analyze data in order to make generalizations and decisions. For example, a political analyst can use data from a portion of the voting population to predict the political preferences of the entire voting population. In this chapter we introduce some basic terminology so that the various meanings of the word "statistics" will become clear.

The nature of statistics

CHAPTER   OUTLINE

1.1 Two kinds of statistics   Discusses the two major types of statistics—descriptive statistics and inferential statistics.

1.2 Classification of statistical studies   Presents several examples that illustrate how to classify a statistical study as either descriptive or inferential.

1.3 The development of inferential statistics   Gives a brief sketch of the history and development of inferential statistics.

1.1   Two kinds of statistics

You probably feel that you already know something about statistics. If you read newspapers, watch the news on television, or follow sports, then you see and hear the word "statistics" frequently. In this section we will use familiar examples such as baseball statistics and voter polls to introduce the two major types of statistics—**descriptive statistics** and **inferential statistics.**

Each spring in the late 1940s, the major-league baseball season was officially opened when President Harry S. Truman threw out the "first ball" of the season at the opening game of the Washington Senators. Both President Truman and the Washington Senators had reason to be interested in statistics. Consider, for instance, the year 1948.

EXAMPLE   1.1   **Illustrates descriptive statistics**

In 1948 the Washington Senators played 153 games, winning 56 and losing 97. They finished seventh in the American League and were led in hitting by B. Stewart, whose batting average was .279. These and many other statistics were compiled by baseball statisticians, who took the complete records for each game of the season and organized that great mass of information effectively and efficiently.

Although baseball fans take these statistics for granted, a great deal of time and effort is required to gather and organize them. Moreover, without such statistics baseball would be much harder to understand. For instance, picture yourself trying to select the best hitter in the American League with only the official score sheets for each game. [More than 600 games were played in 1948; the best hitter was Ted Williams, who led the league with a batting average of .369.]   ■

The work of baseball statisticians provides us with a good illustration of descriptive statistics. Below we give a formal definition of descriptive statistics.

DEFINITION 1.1 Descriptive statistics

Descriptive statistics consists of methods for organizing and summarizing information in a clear and effective way.

Among other things, descriptive statistics includes the construction of graphs, charts, and tables, and the calculation of various descriptive measures, such as averages and percentiles. We will discuss descriptive statistics in detail in Chapters 2 and 3. Descriptive statistics is one of the two major types of statistics. The other major type, inferential statistics, is illustrated in the next example.

EXAMPLE   1.2   Illustrates inferential statistics

In the fall of 1948, President Truman was also concerned about statistics. The Gallup Poll just before the election predicted that he would win only 44.5% of the vote and be defeated by the Republican nominee, Thomas E. Dewey. This time, however, the statisticians had predicted incorrectly. Truman won more than 49% of the vote, and with it the presidency. The Gallup Organization modified some of its procedures and has correctly predicted the winner since.   ∎

Political polling provides us with an example of inferential statistics. It would be tremendously expensive to interview all Americans on their voting preferences. Statisticians who wish to gauge the sentiment of the entire **population** of American voters can afford to interview only a carefully chosen group of a few thousand voters. This group is referred to as a **sample** of the population. Statisticians analyze the information obtained from the sample to make inferences (draw conclusions) about the preferences of the entire voting population. Inferential statistics provides methods for making such inferences.

The terminology introduced above in the context of political polling is used in general in statistics. Specifically, we have the following definitions:

DEFINITION 1.2 Population and sample

Population: The collection of all individuals or items under consideration in a statistical study.
Sample: That part of the population from which information is collected.

Figure 1.1 provides a graphical display of the relationship between a population and a sample from the population.

FIGURE 1.1
Population and sample.

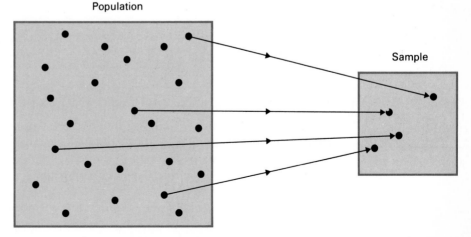

With Definition 1.2 in mind, we now present the definition of inferential statistics.

DEFINITION 1.3 Inferential statistics

Inferential statistics consists of methods of drawing conclusions about a population based on information obtained from a sample of the population.

Exercises 1.1

1.1   Define the following terms:
  a) population
  b) sample

1.2   What are the two major types of statistics? Describe them in detail.

1.2   Classification of statistical studies

In later chapters we will conduct a thorough examination of both descriptive and inferential statistics. At this point, however, you need only classify statistical studies as either descriptive or inferential. The examples below are intended to give you some practice. In each example, we have presented the result of a statistical study and have classified the study as descriptive or inferential. You should attempt to classify each study yourself before reading our explanation.

EXAMPLE   1.3   Illustrates the classification of statistical studies

The study   Table 1.1 gives the voting results for the 1948 presidential election.

TABLE 1.1
Final results of the
1948 presidential election.

Ticket	Votes	Percent
Truman–Barkley (Democratic)	24,179,345	49.7%
Dewey–Warren (Republican)	21,991,291	45.2%
Thurmond–Wright (States Rights)	1,176,125	2.4%
Wallace–Taylor (Progressive)	1,157,326	2.4%
Thomas–Smith (Socialist)	139,572	0.3%

Classification   This study is descriptive. It is a summary of the votes cast by the entire population of voters.   ∎

EXAMPLE   1.4   Illustrates the classification of statistical studies

The study   For the 101 years preceding 1977, the baseballs used by the major leagues were purchased from the Spalding Company. In 1977 that company stopped manufacturing major-league baseballs, and the major leagues arranged to buy their baseballs from the Rawlings Company.

Early in the 1977 season, pitchers began to complain that the Rawlings ball was "livelier" than previous balls. They claimed that it was harder, bounced farther and faster, and gave an unfair advantage to hitters. There was some evidence for this. For instance, in the first 616 games of 1977, there were 1033 home runs, compared to only 762 home runs in the first 616 games of the previous year.

Sports Illustrated magazine sponsored a careful study of the liveliness question, and the results appeared in the June 13, 1977 issue. In this study, an independent testing company randomly selected 85 baseballs from the current (1977) supplies of various major league teams. The bounce, weight, and hardness of the baseballs chosen were carefully measured. These measurements were then compared with measurements obtained previously from similar tests on baseballs used in 1952, 1953, 1961, 1963, 1970, and 1973. The conclusion, given on page 24 of the *Sports Illustrated* article, was as follows: ". . . the 1977 Rawlings ball is livelier than the 1976 Spalding, but not as lively as it could be under big league rules, or as the ball has been in the past."

Classification This is an inferential study. The independent testing company used a sample of 85 baseballs from the 1977 supplies of major league teams to make an inference about the population of all such baseballs. [It has been estimated that approximately 360,000 baseballs were used by the major leagues in 1977.]   ■

The *Sports Illustrated* study also provides an excellent illustration of a situation in which it is not feasible to obtain data for the entire population. Indeed, after the bounce and hardness tests, all of the baseballs sampled were taken to a butcher in Plainfield, New Jersey to be sliced in half so that researchers could look inside them. Clearly, it would not be practical to test every baseball this way.

EXAMPLE   1.5   Illustrates the classification of statistical studies

The study In the late 1940s and early 1950s there was great public concern over epidemics of polio. In an attempt to alleviate this serious problem, Jonas Salk of the University of Pittsburg developed a vaccine for polio. Various preliminary experiments indicated that the vaccine was safe and potentially effective. Nonetheless, it was deemed necessary to conduct a large-scale study to determine whether the vaccine would truly work.

A test was devised involving nearly two million grade-school children. All of the children were inoculated, but only half received the Salk vaccine. The other half were given a placebo, a harmless solution. An evaluation center kept records of who received the Salk vaccine and who did not. The center found that the incidence of polio was far less among the children inoculated with the Salk vaccine. From that information it was concluded that the Salk vaccine would be effective in preventing polio for all American schoolchildren. The Salk vaccine was then made available for general use.

Classification This is an inferential study. The group of children inoculated may have been quite large, but it was still just a sample; and the information obtained from that sample was used to make an inference about the effectiveness of the Salk vaccine in preventing polio among the population of *all* American schoolchildren.   ■

Finally, we should emphasize that it is possible to perform a descriptive study on a sample, as well as on a population. It is only when inferences are made about the population from information obtained from the sample that the study becomes inferential.

Exercises 1.2

In Exercises 1.3 through 1.10, classify each of the studies as either descriptive or inferential.

1.3 The A.C. Nielsen Company collects and publishes information on the television-viewing habits of Americans. Data collected in 1983 from a sample of Americans yielded the following estimates of average TV viewing time per week for all Americans. The times are given in hours and minutes.

Group (by age)		Time	Group (by age)		Time
Average all persons		30:47			
Women	Total 18+	34:36	Teens	Female	24·16
	18–24	27:26		Male	25:17
	55+	41:13	Children	2–5	27:09
Men	Total 18+	30:13		6–11	24:50
	18–24	24:52			
	55+	35:53			

1.4 In 1936, the voters of North Carolina cast their presidential votes as follows:

Candidate	Number of votes
Roosevelt, Democratic	616,414
Landon, Republican	223,283
Thomas, Socialist	21
Browder, Communist	11
Lemke, Union	2

1.5 The U.S. National Center for Health Statistics made the following estimates of the leading causes of death in the United States for 1983. The estimates are based on a 10% sampling of all death certificates for the 11 months from January through November.

Cause	Number
Major cardiovascular diseases	887,460
Malignant neoplasms (cancers)	403,190
Accidents	82,560
Chronic obstructive pulmonary diseases	60,390
Influenza and pneumonia	48,880

1.6 The U.S. National Institute on Drug Abuse compiled the following table on drug use by young adults, by type of drug. The percentages given are estimates based on information obtained from national samples.

Type of drug	Percent of young adults			
	Ever used		Current user	
	1974	1982	1974	1982
Marijuana	52.7	64.1	25.2	27.4
Inhalants	9.2	(NA)	(z)	(NA)
Hallucinogens	16.6	21.1	2.5	1.7
Cocaine	12.7	28.3	3.1	6.8
Heroin	4.5	1.2	(z)	(z)
Analgesics	(NA)	12.1	(NA)	4.7
Stimulants[1]	17.0	18.0	3.7	2.6
Sedatives[1]	15.0	18.7	1.6	1.6
Tranquilizers[1]	10.0	15.1	1.2	1.0
Alcohol	81.6	94.6	69.3	67.9
Cigarettes	68.8	76.9	48.8	39.5

NA = Not available.   z = Less than .5 percent.
[1] = Prescription drugs.

1.7 The table below displays the 1983 attendance figures for selected spectator sports. The data are given in thousands, rounded to the nearest thousand.

Sport	Attendance (thousands)
Baseball, major league	46,269
Basketball	
college	31,471
pro	10,262
Football	
college	36,302
pro	15,958
Horse racing	75,693
Greyhound racing	22,140

1.8 Newspapers publish weather data for cities all over the world. Below are temperature and precipitation readings for some selected cities in Canada and Mexico on September 9, 1985.

Canada			
	H	L	P
Calgary	43	40	.01
Edmonton	49	41	
Montreal	76	57	
Ottawa	—	—	—
Regina	51	45	
Toronto	81	69	.01
Vancouver	63	53	.01
Winnipeg	58	37	

Mexico			
	H	L	P
Acapulco	89	73	
Guadalajara	78	60	.50
Mazatlan	89	78	
Mexico City	76	49	.02

1.9 The New York Stock Exchange keeps records of the selling prices for seats on the Exchange. Below are the high and low prices for some years in this century.

Year	High price	Low price
1900	$47,500	$37,500
1920	115,000	85,000
1929	625,000	550,000
1935	140,000	65,000
1940	60,000	33,000
1960	162,000	135,000
1970	320,000	130,000
1982	340,000	190,000

1.10 A 1984 study conducted by the Gallup Organization concluded that an estimated 83% of American households are involved in at least one form of indoor or outdoor gardening. [This result shows that gardening is the number one American leisure activity.]

1.11 For each of the situations below, decide whether the indicated study is descriptive or inferential. Give a reason for each of your answers.
a) A tire manufacturer estimates the average life of a new type of steel-belted radial.
b) A sports writer lists the winning times for all swimming events in the 1984 Olympics.
c) A politician obtains the exact number of votes that were cast for her opponent in 1986.
d) A medical researcher tests an anticancer drug that may have harmful side effects.
e) A candidate for governor estimates the percentage of voters that will vote for him in the upcoming election.
f) An economist estimates the average income of all California residents.
g) The owner of a small business determines the average salary of his 20 employees.

1.12 The chairperson of the mathematics department at a large state university wants to estimate the average final exam score for the 2476 students in basic algebra. She randomly selects 50 exams from the 2476 and finds that the average score on the 50 exams chosen is 78.3%. From this she estimates that the average score for all 2476 students is 78.3%.
a) What kind of study has the chairperson done?
b) What kind of study would she have done had she averaged all 2476 exam scores?

1.3 The development of inferential statistics

As you know, the science of statistics includes both descriptive and inferential statistics. Descriptive statistics appeared first—censuses were taken as long ago as Roman times. Over the years, records of such things as births, deaths, and taxes have led naturally to the development of descriptive statistics.

Inferential statistics is a newer arrival. Major developments began to occur with the research of Karl Pearson and Ronald Fisher, who published their results in the early years of this century. Since the work of Pearson and Fisher, inferential statistics has evolved rapidly and is now applied in a wide range of subject areas. In fact, an understanding of its fundamental concepts has become mandatory for virtually every professional.

Familiarity with statistics will also help you make more sense out of many things you read in newspapers and magazines. For instance, in the description of the *Sports Illustrated* baseball test in Example 1.4, it may have struck you as unreasonable that a sample of only 85 baseballs could be used to draw a conclusion about a population of roughly 360,000 baseballs. By the time you have completed Chapter 9 of this book, you will understand why such inferences are not unreasonable.

The primary objective of this text is to present the fundamentals of inferential statistics. However, since almost any inferential study involves aspects of descriptive statistics, we will first consider the basic principles and methods of descriptive statistics.

◆ Chapter Review

KEY TERMS

descriptive statistics, 2   population, 3
inferential statistics, 4   sample, 3

YOU SHOULD BE ABLE TO . . .

1 classify statistical studies as either descriptive or inferential.

2 identify the population and the sample in an inferential study.

REVIEW TEST

1 Give an example of
  a) a descriptive study.
  b) an inferential study.

2 At the end of Section 1.3, we stated that almost any inferential study involves aspects of descriptive statistics. Why do you think that is true?

In Problems 3 through 6 classify each study as either descriptive or inferential.

3 A metropolitan newspaper displayed the following college football scores on the front page of the Sunday paper on September 15, 1985.

COLLEGE FOOTBALL			
Michigan St.	12	Arizona	12
ASU	3	Washington St.	7
BYU	31	Rutgers	28
Washington	3	Florida	28
Michigan	20	UCLA	26
Notre Dame	12	Tennessee	26

4 The U.S. Bureau of Justice Statistics (BJS) conducts monthly surveys of about 60,000 households in the United States. These monthly surveys are then used to obtain annual estimates on criminal victimization. For example, the BJS reported the following victimization rate estimates for crimes against households between 1976 and 1982. Rates are per 1000 households.

Year	Burglary	Larceny	Motor vehicle theft
1976	89	124	16
1977	89	123	17
1978	86	120	18
1979	84	134	18
1980	84	127	17
1981	88	121	17
1982	78	114	16

5 On February 14, 1985, the U.S. Bureau of the Census released a survey indicating that "fewer Americans have health insurance coverage than previously thought." The results of the survey, based on a sample of 20,000 households, showed that 85% of the population is covered by health insurance—a far cry from the 97.3% figure found in a 1978 survey by the Department of Health and Human Services.

6 A newspaper reporter who was doing library research for an article on civil aviation obtained the following data on fatalities. (Source: U.S. Federal Aviation Administration.)

Year	Fatalities	Year	Fatalities
1975	1477	1981	1414
1978	1921	1982	1480
1979	1731	1983	1135
1980	1392		

7 A National Institute of Mental Health survey concluded that "about 20% of adult Americans suffer from at least one psychiatric disorder." This and other national estimates were obtained from results of interviews with thousands of Americans in St. Louis, Baltimore, and New Haven, Connecticut. Is this study descriptive or inferential? Explain your answer.

As we discovered in Chapter 1, *descriptive statistics* consists of methods for organizing and summarizing information in a clear and effective way. We will now begin our study of descriptive statistics. We will learn how to classify data by type, organize data into tables, and summarize data with graphical displays. The methods discussed here by no means exhaust all of those available, but they do represent some of the most commonly used techniques.

Organizing data

CHAPTER OUTLINE

2.1 Data

The information collected, organized, and analyzed by statisticians is called **data.** There are several different types of data, and the statisticians' choice of methodology is partly determined by the type of data being considered. In this section we will discuss some of the more important types of data.

EXAMPLE 2.1   Illustrates three major types of data

At noon on April 15, 1985, more than 5500 men and women set out to run from Hopkinton Center to the Prudential Plaza in Boston. Their run, covering 26 miles and 385 yards, would be watched by thousands of people lining Boston streets and by millions more on television news reports. It was the eighty-ninth running of the Boston Marathon.

A great deal of information was accumulated that afternoon. The men's competition was won by Geoff Smith of Great Britain with a time of two hours, fourteen minutes, and five seconds. The winner of the women's competition was Lisa Larson-Weidenbach of Marblehead, Massachusetts; her time was two hours, thirty-four minutes, and six seconds. There were 3468 men and 460 women who finished before the official cutoff time of four hours. [After four hours no official times are recorded, although some runners may still be on the course.]

The Boston Marathon provides us with examples of three major types of data. The simplest type is illustrated by the information that classifies each entrant as either male or female. Such data, which give qualitative information about an individual or item, are called **qualitative data.** For instance, the information that Lisa Larson-Weidenbach is a female is qualitative data.

Most racing fans are interested in the places of the finishers. Information on place is an example of **ordinal data**—data about order or rank. The information that Geoff Smith and Gary Tuttle finished first and second among the men while Steve Allen and Rick Ironside finished 500th and 501st is ordinal data.

Ordinal data give information about place but do not measure the difference in performance between places. For instance, Geoff Smith finished five minutes and six seconds ahead of Gary Tuttle, but Steve Allen beat Rick Ironside by only three seconds. More can be learned about what happened in a race by looking at the times of the finishers. Differences between times indicate exactly how far apart two runners finished, whereas differences in places do not. Information on time is an example of **metric data**—data obtained from measurement. The information that Geoff Smith ran his race in 2:14:05 is metric data.[†]   ∎

Below we summarize the definitions of the three types of data discussed in Example 2.1.

DEFINITION 2.1 Qualitative, ordinal, and metric data

Qualitative data: Data that give non-numerical information such as gender, eye color, and blood type.
Ordinal data: Data about order or rank on a scale such as 1, 2, 3, . . . or A, B, C, ...
Metric data: Data obtained from the measurement of quantities such as time, height, and weight.

Another important type of data is **frequency data** (also called **count data**). Counting the number of individuals or items that fall into categories such as "male" and "female" yields frequency data. For instance, the information that 3468 men and 460 women finished the Boston Marathon in under four hours is frequency data.

DEFINITION 2.2 Frequency data

Frequency data: Data on the number of individuals or items falling in various categories.

The next three examples provide additional illustrations of qualitative, ordinal, metric, and frequency data.

EXAMPLE   2.2   Illustrates types of data

Humans are classified as having one of the four blood types A, B, AB, and O.
 a) What kind of data do you receive when you are told your blood type?
 b) Geneticists and anthropologists record the number of individuals of each blood type. What kind of data are they collecting?

S O L U T I O N
 a) Your blood type is qualitative data. It places you in one of four non-numerical categories—A, B, AB, or O.
 b) Recording the number of individuals in each of the four blood-type categories gives frequency data.   ∎

[†]The term *metric* refers to *measurement* and not to the metric system. Metric data can be given in units other than those in the metric system (e.g., inches).

EXAMPLE   2.3   **Illustrates types of data**

At most colleges and universities, students completing a course receive a grade of A, B, C, D, or F.

a) What type of data is the information that Carol Scott received a grade of A in Professor H's statistics class?

b) What type of data is provided by the information that the final grades in Professor H's statistics class were 17 As, 16 Bs, and 8 Cs, with no Ds or Fs?

SOLUTION

a) The information that Carol Scott received an A is ordinal data, since the grades A, B, C, D, and F rank students' performance.

b) The information on the number of As, Bs, Cs, Ds, and Fs gives frequency data, obtained by counting the number of students falling into each of the five grade categories. [By the way, Professor H is real, and we actually did this count using his final grades.]   ∎

Note that Professor H's grade distribution, given in Example 2.3, is somewhat higher than the grades usually observed in statistics classes. This illustrates that not all ranking schemes are the same. Professor H's grade of A may have been easier to get than Professor W's. Thus, we see that ordinal data may have different meanings when the ranking is done by different people.

EXAMPLE   2.4   **Illustrates types of data**

The *Information Please Almanac* lists the world's highest waterfalls.

a) The list shows that Angel Falls in Venezuela is 3281 feet high, more than twice as high as Ribbon Falls at Yosemite, California, which is 1612 feet high. What kind of data are these heights?

b) From the list, we also find that of the world's 40 highest waterfalls, four are more than 1700 feet high, five are between 1000 and 1700 feet high, and 31 are less than 1000 feet high. What type of data do these counts provide?

SOLUTION

a) The waterfall heights are metric data, determined by taking precise measurements.

b) The counts of the number of waterfalls in the three height categories (more than 1700 ft, 1000–1700 ft, less than 1000 ft) are frequency data.   ∎

Discrete and continuous data

Qualitative and ordinal data are referred to as **discrete,** because they sort individuals or items into separate, or discrete, classes. For instance, track and field authorities classify entrants for competition as either male or female (qualitative data) and assign the first two finishers in each category places 1 and 2 (ordinal data). There is no way to be classified between male and female or to be assigned place 1.357.

On the other hand, most metric data are called **continuous,** since they involve measurement on a continuous scale. For example, the time that it takes a runner to complete a marathon can conceptually be *any* positive number.

Classification and the choice of a statistical method

Statisticians make other distinctions in classifying data, but the types just presented are sufficient for most applications. Data classification is sometimes difficult—statisticians themselves will often disagree over data type. For example, some consider data involving amounts of money to be metric data, while others consider it to be frequency data. In most cases, however, the classification of data is fairly clear and serves as an aid in the choice of the correct statistical method.

Exercises **2.1**

2.1 Give a reason why the classification of data is important.

Classify the data given in Exercises 2.2 through 2.10 as qualitative data, ordinal data, metric data, or frequency data.

2.2 According to Sidney S. Culbert of the University of Wisconsin, the principal languages of the world in 1984 were as follows:

Rank	Language	Speakers (millions)
1	Mandarin (China)	740
2	English	403
3	Russian	277
4	Spanish	266
5	Hindustani	264
6	Arabic	160
7	Bengali	155

a) What type of data is given in the first column of the table?
b) What type of data is provided by the information that Neil Armstrong speaks English?
c) What type of data is provided by the information in the last column of the table?

2.3 Tobacco production in the United States for the years 1978–1983 is given in the table at the top of the next column. (Source: U.S. Department of Agriculture.)

What type of data is given in the second column of this table?

Year	Pounds (millions)
1978	2025
1979	1527
1980	1786
1981	2064
1982	1994
1983	1428

2.4 On May 4, 1961, Commander Malcolm Ross, USNR, ascended 113,739.9 feet in a free balloon. What kind of data is the height given here?

2.5 What type of data do the numbers in the table below provide? (Reprinted by permission from the October issue of *Science 85.* Copyright © 1985 by the American Association for the Advancement of Science.)

Experts and lay people ranked the risk of dying in any year from various activities and technologies. The experts' ranking closely matches known fatality statistics.

Public		Experts
1	Nuclear power	20
2	Motor vehicles	1
3	Handguns	4
4	Smoking	2
5	Motorcycles	6
6	Alcoholic beverages	3
7	General (private) aviation	12
8	Police work	17
9	Pesticides	8

(cont.)

10	Surgery	5
11	Fire fighting	18
12	Large construction	13
13	Hunting	23
14	Spray cans	26
15	Mountain climbing	29
16	Bicycles	15
17	Commercial aviation	16
18	Electric power (nonnuclear)	9
19	Swimming	10
20	Contraceptives	11
21	Skiing	30
22	X rays	7
23	High school and college football	27
24	Railroads	19
25	Food preservatives	14
26	Food coloring	21
27	Power mowers	28
28	Prescription antibiotics	24
29	Home appliances	22
30	Vaccinations	25

2.6 Below are some figures on U.S. industrial employment for 1982. (Source: U.S. Bureau of Labor Statistics.)

Industry	Employees (thousands)
Agriculture	2,815
Mining	742
Construction	5,491
Manufacturing	19,234
Transportation	3,103
Communications	1,420
Trade	22,536
Services	22,617

What kind of data is given by the employee numbers?

2.7 The following continental statistics were compiled by the National Geographic Society, Washington, D.C.

Continent	Area (sq. mi.)	Population (est.)
Asia	16,999,000	2,757,383,000
Africa	11,688,000	513,000,000
North America	9,366,000	390,000,000
South America	6,881,000	259,000,000
Europe	4,017,000	692,879,000
Australia	2,966,000	15,300,000
Antarctica	5,100,000	

a) What type of data is provided by the area figures?

b) What type of data is contained in the statement that "Africa is second largest in area and third largest in population"?

c) What type of data is given by the population figures?

d) What type of data do we obtain from the fact that Madame Curié was born in Europe?

2.8 The following table gives the rank by population and the population of the 10 largest metropolitan areas in the U.S. in 1970 and 1980. (Source: U.S. Bureau of the Census.)

Metropolitan area	1980		1970	
	Rank	Population	Rank	Population
New York, NY–NJ	1	9,120,346	1	9,973,716
Los Angeles–Long Beach, CA	2	7,477,503	2	7,041,980
Chicago, IL	3	7,103,624	3	6,974,755
Philadelphia, PA–NJ	4	4,716,818	4	4,824,110
Detroit, MI	5	4,353,413	5	4,435,051
San Francisco–Oakland, CA	6	3,250,630	6	3,109,249
Washington, DC–MD–VA	7	3,069,922	7	2,910,111
Dallas–Fort Worth, TX	8	2,974,805	12	2,377,623
Houston, TX	9	2,905,353	16	1,999,316
Boston, MA	10	2,763,357	8	2,899,101

a) What type of data is provided by the statement that "in 1980 Houston, Texas was the ninth largest metropolitan area in the United States"?

b) What type of data is given in the last column of the table?

c) What type of data is the information that Theodore Roosevelt was born in New York?

2.9 The five largest U.S. commercial banks (by deposits as of Dec. 31, 1983) are listed below.

Bank	Rank	Deposits
Bank of America NT&SA, San Francisco	1	$90,250,018,000
Citibank NA, New York	2	78,392,000,000
Chase Manhattan Bank NA, New York	3	59,824,598,000
Manufacturers Hanover Trust Co., New York	4	42,186,468,000
Morgan Guaranty Trust Co., New York	5	39,419,026,000

a) What kind of data is given in the second column of the table?

b) What kind of data is given in the third column of the table?

2.10   What kinds of data would be collected in each of the following situations?

a) A quality-control engineer measures the lifetimes of electric light bulbs.

b) A businessperson wants to know the number of families with preteen children in Pueblo, CO.

c) A sporting goods manufacturer is going to classify each major league baseball player as right-handed or lefthanded and count the number in each category.

d) A sociologist needs to estimate the average annual income of residents of Ossining, New York.

e) A pollster plans to classify each individual in a sample of voters as Democratic or Republican and count the total number in each group.

f) An administrator at a community college needs to know how many men and women participated in varsity sports during the spring semester and how much money was spent on men's sports and on women's sports.

2.11   Of the data types qualitative, metric, and frequency, only one involves non-numerical data. Which one is that?

2.12   Of the data types qualitative, ordinal, metric, and frequency, only one involves continuous data. Which one is that?

2.13   Below are several items pertaining to individuals.

a) height
b) weight
c) age
d) sex
e) number of siblings
f) religion
g) place of birth
h) high school class rank

Which involve continuous data?

2.2   Grouping data

The data collected in a real-world situation can sometimes be overwhelming. For example, a list of American colleges and universities with information on enrollment, number of teachers, highest degree offered, and governing official can be found in *The World Almanac.* These data occupy 28 pages of small type!

By suitably organizing data, it is often possible to make a rather complicated set of data easier to understand. In this section we will discuss **grouping,** one of the most common methods for organizing data.

EXAMPLE   2.5   Illustrates grouping data

Table 2.1 gives the number of days to maturity for 40 short-term investments. The data are from *Barron's National Business and Financial Weekly.*

TABLE 2.1

Days to maturity for 40 short-term investments.

70	64	99	55	64	89	87	65
62	38	67	70	60	69	78	39
75	56	71	51	99	68	95	86
57	53	47	50	55	81	80	98
51	36	63	66	85	79	83	70

It is somewhat difficult to get a clear picture of the data in Table 2.1. By grouping the data into categories, or **classes,** we can make it much simpler to comprehend.

The first step is to decide on the classes. One convenient way to group this data is by tens. Since the smallest piece of data is 36 and the largest is 99, grouping by tens results in the classes 30–39, 40–49, and so on up to 90–99. These classes are given in the first column of Table 2.2.

TABLE 2.2

Days to maturity	Tally	Number of investments
30–39	|||	3
40–49			1
50–59	⊬⊬ |||	8
60–69	⊬⊬ ⊬⊬	10
70–79	⊬⊬ ||	7
80–89	⊬⊬ ||	7
90–99	||||	4
		40

The second (and final) step in grouping the data is to determine how many investments are in each class. We do this by going through the data in Table 2.1 and making a tally mark in the appropriate line of Table 2.2 for each investment. For instance, the first investment has a 70-day maturity period. This calls for a tally mark on the line for the class 70–79. The results of the tallying procedure are shown in the second column of Table 2.2. Now we count the tallies for each class and record the totals in the third column of Table 2.2.

By simply glancing at Table 2.2 we can obtain various pieces of useful information. For instance, we see that there are more investments in the 60–69 days-to-maturity range than in any other. Comparing Tables 2.1 and 2.2, we see that grouping the data makes it much easier to read and understand.   ∎

In Example 2.5 we used a common-sense approach to grouping data into classes. Some of that common sense can be written down as guidelines for grouping. Three of the most important guidelines are the following:

1 *There should be between five and twenty classes.*
    In Example 2.5, seven classes are used. Generally, the number of classes should be small enough to provide an effective summary, but large enough to display the relevant characteristics of the data.

2 *Each piece of data must belong to one, and only one, class.*
    Careless planning in Example 2.5 could have led to classes like 30–40, 40–50, 50–60, and so on. Then, for instance, to which class would the investment with a 50-day maturity period belong? The classes in Table 2.2 do not cause such confusion—they cover all maturity periods and do not overlap.

3 *Whenever feasible, all classes should have the same width.*
    The classes in Table 2.2 all have a width of 10 days. Among other things, choosing classes of equal width facilitates the graphical display of the data.

The list of guidelines could go on, but that would be artificial. The purpose of grouping is to organize the data into a reasonable number of classes in order to make it more accessible and understandable.

Frequency and relative-frequency distributions

The number of pieces of data that fall into a particular class is called the **frequency** of that class. For example, as we see from Table 2.2, the frequency of the class 50–59 is eight, since there are eight investments in the 50–59 days-to-maturity range. A table listing all classes and their frequencies is called a **frequency distribution.** The first and third columns of Table 2.2 constitute a frequency distribution for the days-to-maturity data.

In addition to the frequency of a class, we are often interested in the **percentage** of a class. We find the percentage by dividing the frequency of the class by the total number of pieces of data. Referring again to Table 2.2, we see that the percentage of investments in the class 50–59 is

$$\begin{array}{c}\text{Frequency of}\\\text{class }50\text{--}59 \longrightarrow\end{array} \frac{8}{40} = 0.20 \text{ or } 20\%$$
$$\begin{array}{c}\text{Total number of}\\\text{pieces of data}\end{array} \nearrow$$

In other words, 20% of the investments have a maturity period of between 50 and 59 days, inclusive.

The percentage of a class, expressed as a decimal, is called the **relative frequency** of the class. For the class 50–59, the relative frequency is 0.20. A table listing all classes and their relative frequencies is called a **relative-frequency distribution.** Table 2.3 displays a relative-frequency distribution for the days-to-maturity data.

TABLE 2.3
Relative-frequency
distribution for the
days-to-maturity data.

Days to maturity	Relative frequency	
30–39	0.075	⟵ 3/40
40–49	0.025	⟵ 1/40
50–59	0.200	⟵ 8/40
60–69	0.250	⟵ 10/40
70–79	0.175	⟵ 7/40
80–89	0.175	⟵ 7/40
90–99	0.100	⟵ 4/40
	1.000	

Note that the relative frequencies must always add up to 1 (100%). Why?

Terminology

Although the basic ideas of grouping use common sense, there is quite a bit of associated terminology. We have already discussed several of the terms used in grouping. To introduce some additional ones, let us return to the days-to-maturity data.

Consider, for example, the class 50–59. The smallest maturity period that can go in the class is 50. This is called the **lower class limit** of the class. The largest maturity period that can go in the class is 59. This is called the **upper class limit** of the class.

The midpoint of the class 50–59 is $(50 + 59)/2 = 54.5$, and this is called the **class mark.** Finally, the width of the class, obtained by subtracting its lower class limit from the lower class limit of the next higher class, is $60 - 50 = 10$. This is called the **class width** of the class. Definition 2.3 summarizes the terminology of grouping.

DEFINITION 2.3 Terms used in grouping

Classes: Categories for grouping data.
Frequency: The number of data values in a class.
Relative frequency: The ratio of the frequency of a class to the total number of pieces of data.
Frequency distribution: A listing of classes and their frequencies.
Relative-frequency distribution: A listing of classes and their relative frequencies.
Lower class limit: The smallest value that can go in a class.
Upper class limit: The largest value that can go in a class.
Class mark: The midpoint of a class.
Class width: The difference between the lower class limit of the given class and the lower class limit of the next higher class.

A table giving the classes, frequencies, relative-frequencies, and class marks for a data set is called a **grouped-data table.** A grouped-data table for the days-to-maturity data is presented in Table 2.4.

TABLE 2.4
Grouped-data table for days-to-maturity data.

Days to maturity	Frequency	Relative frequency	Class mark
30–39	3	0.075	34.5
40–49	1	0.025	44.5
50–59	8	0.200	54.5
60–69	10	0.250	64.5
70–79	7	0.175	74.5
80–89	7	0.175	84.5
90–99	4	0.100	94.5
	40	1.000	

In the next example, we will find the grouped-data table for another data set.

EXAMPLE 2.6 **Illustrates grouped-data tables**

A pediatrician began testing cholesterol levels of young patients and was alarmed to find that a large number had levels over 200 milligrams per 100 milli-liters. The readings of 20 patients with high levels are presented in Table 2.5.

TABLE 2.5
Cholesterol levels for 20 high-level patients.

210	209	212	208
217	207	210	203
208	210	210	199
215	221	213	218
202	218	200	214

Construct a grouped-data table. Use a class width of five and start at 195.

S O L U T I O N
Since we are to use a class width of five and start at 195, the first class will be 195–199. From Table 2.5 we see that the highest cholesterol level is 221. Thus, we choose the classes displayed in the first column of Table 2.6.

TABLE 2.6

Cholesterol level	Tally	Frequency
195–199			1
200–204	|||	3
205–209	||||	4
210–214	⊞ ||	7
215–219	||||	4
220–224			1
		20

The results of tallying the data in Table 2.5 are shown in the second and third columns of Table 2.6. From Table 2.6 we obtain the grouped-data table given in Table 2.7.

TABLE 2.7

Grouped-data table for cholesterol-level data.

Cholesterol level	Frequency	Relative frequency	Class mark
195–199	1	0.05	197
200–204	3	0.15	202
205–209	4	0.20	207
210–214	7	0.35	212
215–219	4	0.20	217
220–224	1	0.05	222
	20	1.00	

To illustrate some typical computations for the third and fourth columns, consider the class 210–214. We have

$$\text{Relative frequency} = \frac{7}{20} = 0.35$$

and

$$\text{Class mark} = \frac{210 + 214}{2} = 212$$

■

Single-value grouping

Up to this point, each class we have used for grouping data represents several possible numerical values. For instance, as we can see from Table 2.7, each class in Example 2.6 represents five possible cholesterol levels. The first class, 195–199, is for levels of 195, 196, 197, 198, or 199; the second class, 200–204, is for levels of 200, 201, 202, 203, or 204; and so on.

In some cases, however, it is more appropriate to use classes that each represent a single possible numerical value. For instance, it is often preferable to group *discrete data* using such classes. Here is an example.

EXAMPLE   2.7   Illustrates single-value grouping

A planner is collecting data on the number of school-age children per family in a small town. Thirty families are selected at random. Table 2.8 displays the number of school-age children in each of the 30 families chosen.

TABLE 2.8

Number of school-age children in each of 30 families.

0	3	0	0	3	0
2	2	0	1	2	1
0	0	1	2	4	0
4	2	1	0	1	0
0	2	0	1	3	2

a) Group these data using classes that each represent a single numerical value.
b) Identify the class limits and the class marks.
c) Construct a grouped-data table.

S O L U T I O N

a) We first note that since each class is to represent a single numerical value, the classes must be 0, 1, 2, 3, and 4. These are displayed in the first column of Table 2.9.

TABLE 2.9

Frequency and relative-frequency distributions for the number of school-age children.

Number of school-age children	Frequency	Relative frequency
0	12	0.400
1	6	0.200
2	7	0.233
3	3	0.100
4	2	0.067
	30	1.000

Applying the tallying procedure to the data in Table 2.8, we obtain the frequencies in the second column of Table 2.9. Dividing each frequency by the total number of pieces of data, 30, gives the relative frequencies shown in the third column of Table 2.9. The table indicates, for example, that seven of the 30 families, or 23.3%, have two school-age children.

b) For this part we are to identify the class limits and class marks. Consider, for instance, the class "3" (i.e., three school-age children). We have

Lower class limit = 3 (the smallest value that can go in the class)
Upper class limit = 3 (the largest value that can go in the class)

and

$$\text{Class mark} = \frac{3 + 3}{2} = 3 \text{ (the midpoint of the class)}$$

Thus, for the class "3," the lower class limit, the upper class limit, and the class mark are all equal to 3. A similar statement is true for the remaining classes.

c) Finally, we are to construct a grouped-data table. This requires us to append a class-mark column to Table 2.9. However, from part (b) we know that the class mark for each class is the same as the class itself. Thus, it is unnecessary to add a class-mark column to Table 2.9, since such a column would be identical to the first column. In other words, Table 2.9 can serve as a grouped-data table. ∎

We now summarize the important points made about single-value grouping in Example 2.7. When a class for grouping data represents a single numerical value, then that value is the lower class limit, the upper class limit, and the class mark for the class. If every class for grouping a data set is based on a single value, then the class-mark column may be omitted from the grouped-data table because the first column gives the class marks as well as the classes.

Another way of grouping data

It is often convenient to group data using classes that consist of values from one number up to, but not including, another number. This is particularly true when dealing with *continuous data* or *decimal data*. Consider Example 2.8.

EXAMPLE   2.8   **Illustrates another way to group data**

The U.S. National Center for Health Statistics gathers data on weights of American adults by age and sex. The weights in Table 2.10 were obtained from a sample of 18–24-year-old males. Data are given to the nearest tenth of a pound.

TABLE 2.10

Weights of 37 males, aged 18–24 years.

129.2	185.3	218.1	182.5	142.8	155.2	170.0	151.3
187.5	145.6	167.3	161.0	178.7	165.0	172.5	191.1
150.7	187.0	173.7	178.2	161.7	170.1	165.8	
214.6	136.7	278.8	175.6	188.7	132.1	158.5	
146.4	209.1	175.4	182.0	173.6	149.9	158.6	

We will group these weights using the classes "120–under 140," "140–under 160," and so on. The class "120–under 140" is for weights of *at least* 120 lb, but *less than* 140 lb; the class "140–under 160" is for weights of *at least* 140 lb, but *less than* 160 lb; and so forth. These classes are displayed in the first column of Table 2.11.

TABLE 2.11

Grouped-data table for the weights of 37 males, aged 18–24 years.

Weight (lb)	Frequency	Relative frequency	Class mark
120–under 140	3	0.081	130
140–under 160	9	0.243	150
160–under 180	14	0.378	170
180–under 200	7	0.189	190
200–under 220	3	0.081	210
220–under 240	0	0.000	230
240–under 260	0	0.000	250
260–under 280	1	0.027	270
	37	0.999	

Applying the tallying procedure to the data in Table 2.10, we obtain the frequencies in the second column of Table 2.11. The relative frequencies, given in the third column of Table 2.11, are then found in the usual manner. [Note that here the sum of the relative frequencies is given as 0.999. Actually, the sum of the relative frequencies is exactly equal to 1, but since we rounded to three decimal places the sum is off by a little.]

Finally, let us discuss the class marks. For a class of the form *"a–under b,"* we define the class mark to be $(a + b)/2$. For instance, the class mark for the class

"120–under 140" is $(120 + 140)/2 = 130$. Similar calculations give the remaining class marks shown in the fourth column of Table 2.11. ∎

Frequency and relative-frequency distributions for qualitative data

The concepts of class limits and class marks are applicable to metric data and certain kinds of ordinal data. However, they are not appropriate for qualitative data. For instance, with data that categorize runners as male or female, the classes are "male" and "female." For qualitative-data classes such as these, it makes no sense to look for class limits or class marks.

We can, of course, still compute frequencies and relative frequencies for qualitative data. Example 2.9 provides an illustration.

EXAMPLE 2.9 **Illustrates frequency and relative-frequency distributions for qualitative data**

In the fall semester of 1985, Professor W asked his introductory statistics students to state their political party affiliations as Democratic, Republican, or Other. The responses are depicted in Table 2.12. [D = Democratic, R = Republican, O = Other.]

TABLE 2.12
Political party affiliations of the students in Professor W's fall, 1985 introductory statistics class.

D	R	O	R	R	R	R	R
D	O	R	D	O	O	R	D
D	R	O	D	R	R	O	R
D	O	D	D	D	R	O	D
O	R	D	R	R	R	R	D

Determine the frequency and relative-frequency distributions for the data.

SOLUTION
The classes for grouping the data are "Democratic," "Republican," and "Other." Tallying the data in Table 2.12, we obtain the frequency distribution displayed in the first two columns of Table 2.13.

TABLE 2.13
Frequency and relative-frequency distributions for the political party affiliations of students in Professor W's fall, 1985 introductory statistics class.

Party	Frequency	Relative frequency
Democratic	13	0.325
Republican	18	0.450
Other	9	0.225
	40	1.000

Dividing each frequency in the second column by the total number of students (40), we get the relative frequencies in the third column of Table 2.13. The first and third columns of the table constitute the relative-frequency distribution for the data. ∎

Exercises 2.2

2.14   What is one of the main reasons for grouping data?

2.15   Do class limits and class marks make sense for qualitative-data classes? Explain your answer.

2.16   State the three most important guidelines in choosing the classes for grouping a data set.

2.17   Explain the difference between each of the following pairs of terms:
a) frequency and relative frequency
b) percentage and relative frequency

2.18   A research physician conducted a study on the ages of persons with diabetes. The following data were obtained for the ages of a sample of 35 diabetics.

48	41	57	83	41	55	59
61	38	48	79	75	77	7
54	23	47	56	79	68	61
64	45	53	82	68	38	70
10	60	83	76	21	65	47

Construct a grouped-data table for these ages. Use classes of equal width, beginning with the class 0–9.

2.19   A soft-drink bottler sells "one-liter" bottles of soda. A consumer group was concerned that the bottler may have been short-changing the customers. Thirty bottles of soda were randomly selected. The contents, in ml, of the bottles chosen are shown below.

1025	977	1018	975	977
990	959	957	1031	964
986	914	1010	988	1028
989	1001	984	974	1017
1060	1030	991	999	997
996	1014	946	995	987

Construct a grouped-data table for these soft-drink data. Use classes of equal width, starting with the class 910–929.

2.20   The U.S. Bureau of Economic Analysis gathers information on the length of stay in Europe and the Mediterranean of U.S. travelers. A sample of 36 U.S. residents who have traveled to Europe and the Mediterranean this year yielded the following data on length of stay. The data are in days.

41	16	6	21	1	21
5	31	20	27	17	10
3	32	2	48	8	12
21	44	1	56	5	12
3	13	15	10	18	3
1	11	14	12	64	10

Construct a grouped-data table using classes of equal width, beginning with the class 1–7.

2.21   The U.S. Energy Information Administration collects and publishes data on energy production and consumption. Data on last year's energy consumption for a sample of 50 households in the South are given below. Data are in millions of BTU.

130	55	45	64	155	66	60	80	102	62
58	101	75	111	151	139	81	55	66	90
97	77	51	67	125	50	136	55	83	91
54	86	100	78	93	113	111	104	96	113
96	87	129	109	69	94	99	97	83	97

Construct a grouped-data table for these energy consumption figures. Use classes of equal width, starting with the class 40–49.

2.22   The U.S. Bureau of the Census conducts nationwide monthly surveys to obtain data on characteristics of American households. Below are data on the number of persons per household for a sample of 40 households.

2	5	2	1	1	2	3	4
1	4	4	2	1	4	3	3
7	1	2	2	3	4	2	2
6	5	2	5	1	3	2	5
2	1	3	3	2	2	3	3

Construct a grouped-data table for these household sizes using classes based on a single value.

2.23   A car salesperson keeps track of the number of cars she sells per week. The number of cars she sold per week last year is as follows:

1	0	3	3	1	0	2	1	4	0	4	1	2
3	6	4	3	0	2	2	1	1	2	2	2	3
5	1	0	2	5	3	1	3	1	1	1	1	2
2	3	0	4	4	1	0	1	1	3	2	5	2

Construct a grouped-data table for the number of sales per week. Use classes based on a single value.

2.24   An eight-question quiz on fractions was given to a fifth-grade class. The following data provide the number of incorrect answers for each of the 25 students in the class:

```
3   4   2   3   1
1   1   6   1   2
0   5   1   2   1
0   3   4   0   3
2   0   3   1   1
```

Construct an appropriate grouped-data table.

2.25 Cudahey Masonry, Inc. employs 80 bricklayers. Records are kept on the number of days missed by each of the employees. The absentee records for the past year are shown below.

```
2   3   6   2   6   5   5   2   4   7   5   3   6   4   4   4
2   2   4   5   4   0   2   1   6   3   5   3   6   6   4   7
5   2   5   0   5   6   5   2   4   2   6   2   4   3   5   4
2   4   4   3   3   4   0   5   6   3   5   5   2   4   4   2
0   7   5   5   7   6   1   5   3   3   4   7   7   2   5   5
```

Construct an appropriate grouped-data table for these absentee data.

2.26 The Food and Nutrition Board of the National Academy of Sciences states that the recommended daily allowance (RDA) of iron for adult females under the age of 51 is 18 mg. The amounts of iron intake during a 24-hour period for a sample of 45 such females are as follows. Data are in mg.

```
15.0   18.1   14.4   14.6   10.9   18.1   18.2   18.3   15.0
16.0   12.6   16.6   20.7   19.8   11.6   12.8   15.6   11.0
15.3    9.4   19.5   18.3   14.5   16.6   11.5   16.4   12.5
14.6   11.9   12.5   18.6   13.1   12.1   10.7   17.3   12.4
17.0    6.3   16.8   12.5   16.3   14.7   12.7   16.3   11.5
```

Construct a grouped-data table for these iron intakes. Use classes of equal width, beginning with the class 6–under 8.

2.27 The National Education Association does surveys on starting salaries for college graduates. A sample of 35 liberal arts graduates yielded the following starting annual salaries. Data are in thousands of dollars, rounded to the nearest hundred dollars.

```
20.0   16.8   21.3   20.6   21.0   18.7   23.8
18.3   17.7   18.0   19.1   21.1   19.6   19.0
18.7   20.8   20.4   17.1   19.5   19.9   19.2
19.1   17.2   18.3   22.7   20.0   19.2   20.9
19.1   20.8   20.5   21.4   16.3   16.3   20.5
```

Using classes of equal width and starting with the class 16–under 17, construct a grouped-data table for these starting annual salaries.

The following article appeared on September 15, 1985. We will use the article in Exercises 2.28 and 2.29. (Used by permission of the Associated Press.)

DOLLAR LEADERS

NEW YORK (AP) — The following is a list of the most active NYSE stocks based on the dollar volume. The total is based on the median price of the stock traded multiplied by the shares traded.

	Tot. ($1000)	Sales (hds.)	Last
Reyl pfC	$100,9720	80536	127
IBM	$705,516	55011	127½
RchVck	$523,317	10755	48¾
SCM	$358,257	49245	72⅝
Revlon	$271,385	63113	43
WstgE	$261,517	68148	38⅜
CocaCl	$259,958	37607	68⅜
GnFds	$251,132	29545	83¼
Digital	$218,483	20636	105⅜
RckCtr n	$217,851	09611	19⅞
CessAir	$206,341	73366	29½
GMot	$193,443	28500	68¼
PhilMr	$173,890	22258	75⅝
AmExp	$158,911	37949	42
McDnld	$148,393	22273	65
MCA	$143,096	20190	68
GenEl	$142,523	23509	60
Exxon	$140,513	27087	51⅛
AtlRich	$138,506	23181	59
MartM s	$137,368	38832	33⅝
Morgn s	$126,692	26672	46⅝
Burrgh	$126,061	19469	63⅛
BeafCo	$125,221	36695	34⅜
CBS	$120,761	10234	117⅛

2.28 The column headed "Tot." in the article shows the total dollar volumes for the most active New York Stock Exchange (NYSE) stocks. The totals are given in thousands of dollars.

a) Group these totals into a grouped-data table with classes as given in the first column of the following table. Note that the classes use values in millions of dollars. For instance the class "100–under 150" is for dollar volumes of at least $100 million, but less than $150 million.

b) Why is there no class mark for the last class?

Dollar volume ($millions)	Frequency	Relative frequency	Class mark
100–under 150			
150–under 200			
200–under 250			
250–under 300			
300–under 350			
350–under 400			
400 and over			

2.29 The second column of the Associated Press article displays the numbers of shares traded for the most active New York Stock Exchange (NYSE) stocks. The sales are given in hundreds. Group these sales into a grouped-data table using classes of equal width. Begin with the class 0–under 1, where the values are in millions of sales.

2.30 The following table gives the all-time top television programs by number of viewers as of January, 1984. (Source: Nielsen Media Research.)

Program	Date	Network	Households
M*A*S*H Special	2/28/83	CBS	50,150,000
Dallas	11/21/80	CBS	41,470,000
Super Bowl XVII	1/30/83	NBC	40,480,000
Super Bowl XVI	1/24/82	CBS	40,020,000
Super Bowl XVIII	1/22/84	CBS	38,800,000
The Day After	11/20/83	ABC	38,550,000
Roots	1/30/77	ABC	36,380,000
Thorn Birds	3/29/83	ABC	35,990,000
Thorn Birds	3/30/83	ABC	35,900,000
Thorn Birds	3/28/83	ABC	35,400,000
Super Bowl XIV	1/20/80	CBS	35,330,000
Super Bowl XIII	1/21/79	NBC	35,090,000
CBS NFL Championship Game	1/10/82	CBS	34,960,000
Super Bowl XV	1/25/81	NBC	34,540,000
Super Bowl XII	1/15/78	CBS	34,410,000
Winds of War	2/13/83	ABC	34,150,000
Gone With The Wind Pt. 1	11/7/76	NBC	33,960,000
Gone With The Wind Pt. 2	11/8/76	NBC	33,750,000
Winds of War	2/7/83	ABC	33,490,000
Thorn Birds	3/27/83	ABC	32,900,000
Roots	1/28/77	ABC	32,680,000
Winds of War	2/6/83	ABC	32,570,000
Roots	1/27/77	ABC	32,540,000
Winds of War	2/9/83	ABC	32,490,000
Winds of War	2/8/83	ABC	32,240,000
Roots	1/25/77	ABC	31,900,000
World Series Game 7	10/20/82	NBC	31,820,000

Construct a frequency and relative-frequency distribution for the network data. *Hint:* The classes are "ABC," "CBS," and "NBC."

2.31 The National Collegiate Athletic Association (NCAA) wrestling champions for the years 1963–1984 are as follows. (Source: *The World Almanac, 1985.*)

Year	Champion	Year	Champion
1963	Oklahoma	1974	Oklahoma
1964	Oklahoma State	1975	Iowa
1965	Iowa State	1976	Iowa
1966	Oklahoma State	1977	Iowa State
1967	Michigan State	1978	Iowa
1968	Oklahoma State	1979	Iowa
1969	Iowa State	1980	Iowa
1970	Iowa State	1981	Iowa
1971	Oklahoma State	1982	Iowa
1972	Iowa State	1983	Iowa
1973	Iowa State	1984	Iowa

Construct a frequency and relative-frequency distribution for the champions.

2.32 The exam scores for the students in an introductory statistics class are as follows:

88	82	89	70	85
63	100	86	67	39
90	96	76	34	81
64	75	84	89	96

a) Group these exam scores using the classes 30–39, 40–49, 50–59, 60–69, 70–79, 80–89, and 90–100.
b) What are the widths of the classes?
c) If you wanted all of the classes to have equal widths, what classes would you use?

The methods we have examined in this section apply to grouping data that involve *one* characteristic of the members of a population or sample. Such data are called **univariate.** For instance, in Example 2.8 we considered data on the single characteristic "weight" for a sample of 18–24-year-old males. Thus, those data are univariate. We could have considered not only the weights of these males, but also their heights. Then we would have data on two characteristics—"height" and "weight." Data that involve *two* characteristics of the members of a population or sample are called **bivariate.** Bivariate data can be grouped using tables called **contingency tables.** Exercises 2.33 and 2.34 deal with the grouping of bivariate data.

2.33 The bivariate data on age and sex shown at the top of the following page were obtained from the students in a freshman calculus course. The data show, for example, that the first student on the list is 21 years old and is a male.

Age	Sex	Age	Sex	Age	Sex	Age	Sex	Age	Sex
21	M	29	F	22	M	23	F	21	F
20	M	20	M	23	M	44	M	28	F
42	F	18	F	19	F	19	M	21	F
21	M	21	M	21	M	21	F	21	F
19	F	26	M	21	F	19	M	24	F
21	F	24	F	21	F	25	M	24	F
19	F	19	M	20	F	21	M	24	F
19	M	25	M	20	F	19	M	23	M
23	M	19	F	20	F	18	F	20	F
20	F	23	M	22	F	18	F	19	M

We will discuss the grouping of these data into the following contingency table:

Age (yrs)

		Under 21	21–25	Over 25	Total
Sex	Male		│		
	Female				
	Total				

a) To tally the data for the first student, we place a tally mark in the box labelled by the "21–25"-column and the "Male"-row, as indicated. Tally the data for the remaining 49 students.

b) Construct a table similar to the one above, but with the frequencies replacing the tally marks. Also, add the frequencies in each row and column of your table and record the sums in the appropriate "Total" boxes.

c) What do the row and column totals represent?

d) Add the row totals and the column totals. Why are these two sums equal, and what does their common value represent?

e) Construct a table giving the relative frequencies for the data. [Divide each frequency by the grand total of 50 students.]

f) Interpret the entries in your table from part (e) in terms of percentages.

2.34 The heights and weights of the students in Exercise 2.33 are presented in the following table.

Height	Weight	Height	Weight	Height	Weight
68	140	63	130	73	180
67	129	64	105	68	160
64	127	76	200	69	178
72	185	74	215	70	170
61	120	68	160	63	105
62	103	66	115	64	130
66	110	75	175	67	130
69	155	69	170	69	145
70	215	69	135	60	95
68	135	63	105	63	142
67	155	64	130	62	127
72	145	67	120	66	130
68	135	65	115	64	122
73	180	67	140	72	185
65	145	72	275	65	132
64	125	74	170	71	169
75	185	64	130		

Repeat parts (a)–(f) of Exercise 2.33 for the bivariate data on height and weight. Use the following contingency table:

Height (in)

		60–65	66–71	72–77	Total
Weight (lb)	90–129				
	130–169				
	170–209				
	210–249				
	250–289				
	Total				

2.3 Graphs and charts

Besides grouping, another method for organizing and summarizing data is to draw a picture of some kind. The old saying "a picture is worth a thousand words" has particular relevance in statistics—a graph or chart of a data set often provides the simplest and most efficient display. In this section, we will examine various techniques for organizing and summarizing data using graphs and charts. We begin by discussing *histograms*.

EXAMPLE 2.10   Introduces histograms

Table 2.4 gives a grouped-data table for the number of days to maturity for 40 short-term investments. The first three columns of that table are repeated in Table 2.14.

TABLE 2.14
Frequency and relative-frequency distributions for days-to-maturity data.

Days to maturity	Frequency (no. of investments)	Relative frequency
30–39	3	0.075
40–49	1	0.025
50–59	8	0.200
60–69	10	0.250
70–79	7	0.175
80–89	7	0.175
90–99	4	0.100
	40	1.000

One way to display the data pictorially is to construct a graph with the classes depicted on the horizontal axis and the frequencies on the vertical axis. This can be done using a **frequency histogram** as shown in Figure 2.1.

FIGURE 2.1
Frequency histogram for days-to-maturity data.

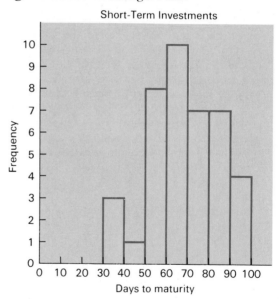

Data from *Barron's National Business and Financial Weekly*

The height of each bar is equal to the frequency of the class it represents. Note that each bar extends from the lower class limit of a class to the lower class limit of the next higher class.[†]

Some other important observations about Figure 2.1 are as follows. Each axis of the frequency histogram has a label, and the frequency histogram as a whole has a title. These are common elements that all such graphical displays should

---

[†] This is only one of several methods for depicting the classes. Additional methods are discussed in Exercises 2.54 and 2.55.

possess. Additionally, the source of the data is given. This should be included whenever possible.

A frequency histogram displays the frequencies of the classes. To display the relative frequencies (or percentages), we can use a **relative-frequency histogram,** which is quite similar to a frequency histogram. In fact, the only difference is that the height of each bar in a relative-frequency histogram is equal to the *relative frequency* of the class instead of the frequency. The relative-frequency histogram for the days-to-maturity data is shown in Figure 2.2.

FIGURE 2.2
Relative-frequency histogram
for days-to-maturity data.

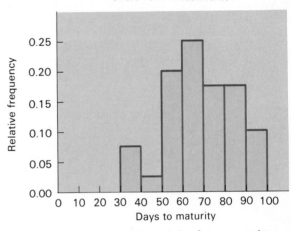

Data from *Barron's National Business and Financial Weekly*

Note that the relative-frequency histogram in Figure 2.2 and the frequency histogram in Figure 2.1 have identical shapes. This is because the frequencies and relative frequencies are in the same proportion.

Finally, we present the following word of caution. We have used the lower class limits 30, 40, 50, . . . to label the horizontal axes of the histograms in Figures 2.1 and 2.2, since it seems natural to do so. However, care must be taken when reading the histograms. For instance, the bar for 80 to 90 represents only maturity periods in the 80s. A maturity period of 90 days is included in the next bar to the right. ∎

Below we present the definition of a frequency histogram and the definition of a relative-frequency histogram.

DEFINITION 2.4 Frequency and relative-frequency histograms

Frequency histogram: A graph that displays the classes on the horizontal axis and the frequencies of the classes on the vertical axis. The frequency of each class is represented by a vertical bar whose height is equal to the frequency of the class.

Relative-frequency histogram: A graph that displays the classes on the horizontal axis and the relative frequencies of the classes on the vertical axis. The relative frequency of each class is represented by a vertical bar whose height is equal to the relative frequency of the class.

MTB
SPSS

Histograms for single-value grouping

For the days-to-maturity data, each class represents 10 possible days to maturity, and the histogram bar for each class extends over those 10 possible days (see Figures 2.1 and 2.2). When data are grouped using classes based on a single value, we proceed somewhat differently. In that case, each bar is placed directly over the only possible numerical value in the class, as illustrated in the next example.

EXAMPLE 2.11   **Illustrates histograms for single-value grouped data**

In Example 2.7 on pages 20–21 we considered data on the number of school-age children in each of 30 families. We grouped the data using classes based on a single value. The frequency and relative-frequency distributions are given in Table 2.9 and are repeated here in Table 2.15.

TABLE 2.15
Frequency and relative-frequency distributions for number of school-age children.

Number of school-age children	Frequency	Relative frequency
0	12	0.400
1	6	0.200
2	7	0.233
3	3	0.100
4	2	0.067
	30	1.000

Construct a frequency and relative-frequency histogram for this grouped data.

S O L U T I O N
As we just mentioned, for single-value grouping such as in Table 2.15, we place the middle of each histogram bar directly over the single numerical value represented by the class. Hence, the frequency and relative-frequency histograms for the grouped data in Table 2.15 are as pictured in Figures 2.3 and 2.4.   ∎

MTB
SPSS

FIGURE 2.3
Frequency histogram
for number of school-age
children in each of 30
families.

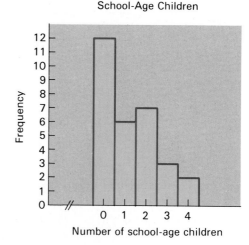

FIGURE 2.4
Relative-frequency
histogram for number
of school-age children
in each of 30 families.

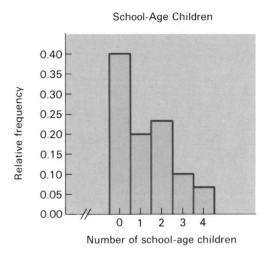

Note: In Figures 2.3 and 2.4 the symbol "//" has been placed on the horizontal axis to indicate that the zero point (0) on that axis is not in its usual position at the intersection of the horizontal and vertical axes. Whenever any such modification is made, whether on the horizontal or vertical axis, the symbol "//," or some similar symbol, should be used to indicate that fact.

Dotplots

Another kind of graphical display of numerical data is the **dotplot.** Dotplots are particularly useful for showing the relative positions of the data in a data set or for comparing two or more data sets.

EXAMPLE 2.12 Illustrates dotplots

A farmer is interested in a newly developed fertilizer that supposedly will increase his yield of oats. He uses the fertilizer on a sample of 15 one-acre plots. The yields in bushels are depicted in Table 2.16.

TABLE 2.16
Oat yields, in bushels,
for 15 one-acre plots.

67	65	55	57	58
61	61	61	64	62
62	60	62	60	67

To construct a dotplot for the data in Table 2.16, we begin by drawing a horizontal axis that displays the possible oat yields. Then we go through the data and record each yield by placing a dot over the appropriate value on the horizontal axis. For instance, the first yield is 67 bushels. This calls for a dot over the "67" on the horizontal axis. The dotplot for the data in Table 2.16 is pictured in Figure 2.5 on the top of page 32. ∎

MTB
SPSS

FIGURE 2.5
Dotplot for oat yields.

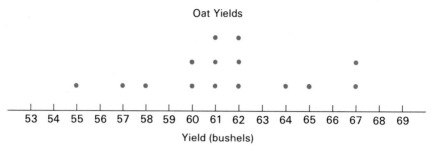

As you can see, dotplots are quite similar to histograms. In fact, when data are grouped in classes based on a single value, a dotplot and frequency histogram are essentially identical. However, dotplots are generally more convenient to use than histograms when dealing with single-value grouped data that involve decimals.

Graphical displays for qualitative data

Histograms and dotplots are designed for use with numerical data. Qualitative data are portrayed using different techniques. Two common methods for portraying qualitative data graphically are *pie charts* and *bar graphs*.

EXAMPLE 2.13   Illustrates pie charts and bar graphs

The U.S. Bureau of the Census divides the United States into four *regions*—Northeast, Midwest, South, and West. In Table 2.17 we have presented a frequency distribution and relative-frequency distribution for the 1983 resident population of the United States, by region.

TABLE 2.17
Resident population of the United States, by region, for 1983.

Region	Frequency (thousands)	Relative frequency
Northeast	49,519	0.212
Midwest	58,953	0.252
South	79,539	0.340
West	45,970	0.196
	233,981	1.000

Data from U.S. Bureau of the Census

The table shows, for instance, that 21.2% of the 1983 U.S. population resided in the Northeast region.

Note that we are dealing here with qualitative-data classes—namely, "Northeast," "Midwest," "South," and "West." We can use a *pie chart* to display the relative-frequency distribution given in the first and third columns of Table 2.17. A **pie chart** is a disk divided into pie-shaped pieces proportional to the relative frequencies. In this case, we need to divide a disk into four pie-shaped pieces comprising 21.2%, 25.2%, 34.0%, and 19.6% of the disk. We can do this by using a protractor and the fact that there are 360° in a circle. Thus, for instance, the first piece of the disk is obtained by marking off 76.32° (21.2% of 360°). The pie chart for the relative-frequency distribution in Table 2.17 is shown in Figure 2.6.

FIGURE 2.6
Pie chart of relative-frequency distribution of 1983 U.S. population by region.

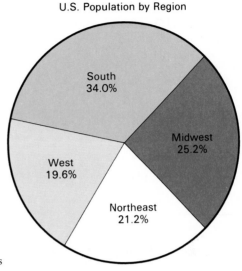

U.S. Population by Region

Data from U.S. Bureau of the Census

We can also use a *bar graph* to portray the same information. A **bar graph** is like a histogram except that its bars are separated. The bar graph for the relative-frequency distribution in Table 2.17 is pictured in Figure 2.7. ∎

FIGURE 2.7
Bar graph of relative-frequency distribution of 1983 U.S. population by region.

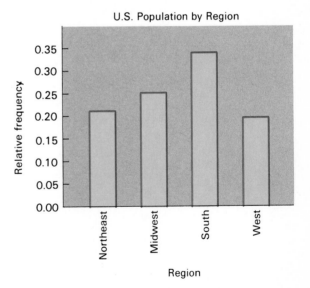

U.S. Population by Region

Data from
U.S. Bureau of the Census

Pie charts are generally preferable to bar graphs for qualitative data, since people are accustomed to having the horizontal axis of a graph show order. For example, someone might infer from Figure 2.7 that the South is "less than" the West, because South is shown to the left of West on the horizontal axis. Pie charts do not lead to such inferences.

Other graphical displays

Histograms, pie charts, and bar graphs are only a few of the countless ways that data can be portrayed pictorially. Fortunately, most graphical displays are based on common sense and are easy to understand. Pictograms, line graphs, and bar charts, shown in Figures 2.8–2.10, are additional methods for portraying data graphically.

FIGURE 2.8   Pictograms showing changes in farming, 1940–1983.

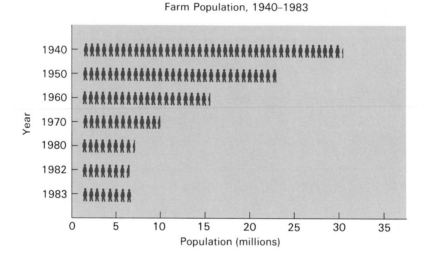

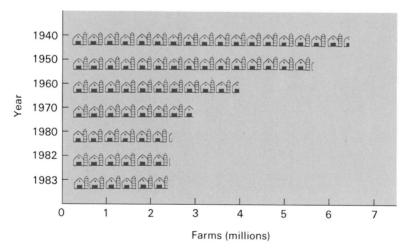

Data from U.S. Department of Agriculture

FIGURE 2.9   Line graphs giving selected crime rates, 1972–1983.

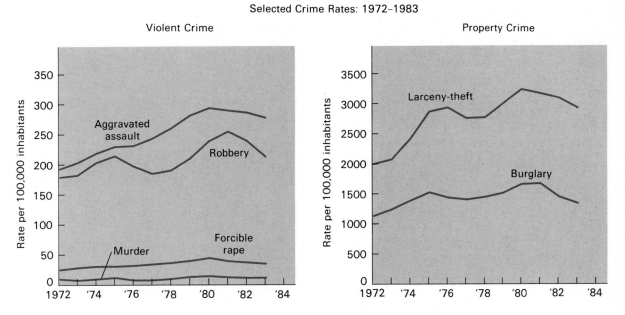

Chart prepared by U.S. Bureau of the Census. Data from U.S. Federal Bureau of Investigation.

FIGURE 2.10   Bar charts portraying percent distributions for the sources of federal receipts, 1970 and 1983.

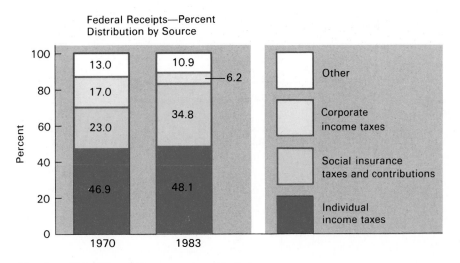

Chart from U.S. Office of Management and Budget

Exercises 2.3

2.35 What is the difference between a frequency histogram and a relative-frequency histogram?

In Exercises 2.36 through 2.45, we have given the frequency and relative-frequency distributions obtained in Exercises 2.18 through 2.27. For each exercise,
a) construct a frequency histogram.
b) construct a relative-frequency histogram.

2.36 The ages of a sample of 35 diabetics:

Age (yrs)	Frequency	Relative frequency
0–9	1	0.029
10–19	1	0.029
20–29	2	0.057
30–39	2	0.057
40–49	7	0.200
50–59	6	0.171
60–69	7	0.200
70–79	6	0.171
80–89	3	0.086

2.37 The contents of a sample of 30 "one-liter" bottles of soda:

Contents (ml)	Frequency	Relative frequency
910–929	1	0.033
930–949	1	0.033
950–969	3	0.100
970–989	9	0.300
990–1009	7	0.233
1010–1029	6	0.200
1030–1049	2	0.067
1050–1069	1	0.033

2.38 The length of stay in Europe and the Mediterranean for a sample of 36 U.S. travelers:

Length of stay (days)	Frequency	Relative frequency
1–7	10	0.278
8–14	10	0.278
15–21	8	0.222
22–28	1	0.028
29–35	2	0.056
36–42	1	0.028
43–49	2	0.056
50–56	1	0.028
57–63	0	0.000
64–70	1	0.028

2.39 Last year's energy consumption for a sample of 50 households in the South:

Energy consumption (millions of BTU)	Frequency	Relative frequency
40–49	1	0.02
50–59	7	0.14
60–69	7	0.14
70–79	3	0.06
80–89	6	0.12
90–99	10	0.20
100–109	5	0.10
110–119	4	0.08
120–129	2	0.04
130–139	3	0.06
140–149	0	0.00
150–159	2	0.04

2.40 The number of persons per household for a sample of 40 American households:

Number of persons	Frequency	Relative frequency
1	7	0.175
2	13	0.325
3	9	0.225
4	5	0.125
5	4	0.100
6	1	0.025
7	1	0.025

2.41 The number of cars sold per week last year by a car salesperson:

Number of cars sold	Frequency	Relative frequency
0	7	0.135
1	15	0.288
2	12	0.231
3	9	0.173
4	5	0.096
5	3	0.058
6	1	0.019

2.42 The number of incorrect answers on an eight-question fraction quiz taken by 25 fifth-graders:

Number incorrect	Frequency	Relative frequency
0	4	0.16
1	8	0.32
2	4	0.16
3	5	0.20
4	2	0.08
5	1	0.04
6	1	0.04

2.43 The number of days missed last year by each of the bricklayers employed at Cudahey Masonry:

Number of days missed	Frequency	Relative frequency
0	4	0.050
1	2	0.025
2	14	0.175
3	10	0.125
4	16	0.200
5	18	0.225
6	10	0.125
7	6	0.075

2.44 The 24-hour iron intake of a sample of 45 women under the age of 51:

Iron intake (mg)	Frequency	Relative frequency
6–under 8	1	0.022
8–under 10	1	0.022
10–under 12	7	0.156
12–under 14	9	0.200
14–under 16	9	0.200
16–under 18	9	0.200
18–under 20	8	0.178
20–under 22	1	0.022

2.45 The starting annual salaries of a sample of 35 liberal arts graduates:

Starting salary ($thousands)	Frequency	Relative frequency
16–under 17	3	0.086
17–under 18	3	0.086
18–under 19	5	0.143
19–under 20	9	0.257
20–under 21	9	0.257
21–under 22	4	0.114
22–under 23	1	0.029
23–under 24	1	0.029

In Exercises 2.46 through 2.49, construct a dotplot for each of the data sets.

2.46 The exam scores for the students in an introductory statistics class are as follows:

88	82	89	70	85
63	100	86	67	39
90	96	76	34	81
64	75	84	89	96

2.47 A paint manufacturer claims that the average drying time for his new latex paint is two hours. To test this claim, the drying times are obtained for a sample of 20 cans of paint. The results are displayed below in minutes.

123	109	115	121	130
127	106	120	116	136
131	128	139	110	133
122	133	119	135	109

2.48 A consumer advocacy group suspects that Wheat Flakes brand cereal contains, on the average, less than the advertised weight of 15 ounces per box. A sample of 40 boxes of Wheat Flakes gives the following weights:

15.8	15.1	15.2	15.4	14.8	15.6	15.7	14.5
14.8	15.4	15.3	15.5	15.2	14.6	15.4	15.4
15.5	14.7	14.7	15.1	14.7	15.3	15.3	15.5
14.0	14.2	14.6	15.0	15.1	14.9	14.9	15.8
15.0	14.4	15.4	14.3	15.4	15.9	15.2	15.6

2.49 The Motor Vehicle Manufacturers Association gathers information on the ages of cars and trucks in use. A sample of 37 trucks yields the following ages, in years:

8	12	14	16	15	5	11	13
4	12	12	15	12	3	10	9
11	3	18	4	9	11	17	
7	4	12	12	8	9	10	
9	9	1	7	6	9	7	

In Exercises 2.50 and 2.51 we have given the frequency and relative-frequency distributions obtained in Exercises 2.30 and 2.31, respectively. For each exercise,
a) draw a pie chart for the relative-frequencies.
b) construct a bar graph for the relative-frequencies.

2.50 The network data for the all-time top television programs by number of viewers as of January, 1984:

Network	Frequency	Relative frequency
ABC	14	0.519
CBS	7	0.259
NBC	6	0.222

2.51 The winners of the NCAA wrestling championships for the years 1963–1984:

Champion	Frequency	Relative frequency
Oklahoma	2	0.091
Oklahoma State	4	0.182
Iowa State	6	0.273
Michigan State	1	0.045
Iowa	9	0.409

2.52 The U.S. Internal Revenue Service groups income-tax returns by adjusted gross income. The following is a relative-frequency histogram for the 1982 federal individual income-tax returns showing an adjusted gross income less than $50,000.

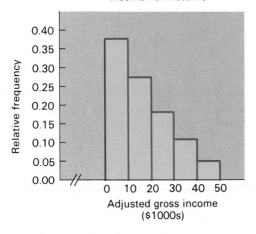

Income Tax Returns

Data from United States Internal Revenue Service

a) About what percentage of the returns had an adjusted gross income between $10,000 and $19,999, inclusive?

b) About what percentage had an adjusted gross income less than $30,000?

c) Given that there were 89,928,000 returns with an adjusted gross income less than $50,000, about how many had an adjusted gross income between $30,000 and $49,999, inclusive?

2.53 The following graph displays a relative-frequency histogram for the cholesterol-level data discussed in Example 2.6:

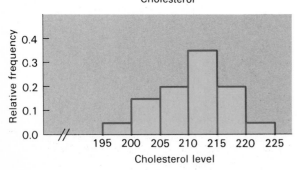

Cholesterol

Answer the following questions using only the graph:

a) What percentage of the patients have cholesterol levels between 205 and 209, inclusive?

b) What percentage have levels of 215 or higher?

c) Given that the number of patients is 20, how many have levels between 210 and 214?

Another method of depicting the classes on the horizontal axis of a histogram is to use *class boundaries*. The **lower class boundary** of a class is the number halfway between the lower class limit of the class and the upper class limit of the next lower class. Similarly, the **upper class boundary** of a class is the number halfway between the upper class limit of the class and the lower class limit of the next higher class. For instance, consider the class 50–59 of the days-to-maturity data (see Table 2.14 on page 28). We have

$$\text{Lower class boundary} = \frac{49 + 50}{2} = 49.5$$

$$\text{Upper class boundary} = \frac{59 + 60}{2} = 59.5$$

2.54 Refer to Exercise 2.36.

a) Append a "lower class boundary" column and an "upper class boundary" column to the table given in Exercise 2.36. Then fill in these columns by computing the lower and upper class boundaries for all the classes.

b) Construct a frequency histogram for the data using class boundaries on the horizontal axis instead of class limits.

c) What are the advantages of using class boundaries instead of class limits? What are the disadvantages?

2.55 Refer to Exercise 2.37.

a) Append a "lower class boundary" column and an "upper class boundary" column to the table given in Exercise 2.37, and fill in those columns by computing the lower and upper class boundaries for all the classes.

b) Construct a frequency histogram for the data using class boundaries on the horizontal axis instead of class limits.

c) What are the advantages of using class boundaries instead of class limits? What are the disadvantages?

In Exercises 2.56 and 2.57 we will discuss the *relative-frequency polygon*—a commonly used graphical display that is similar to the relative-frequency histogram. In a **relative-frequency polygon,** a point is plotted above each *class mark* at a height equal to the relative frequency of the class. Then the points are joined with connecting lines. For instance, the relative-frequency polygon for the days-to-maturity data is as follows:

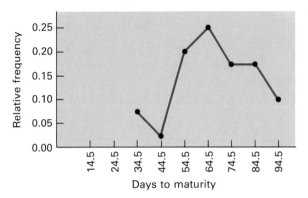

Short-Term Investments

2.56 Construct a relative-frequency polygon for the length-of-stay data given in Exercise 2.38.

2.57 Construct a relative-frequency polygon for the energy-consumption data given in Exercise 2.39.

Cumulative information can be portrayed using a graphical display called an *ogive* ($\overline{o}'$ jīv). To construct an ogive, make a table showing cumulative frequencies and cumulative relative frequencies. Table 2.18 provides a table for the days-to-maturity data.

TABLE 2.18

Less than	Cumulative frequency	Cumulative relative frequency
30	0	0.000
40	3	0.075
50	4	0.100
60	12	0.300
70	22	0.550
80	29	0.725
90	36	0.900
100	40	1.000

The first column of the table gives the class limits. The second column gives the cumulative frequencies. A **cumulative frequency** is obtained by summing the frequencies of all classes representing values *less than* the specified class limit. For instance, by referring to Table 2.14 on page 28, we can find the cumulative frequency of investments with a maturity period of less than 50 days:

Cumulative frequency $= 3 + 1 = 4$

The third column of Table 2.18 gives the cumulative relative frequencies. A **cumulative relative frequency** is found by dividing the corresponding cumulative frequency by the total number of pieces of data. For instance, the cumulative relative frequency of investments with a maturity period of less than 50 days is obtained as follows:

$$\text{Cumulative relative frequency} = \frac{4}{40} = 0.100$$

This means that 10% of the investments have a maturity period of less than 50 days.

Using Table 2.18 we can now construct an ogive for the days-to-maturity data. In an **ogive,** a point is plotted above each *class limit* at a height equal to the cumulative relative frequency. Then the points are joined with connecting lines. Thus, the ogive for the days-to-maturity data is as follows:

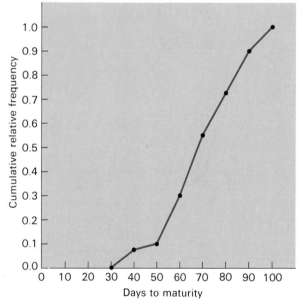

Short-Term Investments

2.58 Refer to Exercise 2.38.
   a) Construct a table similar to Table 2.18 for the length-of-stay data given in Exercise 2.38. Interpret the values in your table.
   b) Draw an ogive for the data.

2.59 Refer to Exercise 2.39.
   a) Construct a table similar to Table 2.18 for the energy-consumption data given in Exercise 2.39. Interpret the values in your table.
   b) Draw an ogive for the data.

2.4   Stem-and-leaf diagrams

New ways of displaying data are constantly being invented. One recently developed method is called a *stem-and-leaf diagram*. This ingenious diagram, invented in the late 1960s by Professor John Tukey of Princeton University, is often easier to construct than either a frequency distribution or a histogram, and generally displays more information. To illustrate stem-and-leaf diagrams, we return once more to the days-to-maturity data.

EXAMPLE 2.14   **Illustrates stem-and-leaf diagrams**

The number of days to maturity for 40 short-term investments, given in Table 2.1, is repeated here in Table 2.19.

TABLE 2.19 Days to maturity for 40 short-term investments.							
70	64	99	55	64	89	87	65
62	38	67	70	60	69	78	39
75	56	71	51	99	68	95	86
57	53	47	50	55	81	80	98
51	36	63	66	85	79	83	70

We grouped these data using the classes 30–39, 40–49, . . . , 90–99 (Table 2.2, page 17), and we portrayed the data graphically with a frequency histogram (Figure 2.1, page 28). Now we will construct a stem-and-leaf diagram for the data. As you will see, with a stem-and-leaf diagram we can simultaneously group the data and obtain a graphical display similar to a histogram. A stem-and-leaf diagram for the days-to-maturity data can be constructed as follows:

STEP 1 *Select the leading digits from the data in Table 2.19. This gives the numbers 3, 4, . . . , 9.*

STEP 2 *List the leading digits from Step 1 on the lefthand side of a page. [See the colored numbers in Figure 2.11.]*

STEP 3 *Go through the data in Table 2.19 and write the final digit of each number to the right of the appropriate leading digit. For instance, the first investment has a maturity period of 70 days. This calls for a "0" to the right of the "7" in colored type. The second investment has a maturity period of 62 days, so this calls for a "2" to the right of the "6" in colored type. Continuing in this manner, we obtain the diagram displayed in Figure 2.11.*

FIGURE 2.11
Stem-and-leaf diagram for days-to-maturity data.

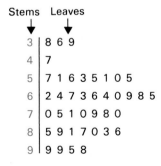

Stems   Leaves

3	8 6 9
4	7
5	7 1 6 3 5 1 0 5
6	2 4 7 3 6 4 0 9 8 5
7	0 5 1 0 9 8 0
8	5 9 1 7 0 3 6
9	9 9 5 8

As indicated in the figure, the leading digits are called **stems,** the final digits are called **leaves,** and the entire diagram is called a **stem-and-leaf diagram.**

The stem-and-leaf diagram for the days-to-maturity data is similar to the frequency histogram for that data, since the length of the row of leaves for a class equals the frequency of the class. [Turn the stem-and-leaf diagram 90° counterclockwise and compare it with the frequency histogram in Figure 2.1.] By shading each row of leaves, as in Figure 2.12, we get a diagram that looks even more like the frequency histogram of the data.

FIGURE 2.12
Shaded stem-and-leaf diagram
for days-to-maturity data.

```
3 | 8 6 9
4 | 7
5 | 7 1 6 3 5 1 0 5
6 | 2 4 7 3 6 4 0 9 8 5
7 | 0 5 1 0 9 8 0
8 | 5 9 1 7 0 3 6
9 | 9 9 5 8
```

The diagram in Figure 2.12 is appropriately called a **shaded stem-and-leaf diagram.** Because the numbers in the diagram are not shaded over completely, the shaded stem-and-leaf diagram exhibits the original, or raw, data as well as providing a graphical display of the frequency distribution. On the other hand, although a frequency histogram gives a graphical display of the frequency distribution, it is generally not possible to recover the raw data from the frequency histogram.

Another form of stem-and-leaf diagram is the **ordered stem-and-leaf diagram.** For this type of stem-and-leaf diagram, the leaves in each row are ordered from smallest to largest. This makes it easier to comprehend the data and also facilitates the computation of descriptive measures such as the median (to be discussed in Chapter 3). The ordered stem-and-leaf diagram for the days-to-maturity data is presented in Figure 2.13.

FIGURE 2.13
Ordered stem-and-leaf
diagram for days-to-maturity
data.

```
3 | 6 8 9
4 | 7
5 | 0 1 1 3 5 5 6 7
6 | 0 2 3 4 4 5 6 7 8 9
7 | 0 0 0 1 5 8 9
8 | 0 1 3 5 6 7 9
9 | 5 8 9 9
```

∎

EXAMPLE 2.15 **Illustrates stem-and-leaf diagrams**

The cholesterol-level data from Example 2.6 are repeated in Table 2.20.

TABLE 2.20
Cholesterol levels of 20
patients, in milligrams per
100 milliliters.

210	209	212	208
217	207	210	203
208	210	210	199
215	221	213	218
202	218	200	214

Since these data are three-digit numbers, we use the first *two* digits as the stems and the third digits as the leaves. A stem-and-leaf diagram for the cholesterol levels is displayed in Figure 2.14.

FIGURE 2.14
Stem-and-leaf diagram for
cholesterol-level data.

```
19 | 9
20 | 8 2 9 7 0 8 3
21 | 0 7 5 0 8 2 0 0 3 8 4
22 | 1
```

The stem-and-leaf diagram in Figure 2.14 is only moderately helpful because there are so few stems. We can construct a better stem-and-leaf diagram by using two lines for each stem, with the first line for the leaf digits 0–4 and the second line for the leaf digits 5–9. This stem-and-leaf diagram is shown in Figure 2.15.

FIGURE 2.15
Stem-and-leaf diagram for
cholesterol-level data using
two lines per stem.

```
19 |
19 | 9
20 | 2 0 3
20 | 8 9 7 8
21 | 0 0 2 0 0 3 4
21 | 7 5 8 8
22 | 1
22 |
```

■

MTB
SPSS

As we have seen, stem-and-leaf diagrams have several advantages over the more classical techniques for grouping and graphing. However, stem-and-leaf diagrams do have some drawbacks. For instance, they are sometimes awkward to use with decimal data. Nonetheless, stem-and-leaf diagrams offer a viable alternative for organizing and summarizing data.

Exercises 2.4

In Exercises 2.60 through 2.63,
a) construct a stem-and-leaf diagram.
b) construct an ordered stem-and-leaf diagram.

2.60 The data from Exercise 2.18 on the ages of a sample of 35 diabetics:

48	41	57	83	41	55	59
61	38	48	79	75	55	7
54	23	47	56	79	68	61
64	45	53	82	68	38	70
10	60	83	76	21	65	47

2.61 The data from Exercise 2.19 on the contents of a sample of 30 "one-liter" bottles of soda:

1025	977	1018	975	977
990	959	957	1031	964
986	914	1010	988	1028
989	1001	984	974	1017
1060	1030	991	999	997
996	1014	946	995	987

[Use the stems 91, 92, . . . , 106.]

2.62 The data from Exercise 2.20 on the length of stay in Europe and the Mediterranean for a sample of 36 U.S. travelers:

41	16	6	21	1	21
5	31	20	27	17	10
3	32	2	48	8	12
21	44	1	56	5	12
3	13	15	10	18	3
1	11	14	12	64	10

2.63 The data from Exercise 2.21 on last year's energy consumption for a sample of 50 households in the South:

130	55	45	64	155	66	60	80	102	62
58	101	75	111	151	139	81	55	66	90
97	77	51	67	125	50	136	55	83	91
54	86	100	78	93	113	111	104	96	113
96	87	129	109	69	94	99	97	83	97

2.64 According to the U.S. Bureau of the Census, the percentage of the population in each state that has completed high school is as follows:

State	Percent	State	Percent	State	Percent
AL	57	LA	58	OH	67
AK	83	ME	69	OK	66
AZ	72	MD	67	OR	76
AR	56	MA	72	PA	65
CA	74	MI	68	RI	61
CO	79	MN	73	SC	54
CT	70	MS	55	SD	68
DE	69	MO	64	TN	56
FL	67	MT	74	TX	63
GA	56	NE	73	UT	80
HI	74	NV	76	VT	71
ID	74	NH	72	VA	62
IL	67	NJ	67	WA	78
IN	66	NM	69	WV	56
IA	72	NY	66	WI	70
KS	73	NC	55	WY	78
KY	53	ND	66		

Construct a stem-and-leaf diagram for the percentages
a) using one line per stem.
b) using two lines per stem.

2.65 The U.S. Federal Bureau of Investigation (FBI) keeps records on crime rates, by state. In 1983, the number of crimes per 1000 population for each state was as shown in the table at the top of the next column.

State	Crimes per 1000 pop.	State	Crimes per 1000 pop.
AL	41	MT	46
AK	60	NE	38
AZ	64	NV	67
AR	35	NH	34
CA	67	NJ	52
CO	66	NM	63
CT	50	NY	59
DE	55	NC	42
FL	68	ND	27
GA	45	OH	45
HI	58	OK	49
ID	39	OR	63
IL	52	PA	32
IN	41	RI	50
IA	39	SC	48
KS	45	SD	25
KY	34	TN	40
LA	50	TX	59
ME	37	UT	51
MD	54	VT	41
MA	50	VA	40
MI	65	WA	61
MN	40	WV	24
MS	32	WI	43
MO	45	WY	40

Construct a stem-and-leaf diagram for the crime rates
a) using one line per stem.
b) using two lines per stem.

2.66 The annual average maximum temperatures for selected cities in the United States are as shown on the following page. (Source: U.S. National Oceanic and Atmospheric Administration.)

Construct a stem-and-leaf diagram for the temperatures
a) using two lines per stem.
b) using five lines per stem.

2.67 The annual average minimum temperatures for selected cities in the United States are as shown on the following page. (Source: U.S. National Oceanic and Atmospheric Administration.)

Construct a stem-and-leaf diagram for the temperatures
a) using two lines per stem.
b) using five lines per stem.

Data for Exercise 2.66:

City	Annual avg. max. temp.	City	Annual avg. max. temp.
Mobile, AL	77	Reno, NV	67
Juneau, AK	47	Concord, NH	57
Phoenix, AZ	85	Atlantic City, NJ	63
Little Rock, AR	73	Albuquerque, NM	70
Los Angeles, CA	70	Albany, NY	58
Sacramento, CA	73	Buffalo, NY	56
San Francisco, CA	65	New York, NY	62
Denver, CO	64	Charlotte, NC	71
Hartford, CT	60	Raleigh, NC	70
Wilmington, DE	64	Bismarck, ND	54
Washington, DC	67	Cincinnati, OH	64
Jacksonville, FL	79	Cleveland, OH	59
Miami, FL	83	Columbus, OH	62
Atlanta, GA	71	Oklahoma City, OK	71
Honolulu, HI	84	Portland, OR	62
Boise, ID	63	Philadelphia, PA	63
Chicago, IL	59	Pittsburgh, PA	60
Peoria, IL	60	Providence, RI	59
Indianapolis, IN	62	Columbia, SC	75
Des Moines, IA	59	Sioux Falls, SD	57
Wichita, KS	68	Memphis, TN	72
Louisville, KY	66	Nashville, TN	70
New Orleans, LA	78	Dallas-Forth Worth, TX	77
Portland, ME	55	El Paso, TX	78
Baltimore, MD	65	Houston, TX	79
Boston, MA	59	Salt Lake City, UT	64
Detroit, MI	58	Burlington, VT	54
Sault Ste. Marie, MI	49	Norfolk, VA	68
Duluth, MN	48	Richmond, VA	69
Minneapolis-St. Paul, MN	54	Seattle-Tacoma, WA	59
Jackson, MS	76	Spokane, WA	57
Kansas City, MO	64	Charleston, WV	66
St. Louis, MO	66	Milwaukee, WI	55
Great Falls, MT	56	Cheyenne, WY	58
Omaha, NE	62	San Juan, PR	86

Data for Exercise 2.67:

City	Annual avg. min. temp.	City	Annual avg. min. temp.
Mobile, AL	58	Reno, NV	32
Juneau, AK	33	Concord, NH	33
Phoenix, AZ	57	Atlantic City, NJ	43
Little Rock, AR	51	Albuquerque, NM	42
Los Angeles, CA	55	Albany, NY	37
Sacramento, CA	48	Buffalo, NY	39
San Francisco, CA	48	New York, NY	47
Denver, CO	36	Charlotte, NC	49
Hartford, CT	40	Raleigh, NC	48
Wilmington, DE	45	Bismarck, ND	29
Washington, DC	49	Cincinnati, OH	45
Jacksonville, FL	57	Cleveland, OH	41
Miami, FL	69	Columbus, OH	42
Atlanta, GA	51	Oklahoma City, OK	49
Honolulu, HI	70	Portland, OR	44
Boise, ID	39	Philadelphia, PA	45
Chicago, IL	40	Pittsburgh, PA	41
Peoria, IL	41	Providence, RI	41
Indianapolis, IN	42	Columbia, SC	51
Des Moines, IA	40	Sioux Falls, SD	34
Wichita, KS	45	Memphis, TN	52
Louisville, KY	46	Nashville, TN	49
New Orleans, LA	59	Dallas-Forth Worth, TX	55
Portland, ME	35	El Paso, TX	49
Baltimore, MD	45	Houston, TX	57
Boston, MA	44	Salt Lake City, UT	39
Detroit, MI	39	Burlington, VT	35
Sault Ste. Marie, MI	31	Norfolk, VA	51
Duluth, MN	29	Richmond, VA	47
Minneapolis-St. Paul, MN	35	Seattle-Tacoma, WA	44
Jackson, MS	53	Spokane, WA	37
Kansas City, MO	44	Charleston, WV	44
St. Louis, MO	45	Milwaukee, WI	38
Great Falls, MT	33	Cheyenne, WY	33
Omaha, NE	40	San Juan, PR	73

2.5   Misleading graphs*

Graphs and charts are frequently constructed in a manner that causes them to be misleading. Sometimes this is intentional and sometimes it is not. Regardless of the intent, it is important to read and interpret graphs and charts with a great deal of care. Consider the following example.

EXAMPLE 2.16   Illustrates truncated graphs

Figure 2.16 on the following page shows a bar graph from an article in a major metropolitan newspaper. The graph displays the unemployment rates in the United States from September of one year to March of the next year.

FIGURE 2.16

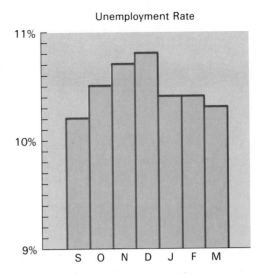

Unemployment Rate

A quick look at the graph might lead you to the conclusion that the unemployment rate dropped by roughly one-fourth between December and March, since the bar for March is about one-fourth smaller than the bar for December. In reality, however, the unemployment rate dropped less than one-twentieth—from 10.8% to 10.3%. Consequently, a more careful analysis of the graph is required than first meets the eye.

The unemployment-rate graph in Figure 2.16 is an example of a **truncated graph** because the vertical axis, which should start at 0%, starts at 9% instead. Thus, the part of the graph from 0% to 9% has been cut off, or *truncated.* This truncation causes the bars to be out of correct proportion and hence creates a misleading impression, as we discovered earlier. The graph would be even more misleading if it started at 10%. To see this, slide a piece of paper over the bottom of Figure 2.16 so that the bars begin at 10%. By how much does it now appear that the unemployment rate dropped between December and March?

As we have observed, the truncated graph in Figure 2.16 is potentially misleading. However, it is probably safe to say that the truncation was done to give a picture of the "ups" and "downs" in the unemployment-rate pattern, rather than to intentionally mislead the reader.

A version of Figure 2.16 that is not truncated is shown in Figure 2.17 at the top of page 46. Although Figure 2.17 provides a correct graphical display of the unemployment-rate data, the "ups" and "downs" are not as easy to spot as in the truncated graph in Figure 2.16.   ■

Truncated graphs have been a target of statisticians for a long time. Many statistics books discuss them and warn against their use. Nonetheless, as we have seen in Example 2.16, truncated graphs are still used today, even in reputable publications.

On the other hand, Example 2.16 also suggests that it may be desirable to cut off part of the vertical axis of a graph in order to more easily convey certain relevant information. In these cases, a truncated graph should *not* be used. Instead, a

FIGURE 2.17
"Untruncated" version
of Figure 2.16.

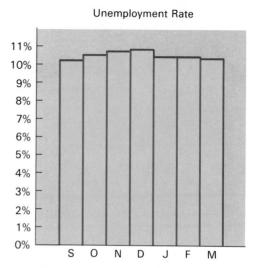

special symbol such as "//" should be employed to signify that the vertical axis
has been modified. The two graphs in Figure 2.18 provide an excellent illustration.

FIGURE 2.18
New single-
family home sales.

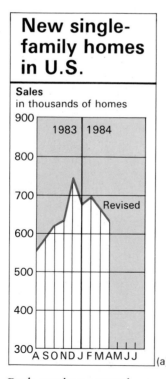

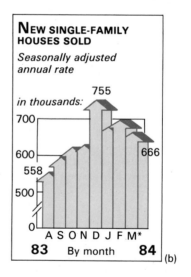

a) Reprinted by permission Tribune Media Services

b) Data from U.S. Department of Commerce, U.S. Department
of Housing and Urban Development

Both graphs portray the number of new single-family homes sold per month during part of 1983 and 1984. The first graph, Figure 2.18(a), is a truncated graph—truncated most likely in an attempt to present the reader with a clear visual display of the variation in sales. The second graph, Figure 2.18(b), accomplishes the same result, but is less subject to misinterpretation. Indeed, the reader is aptly warned by the slashes that part of the vertical axis between 0 and 500 has been removed.

Misleading graphs and charts can also result because of **improper scaling.** The following example shows how this can happen.

EXAMPLE 2.17 Illustrates improper scaling

A developer is preparing a brochure to attract investors for a new shopping center that is to be built in an area of Denver, Colorado. The area is growing rapidly—this year twice as many homes will be built there as last year. To illustrate this fact, the developer draws a pictogram as in Figure 2.19.

FIGURE 2.19

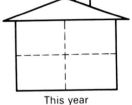

Last year   This year

The house on the left represents the number of homes built last year. Since the number of homes that will be built this year is double the number built last year, the developer makes the house on the right twice as tall and twice as wide as the house on the left. However, this scaling is improper because it gives the impression that *four times* as many homes will be built this year as last. So, the developer's brochure may mislead the unwary investor. ∎

There are countless ways in which graphs and charts can be misleading, besides the two ways we have discussed. Many more can be found in the entertaining and classic book *How to Lie with Statistics* by Darrell Huff (New York: W. W. Norton & Co., 1955). The main purpose of this section has been to show that graphs and charts should be read and constructed with care.

Exercises 2.5

2.68 Find two examples of graphs in a current newspaper or magazine that might be misleading. Explain why you think the graphs you obtain are potentially misleading.

2.69 Each year, the director of the reading program in a school district administers a standard test of reading skills. Then the director compares the average score for his district with the national average. The graph at the right (Figure 2.20) was presented to the school board in 1985.

a) Observe a truncated version of Figure 2.20 by sliding a piece of paper over the bottom of the graph so that the bars start at 16.

FIGURE 2.20 Average reading scores

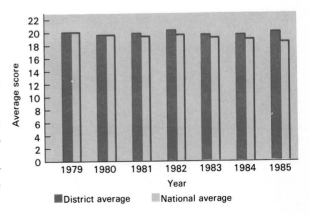

b) Observe a truncated version of Figure 2.20 by sliding a piece of paper over the bottom of the graph so that the bars start at 18.

c) What misleading impression about the 1985 scores is given by the truncated graphs that you observed in parts (a) and (b)?

2.70 The newspaper article below includes a bar graph on personal income for the first seven months of 1984. (Reprinted with permission of United Press International, Inc.)

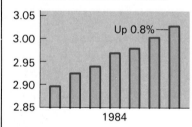

Today

in business

Personal Income

Seasonally adjusted annual rates in trillions of dollars

Consumers cautious

WASHINGTON (UPI) — Americans' personal income climbed 0.8 of a percent in July but spending slowed for the second month, actually declining for cars and heavy appliances, the Commerce Department reported.

The figures strengthened the view that consumers are being more cautious, even while employment and incomes remain strong.

Income was up nearly as much as the 0.9 of a percent increase in June and far more than May's anemic 0.4 of a percent rise, the department said.

White House spokesman Larry Speakes said the personal income figures indicate "a steady growth, as we wish, for the economy, and it doesn't show any evidence of overheating."

a) What is wrong with the bar graph?

b) Draw a version of the bar graph with an untruncated and unmodified vertical axis.

c) Draw a version of the bar graph in which the vertical axis is modified in an acceptable manner.

2.71 The following bar graph was taken from a 1985 newspaper article entitled "Immigrants add seasoning to America's melting pot." (Used with permission from American Demographics, © 1985, Ithaca, NY.)

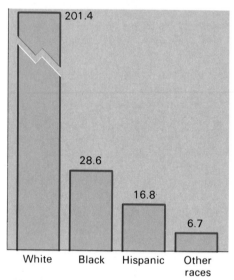

Race and Ethnicity in America
(in millions)

Data from Census Bureau July 1984 Estimates

a) Explain why a break is shown in the first bar on the left.

b) Why was it necessary to construct the graph with a broken bar?

c) Do you think this bar graph is potentially misleading? Why?

2.72 Refer to Example 2.17 on page 47. Indicate a correct way in which the developer can illustrate the fact that twice as many homes will be built in the area this year as last year.

2.73 A manufacturer of golf balls has determined that a newly developed process results in a ball lasting roughly twice as long as a ball produced using the current process. To illustrate this graphically, she designs a brochure showing a "new" ball with twice the radius of the "old" ball, as shown on the following page.

Old ball

New ball

a) *What is wrong with this? Hint:* The volume of a sphere is proportional to the cube of its radius.
b) Indicate a correct way in which the manufacturer can illustrate the fact that the "new" ball lasts twice as long as the "old" ball.

2.6 Computer packages*

Computers are ideal for performing statistical calculations and analyses. It is rarely necessary to write your own computer programs, since programs already exist for almost all aspects of statistics. The most commonly used programs for statistical work are taken from **computer packages**—collections of statistical computer programs written by some organization or individual. Nowadays, computer packages are available for use on mainframes, minicomputers, and microcomputers.

Computer packages are generally known by acronyms such as BMDP, Minitab, SAS, and SPSS. In this book we will give examples of printouts from the widely used package Minitab. Our objective here is to present a brief description of the Minitab programs corresponding to statistical procedures and analyses that we have discussed.[†]

In this chapter, we considered methods of grouping data into tables and displaying data graphically. The following examples show how Minitab can be employed to perform those tasks for us.

HISTOGRAM

Minitab has a program called **HISTOGRAM** that will simultaneously group a data set and display a frequency histogram for the data. We will illustrate the use of HISTOGRAM for the days-to-maturity data of Example 2.5.

EXAMPLE 2.18 Illustrates the use of HISTOGRAM

The number of days to maturity for 40 short-term investments, given in Table 2.1, are repeated here in Table 2.21.

TABLE 2.21
Days to maturity for 40
short-term investments.
(Source: *Barron's.*)

70	64	99	55	64	89	87	65
62	38	67	70	60	69	78	39
75	56	71	51	99	68	95	86
57	53	47	50	55	81	80	98
51	36	63	66	85	79	83	70

We grouped these data into a frequency distribution in Table 2.2 on page 17 and displayed them graphically with a frequency histogram in Figure 2.1 on page 28. Using Minitab, we can accomplish both of these tasks with relatively little effort.

---

† For more details, consult the *Minitab Supplement* by Neil Weiss (Addison-Wesley, 1989).

To begin, we enter the data from Table 2.21 into the computer (procedures for entering data are discussed in the Minitab supplement). Then we type the command HISTOGRAM followed by the location of the sample data (C1). On the next line we type the subcommand INCREMENT = 10 to specify that the class width is to be 10, and on the following line the subcommand START = 34.5 to specify that the first class mark is to be 34.5. [The two subcommands are optional. We have used them to obtain a frequency distribution and histogram that are comparable to the ones constructed earlier for these data by hand. The two subcommands instruct Minitab to use a class width of 10 and begin with the class 30–39.] Printout 2.1 shows these commands along with the resulting output.

PRINTOUT 2.1
Minitab output for
HISTOGRAM

```
MTB > HISTOGRAM C1;
SUBC> INCREMENT=10;
SUBC> START=34.5.

Histogram of C1   N = 40

Midpoint   Count
    34.5       3   ***
    44.5       1   *
    54.5       8   ********
    64.5      10   **********
    74.5       7   *******
    84.5       7   *******
    94.5       4   ****
```

The first column of the printout, headed Midpoint, gives the class marks corresponding to the classes 30–39, 40–49, . . . , 90–99. The second column, headed Count, displays the frequencies for those classes. Finally, by turning the printout 90° counterclockwise, we obtain, from the stars, Minitab's version of the frequency histogram in Figure 2.1 on page 28. ∎

DOTPLOT

On pages 31–32 we discussed dotplots. Recall that a dotplot for a data set is constructed by displaying the possible numerical values on a horizontal axis and by recording each piece of data with a dot over the appropriate value on the horizontal axis. The Minitab program for obtaining dotplots is called **DOTPLOT.**

EXAMPLE 2.19   **Illustrates the use of DOTPLOT**

A farmer is interested in a newly developed fertilizer that supposedly will increase his yield of oats. He uses the fertilizer on a sample of 15 one-acre plots. The yields, in bushels, are depicted in Table 2.22.

TABLE 2.22
Oat yields, in bushels,
for 15 one-acre plots.

67	65	55	57	58
61	61	61	64	62
62	60	62	60	67

In Example 2.12 we constructed a dotplot for these data by hand. To have Minitab construct the dotplot, we begin by entering the data from Table 2.22 into the computer. Then we type the command DOTPLOT followed by the storage location of the data (C2). Finally, on the next line, we type the (optional) subcommand INCREMENT = 10 to specify that we want 10 spaces between tick (+) marks. These commands and the resulting output are given in Printout 2.2.

PRINTOUT 2.2
Minitab output
for DOTPLOT

```
MTB > DOTPLOT C2;
SUBC> INCREMENT=10.
```

You should compare Minitab's dotplot in Printout 2.2 to the one we obtained by hand in Figure 2.5 on page 32. ∎

STEM-AND-LEAF

Minitab also has a program, called **STEM-AND-LEAF,** that will construct a stem-and-leaf diagram for a set of data. Example 2.20 illustrates the use of STEM-AND-LEAF in obtaining a stem-and-leaf diagram for the days-to-maturity data.

EXAMPLE 2.20 **Illustrates the use of STEM-AND-LEAF**

The number of days to maturity for 40 short-term investments are given in Table 2.21 on page 49. To have Minitab construct a stem-and-leaf diagram for those data, we first enter the data into the computer. Next, we type the command STEM-AND-LEAF followed by the storage location of the sample data (C1). Finally, we type the (optional) subcommand INCREMENT = 10 to instruct Minitab to use a distance of 10 units between stems. Printout 2.3 displays these commands along with the resulting output.

PRINTOUT 2.3
Minitab output for
STEM-AND-LEAF

```
MTB > STEM-AND-LEAF C1;
SUBC> INCREMENT=10.

Stem-and-leaf of C1        N = 40
Leaf Unit = 1.0

    3      3 689
    4      4 7
   12      5 01135567
  (10)     6 0234456789
   18      7 0001589
   11      8 0135679
    4      9 5899
```

On the first line of the output we find a verbal description of what follows (Stem-and-leaf of C1) and the number of pieces of data (N = 40). The second line of the output (Leaf Unit = 1.0) indicates where the decimal point goes, in this case directly after each leaf digit.

Let us now discuss in detail the stem-and-leaf diagram in Printout 2.3. The second column of numbers gives the stems, and to the right of the stems are the leaves. Since the leaves on each line are ordered, we see that Printout 2.3 actually provides us with an *ordered* stem-and-leaf diagram.

The numbers in the first column of Printout 2.3, called **depths,** are used to display cumulative frequencies. Starting from the top, the depths indicate the number of leaves (pieces of data) that lie in the given row or before. For instance, the "12" in the third row shows that there are a total of 12 leaves in the first three rows.

When the row in which the middle observations lie is reached, the cumulative frequency is replaced by the number of leaves in that row enclosed by parentheses. Thus, in this case, we see that the middle observations lie in the 60s and that there are 10 data values in the 60s.

Finally, the depths following the row where the middle observations lie indicate the number of leaves that lie in the given row or after. For instance, the "11" in the sixth row shows that there are a total of 11 leaves in the last two rows. ■

◆ Chapter Review

KEY TERMS

bar graph, 33
class mark, 19
class width, 19
classes, 19
computer packages*, 49
continuous, 14
data, 11
depths*, 52
discrete, 13
dotplot, 31
DOTPLOT*, 50
frequency, 19
frequency data, 12
frequency distribution, 19
frequency histogram, 29
grouped-data table, 19
HISTOGRAM*, 49
improper scaling*, 47

leaves, 41
lower class limit, 19
metric data, 12
ordered stem-and-leaf diagram, 41
ordinal data, 12
percentage, 18
pie chart, 32
qualitative data, 12
relative-frequency distribution, 19
relative-frequency histogram, 29
relative frequency, 19
shaded stem-and-leaf diagram, 41
STEM-AND-LEAF*, 51
stem-and-leaf diagram, 41
stems, 41
truncated graph*, 45
upper class limit, 19

YOU SHOULD BE
ABLE TO . . .

1   classify data as qualitative, ordinal, metric, or frequency data.

2   distinguish between discrete data and continuous data.

3   group data into a frequency distribution and a relative-frequency distribution.

4   construct a grouped-data table.

5 draw a frequency and relative-frequency histogram.

6 construct a dotplot.

7 draw a pie chart and a bar graph.

8 construct a stem-and-leaf, shaded stem-and-leaf, and ordered stem-and-leaf diagram.

9* identify and correct misleading graphs.

REVIEW TEST

1 According to the Bureau of Reclamation, the world's five largest hydroelectric plants, based on ultimate capacity, are as follows. Capacities are in megawatts.

Rank	Name	Country	Capacity
1	Itaipu	Brazil/Paraguay	12,600
2	Grand Coulee	U.S.A.	10,080
3	Guri	Venezuela	10,060
4	Tucurui	Brazil	6,480
5	Sayano-Shushensk	USSR	6,400

a) What type of data is given in the left-hand column of the table?

b) What type of data is given in the right-hand column of the table?

c) What type of data is the information that Grand Coulee is in the United States?

d) What type of data is the information that three of the five largest hydroelectric plants in the world are in South America?

2 The ages at inauguration for the first 40 Presidents of the United States are as follows. (Source: *The World Almanac, 1985*.)

George Washington 57	Chester A. Arthur 50
John Adams 61	Grover Cleveland 47
Thomas Jefferson 57	Benjamin Harrison 55
James Madison 57	Grover Cleveland 55
James Monroe 58	William McKinley 54
John Quincy Adams 57	Theodore Roosevelt 42
Andrew Jackson 61	William Howard Taft 51
Martin Van Buren 54	Woodrow Wilson 56
William H. Harrison 68	Warren G. Harding 55
John Tyler 51	Calvin Coolidge 51
James K. Polk 49	Herbert C. Hoover 54
Zachary Taylor 64	Franklin D. Roosevelt 51
Millard Fillmore 50	Harry S. Truman 60
Franklin Pierce 48	Dwight D. Eisenhower 62
James Buchanan 65	John F. Kennedy 43
Abraham Lincoln 52	Lyndon B. Johnson 55
Andrew Johnson 56	Richard M. Nixon 56
Ulysses S. Grant 46	Gerald Rudolph Ford 61
Rutherford B. Hayes 54	Jimmy Carter 52
James A. Garfield 49	Ronald Reagan 69

a) Construct a grouped-data table for these inauguration ages using equal-width classes and beginning with the class 40–44.

b) Draw a frequency histogram for the ages at inauguration based on your grouping in part (a).

3 Refer to Problem 2. Construct a dotplot for the ages at inauguration of the first 40 U.S. Presidents.

4 Refer to Problem 2. Construct an *ordered* stem-and-leaf diagram for the inauguration ages of the first 40 U.S. Presidents

a) using one line per stem.

b) using two lines per stem.

c) Which of your stem-and-leaf diagrams corresponds to the frequency distribution of Problem 2 (a)?

5 The Prescott National Bank has six tellers available to serve customers. The data below give the number of busy tellers observed at 25 spot checks.

6	5	2	5
6	5	1	3
3	4	5	4
4	4	4	6
3	5	6	
5	2	3	
1	0	5	

a) Construct a grouped-data table for these data using single-value grouping.

b) Draw a relative-frequency histogram for the data.

6 According to the National Safety Council, the number of accidental deaths by type in the United States for 1982 was as follows:

Type	Frequency
Motor vehicle	46,000
Work	11,200
Home	21,000
Public	19,500

a) Draw a pie chart of the data that displays the percentage of accidental deaths in each of the four categories.

b) Draw a bar graph of the data that displays the relative frequency of accidental deaths in each of the four categories.

7 The following table gives the highs for the Dow Jones Industrial Averages from 1950–1984. (Source: *Barron's.*)

Year	High	Year	High	Year	High
1950	235.47	1962	726.01	1974	891.66
1951	276.37	1963	767.21	1975	881.81
1952	292.00	1964	891.71	1976	1014.79
1953	293.79	1965	969.26	1977	999.75
1954	404.39	1966	995.15	1978	907.74
1955	488.40	1967	943.08	1979	897.61
1956	521.05	1968	985.21	1980	1000.17
1957	520.77	1969	968.85	1981	1024.05
1958	583.65	1970	842.00	1982	1071.55
1959	679.36	1971	950.82	1983	1287.20
1960	685.47	1972	1036.27	1984	1286.64
1961	734.91	1973	1051.70		

a) Construct a grouped-data table for the highs using equal-width classes, starting with the class "200–under 400."

b) Draw a relative-frequency histogram for the highs based on your relative-frequency distribution from part (a).

8* The graph at the top of the next column is taken from a newspaper article entitled "Hand that rocked cradle turns to work as women reshape U.S. labor force." (Used with permission from American Demographics, © 1985, Ithaca, NY.)

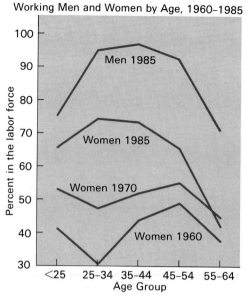

Working Men and Women by Age, 1960–1985

Data from Bureau of Labor Statistics

a) Cover up the numbers on the vertical axis of the graph with a piece of paper.

b) Look at the 1970 and 1985 graphs for women, and concentrate on the 25–34-year-old age group. What impression does the graph convey regarding the ratio of the percentages of women in the labor force for 1985 and 1970?

c) Now remove your piece of paper from the graph. Using the vertical scale, find the actual ratio of the percentages of 25–34-year-old women in the labor force for 1985 and 1970.

d) Why is the graph potentially misleading?

e) What can be done to improve the graph?

In Chapter 2 we began our study of descriptive statistics. We learned how to organize data into tables and summarize data using graphical displays. Another method of summarizing data is to compute a number, such as an average, that describes the data set. Numbers used to describe data sets are called **descriptive measures.** In this chapter we will continue our study of descriptive statistics by examining some of the most important descriptive measures.

Descriptive measures

CHAPTER OUTLINE

3.1

Measures of central tendency

Descriptive measures that indicate where the center or most typical value of a data set lies are called **measures of central tendency,** often more simply referred to as averages. In this section we will discuss the two most important measures of central tendency, the *mean* and the *median.*

The mean

The most commonly used measure of central tendency is the *mean.* When people speak of taking an average, it is the mean that they are most often referring to. The definition of the mean is as follows:

DEFINITION 3.1 Mean of a data set

The **mean** of a data set is defined to be the sum of the data divided by the number of pieces of data:

$$\text{Mean} = \frac{\text{Sum of the data}}{\text{Number of pieces of data}}$$

EXAMPLE   3.1   Illustrates Definition 3.1

One of the authors spent a summer working for a small mathematical consulting firm. The firm employed a few senior consultants who made between $700 and $950 per week, a few junior consultants who made between $300 and $350 per week, and several clerical helpers who made $200 per week. There was more work in the first half of the summer than in the second half, so there were more employees during the first half. Tables 3.1 and 3.2 give typical lists of weekly earnings for the two halves of the summer.

TABLE 3.1
Data Set I (week
ending July 15)

TABLE 3.2
Data Set II (week
ending August 12)

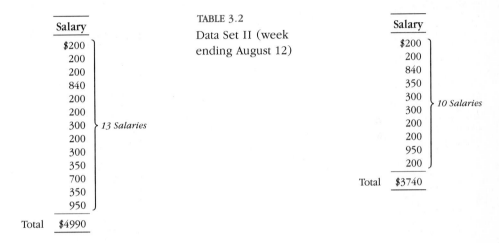

Salary	
$200	
200	
200	
840	
200	
200	
300	} 13 Salaries
200	
300	
350	
700	
350	
950	

Total $4990

Salary	
$200	
200	
840	
350	
300	} 10 Salaries
300	
200	
200	
950	
200	

Total $3740

Determine the mean of each of the two data sets.

SOLUTION
According to Definition 3.1, the mean of a data set is obtained by summing all the data and then dividing that sum by the total number of pieces of data. Consequently,

$$\text{Mean of Data Set I} = \frac{\$4990}{13} = \$383.85 \text{ (to the nearest cent)}$$

$$\text{Mean of Data Set II} = \frac{\$3740}{10} = \$374.00$$

MTB
SPSS

Thus, the mean salary of the 13 employees in Data Set I is $383.85 and that of the 10 employees in Data Set II is $374.00.   ■

The median

Another frequently used measure of central tendency is the *median*. Essentially, the median of a data set is the number that divides the bottom 50% of the data from the top 50%. To obtain the median of a data set, we arrange the data in increasing order and then determine the middle value in the ordered list. More precisely, we have the following definition.

DEFINITION 3.2 Median of a data set

The **median** of a data set is defined to be
1 the data value exactly in the middle of its ordered list if the number of pieces of data is *odd.*
2 the mean of the two middle data values in its ordered list if the number of pieces of data is *even.*

EXAMPLE   3.2   Illustrates Definition 3.2

Consider again the two sets of salary data given in Tables 3.1 and 3.2 on page 57. Determine the median of each of the two data sets.

S O L U T I O N
The number of pieces of data in Data Set I is 13, an *odd* number. Consequently, the median of that data set is the data value exactly in the middle of its ordered list. We have arranged the salaries in Data Set I in increasing order and have obtained the median, which is the seventh salary in the ordered list:

six lowest salaries							six highest salaries					
200	200	200	200	200	200	300	300	350	350	700	840	950

$$\text{Median} = 300$$

Thus, the median salary of the 13 employees in Data Set I is $300.

The number of pieces of data in Data Set II is 10, an *even* number. Consequently, the median of that data set is the mean of the two middle data values in its ordered list. We have arranged the salaries in Data Set II in increasing order and have obtained the median, which is the mean of the fifth and sixth salaries in the ordered list:

five lowest salaries					five highest salaries				
200	200	200	200	200	300	300	350	840	950

$$\text{Median} = \frac{200 + 300}{2} = 250$$

MTB
SPSS

Thus, the median salary of the 10 employees in Data Set II is $250.   ■

Comparison of the mean and the median

The mean and the median of a data set are frequently different. Table 3.3 summarizes the salary "averages" computed in Examples 3.1 and 3.2.

TABLE 3.3 Means and medians of salaries in Data Set I and Data Set II.

Measure of central tendency	Definition	Data Set I	Data Set II
Mean	Sum of the data / Number of pieces of data	$383.85	$374.00
Median	Middle value in ordered list	$300.00	$250.00

In both Data Sets I and II, the mean is larger than the median. This is because the mean is affected strongly by the few large salaries in each data set.

In general, the mean is sensitive to very large or very small data values, whereas the median is not. Consequently, the median is ordinarily preferred for data sets that have exceptional (very large or small) values. On the other hand, the mean takes into account the numerical value of every piece of data in a data set, whereas the median does not.

Both the mean and the median have other advantages and disadvantages. Consequently, it is not surprising that no simple rules exist for deciding which one of these measures of central tendency should be used in a given situation. You will attain skill in making such decisions through practice. However, even experts may disagree on whether the mean or the median is the more suitable measure of central tendency for a particular data set. The next example discusses two data sets and suggests which measure of central tendency is probably more appropriate for each.

EXAMPLE 3.3 **Illustrates the selection of the appropriate measure of central tendency**

a) A student takes four exams in a biology class. His grades are 88, 75, 95, and 100. If asked for his average, which measure of central tendency is the student likely to report?

b) The National Association of REALTORS® publishes data on resale prices of homes in the United States. Which measure of central tendency is more appropriate for such resale prices?

SOLUTION

a) Chances are that the student would report the mean of his four exam scores, which is 89.5. The mean is probably the more reasonable measure of central tendency to use since it takes into account the numerical value of each score and, therefore, indicates total overall performance.

b) The more appropriate measure of central tendency for resale home prices is the median. This is because the median is not affected strongly by the relatively few homes with extremely high or low resale prices, whereas the mean is. Thus, the median provides a better indication of the "typical" resale price than the mean. ∎

There are other measures of central tendency besides the mean and the median, but those are the two most important ones. Some additional measures of central tendency are discussed in Exercises 3.16 and 3.17.

Population and sample averages

We have defined the mean and the median of a data set. Whether these averages are considered population averages or sample averages depends on various things, such as the use to be made of them. For example, suppose the data set consists of the ages of the students in Professor W's introductory statistics class, as shown in Table 3.4.

TABLE 3.4
Ages of students in
Professor W's introductory
statistics class.

21	24	24	24	23	20	22	21
18	20	19	19	21	19	19	23
36	22	20	35	22	23	19	26
22	17	19	20	20	21	19	21
20	20	21	19	24	21	22	21

We computed the mean and the median of this data set and found that

$$\text{Mean} = 21.7 \text{ years}$$
$$\text{Median} = 21.0 \text{ years}$$

Now, the age data in Table 3.4 may constitute the entire population or it may be only a sample from a population. For a researcher interested in the ages of the students in this particular introductory statistics class, the data set is a population, and its mean and median are a population mean and a population median. For another researcher, interested in the ages of students in all introductory statistics classes, the data set is only a sample, and its mean and median are a sample mean and a sample median. Figure 3.1 illustrates the two ways in which the mean of a data set may be interpreted.

FIGURE 3.1
Possible interpretations for the mean of a data set.

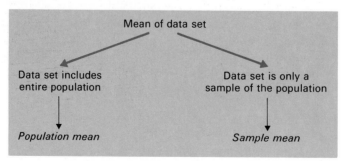

Concluding remarks

Recall that inferential statistics consists of methods of making inferences about a population based on information obtained from a sample of the population. Consequently, the data we actually deal with in inferential statistics is sample data. Since the majority of this text is devoted to inferential statistics, we will concentrate on the calculations required for computing descriptive measures of sample data.

However, keep in mind that descriptive measures of sample data are rarely an end in themselves. Rather they are usually a means of drawing conclusions about the population from which the sample was taken. The relationships between population and sample will be discussed further in Section 3.7.

Exercises 3.1

3.1 What is the purpose of a measure of central tendency?

3.2 Name and describe the two most important measures of central tendency.

Determine the mean and the median for each of the data sets in Exercises 3.3 through 3.10. Round each answer to one more decimal place than the original data.

3.3 The U.S. National Science Foundation collects data on the ages of recipients of science and engineering doctoral degrees. A sample of this year's recipients has the following ages:

37	28	36	33
37	43	41	28
24	44	27	24

3.4 The Health Insurance Association of America publishes figures on the costs to community hospitals per patient per day. Suppose a sample of 10 such costs in New York yields the data below.

$602	539	569	916	768
335	776	887	806	422

3.5 The average retail price for oranges in 1983 was 38.5 cents per pound, as reported by the U.S. Department of Agriculture. Recently, a sample of 15 markets reported the following prices for oranges in cents per pound.

43.0	40.0	42.6	40.2	37.5
44.1	45.2	41.8	35.6	34.6
37.9	44.2	44.5	38.2	42.4

3.6 A biologist is studying the gestation period (duration of pregnancy) of domestic dogs. Fifteen dogs are observed during pregnancy and are found to have the following gestation periods, in days:

62.0	61.4	59.8
62.2	60.3	60.4
59.4	60.2	60.4
60.8	61.8	59.2
61.1	60.4	60.9

3.7 A manufacturer of liquid soap produces a bottle with an advertised contents of 310 ml. A sample of 16 bottles yields the following contents:

297	318	306	300
311	303	291	298
322	307	312	300
315	296	309	311

3.8 A city planner working on bikeways needs information about local bicycle commuters. She designs a questionnaire. One of the questions asks how many minutes it takes the rider to pedal from home to his or her destination. A sample of 22 local bicycle commuters yields the following times:

22	19	24	31
29	29	21	15
27	23	37	31
30	26	16	26
12	23	48	
22	29	28	

3.9 The U.S. National Center for Educational Statistics surveys college and university libraries to obtain information on the number of volumes held. Suppose a sample of seven public colleges and universities hold the number of volumes, in thousands, indicated below.

79	516	24	265	41	15	411

3.10 According to the College Entrance Examination Board, the average verbal score on the Scholastic Aptitude Test in 1983 was 425 points out of a possible 800. A sample of 25 verbal scores for last year yielded the following results:

346	496	352	378	315
491	360	385	500	558
381	303	434	562	496
420	485	446	479	422
494	289	436	516	615

3.11 The U.S. Energy Information Administration conducts annual surveys to estimate the average number of liveable square feet for housing units. In 1982 it was reported that the mean was 1449 sq. ft. and the median was 1222 sq. ft. Which measure of central tendency do you think is more appropriate? Explain your answer.

3.12 The U.S. Bureau of the Census provides figures on the average annual income of all U.S. households. For 1982, the mean household income was $24,309, and the median household income was $20,171. Which measure of central tendency do you think is more appropriate? Explain your answer.

Most statisticians recommend using the median for ordinal data, but some researchers also use the mean. In Examples 3.13 and 3.14 we have presented ordinal data sets. For each exercise,
a) compute the mean of the data.
b) compute the median of the data.
c) decide which measure of central tendency provides a better descriptive summary of the data.

3.13 A distance runner has entered seven marathons. His finishing places in the first six races were 4, 5, 3, 2, 7, and 4. In the seventh race he decided to go all out to win and ran in first place for 20 miles. This tired him out so badly that he ended up walking parts of the last six miles. He did finish, but only in 72nd place. The runner's places give the following ordinal data set:

$$4 \quad 5 \quad 3 \quad 2 \quad 7 \quad 4 \quad 72$$

3.14 Twenty-one algebra students were asked to rate the change in "test anxiety" produced by their algebra course. Negative ratings meant that they worried more about tests at the end of the course than at the beginning, whereas positive ratings meant they worried less at the end than at the beginning. Their ratings were:

0	0	0	−1	−1	1	0
−2	1	0	−2	0	−2	2
−2	1	0	−1	−3	−3	−1

3.15 In some data sets there are values called **outliers** that probably should not be included in the data set for one reason or another. Suppose, for example, you are interested in the ability of high-school algebra students to compute square roots. You decide to give a square-root exam to 10 of these students. Unfortunately, one of the students had a fight with his girlfriend and cannot concentrate—he gets a 0. The 10 scores in order are as follows:

$$0 \quad 58 \quad 61 \quad 63 \quad 67 \quad 69 \quad 70 \quad 71 \quad 78 \quad 80$$

The score of 0 is an *outlier*.

Many data sets contain obvious outliers. Statisticians have a systematic method for avoiding outliers when they calculate means. They compute **trimmed means,** in which high and low values are deleted or "trimmed off" before the mean is calculated. For instance, to compute the 10% trimmed mean of the test-score data, we first delete *both* the top 10% and the bottom 10% of the ordered data. Then we calculate the mean of the remaining data:

10% trimmed mean =

$$\frac{58 + 61 + 63 + 67 + 69 + 70 + 71 + 78}{8} = 67.1$$

[We deleted 0 and 80 from the original data set, since they are the bottom 10% and the top 10% of the data, respectively.]

Below is a set of algebra final-exam scores for a 40-question test.

2	15	16	16	19	21	21	25	26	27
4	15	16	17	20	21	24	25	27	28

a) Do any of the scores look like outliers?
b) Compute the usual mean for the data.
c) Compute the 5% trimmed mean for the data.
d) Compute the 10% trimmed mean for the data.
e) Compare the three means you obtained in parts (b)–(d). Which of the three means do you think provides the best measure of central tendency for the data?

Exercises 3.16 and 3.17 introduce two additional measures of central tendency—the *midrange* and the *mode*.

3.16 The **midrange** of a data set is defined to be the mean of the smallest and largest data values in the data set:

$$\text{Midrange} = \frac{\text{Smallest} + \text{Largest}}{2}$$

For instance, the midrange of the four exam scores 88, 75, 95, and 100 from Example 3.3(a) on page 59 is

$$\text{Midrange} = \frac{75 + 100}{2} = 87.5$$

a) Compute the midrange of the ages in Exercise 3.3.
b) Compute the midrange of the costs in Exercise 3.4.
c) Compute the midrange of the gestation periods in Exercise 3.6.

d) What advantages does the midrange have as a measure of central tendency? What disadvantages does it have?

3.17 The **mode** of a data set is defined to be the data value or values that occur most frequently. Unlike the mean and the median, a data set can have more than one mode.
   a) Determine the mode or modes of the ages in Exercise 3.3.

b) Determine the mode or modes of the costs in Exercise 3.4.
   c) Determine the mode or modes of the gestation periods in Exercise 3.6.
   d) Do you think it is reasonable to use the mode as a measure of central tendency? Give a reason for your answer.

3.2 Summation notation; the sample mean

In Definition 3.1, we defined the mean of a data set with an equation written in words:

$$\text{Mean} = \frac{\text{Sum of the data}}{\text{Number of pieces of data}}$$

It is possible to write such equations more concisely by using mathematical notation. We begin by introducing the mathematical notation for "sum of the data."

EXAMPLE 3.4 **Introduces summation notation**

The exam scores of the student in Example 3.3(a) are repeated here:

$$88 \quad 75 \quad 95 \quad 100$$

In mathematical notation, we let the letter x_i denote the ith value in the data set. For the exam scores above we have

$$\begin{aligned}
x_1 &= \text{score on Exam } 1 = 88 \\
x_2 &= \text{score on Exam } 2 = 75 \\
x_3 &= \text{score on Exam } 3 = 95 \\
x_4 &= \text{score on Exam } 4 = 100
\end{aligned}$$

More simply, we can just write $x_1 = 88$, $x_2 = 75$, $x_3 = 95$, and $x_4 = 100$. The numbers 1, 2, 3, and 4 written below the xs are called **subscripts.** Using this notation, the sum of the exam-score data can be expressed symbolically as

$$x_1 + x_2 + x_3 + x_4$$

Summation notation provides a shorthand for this last sum. The notation uses the upper-case Greek letter Σ (sigma). That letter, which corresponds to the English letter S, stands for the phrase "the sum of." Thus, in place of the lengthy expression

$$x_1 + x_2 + x_3 + x_4$$

we can simply use Σx, read "summation x" or "the sum of the x values:"

$$\Sigma x$$

$$\nearrow \quad \nwarrow$$

the sum of   the x values

For the exam-score data,

$$\Sigma x = x_1 + x_2 + x_3 + x_4 = 88 + 75 + 95 + 100 = 358$$

Since many people use a column to sum numbers, we often present data in a column with an x above and Σx below, as in Table 3.5.

TABLE 3.5
Exam-score data and its sum.

Score x
88
75
95
100
$\Sigma x = 358$

MTB
SPSS

■

Notation for the sample mean

To save us the trouble of writing the phrase "sample mean," we use the symbol $\bar{x}$ (read "x bar"). If we also use the letter n to denote the number of pieces of data, then the definition of the sample mean can be expressed in a very concise form:

$$\text{sample mean} \longrightarrow \bar{x} = \frac{\Sigma x}{n}$$

sum of the data ↙

number of pieces of data ↖

We summarize the above discussion in Definition 3.3.

DEFINITION 3.3 Sample mean

The **sample mean**, $\bar{x}$, of n pieces of sample data is given by

$$\bar{x} = \frac{\Sigma x}{n}$$

EXAMPLE 3.5 Illustrates Definition 3.3

Each year, manufacturers perform mileage tests on new car models and submit the results to the Environmental Protection Agency (EPA). The EPA then tests the vehicles to determine whether the manufacturers are correct. In 1985, one company reported that a particular model equipped with a four-speed manual transmission averaged 29 miles per gallon (mpg) on the highway. Let us suppose that the EPA tested 15 of the cars and obtained the gas mileages given in Table 3.6.

TABLE 3.6   Gas mileages, in miles per gallon, for a sample of 15 cars tested by the EPA.

27.3	31.2	29.4	31.6	28.6
30.9	29.7	28.5	27.8	27.3
25.9	28.8	28.9	27.8	27.6

Determine the sample mean of these gas mileages.

SOLUTION

Summing the gas mileages in Table 3.6 we obtained $\Sigma x = 431.3$. Since the number of pieces of data is $n = 15$, the sample mean gas mileage is

$$\bar{x} = \frac{\Sigma x}{n} = \frac{431.3}{15} = 28.75 \text{ mpg}$$

■

MTB
SPSS

Other important sums

We must often find sums besides the sum of the data, Σx. One such sum is the sum of the *squares* of the data, Σx^2. In the next section, we will need to obtain Σx, Σx^2, and various other sums. So that we can concentrate on the concepts to be presented there, instead of on the computations, we will discuss those sums now.

EXAMPLE   3.6   Illustrates some other important sums

The exam-score data from Example 3.4 are repeated below in the first column of Table 3.7. The remaining columns of the table give some related quantities of importance. [Do not be concerned right now with the meaning of these other quantities. You will see their significance in the next section.]

TABLE 3.7
Exam-score data and related quantities.

	x	x^2	$x - \bar{x}$	$(x - \bar{x})^2$
	88	7,744	−1.5	2.25
	75	5,625	−14.5	210.25
	95	9,025	5.5	30.25
	100	10,000	10.5	110.25
Σ	358	32,394	0	353.00

In Example 3.4 we found that the sum of the exam-score data is $\Sigma x = 358$. This is recorded as the first entry in the row labelled "Σ." The second column of Table 3.7 displays the squares, x^2, of the exam scores along with the sum of these squares, which is $\Sigma x^2 = 32,394$.

To obtain the third column of Table 3.7, we must first compute the mean, $\bar{x}$, of the four exam scores. Since $n = 4$ and $\Sigma x = 358$,

$$\bar{x} = \frac{\Sigma x}{n} = \frac{358}{4} = 89.5$$

Subtracting $\bar{x}$ (=89.5) from each of the four exam scores in the first column of the table, we get the $x - \bar{x}$ values shown in the third column. The sum of those values is $\Sigma(x - \overline{x}) = 0$. Finally, the fourth column of Table 3.7 gives the squares, $(x - \overline{x})^2$, of the $x - \overline{x}$ values. The sum of those squares is $\Sigma(x - \overline{x})^2 = 353$, as indicated. ∎

Exercises 3.2

3.18 Let $x_1 = 1$, $x_2 = 7$, $x_3 = 4$, $x_4 = 5$, and $x_5 = 10$.
   a) Compute Σx.
   b) Find n.
   c) Determine $\overline{x}$.

3.19 Let $x_1 = 12$, $x_2 = 8$, $x_3 = 9$, and $x_4 = 17$.
   a) Compute Σx.
   b) Find n.
   c) Determine $\overline{x}$.

For each data set in Exercises 3.20 through 3.23,
a) compute Σx.
b) find n.
c) determine the sample mean (round your answer to one more decimal place than the original data).

3.20 Atlas Fishing Line, Inc. manufactures a 10-lb test line. A sample of 12 spools is subjected to tensile-strength tests. The results are

9.8	10.2	9.8	9.4
9.7	9.7	10.1	10.1
9.8	9.6	9.1	9.7

3.21 A tire manufacturer needs information about the life of a new steel-belted radial he is going to sell. The results of tests on a sample of these tires are given below (data are in miles).

43,725	37,732	44,473	37,396
40,652	41,868	43,097	42,200

3.22 The U.S. Energy Information Administration reports that in 1983 the average annual motor fuel expenditure per U.S. household was $1317. That same year a random sample of 16 households within metropolitan areas gave the following annual motor fuel expenditures:

$1390	1459	2043	1551
415	1359	1778	1537
1167	1716	1463	904
560	1592	1710	638

3.23 According to the Salt River Project (SRP), a supplier of electricity to the greater Phoenix area, the mean annual electric bill in 1984 was $852.31. An economist wants to estimate the mean for last year. He takes a sample of SRP customers and obtains the following amounts for last year's electric bills.

$1175	778	1506	1040	1130	816
1038	786	1241	908	1094	1128
564	1299	1098	1094	898	868

In Exercises 3.24 through 3.27,
a) compute $\overline{x}$.
b) compute Σx^2, $\Sigma(x - \overline{x})$, and $\Sigma(x - \overline{x})^2$ by constructing a table similar to Table 3.7 on page 65.

3.24 A sample of five families gave the following data on the number of children per family:

2	3	4	4	3

3.25 The amount of money a salesperson earned on six randomly selected days yielded the following data set:

$75	98	130	63	112	107

3.26 According to the R. R. Bowker Company of New York, the mean annual subscription rate to law periodicals was $29.66 in 1983. A sample of this year's law periodicals gives the following subscription rates:

$20	36	34	37
32	28	52	45
42	38	33	44

3.27 A team of medical researchers has developed an exercise program to help reduce hypertension (high blood pressure). To determine whether the program is effective, the team selects a sample of 10 hypertensive individuals and places them in the exercise program for one month. Below are the diastolic blood pressures for the hypertensive individuals *before* beginning the exercise program.

106	118	118	99	109
94	109	95	97	106

In Exercises 3.28 through 3.32, assume that we have two data sets each with n data values. Denote the data values in one data set by $x_1, x_2, \ldots, x_n$ and the data values in the other data set by $y_1, y_2, \ldots, y_n$.

3.28 Explain the difference between $(\Sigma x)^2$ and Σx^2. Give an example to show that generally these two quantities are unequal.

3.29 Explain the difference between Σxy and $\Sigma x \Sigma y$. Give an example to show that generally these two quantities are unequal.

3.30 Prove the algebraic identity

$$\Sigma(x + y) = \Sigma x + \Sigma y$$

Hint: Write $\Sigma(x + y) = (x_1 + y_1) + \cdots + (x_n + y_n)$

and rearrange the terms to obtain $\Sigma x + \Sigma y$.

3.31 If c is a constant, prove that

$$\Sigma cx = c\Sigma x$$

3.32 If c is a constant, prove that

$$\Sigma c = nc$$

3.3   Measures of dispersion; the sample standard deviation

Up to this point, the only descriptive measures that we have discussed are measures of central tendency; specifically, the mean and the median. These descriptive measures indicate where the center or most typical value of a data set lies.

However, two data sets can have the same mean or the same median and yet still be quite different in other respects. For example, consider the heights of the five starting players on each of two men's college basketball teams, as shown in Figure 3.2. The two teams have the same mean heights; namely, 75 inches (6'3''). The median heights for the two teams are also identical; namely, 76 inches (6'4''). Nonetheless, it is clear that the two data sets differ. In particular, there is much more *variation* in the heights of the players on Team II than on Team I. To describe this difference quantitatively, we use a **measure of dispersion**—a descriptive measure that indicates the amount of variation in a data set.

Just as there are several different measures of central tendency, there are also several different measures of dispersion. In this section we will discuss two of the most frequently used measures of dispersion, the *range* and the *standard deviation.*

FIGURE 3.2 Heights of the five starting players on each of two men's college basketball teams.

	Team I					Team II				
Feet and inches	6'	6'1"	6'4"	6'4"	6'6"	5'7"	6'	6'4"	6'4"	7'
Inches	72	73	76	76	78	67	72	76	76	84

The range

We will first discuss the *range,* since it is simpler to understand and compute. Referring to the previous example, we see that the contrast between the two teams becomes clear if we place the shortest player on each team next to the tallest. See Figure 3.3.

FIGURE 3.3   Heights of the shortest and tallest starting players on each men's college basketball team.

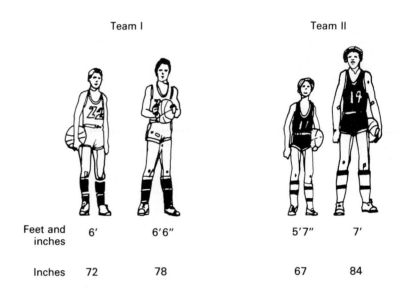

	Team I		Team II	
Feet and inches	6′	6′6″	5′7″	7′
Inches	72	78	67	84

The **range** of a data set is obtained by computing the difference between the largest and smallest data values in the data set. Hence, as we see from Figure 3.3:

$$\text{Team I: Range} = 78'' - 72'' = 6''$$
$$\text{Team II: Range} = 84'' - 67'' = 17''$$

In general, we have the following definition.

DEFINITION 3.4 Range of a data set

The **range** of a data set is defined to be the difference between the largest and smallest data values in the data set:

$$\text{Range} = \text{Largest value} - \text{Smallest value}$$

The range of a data set is quite easy to compute. However, in using the range a great deal of information is ignored—only the largest and smallest data values are considered; the remainder of the data is disregarded.

MTB
SPSS

The sample standard deviation[†]

In contrast to the range, the *standard deviation* takes into account all the data values. For this and other reasons, the standard deviation is preferable to the range as a measure of dispersion. The computations required in determining the standard deviation are more involved than those for the range. However, this is not a serious problem since computers and sophisticated calculators are available to do the necessary computations.

Roughly speaking, the standard deviation measures the variation in a data set by determining how far the data values are from the mean, on the average. If there is a large amount of variation in the data, then on the average, the data values will be far from the mean. Hence, the standard deviation will be large. On the other hand, if there is only a small amount of variation in the data, then on the average, the data values will be close to the mean. Hence, the standard deviation will be small.

The first step in computing the sample standard deviation is to find how far each data value is from the mean—the so-called **deviations from the mean.** We illustrate the calculations of the deviations from the mean in the next example.

EXAMPLE 3.7   Illustrates the deviations from the mean

The heights, in inches, for the five starting players on Team I are

$$72 \quad 73 \quad 76 \quad 76 \quad 78$$

(See Figure 3.2, page 67.) Determine the deviations from the mean.

SOLUTION

The mean height of the starting players on Team I is

$$\bar{x} = \frac{\Sigma x}{n} = \frac{72 + 73 + 76 + 76 + 78}{5} = \frac{375}{5} = 75 \text{ inches}$$

To obtain the deviation from the mean for a particular data value, we simply subtract the mean from that data value; that is, we compute $x - \bar{x}$. For instance, the deviation from the mean for the height of 72 inches is $x - \bar{x} = 72 - 75 = -3$. The deviations from the mean for all five data values are given in the second column of Table 3.8 and are displayed graphically in Figure 3.4.

TABLE 3.8
Deviations from the mean for the heights of the five starting players on Team I. [Recall that $\bar{x} = 75$.]

Height x	Deviation from the mean $x - \bar{x}$
72	−3
73	−2
76	1
76	1
78	3

[†] As we mentioned earlier, most data sets we consider will be sample data. Consequently we will assume that all data sets in this section are sample data. We will discuss population data and the population standard deviation in Section 3.7.

FIGURE 3.4
Deviations from the mean.
(Dots represent data values.)

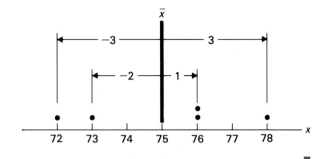

The second step in computing the sample standard deviation is to obtain a measure of the *total deviation* from the mean for all data values. Although the quantities $x - \bar{x}$ represent deviations from the mean, adding them to get a total deviation from the mean is of no value because their sum, $\Sigma(x - \bar{x})$, *always* equals zero. [The second column of Table 3.8 shows that $\Sigma(x - \bar{x}) = 0$ for the height data. See Exercise 3.48 for a verification that $\Sigma(x - \bar{x}) = 0$ in general.]

In the calculation of the standard deviation, the deviations from the mean, $x - \bar{x}$, are *squared* to obtain quantities that do not add up to zero. The sum of the squared deviations from the mean, $\Sigma(x - \bar{x})^2$, is called the **sum of squared deviations** and is a measure of the total deviation of the data values from the mean.

EXAMPLE   3.8   Illustrates the sum of squared deviations

Compute the sum of squared deviations for the heights of the starting players on Team I.

SOLUTION
In Table 3.9 we have appended a column for $(x - \bar{x})^2$ to Table 3.8.

TABLE 3.9
Table for computing the sum of squared deviations for the heights of the starting players on Team I.

Height x	Deviation from mean $x - \bar{x}$	Squared deviation $(x - \bar{x})^2$
72	−3	9
73	−2	4
76	1	1
76	1	1
78	3	9
Σ		24

From the third column of the table we find that

$$\text{Sum of squared deviations} = \Sigma(x - \bar{x})^2 = 24 \text{ inches}^2$$

The third step in computing the sample standard deviation is to take an *average* of the squared deviations. This is accomplished by dividing the sum of squared deviations by $n-1$ (one less than the sample size, n). The resulting quantity is called the **sample variance** and is denoted by s^2.[†] In symbols,

$$s^2 = \frac{\Sigma(x - \overline{x})^2}{n - 1}$$

EXAMPLE 3.9   **Illustrates the sample variance**

Compute the sample variance of the heights of the starting players on Team I.

SOLUTION
From Example 3.8, the sum of squared deviations is $\Sigma(x - \overline{x})^2 = 24$ inches². Also, $n = 5$ since there are five pieces of data. Thus, the sample variance is

$$s^2 = \frac{\Sigma(x - \overline{x})^2}{n - 1} = \frac{24}{5 - 1} = 6 \text{ inches}^2$$

∎

It is important to realize that the sample variance is in units that are the square of the original units. This results from squaring the deviations from the mean. For instance, the sample variance of the heights of the players on Team I is $s^2 = 6$ inches² (Example 3.9). Since it is desirable to have descriptive measures in the original units, the final step in computing the sample standard deviation is to take the square root of the sample variance. In other words, the **sample standard deviation, s,** is

$$s = \sqrt{\frac{\Sigma(x - \overline{x})^2}{n - 1}}$$

EXAMPLE 3.10   **Illustrates the sample standard deviation**

Compute the sample standard deviation of the heights of the starting players on Team I.

SOLUTION
From Example 3.9, the sample variance is $s^2 = \Sigma(x - \overline{x})^2/(n - 1) = 6$ inches². Thus, the sample standard deviation is

$$s = \sqrt{\frac{\Sigma(x - \overline{x})^2}{n - 1}} = \sqrt{6} = 2.4 \text{ inches}$$

∎

We summarize the above discussion regarding the sample standard deviation in Definition 3.5.

[†] If instead of dividing by $n-1$, we divided by n, then the sample variance would be the mean of the squared deviations. Although dividing by n seems more natural, we divide by $n-1$ for the following reason: One of the main uses of the sample variance is to estimate the population variance (to be defined in Section 3.7). Division by n tends to underestimate the population variance, whereas division by $n-1$ does not.

DEFINITION 3.5 Sample standard deviation

The **sample standard deviation, s,** of n pieces of sample data is defined by

$$s = \sqrt{\frac{\Sigma(x - \bar{x})^2}{n - 1}}$$

The steps required for computing the sample standard deviation were illustrated for the heights of the starting players on Team I in Examples 3.7–3.10. The computations were performed in four separate examples so that we could explain the interpretation of the sample standard deviation as well as the calculations involved. However, now that we have done that, we can give a relatively simple procedure for computing the sample standard deviation, s, for any data set: *(1) calculate the sample mean, $\bar{x}$; (2) construct a table (similar to Table 3.9 on page 70) to determine the sum of squared deviations, $\Sigma(x - \bar{x})^2$; and (3) apply Definition 3.5.*

EXAMPLE 3.11   Illustrates Definition 3.5

The heights, in inches, for the five starting players on Team II are

$$67 \quad 72 \quad 76 \quad 76 \quad 84$$

Compute the sample standard deviation for these heights.

S O L U T I O N

We apply the procedure described above. First, we calculate the sample mean, $\bar{x}$:

$$\bar{x} = \frac{\Sigma x}{n} = \frac{67 + 72 + 76 + 76 + 84}{5} = \frac{375}{5} = 75 \text{ inches}$$

Next, we construct a table to determine the sum of squared deviations, $\Sigma(x - \bar{x})^2$:

TABLE 3.10

x	$x - \bar{x}$	$(x - \bar{x})^2$
67	-8	64
72	-3	9
76	1	1
76	1	1
84	9	81
Σ		156

From the third column of Table 3.10 we see that the sum of squared deviations is

$$\Sigma(x - \bar{x})^2 = 156 \text{ inches}^2$$

Finally, we apply Definition 3.5. We have $n = 5$ and $\Sigma(x - \bar{x})^2 = 156$. Consequently, the sample standard deviation of the heights for Team II is

$$s = \sqrt{\frac{\Sigma(x - \bar{x})^2}{n - 1}} = \sqrt{\frac{156}{5 - 1}} = \sqrt{39} = 6.2 \text{ inches}$$

■

In Example 3.10 (page 71) we found that the sample standard deviation for the heights of the players on Team I is $s = 2.4$ inches. In Example 3.11 we found that the sample standard deviation for the heights of the players on Team II is $s = 6.2$ inches. Consequently, we see that Team II, which has more variation in height than Team I, also has a larger standard deviation. This is the way a measure of dispersion is supposed to work.

KEY FACT

The more variation there is in a data set, the larger its standard deviation.

A shortcut formula for *s*

We are going to present an alternative formula for computing the sample standard deviation, *s*. Consequently, it will be convenient to have a name for our original formula

$$s = \sqrt{\frac{\Sigma(x - \overline{x})^2}{n - 1}}$$

Since this is the formula used to define the sample standard deviation, we will call it the **defining formula** for *s*.

The alternative formula for computing the sample standard deviation is given below. We will call this formula the **shortcut formula** for *s*.

FORMULA 3.1 Shortcut formula for *s*

The sample standard deviation, *s*, of *n* pieces of sample data can be computed from the formula

$$s = \sqrt{\frac{n(\Sigma x^2) - (\Sigma x)^2}{n(n - 1)}}$$

The shortcut formula for *s* is equivalent to the defining formula. That is, both formulas give the same result.[†] However, the shortcut formula is generally faster and easier for doing calculations and reduces the chance of roundoff errors.

Before illustrating the shortcut formula for *s*, we should comment on the similar-looking expressions Σx^2 and $(\Sigma x)^2$ that occur in the formula. The expression Σx^2 represents the sum of the squares of the data values. It is obtained by first squaring each data value and then summing those squared values. On the other hand, the expression $(\Sigma x)^2$ represents the square of the sum of the data values; and it is obtained by first summing the data values and then squaring that sum.

EXAMPLE 3.12 **Illustrates Formula 3.1**

In Example 3.11 on page 72, we computed the sample standard deviation of the

---

[†] Differences due to roundoff errors are possible.

heights for the five starting players on Team II using the defining formula for *s*. Compute that sample standard deviation using the shortcut formula for *s*.

S O L U T I O N
To apply the shortcut formula, we need the sums Σx and Σx^2. These are determined in Table 3.11.

TABLE 3.11
Table for computation of *s* using the shortcut formula.

x	x^2
67	4,489
72	5,184
76	5,776
76	5,776
84	7,056
Σ 375	28,281

We have $n = 5$, and from the table we see that $\Sigma x = 375$ and $\Sigma x^2 = 28,281$. Thus by the shortcut formula,

$$s = \sqrt{\frac{n(\Sigma x^2) - (\Sigma x)^2}{n(n-1)}} = \sqrt{\frac{5(28,281) - (375)^2}{5(5-1)}}$$

$$= \sqrt{\frac{780}{20}} = \sqrt{39} = 6.2 \text{ inches}$$

∎

MTB
SPSS

The value for the standard deviation of the heights obtained in Example 3.12 using the shortcut formula is, of course, the same as the value obtained in Example 3.11 using the defining formula. For the height data, both formulas are relatively easy to apply. However, for most data sets, and especially those in which the mean is not a whole number, the shortcut formula is better.

Exercises **3.3**

3.33   What is the purpose of a measure of dispersion?

3.34   Why is the sample standard deviation preferable to the range as a measure of dispersion?

In Exercises 3.35 through 3.42 we have repeated the data from Exercises 3.3 through 3.10. For each exercise,
a) determine the range.
b) compute the sample standard deviation, *s*, using the defining formula.
c) compute the sample standard deviation, *s*, using the shortcut formula.
d) state which formula you found easier to use in computing *s*.

Note: In parts (b) and (c) round your final answers to one more decimal place than the original data.

3.35   The U.S. National Science Foundation collects data on the age of recipients of science and engineering doctoral degrees. Suppose that a sample of this year's recipients has the following ages:

37	28	36	33
37	43	41	28
24	44	27	24

3.36   The Health Insurance Association of America publishes figures on the costs to community hospitals per patient per day. Suppose that a sample of 10 such costs in New York yields the data below.

$602	539	569	916	768
335	776	887	806	422

3.37 The average retail price for oranges in 1983 was 38.5 cents per pound, as reported by the U.S. Department of Agriculture. Recently a sample of 15 markets reported the following prices for oranges in cents per pound:

43.0	40.0	42.6	40.2	37.5
44.1	45.2	41.8	35.6	34.6
37.9	44.2	44.5	38.2	42.4

3.38 A biologist is studying the gestation period of domestic dogs. Fifteen dogs are observed during pregnancy and are found to have the gestation periods, in days, given below.

62.0	61.4	59.8
62.2	60.3	60.4
59.4	60.2	60.4
60.8	61.8	59.2
61.1	60.4	60.9

3.39 A manufacturer of liquid soap produces a bottle with an advertised contents of 310 ml. A sample of 16 bottles yields the following contents:

297	318	306	300
311	303	291	298
322	307	312	300
315	296	309	311

3.40 A city planner working on bikeways needs information about local bicycle commuters. She designs a questionnaire. One of the questions asks how many minutes it takes the rider to pedal from home to his or her destination. A sample of 22 local bicycle commuters yields the following times:

22	19	24	31
29	29	21	15
27	23	37	31
30	26	16	26
12	23	48	
22	29	28	

3.41 The U.S. National Center for Educational Statistics surveys college and university libraries to obtain information on the number of volumes held. A sample of seven public colleges and universities hold the number of volumes, in thousands, indicated below.

79	516	24	265	41	15	411

3.42 According to the College Entrance Examination Board, the average verbal score on the Scholastic Aptitude Test in 1983 was 425 points out of a possible 800. A sample of 25 verbal scores for last year yielded the following results:

346	496	352	378	315
491	360	385	500	558
381	303	434	562	496
420	485	446	479	422
494	289	436	516	615

3.43 TSI, an independent testing agency, has tested the lifetimes of two brands of light bulbs. Light bulb life is defined as the number of hours that a bulb will operate continuously before it burns out. The results of tests on seven bulbs of each brand are displayed in the table below (data in hundreds of hours).

Brand A	Brand B
10.5	11.3
9.1	7.0
10.0	9.7
10.3	9.6
9.4	10.5
9.6	11.8
9.7	8.7

a) Compute $\bar{x}$ for each data set.
b) Determine the median of each data set.
c) Even though the two data sets have the same means and medians, they are quite different in another respect. How are they different?
d) Which data set appears to have less variation?
e) Compute s for each data set.
f) Are your answers in parts (d) and (e) consistent? Why?

3.44 Consider the following data sets.

Data Set I		Data Set II		Data Set III		Data Set IV	
1	5	1	9	5	5	10	4
1	8	1	9	5	5	10	4
2	8	1	9	5	5	4	4
2	9	1	9	5	5	4	4
5	9	1	9	5	5	4	2

a) Compute $\bar{x}$ for each data set.
b) Even though the four data sets have the same means, they are quite different in another respect. How are they different?
c) Which data set appears to have the least variation? the greatest variation?
d) Compute the range of each data set.
e) Compute s for each data set.
f) From your answers to parts (d) and (e), which measure of dispersion better distinguishes the variation in the four data sets, the range or the standard deviation? Explain your answer.
g) Are your answers from parts (c) and (e) consistent? Why?

3.45   Below are 10 IQ scores:

110	122	132	107	101
97	115	91	125	142

Time each of the following calculations on a watch or timer.

a) Calculate the sample standard deviation of the 10 IQs using the defining formula for s, and record the time your calculation requires.

b) Calculate the sample standard deviation of the 10 IQs using the shortcut formula for s, and record the time your calculation requires.

c) Was the shortcut formula really a time-saver?

3.46   Another measure of dispersion is the **mean absolute deviation (MAD).** The mean absolute deviation measures the variation in a data set by determining the mean of the absolute values of the deviations from the mean. That is, the mean absolute deviation is defined by

$$MAD = \frac{\Sigma|x - \bar{x}|}{n}$$

a) Compute the mean absolute deviation of the ages in Exercise 3.35.

b) Compute the mean absolute deviation of the costs in Exercise 3.36.

3.47   In Exercise 3.15 (page 62) we discussed *outliers* —extreme values in a data set. At the top of the next column are two data sets. Data Set II is obtained by removing the outliers from Data Set I.

Data Set I					Data Set II			
0	12	14	15	23	10	14	15	17
0	14	15	16	24	12	14	15	
10	14	15	17		14	15	16	

a) Compute the sample standard deviation, s, for both data sets.

b) Compute the range for both data sets.

c) What effect do outliers have on the variation in a data set? Explain.

3.48   On page 70, we pointed out that the sum of the deviations from the mean is always equal to zero; that is, $\Sigma(x - \bar{x}) = 0$ for any data set. Prove that this is true. *Hint:* Use the results of Exercises 3.30–3.32 on page 67.

3.49   This exercise shows that the shortcut formula for s is equivalent to the defining formula.

a) Verify that

$$\Sigma(x - \bar{x})^2 = \Sigma x^2 - \frac{(\Sigma x)^2}{n}$$

Hint: Expand the square on the left and then apply the results of Exercises 3.30–3.32 on page 67.

b) Conclude from part (a) that

$$\frac{\Sigma(x - \bar{x})^2}{n - 1} = \frac{n(\Sigma x^2) - (\Sigma x)^2}{n(n - 1)}$$

c) Deduce from part (b) that the defining formula and shortcut formula for s are equivalent.

3.4

Interpretation of the standard deviation; Chebychev's theorem

As you know, the standard deviation is a measure of dispersion—it is a descriptive measure used to indicate the amount of variation in a data set. The Key Fact on page 73 states that the standard deviation has an important property that any measure of dispersion must have; namely, *the more variation there is in a data set, the larger its standard deviation.*[†]

Up to this point, we have been concentrating more on the calculation of the standard deviation than on its meaning. In this section we will present material that should give you a better idea of how to interpret the standard deviation.

[†] In the extreme case in which all the data values in a data set are equal (that is, there is no variation), the standard deviation is zero; conversely, if the standard deviation of a data set is zero, then all the data values in the data set must be equal. See Exercise 3.64.

Table 3.12 displays two data sets, each of which has 10 data values. A brief inspection of the table reveals that there is more variation in Data Set II than in Data Set I.

TABLE 3.12

Data Set I	Data Set II
49	37
55	61
54	49
49	20
52	70
54	53
45	48
52	48
55	50
35	64

Using the formulas

$$\bar{x} = \frac{\Sigma x}{n} \quad \text{and} \quad s = \sqrt{\frac{n(\Sigma x^2) - (\Sigma x)^2}{n(n-1)}}$$

we computed the sample mean and sample standard deviation of each data set. The results are given below:

Data Set I	Data Set II
$\bar{x} = 50.0$	$\bar{x} = 50.0$
$s = 6.2$	$s = 14.2$

So, as we would expect, the standard deviation of Data Set II is larger than that of Data Set I.

To enable you to visually compare the variations in the two data sets, we have drawn the graphs in Figures 3.5 and 3.6. On each graph we have marked the data values with dots. In addition, we have located the sample mean and have measured off intervals equal in length to the standard deviation (6.2 for Data Set I and 14.2 for Data Set II).

FIGURE 3.5 Data Set I. $\bar{x} = 50$, $s = 6.2$

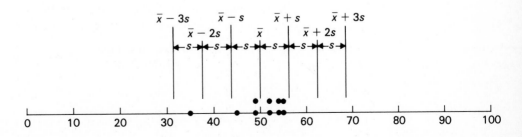

FIGURE 3.6   Data Set II. $\bar{x} = 50$, $s = 14.2$

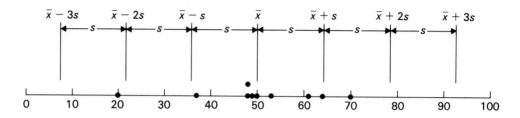

The graphs in Figures 3.5 and 3.6 vividly illustrate that there is more variation in Data Set II than in Data Set I. The graphs also show that, for each data set, all the data values lie within a few standard deviations to either side of the mean. This is no accident—*almost all the data in any data set will lie within three standard deviations to either side of its mean.* A data set with a great deal of variation will have a large standard deviation and, consequently, three standard deviations to either side of its mean will be rather extensive, as in Figure 3.6. On the other hand, a data set with little variation will have a small standard deviation and, hence, three standard deviations to either side of its mean will be rather narrow, as in Figure 3.5.

Chebychev's theorem

As we said, almost all the data in any data set will lie within three standard deviations to either side of its mean. Chebychev's theorem makes this qualitative statement more precise.

KEY FACT Chebychev's theorem

For *any* data set, the proportion of data that lies within k standard deviations to either side of its mean is *at least* $1 - 1/k^2$.

The statement of Chebychev's theorem is rather abstract, so it is helpful to examine some special cases for various values of k. For instance, when $k = 3$,

$$1 - 1/k^2 = 1 - 1/3^2 = 1 - 1/9 = 8/9 = 0.89$$

So by Chebychev's theorem, for any data set, the proportion of data lying within three standard deviations to either side of its mean is at least 0.89. In other words

For any data set, *at least* 89% of the data lies within
three standard deviations to either side of its mean.

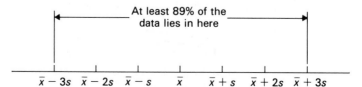

As another special case of Chebychev's theorem, consider $k = 2$. Then

$$1 - 1/k^2 = 1 - 1/2^2 = 1 - 1/4 = 3/4 = 0.75$$

So by Chebychev's theorem, for any data set, the proportion of data lying within two standard deviations to either side of its mean is at least 0.75. In other words

For any data set, *at least* 75% of the data lies within two standard deviations to either side of its mean.

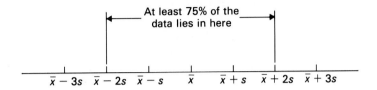

We should emphasize that the value of Chebychev's theorem is not in its precision but in its generality. For example, consider the data set portrayed in Figure 3.5 on page 77. Chebychev's theorem says that *at least* 75% of the data will lie within two standard deviations of the mean. However, as we can see from Figure 3.5, 90% of the data actually lies within two standard deviations of the mean. Similarly, Chebychev's theorem says that *at least* 89% of the data portrayed in Figure 3.5 will lie within three standard deviations of the mean, whereas in fact, 100% of the data lies within three standard deviations of the mean.

Thus, we see that Chebychev's theorem does not necessarily give precise estimates for the percentage of data that lies within k standard deviations to either side of the mean. Its power is its generality—Chebychev's theorem holds for *any* data set. Moreover, Chebychev's theorem permits us to make pertinent statements about a data set when we know only its mean and standard deviation, and frequently that is all we do know. Consider the following example.

EXAMPLE 3.13   **Illustrates Chebychev's theorem**

A sociology instructor tells the 20 students in her class that the mean score on the last 100-point exam is $\bar{x} = 71.4$ with a standard deviation of $s = 9.6$. Using this information only, apply Chebychev's theorem to make some observations about the exam scores.

SOLUTION
Based on the fact that $\bar{x} = 71.4$ and $s = 9.6$, we can construct Figure 3.7. [Note, for instance, that $\bar{x} - 2s = 71.4 - 2(9.6) = 52.2$.]

FIGURE 3.7

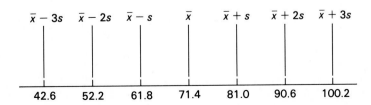

By Chebychev's theorem, with $k = 2$, *at least* 75% of the 20 exams have scores within two standard deviations to either side of the mean. Since 75% of 20 is 15, we can conclude in view of Figure 3.7 that *at least* 15 of the 20 exams have scores between 52.2 and 90.6.

If we apply Chebychev's theorem with $k = 3$, then we can say that *at least* 89% of the 20 exams have scores within three standard deviations to either side of the mean. Now, 89% of 20 is 17.8. Consequently, in view of Figure 3.7, we can conclude that *at least* 18 of the 20 exams have scores between 42.6 and 100.2. [Since the maximum score possible is 100, we can actually conclude that at least 18 of the 20 exam scores are above 42.6.] ■

Exercises 3.4

3.50   Consider the following data sets:

Data Set 1		Data Set 2	
30	20	14	9
16	24	56	32
22	19	13	8
23	13	26	3
18	9	9	16
18	28	31	23

a) In which data set does there appear to be more variation?
b) Compute $\bar{x}$ and s for each data set.
c) Draw graphs similar to Figures 3.5 and 3.6 on pages 77 and 78.
d) Interpret your graphs from part (c).
e) Do most of the data in each data set lie within three standard deviations to either side of the mean?

3.51   Consider the following data sets:

Data Set 3		Data Set 4	
82	78	97	59
85	94	100	100
65	84	30	95
91	86	87	79
81	84	90	93

a) In which data set does there appear to be more variation?
b) Compute $\bar{x}$ and s for each data set.
c) Draw graphs similar to Figures 3.5 and 3.6 on pages 77 and 78.
d) Interpret your graphs from part (c).

e) Do most of the data in each data set lie within three standard deviations to either side of the mean?

3.52   What does Chebychev's theorem say about the percentage of data in a data set that lies within
a) one standard deviation to either side of its mean?
b) 2.5 standard deviations to either side of its mean?
c) 3.75 standard deviations to either side of its mean?

3.53   What does Chebychev's theorem say about the percentage of data in a data set that lies within
a) 1.25 standard deviations to either side of its mean?
b) 3.5 standard deviations to either side of its mean?
c) five standard deviations to either side of its mean?

3.54   Refer to Exercise 3.50.
a) What percentage of the data does Chebychev's theorem say will lie within two standard deviations to either side of the mean? Within three standard deviations to either side of the mean?
b) Using your graph from Exercise 3.50 (c), determine the percentage of the data in Data Set 1 that actually lies within two standard deviations to either side of the mean; within three standard deviations to either side of the mean.
c) Repeat part (b) for Data Set 2.
d) What do parts (b) and (c) illustrate about Chebychev's theorem?

3.55   Refer to Exercise 3.51.
a) What percentage of the data does Chebychev's

theorem say will lie within two standard deviations to either side of the mean? Within three standard deviations to either side of the mean?

b) Using your graph from Exercise 3.51 (c), determine the percentage of the data in Data Set 3 that actually lies within two standard deviations to either side of the mean; within three standard deviations to either side of the mean.

c) Repeat part (b) for Data Set 4.

d) What do parts (b) and (c) illustrate about Chebychev's theorem?

3.56 We stated on page 78 that almost all of the data in any data set will lie within three standard deviations to either side of its mean. But in every example and exercise so far, *all* of the data were within three standard deviations of the mean. Verify that this is not the case for the following data set:

100	28	69	85	85	98
97	100	87	97	96	94
89	97	93	92	95	94
79	83	97	90	96	57
64	74	80	87	89	58

[*Note*: $\Sigma x = 2550$ and $\Sigma x^2 = 224{,}272$.]

In Exercises 3.57 through 3.62 we have given the sample mean, sample standard deviation, and sample size for a data set. For each exercise,

a) construct a graph similar to Figure 3.7 on page 79.

b) apply Chebychev's theorem with $k = 2$ to make some observations about the data.

c) repeat part (b) with $k = 3$.

3.57 The Philadelphia CPA firm Laventhol and Horwath conducts annual surveys to obtain characteristics of full-service and economy lodging establishments. Suppose that the room rates (for a double) for 300 lodging establishments have a mean of $\bar{x} = \$40.97$ with a standard deviation of $s = \$8.66$.

3.58 The U.S. National Center for Health Statistics collects data on cigarette smokers by sex and age. Suppose that a sample of 2760 females who presently smoke has a mean age of $\bar{x} = 38.7$ years, with a standard deviation of $s = 12.6$ years.

3.59 The Gallup Organization conducts annual surveys on home gardening. Results are published by the National Association for Gardening. Suppose a sample of 250 households with vegetable gardens yields a mean garden size of $\bar{x} = 643$ square feet, with a standard deviation of $s = 247$ square feet.

3.60 The Health Insurance Association of America reports that in 1983 the average daily room charge for a semi-private room in U.S. hospitals was $195. In the same year, a sample of 30 Massachusetts hospitals yielded a mean semi-private room charge of $\bar{x} = \$202.68$ with a standard deviation of $s = \$12.77$.

3.61 According to the Internal Revenue Service, the average income tax per *taxable* return was $3603 in 1982. A sample of 30 taxable returns from last year has a mean income tax of $\bar{x} = \$2239.8$ with a standard deviation of $s = \$2215.6$.

3.62 *The World Almanac, 1985* reports that in 1980 the average travel time to work for residents of South Dakota was 13 minutes. For this year, a sample of 35 travel times for South Dakota residents gave a mean of $\bar{x} = 14.0$ minutes with a standard deviation of $s = 13.3$ minutes.

3.63 Suppose you take a final exam with 400 possible points. The instructor tells you that the mean score is 280 and that the standard deviation is 20. He also tells you that you got 350. Did you do well on the exam? Explain.

3.64 Consider a data set with n data values, $x_1, x_2, \ldots, x_n$.

a) Show that if all the data values are equal, then the standard deviation is zero.

b) Show that if the standard deviation of the data set is zero, then all of the data values must be equal.

3.65 How many standard deviations to either side of the mean must we go to be assured that

a) at least 95% of the data lies within?

b) at least 99% of the data lies within?

3.5

Computing $\bar{x}$ and s for grouped data

We have learned how to compute the sample mean, $\bar{x}$, and the sample standard deviation, s, of a data set. Often the data we encounter are grouped into a frequency distribution. In this section we will present methods for computing $\bar{x}$ and s when the data are in grouped form.

Computing $\bar{x}$ for grouped data

In Example 3.1 on page 57, we considered two sets of salary data representing weekly earnings of the employees of a small mathematical consulting firm. Data Set II is repeated here in Table 3.13. A frequency distribution for that data set is given in Table 3.14.

TABLE 3.13
Data Set II (Week ending August 12)

Salary
$200
200
840
350
300
300
200
200
950
200
Σ $3740

TABLE 3.14
Frequency distribution for Data Set II.

Salary ($)	Frequency
200	5
300	2
350	1
840	1
950	1
Σ	10

We computed the mean of Data Set II in Example 3.1. In doing so, we had to determine the sum of the salaries displayed in Table 3.13. That sum can be found more quickly using the frequency distribution given in Table 3.14. To see how, we first express the sum of the salaries in Data Set II as follows:

Sum of salaries $= 200 + 200 + 840 + 350 + 300 + 300 + 200 + 200 + 950 + 200$

$$= \underbrace{200 + 200 + 200 + 200 + 200}_{5} + \underbrace{300 + 300}_{2} + \underbrace{350}_{1} + \underbrace{840}_{1} + \underbrace{950}_{1}$$

In the second sum we combined like salaries and placed the frequency of each of the different salaries above each group. Consequently, the sum of the salaries can be rewritten as shown below:

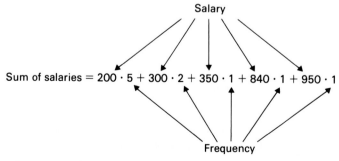

In other words, we can determine the sum of the salaries by multiplying each of the *different* salaries by its corresponding frequency and then adding the results.

This can be done most efficiently by appending a column to the frequency distribution in Table 3.14 in order to perform the necessary computations. (See Table 3.15.)

TABLE 3.15

Salary x	Frequency f	Salary · Frequency xf
200	5	1000
300	2	600
350	1	350
840	1	840
950	1	950
Σ	10	3740 $\longleftarrow$ *Sum of salaries*

We have also introduced some mathematical notation in Table 3.15: x represents salary-class value (not each individual salary, as before); f represents class frequency; and xf represents salary-class value times class frequency. Note that for a given class, xf equals the sum of the salaries in that class.

We will now show how Table 3.15 can be used to compute the mean of the salary data. The sum of all the salaries is Σxf, the sum of the third column in Table 3.15. The number of salaries is Σf, the sum of the second column in Table 3.15. Consequently, the mean of the salaries is

$$\bar{x} = \frac{\text{Sum of salaries}}{\text{Number of salaries}} = \frac{\Sigma xf}{\Sigma f} = \frac{3740}{10} = \$374$$

Note: Previously we have used the letter n to denote the number of pieces of data. As illustrated above, when the data are in grouped form, the number of pieces of data is equal to the sum of the class frequencies; that is, $n = \Sigma f$. To minimize the amount of notation, we will generally use n instead of Σf to denote the number of pieces of data.

The preceding discussion shows how we can compute the mean of a data set when the data are grouped in a frequency distribution. We summarize that discussion in Formula 3.2.

FORMULA 3.2 Computing $\bar{x}$ for grouped data

The sample mean of a data set that is grouped in a frequency distribution can be computed using the formula

$$\bar{x} = \frac{\Sigma xf}{n}$$

where f denotes class frequency and $n \ (= \Sigma f)$ denotes the number of pieces of data.

EXAMPLE 3.14   Illustrates Formula 3.2

Table 3.16 gives a frequency distribution for the salary data in Data Set I (Table 3.1, page 57).

TABLE 3.16
Frequency distribution for
Data Set I.

Salary ($) x	Frequency f
200	6
300	2
350	2
700	1
840	1
950	1

Compute the sample mean of these salaries.

S O L U T I O N
Since the data are in grouped form, we apply Formula 3.2. Appending an xf
column to Table 3.16 we obtain Table 3.17.

TABLE 3.17
Table for calculation of $\overline{x}$.

	x	f	xf
	200	6	1200
	300	2	600
	350	2	700
	700	1	700
	840	1	840
	950	1	950
Σ		13	4990

MTB
SPSS

Using the final row of Table 3.17 we can compute $\overline{x}$:

$$\overline{x} = \frac{\Sigma xf}{n} = \frac{4990}{13} = \$383.85$$

∎

Computing s for grouped data

We have just seen how to compute the sample mean of a data set when it is
grouped into a frequency distribution. Using similar reasoning, we can also obtain a formula for computing the sample standard deviation of a data set when it
is grouped into a frequency distribution. That formula, along with its shortcut
version, is presented below.

FORMULA 3.3 Computing s for grouped data

The sample standard deviation of a data set that is grouped in a frequency distribution can be computed using the formula

$$s = \sqrt{\frac{\Sigma(x - \overline{x})^2 f}{n - 1}}$$

or its shortcut version

$$s = \sqrt{\frac{n(\Sigma x^2 f) - (\Sigma xf)^2}{n(n - 1)}}$$

where f denotes class frequency and $n \ (= \Sigma f)$ denotes the number of pieces of data.

EXAMPLE 3.15 **Illustrates Formula 3.3**

A frequency distribution for the salary data in Data Set I is given in Table 3.16 and is repeated below in the first two columns of Table 3.18. Compute the sample standard deviation of these salaries.

S O L U T I O N
We will apply the shortcut formula presented in Formula 3.3. As you can see from the formula, we will need a table with columns for x, f, xf, x^2, and x^2f. We will also need to sum the second, third, and fifth columns of that table to obtain n, Σxf, and Σx^2f. See Table 3.18.

TABLE 3.18
Table for calculation of s
using the shortcut formula.

x	f	xf	x^2	x^2f
200	6	1200	40,000	240,000
300	2	600	90,000	180,000
350	2	700	122,500	245,000
700	1	700	490,000	490,000
840	1	840	705,600	705,600
950	1	950	902,500	902,500
Σ	13	4990		2,763,100

From the final row of Table 3.18 we find that

$$s = \sqrt{\frac{n(\Sigma x^2 f) - (\Sigma x f)^2}{n(n-1)}} = \sqrt{\frac{13(2,763,100) - (4990)^2}{13(12)}}$$

$$= \sqrt{\frac{35,920,300 - 24,900,100}{156}} = \sqrt{\frac{11,020,200}{156}} = \$265.79$$

MTB
SPSS

∎

Computing $\bar{x}$ and s when the classes are not based on a single value

So far we have considered only the computations of $\bar{x}$ and s for grouped data in which each class is based on a single value. In the case in which each class represents a number of different values, such as the grouped data on page 17, Formulas 3.2 and 3.3 cannot be used to compute the exact values of $\bar{x}$ and s. However, they can be used to compute the approximate values of $\bar{x}$ and s. This is accomplished by substituting the *class marks* for the x-values.

EXAMPLE 3.16 **Illustrates Formulas 3.2 and 3.3**

A grouped-data table for the number of days to maturity for 40 short-term investments is given in Table 2.4 on page 19. The first, second, and fourth columns of that table are repeated here in Table 3.19.

TABLE 3.19

Days to maturity	Frequency	Class mark
30–39	3	34.5
40–49	1	44.5
50–59	8	54.5
60–69	10	64.5
70–79	7	74.5
80–89	7	84.5
90–99	4	94.5
Σ	40	

Use Table 3.19 to compute the approximate values of $\bar{x}$ and s.

SOLUTION

As we said, the approximate values of $\bar{x}$ and s can be obtained by applying Formulas 3.2 and 3.3 and by using the class marks for the x-values. The required sums are obtained in Table 3.20.

TABLE 3.20

x	f	xf	x^2	x^2f
34.5	3	103.5	1190.25	3,570.75
44.5	1	44.5	1980.25	1,980.25
54.5	8	436.0	2970.25	23,762.00
64.5	10	645.0	4160.25	41,602.50
74.5	7	521.5	5550.25	38,851.75
84.5	7	591.5	7140.25	49,981.75
94.5	4	378.0	8930.25	35,721.00
Σ	40	2720.0		195,470.00

From the second, third, and fifth columns of Table 3.20, we find that

$$\bar{x} \approx \frac{\Sigma xf}{n} = \frac{2720}{40} = 68.0 \text{ days}$$

and

$$s \approx \sqrt{\frac{n(\Sigma x^2 f) - (\Sigma xf)^2}{n(n-1)}} = \sqrt{\frac{40(195,470) - (2720)^2}{40(39)}}$$

$$= \sqrt{\frac{420,400}{1560}} = 16.4 \text{ days}$$

MTB
SPSS

■

Since we have the raw (ungrouped) data for the days-to-maturity data in Table 2.1 on page 16, we can compute the exact values for $\bar{x}$ and s using Definition 3.3 (page 64) and Formula 3.1 (page 73). We did this and obtained

$$\bar{x} = 68.3 \text{ days} \quad \text{and} \quad s = 16.7 \text{ days}$$

Comparing these values to the ones found in Example 3.16, we see that the grouped-data formulas really do give only the approximate values for $\bar{x}$ and s when the classes are not based on a single value. The reason for this is that the class mark of each class provides only a typical value for the data values in the class and not the data values themselves.

For the days-to-maturity data we could actually compute the exact values for $\bar{x}$ and s because the raw data were available. In practice, when data are given in a grouped form and the classes are not based on a single value, the raw data are usually not accessible. For such cases, we must be content with approximate values for $\bar{x}$ and s.

Exercises 3.5

In Exercises 3.66 through 3.69 you will be given a frequency distribution for a data set. For each exercise,
a) compute the sample mean, $\bar{x}$, using Formula 3.2 on page 83.
b) compute the sample standard deviation, s, using one of the formulas in Formula 3.3.
Round your answers to one more decimal place than the original data.

3.66 An English professor wants to estimate the average study time per week for students in introductory English courses at her school. To accomplish this, she selects a sample of 25 students and records their weekly study time. The times, to the nearest hour, are as follows:

Study time	Frequency
2	2
3	1
4	3
5	1
6	5
7	5
8	5
9	2
11	1

3.67 The U. S. Bureau of the Census collects and publishes data on family size. Suppose that a sample of 500 American families yields the sizes indicated here:

Size of family	Frequency
2	198
3	118
4	101
5	59
6	12
7	3
8	8
9	1

3.68 The heights to the nearest inch of the basketball players on the 1985–86 Phoenix Suns roster are given in the following frequency distribution. (Source: *Phoenix Suns 1985–86 Media Guide.*)

Height	Frequency
74	1
75	3
76	1
77	2
78	3
79	2
80	2
81	2
82	2
85	1
86	1

3.69 The ages of the students in Professor W's introductory statistics class are given in Table 3.4 on page 60. A frequency distribution for those ages is displayed below.

Age	Frequency
17	1
18	1
19	8
20	7
21	8
22	5
23	3
24	4
26	1
35	1
36	1

In Exercises 3.70 and 3.71 you will be given raw (ungrouped) data. For each data set,
a) compute $\bar{x}$ and s by applying Definition 3.3 (page 64) and Formula 3.1 (page 73).
b) group the data into a frequency distribution in which each class is based on a single value.

c) compute $\bar{x}$ and s using your frequency distribution from part (b) and the formulas discussed in this section.

d) compare your answers from parts (a) and (c).

3.70   The Prescott National Bank has six tellers available to serve customers. The data below give the number of busy tellers observed at 25 spot checks.

6	5	4	1	5
6	1	5	5	5
3	5	2	4	3
4	5	0	6	4
3	4	2	3	6

3.71   A wheat farmer tests a new fertilizer on a sample of 30 one-acre plots. The yields of wheat, in bushels, are as follows:

39	41	38	39	38
41	40	38	37	39
39	41	43	38	40
39	39	43	39	37
38	37	40	43	40
42	40	39	36	43

Compute the approximate values of $\bar{x}$ and s for the grouped data in Exercises 3.72 through 3.75.

3.72   The U. S. National Center for Education Statistics estimates costs for various aspects of higher education. One such cost is the annual tuition and fees in private, four-year universities. Suppose the annual tuitions for a sample of 35 private, four-year universities yield the following frequency distribution:

Annual tuition	Frequency
$2000–$2999	2
$3000–$3999	2
$4000–$4999	6
$5000–$5999	7
$6000–$6999	7
$7000–$7999	6
$8000–$8999	4
$9000–$9999	1

3.73   A study by Lewis M. Terman, published in his book *The Intelligence of School Children* (Boston: Houghton Mifflin, 1946), gave the following frequency distribution for the IQs of 112 children attending kindergarten in San Jose and San Mateo, California:

IQ	Frequency
60–69	1
70–79	5
80–89	13
90–99	22
100–109	28
110–119	23
120–129	14
130–139	3
140–149	2
150–159	1

3.74   Calcium is the most abundant and one of the most important minerals in the body. It works with phosphorus in building and maintaining bones and teeth. According to the Food and Nutrition Board of the National Academy of Sciences, the recommended daily allowance (RDA) of calcium for adults is 800 milligrams (mg). A nutritionist thinks that people with incomes below the poverty level average *less* than the RDA of 800 mg. To test her suspicion, the daily intakes of calcium are determined for a sample of 50 people with incomes below the poverty level. The results are displayed in the following frequency distribution:

Intake (mg)	Frequency
Under 200	1
200–under 400	1
400–under 600	12
600–under 800	16
800–under 1000	12
1000–under 1200	7
1200–under 1400	1

3.75   *Runner's World* magazine conducts statistical studies on the finishing times of runners in the New York City 10-km run. Suppose a sample of 40 finishing times yields the following frequency distribution:

Finishing time (minutes)	Frequency
30–under 40	1
40–under 50	2
50–under 60	14
60–under 70	13
70–under 80	8
80–under 90	2

In each of Exercises 3.76 and 3.77 you will be given a set of raw data along with a frequency distribution for that data. For each data set,

a) compute $\bar{x}$ and s by applying Definition 3.3 (page 64) and Formula 3.1 (page 73) to the raw data.

b) compute the approximate values of $\bar{x}$ and s using the frequency distribution and the formulas discussed in this section.

c) compare your answers from parts (a) and (b).

3.76 The U.S. Bureau of Economic Analysis gathers information on the length of stay in Europe and the Mediterranean by U.S. travelers. A sample of 36 U.S. residents who have traveled to Europe and the Mediterranean this year yielded the following data on length of stay. The data are in days.

Raw data

41	16	6	21	10	21
5	31	20	27	17	10
3	32	2	48	8	12
21	44	1	56	5	12
3	13	15	10	18	3
1	11	14	12	64	10

$[\Sigma x = 643, \Sigma x^2 = 20{,}185]$

Grouped Data

Length of stay (days)	Frequency
1–7	10
8–14	10
15–21	8
22–28	1
29–35	2
36–42	1
43–49	2
50–56	1
57–63	0
64–70	1

3.77 The U.S. Energy Information Administration collects and publishes data on energy production and consumption. Data on last year's energy consumption for a sample of 50 households in the South are given here. (Data are in millions of BTU.)

Raw data

130	55	45	64	155
66	60	80	102	62
58	101	75	111	151
139	81	55	66	90
97	77	51	67	125

(continued at the top of the next column)

50	136	55	83	91
54	86	100	78	93
113	111	104	96	113
96	87	129	109	69
94	99	97	83	97

$[\Sigma x = 4486, \Sigma x^2 = 438{,}942]$

Grouped Data

Energy consumption (million BTU)	Frequency
40–49	1
50–59	7
60–69	7
70–79	3
80–89	6
90–99	10
100–109	5
110–119	4
120–129	2
130–139	3
140–149	0
150–159	2

3.78 If data are grouped into a frequency distribution using classes based on a single value, is it possible to compute $\bar{x}$ and s without using the grouped-data formulas of this section? Explain your answer.

3.79 If data are grouped into a frequency distribution using classes that are *not* based on a single value, why do the grouped data formulas of this section give only the approximate values of $\bar{x}$ and s?

Formula 3.2 for computing the mean of grouped data is $\bar{x} = (\Sigma xf)/n$. Using algebra, we can rewrite this formula as

$$\bar{x} = \Sigma x \cdot \frac{f}{n}$$

Since f/n represents the relative frequency of a class, we have

$$\bar{x} = \Sigma x \cdot \frac{f}{n}$$
(1)
$$= \Sigma[(\text{class mark}) \cdot (\text{relative frequency of class})]$$

This formula allows us to compute the mean of a data set when it is grouped in a relative-frequency distribution. We will apply this formula in Exercises 3.80 and 3.81.

3.80 Refer to Exercise 3.66.

a) Construct a relative-frequency distribution for the data.

b) Use Formula (1) to compute $\bar{x}$, and compare your answer to the one obtained in Exercise 3.66.

3.81 Refer to Exercise 3.67.
  a) Construct a relative-frequency distribution for the data.
  b) Use Formula (1) to compute $\bar{x}$, and compare your answer to the one obtained in Exercise 3.67.

3.82 At the top of the next column is a set of 20 scores on a 100-point, true-false psychology quiz in which each question was worth 10 points.

70	70	80	80	80
70	70	80	80	80
70	70	80	80	90
70	70	80	80	90

a) Compute the sample mean, $\bar{x}$, of the exam scores from the raw data.
b) Construct a frequency distribution for the data using the classes 70–79, 80–89, and 90–99.
c) Use your frequency distribution from part (b) to compute the approximate value of $\bar{x}$.
d) Compare your answers from parts (a) and (c). Why is the approximation in part (c) so poor?

3.6 Percentiles; box-and-whisker diagrams

As we learned in Section 3.1, the *median* of a data set divides the data into two equal parts, a bottom 50% and a top 50%. Frequently it is useful, for both descriptive and inferential purposes, to divide a data set into a larger number of parts. **Quartiles** divide the data into quarters, or four equal parts, **deciles** divide the data into tenths, or 10 equal parts, and **percentiles** divide the data into hundredths, or 100 equal parts.

Quartiles

A data set has *three* quartiles, which we denote by Q_1, Q_2, and Q_3. Roughly speaking, the first quartile, Q_1, is the number that divides the bottom 25% of the data from the top 75%; the second quartile, Q_2, is the median which, as we know, is the number that divides the bottom 50% of the data from the top 50%; and, finally, the third quartile, Q_3, is the number that divides the bottom 75% of the data from the top 25%. In the next example, we will determine the three quartiles for a set of exam scores.

EXAMPLE 3.17 Illustrates quartiles

The exam scores for the students in an introductory statistics class are displayed in Table 3.21.

TABLE 3.21
Exam scores for introductory statistics class.

88	67	64	76	86
85	82	39	75	34
90	63	89	89	84
81	96	100	70	96

Determine the quartiles for these data.

SOLUTION
To determine the quartiles we use the following procedure:

STEP 1 *Arrange the data in increasing order.*

STEP 2 *Divide the data into quarters.*
STEP 3 *Find the numbers dividing the quarters.*

The results of applying these steps are:

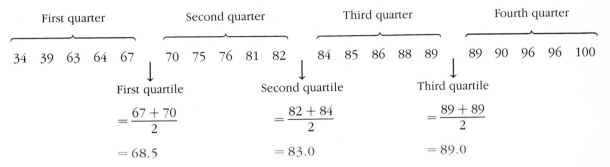

Thus, we see that $Q_1 = 68.5$, $Q_2 = 83.0$, and $Q_3 = 89.0$. ∎

It was easy to find the quartiles for the exam-score data because the number of exam scores (20) is divisible by four. When the number of pieces of data is not divisible by four, the determination of quartiles is more complicated. Many statistics books present general rules for obtaining quartiles. However, these complex rules tend to obscure a simple concept: quartiles divide a data set into four quarters. If your data do not divide evenly into four equal parts, use common sense to get as close to quarters as possible, and then provide an explanation of what you did.

**MTB
SPSS**

Deciles and percentiles

We will now briefly discuss deciles and percentiles. As we said, deciles divide a data set into tenths, or 10 equal parts. A data set has *nine* deciles, which we denote by $D_1, D_2, \ldots, D_9$. Basically, the first decile, D_1, is the number that divides the bottom 10% of the data from the top 90%; the second decile, D_2, is the number that divides the bottom 20% of the data from the top 80%; and so on. To obtain the deciles of a data set, we divide the ordered data into tenths and then determine the numbers dividing the tenths.

The percentiles of a data set divide it into hundredths, or 100 equal parts. A data set has 99 percentiles, denoted by $P_1, P_2, \ldots, P_{99}$. The first percentile, P_1, is the number that divides the bottom 1% of the data from the top 99%; the second percentile, P_2, is the number that divides the bottom 2% of the data from the top 98%; and so forth. To obtain the percentiles of a data set, we divide the ordered data into hundredths and then determine the numbers dividing the hundredths.

The second quartile, fifth decile, and fiftieth percentile of a data set are all the same and all equal the median. In symbols,

$$\text{Median} = Q_2 = D_5 = P_{50}$$

Similarly, we have equalities such as $Q_1 = P_{25}$, $D_1 = P_{10}$, and $Q_3 = P_{75}$.

In practice, quartiles, deciles, and percentiles are most useful with large data sets, and for such data sets the required calculations are almost always done by computer. Our purpose here has been simply to present the basic concepts so that you will be familiar with the terminology and understand the principles behind the computations.

Box-and-whisker diagrams

Box-and-whisker diagrams, or **boxplots,** are used to provide a graphical display of the variation in a data set. These diagrams, like stem-and-leaf diagrams, were invented by Professor John Tukey. To illustrate the construction of box-and-whisker diagrams, we return to the exam-score data of Example 3.17.

EXAMPLE 3.18   Illustrates box-and-whisker diagrams

The exam scores for the students in an introductory statistics class, given in Table 3.21, are repeated here in Table 3.22.

TABLE 3.22
Exam scores for introductory statistics class.

88	67	64	76	86
85	82	39	75	34
90	63	89	89	84
81	96	100	70	96

To construct a box-and-whisker diagram, we proceed as follows:

STEP 1 *Arrange the data in increasing order and determine the median.*

34 39 63 64 67 70 75 76 81 82 84 85 86 88 89 89 90 96 96 100

$$\text{Median} = \frac{82 + 84}{2} = 83$$

STEP 2 *Find the smallest and largest data values.*

$$\text{Low} = 34 \qquad \text{High} = 100$$

STEP 3 *Determine the median of the bottom half of the data set and the median of the top half of the data set.* [The medians of the bottom and top halves of the data set are called **hinges.**[†]]

Bottom half: 34   39   63   64   67   70   75   76   81   82

$$\text{Hinge} = \frac{67 + 70}{2} = 68.5$$

Top Half: 84   85   86   88   89   89   90   96   96   100

$$\text{Hinge} = \frac{89 + 89}{2} = 89$$

[†] The *hinges* are just the first and third quartiles if the number of pieces of data is even, but not necessarily if the number of pieces of data is odd.

STEP 4 *Draw a horizontal axis on which the values obtained in Steps 1-3 can be located. Above this axis, mark the median, the low and high, and the hinges with vertical lines (see Figure 3.8). Finally, connect the hinges to each other to make a box, and then to the low and high by lines.* [These connecting lines are called **whiskers.**]

FIGURE 3.8
Box-and-whisker diagram for
exam–score data.

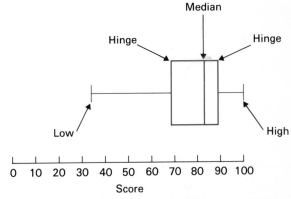

Figure 3.8 is the box-and-whisker diagram for the exam-score data. Note that the two boxes in the box-and-whisker diagram indicate the spread of the two middle quarters of the data and that the two whiskers indicate the spread of the first and fourth quarters. Thus, we see that there is less variation in the top two quarters of the exam-score data than in the bottom two, and that the first quarter has the greatest variation of all. ∎

MTB
SPSS

Exercises 3.6

Determine the quartiles for each of the data sets in Exercises 3.83 through 3.86.

3.83 The A.C. Nielsen Company collects data on the TV viewing habits of Americans by various characteristics. Suppose a sample of 20 married men gives the following weekly viewing times, in hours:

21	41	27	32	43
56	35	30	15	4
34	26	31	36	16
30	38	30	20	21

3.84 A water-heater manufacturer guarantees the electric heating element for a period of five years. The lifetimes, in months, for a sample of 20 such elements are

49.3	79.3	86.4	68.4	62.6
64.1	53.2	30.0	66.6	66.9
65.1	67.8	95.9	40.1	79.3
92.1	92.0	93.0	48.9	63.7

3.85 The U.S. Energy Information Administration publishes figures on residential energy consumption and expenditures. Suppose that a sample of 36 households using electricity as their primary energy source yields the following data on last year's energy expenditures:

$1376	1452	1235	1480	1185	1327
1059	1400	1227	1102	1168	1070
1180	1221	1351	1014	1461	1102
976	1394	1379	987	1002	1532
1450	1177	1150	1352	1266	1109
949	1351	1259	1179	1393	1456

3.86 The hourly temperatures in Colorado Springs, Colorado for Tuesday, August 22, 1978 are shown below. Data are in degrees Fahrenheit.

69	63	61	74	88	87	74	68
66	61	64	79	87	85	71	65
65	60	70	84	85	73	70	62

3.87   Refer to Exercise 3.83. Determine the deciles for the weekly TV viewing times of the sample of 20 married men.

3.88   Refer to Exercise 3.84. Determine the deciles for the lifetimes of the sample of 20 electric heating elements.

In Exercises 3.89 through 3.92, construct a box-and-whisker diagram for the specified data sets.

3.89   The weekly TV viewing times for a sample of 20 married men given in Exercise 3.83.

3.90   The lifetimes, in months, shown in Exercise 3.84 for a sample of 20 electric heating elements.

3.91   The energy expenditures from Exercise 3.85 for a sample of 36 households using electricity as their primary energy source.

3.92   The hourly temperatures in Colorado Springs, Colorado for Tuesday, August 22, 1978 given in Exercise 3.86.

3.93   In January 1984, the U.S. Department of Agriculture estimated that a typical American family of four with an intermediate budget would spend about $92 per week for food. A consumer researcher in Kansas suspected that the median weekly cost was less in her state. She took a sample of 10 Kansas families of four, each with an intermediate budget, and obtained the following weekly food costs:

$78	104	84	70	96
73	87	85	76	94

a) Arrange the data in increasing order.
b) Find the median.
c) Although the data do not divide evenly into quarters, use common sense to determine the quartiles.

3.7   The population mean and population standard deviation; use of samples

Up to now we have concentrated on descriptive measures for sample data (data obtained from a sample of a population). This is because in practice the data we work with is frequently sample data. As a matter of fact, in inferential studies the data analyzed is almost always sample data. However, although we generally deal with sample data in inferential studies, the objective is to describe the entire population.

In this section we will define the **population mean** and the **population standard deviation.** We will also examine the role of samples with regard to those two descriptive measures for a population.

The population mean

Recall that for a sample of size n, the sample mean, $\bar{x}$, is defined by

$$\bar{x} = \frac{\Sigma x}{n}$$

The mean of a finite population is computed in the same way—sum the data and then divide by the total number of pieces of data. However, to distinguish a population mean from a sample mean, we use the lower-case Greek letter μ (pronounced "mew") to denote the mean of a population. Furthermore, we use the upper-case English letter N to represent the size of a population. Table 3.23 summarizes the notations used for both a sample and a population.

TABLE 3.23
Notations used for a sample
and for a population.

	Size	Mean
Sample	n	$\bar{x}$
Population	N	μ

Using the notation displayed in the second row of Table 3.23, we now define the mean of a finite population.

DEFINITION 3.6 Population mean

The **population mean, μ,** for a finite population of size N is defined by

$$\mu = \frac{\Sigma x}{N}$$

In inferential studies, the size of the population is ordinarily quite large and the mean of the population is usually difficult or impossible to determine. However, for the sake of illustration, we will sometimes consider small populations for which the mean, μ, can easily be computed.

EXAMPLE 3.19 **Illustrates Definition 3.6**

Table 3.24 presents some data for the starting offense of the 1985–1986 Los Angeles Rams football team.

TABLE 3.24
Los Angeles Rams offense,
1985–86.

Name	Position	Height	Weight (lbs)
Irv Pankey	LT	6'4"	267
Kent Hill	LG	6'5"	260
Doug Smith	C	6'3"	265
Dennis Harrah	RG	6'5"	265
Jackie Slater	RT	6'4"	271
David Hill	TE	6'2"	240
Dieter Broch	QB	6'	195
Erich Dickerson	RB	6'3"	220
Ron Brown	WR	6'1"	185
Henry Ellard	WR	5'11"	170
Tony Hunter	TE	6'4"	240

Compute the population mean weight of these 11 players.

SOLUTION
The sum of the weights in Table 3.24 is $\Sigma x = 2578$ lb. Since there are 11 players, $N = 11$. Thus, the population mean weight of the players is

$$\mu = \frac{\Sigma x}{N} = \frac{2578}{11} = 234.4 \text{ lb}$$

MTB
SPSS

If we were interested in the mean weight of *all* starting offensive players in professional football, then the 11 starting players on the Rams would be considered a sample instead of a population. The mean weight of those 11 players would then be a sample mean instead of a population mean, and we would write $\bar{x} = 234.4$ lb instead of $\mu = 234.4$ lb. Consequently, we see that whether a data set is considered population data or sample data often depends on the interests of the researcher conducting the study.

As we have mentioned, although we usually deal with sample data in inferential studies, the objective is to describe the entire population. The reason for resorting to a sample is that it is generally more practical. We illustrate this last point in Example 3.20.

EXAMPLE 3.20   **Illustrates a use of the sample mean**

The U.S. Bureau of the Census annually publishes figures on the mean income of households in the United States. To obtain the complete population data—the incomes of *all* American households—would be extremely expensive and time-consuming. It is also unnecessary, as we will see in Chapter 8, because accurate estimates for the mean income of all American households can be obtained from the mean income of a sample of such households.

In reality, the Census Bureau samples 72,000 out of a total of more than 83 million American households. The population of interest here is comprised of the incomes of all American households, and the mean of those incomes is the population mean, μ. The sample consists of the incomes of the 72,000 households sampled by the Census Bureau, and the mean of those incomes is the sample mean, $\bar{x}$. We summarize the situation in Figure 3.9.

FIGURE 3.9
Population and sample for incomes of American households.

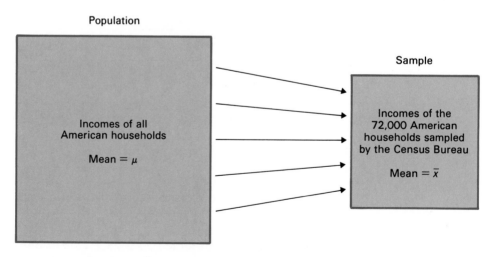

Once the sample is taken, the Census Bureau can compute the sample mean income, $\bar{x}$, of the 72,000 households obtained. Using $\bar{x}$ the Census Bureau can then make an inference about the population mean income, μ, of *all* American households. We will study these kinds of inferences in Chapter 8.   ■

The population standard deviation

We next discuss the standard deviation of a population. Recall that for a sample of size n, the sample standard deviation, s, is defined by

$$s = \sqrt{\frac{\Sigma(x - \bar{x})^2}{n - 1}}$$

The standard deviation of a population is denoted by the lower-case Greek letter σ (pronounced "sigma") and is defined as follows:

DEFINITION 3.7 Population standard deviation

The **population standard deviation, σ,** for a finite population of size N is defined by

$$\sigma = \sqrt{\frac{\Sigma(x - \mu)^2}{N}}$$

and can also be computed using the **shortcut formula**

$$\sigma = \sqrt{\frac{\Sigma x^2}{N} - \mu^2}$$

Just as s^2 is called the sample variance, σ^2 is called the **population variance.**

Note that in the defining formula for the population standard deviation, σ, we divide by N, the size of the population. On the other hand, in the defining formula for the sample standard deviation, s, we divide by $n - 1$, one less than the size of the sample. This is because in inferential studies, the main use of s is as an estimate for σ. Dividing by $n - 1$ instead of n in the formula for s results in a better estimate for σ, on the average.

EXAMPLE 3.21 **Illustrates Definition 3.7**

The weights, to the nearest pound, of the starting offense of the 1985–1986 Los Angeles Rams football team are given in Table 3.24. We have repeated those weights in the first column of Table 3.25 at the top of the next page. Compute the population standard deviation of the weights of these 11 players.

S O L U T I O N
We will apply the shortcut formula

$$\sigma = \sqrt{\frac{\Sigma x^2}{N} - \mu^2}$$

The required sums are shown in Table 3.25. Recalling that $\mu = \Sigma x/N$, we see from the last row of Table 3.25 that

$$\sigma = \sqrt{\frac{\Sigma x^2}{N} - \mu^2} = \sqrt{\frac{617{,}530}{11} - \left(\frac{2578}{11}\right)^2} = 34.8 \text{ lb}$$

TABLE 3.25
Table for computation of σ for the weights of the starting offensive players on the 1985–86 Los Angeles Rams football team.

x	x^2
267	71,289
260	67,600
265	70,225
265	70,225
271	73,441
240	57,600
195	38,025
220	48,400
185	34,225
170	28,900
240	57,600
Σ 2578	617,530

MTB
SPSS

■

In the next example we will illustrate the use of the sample standard deviation to estimate a population standard deviation.

EXAMPLE 3.22   **Illustrates a use of the sample standard deviation**

A hardware manufacturer produces 10-millimeter bolts. The manufacturer knows that the diameters of the bolts produced vary somewhat from 10 mm and also from each other. But even if he is willing to accept some variation in bolt diameters, he cannot tolerate too much variation. For if the variation is too large, then too many of the bolts produced will be unusable.

Thus the manufacturer must make sure that the standard deviation, σ, of the bolt diameters is not unduly large. Since, in this case, it is not possible to determine σ exactly, inferential statistics must be employed.

FIGURE 3.10
Population and sample for bolt diameters.

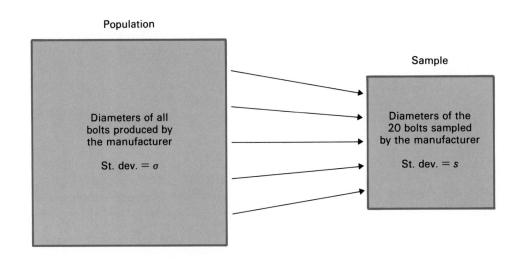

Suppose the manufacturer decides to estimate σ by taking a sample of 20 bolts. The population of interest here consists of the diameters of all bolts that have been or will ever be produced by the manufacturer, and the standard deviation of these diameters is the population standard deviation, σ. The sample consists of the diameters of the 20 bolts sampled by the manufacturer, and the standard deviation of those diameters is the sample standard deviation, s. Figure 3.10 summarizes the above discussion.

After the sample is taken, the manufacturer can compute the sample standard deviation, s, of the diameters of the 20 bolts obtained. Using that information, he can then make an inference about the population standard deviation, σ, of the diameters of *all* bolts being produced. We will examine these types of inferences in Chapter 11. ∎

Parameter and statistic

There is statistical terminology that helps us to distinguish between descriptive measures for populations and descriptive measures for samples.

DEFINITION 3.8 Parameter and statistic

Parameter: A descriptive measure for a population.
Statistic: A descriptive measure for a sample.

Thus, for example, μ and σ are *parameters,* whereas $\bar{x}$ and s are *statistics.*

Formulas for grouped population data

There are formulas for computing the mean and standard deviation of a population when the data are grouped in a frequency distribution. These formulas are derived in the same way as the grouped-data formulas for the sample mean and sample standard deviation (Section 3.5).

FORMULA 3.4 Computing μ for grouped data

The mean of a finite population whose data are grouped in a frequency distribution can be computed using the formula

$$\mu = \frac{\Sigma xf}{N}$$

where f denotes class frequency and $N\,(=\Sigma f)$ denotes the size of the population.

FORMULA 3.5 Computing σ for grouped data

The standard deviation of a finite population whose data are grouped in a frequency distribution can be computed using the formula

$$\sigma = \sqrt{\frac{\Sigma(x-\mu)^2 f}{N}}$$

or its shortcut version

$$\sigma = \sqrt{\frac{\Sigma x^2 f}{N} - \mu^2}$$

where f denotes class frequency and $N (= \Sigma f)$ denotes the size of the population.

EXAMPLE 3.23   Illustrates Formulas 3.4 and 3.5

CJ² Business Services is a small company specializing in word processing. The company employs eight people. A frequency distribution for the weekly salaries of these eight employees is presented in Table 3.26.

TABLE 3.26
Frequency distribution for
weekly salaries of CJ²
employees.

Salary x	Frequency f
$240	3
320	2
450	1
600	2

Regarding these salaries as a population of interest, compute the population mean and population standard deviation.

SOLUTION
Since the population data are in grouped form, we apply Formulas 3.4 and 3.5 (we will use the shortcut formula to compute σ). As the formulas indicate, we will need a table with columns for x, f, xf, x^2, and $x^2 f$. See Table 3.27.

TABLE 3.27

	x	f	xf	x^2	$x^2 f$
	240	3	720	57,600	172,800
	320	2	640	102,400	204,800
	450	1	450	202,500	202,500
	600	2	1200	360,000	720,000
Σ		8	3010		1,300,100

From the second and third columns of Table 3.27 we see that

$$\mu = \frac{\Sigma xf}{N} = \frac{3010}{8} = \$376.25$$

MTB
SPSS

Now using the fifth column we find that

$$\sigma = \sqrt{\frac{\Sigma x^2 f}{N} - \mu^2} = \sqrt{\frac{1,300,100}{8} - (376.25)^2} = \$144.74$$
∎

The frequency distribution in Table 3.26 uses classes based on a single value. When population data are grouped into a frequency distribution in which each class represents several values, we use the class marks for the x-values in Formulas 3.4 and 3.5. However, we must keep in mind that the values for μ and σ so obtained are only approximately equal to the actual population mean and standard deviation.

The population mean and population standard deviation in inferential studies

We would like to emphasize again that in inferential studies the population mean, μ, and population standard deviation, σ, are rarely known. In fact, two major topics in inferential statistics are

1. Using the mean, $\bar{x}$, of a sample to make inferences about the population mean, μ, and
2. Using the standard deviation, s, of a sample to make inferences about the population standard deviation, σ.

Topic 1 will be examined in Chapters 8 and 9 and Topic 2 in Chapter 11.

Exercises 3.7

3.94 Although we generally deal with sample data in inferential studies, what is the ultimate objective?

3.95 In Section 3.3 we considered the heights of the starting five players on each of two men's basketball teams. The heights, in inches, for the players on Team I are 72, 73, 76, 76, and 78. Considering the heights as a sample of the heights of all male starting college basketball players,
a) compute the sample mean height, $\bar{x}$.
b) compute the sample standard deviation, s, of the heights.
Considering the heights now as the population of interest,
c) compute the population mean height, μ.
d) compute the population standard deviation, σ.
Comparing your answers from parts (a) and (c) and from parts (b) and (d),
e) why are the values for the sample mean, $\bar{x}$, and population mean, μ, equal?
f) why are the values for the sample standard deviation, s, and the populaton standard deviation, σ, different?

3.96 In Section 3.3 we considered the heights of the starting five players on each of two men's basketball teams. The heights, in inches, for the players on Team II are 67, 72, 76, 76, and 84. Considering the heights as a sample of the heights of all male starting college basketball players,
a) compute the sample mean height, $\bar{x}$.
b) compute the sample standard deviation, s, of the heights.
Considering the heights now as the population of interest,
c) compute the population mean height, μ.
d) compute the population standard deviation, σ.
Comparing your answers from parts (a) and (c) and from parts (b) and (d),
e) why are the values for the sample mean, $\bar{x}$, and population mean, μ, equal?
f) why are the values for the sample standard deviation, s, and the population standard deviation, σ, different?

In Exercises 3.97 through 3.100 consider the given data set the population of interest. For each exercise,
a) compute the population mean, μ.
b) compute the population standard deviation, σ.

3.97 Last year's monthly electric bills for a family living in the southeast are given on the following page. The data are in dollars, rounded to the nearest dollar.

$87	54	125	89
92	68	142	63
65	95	140	84

3.98 Last year's monthly phone bills for the family in Exercise 3.97 are presented below. The data are in dollars, rounded to the nearest dollar.

$81	35	75	48
55	46	61	41
24	25	33	32

3.99 According to the U.S. Department of Agriculture, the acreages of tobacco harvested in 1983 by each of the tobacco producing states are as follows (data are in thousands of acres).

CT 2	MD 27	SC 54
FL 8	NC. 278	TN 73
GA. 44	OH 12	VA 54
IN 8	PA 12	WI 8
KY 203		

3.100 The U.S. Agency for International Development publishes annual data on U.S. foreign aid commitments for economic assistance. In 1983 the commitments to Latin American countries were as follows (data in millions of dollars).

Bolivia 11	Haiti 27
Costa Rica 184	Honduras 87
Dominican Rep 35	Jamaica 82
Ecuador 22	Panama 6
El Salvador. 199	Peru. 36
Guatemala 22	

3.101 In Example 3.19 we found that the population mean weight of the starting offensive players on the 1985–1986 Los Angeles Rams football team is $\mu = 234.4$ lb. In this context, is the number 234.4 a parameter or a statistic? Explain your answer.

3.102 In Example 3.23 we determined that the population standard deviation of the weekly salaries for the employees of CJ² Business Services is $\sigma = \$144.74$. In this context, is the number $144.74 a parameter or a statistic? Explain your answer.

3.103 The ages of the students in Professor W's introductory statistics class are given in Table 3.4 on page 60. A frequency distribution for those ages is given at the top of the next column.

Regarding the ages of these students as the population of interest,
a) compute the population mean age, μ.

Age	Frequency
17	1
18	1
19	8
20	7
21	8
22	5
23	3
24	4
26	1
35	1
36	1

b) compute the population standard deviation, σ, of the ages.

3.104 In a fifth-grade class, an eight-question quiz on fractions was given. A frequency distribution for the number of incorrect answers is

Number incorrect x	Frequency f
0	4
1	8
2	4
3	5
4	2
5	1
6	1

a) Compute the population mean number of questions missed.
b) Compute σ.

3.105 The U.S. Bureau of the Census collects and publishes annual figures for state expenditures on highways. Here is a frequency distribution of the 1982 highway expenditures of the 50 states.

Expenditures ($millions)	Number of states
0–under 200	13
200–under 400	11
400–under 600	12
600–under 800	5
800–under 1000	3
1000–under 1200	2
1200–under 1400	2
1400–under 1600	1
1600–under 1800	0
1800–under 2000	1

a) Compute the approximate value of the mean 1982 highway expenditure, μ, per state.

b) Compute the approximate value of the standard deviation, σ, of the 1982 state highway expenditures.

c) Why are your values for μ and σ only approximately correct?

3.106 The following table gives the 1983 age distribution for Americans between 3 and 34 years old who are enrolled in school. (Source: U. S. Bureau of the Census.)

Age (yrs)	Frequency (thousands)
3–4	2,624
5–6	6,214
7–13	23,278
14–15	7,093
16–17	6,698
18–19	3,938
20–21	2,609
22–24	2,111
25–29	1,976
30–34	1,203

a) Compute the approximate value of the mean age, μ, for Americans between 3 and 34 years old who are enrolled in school.

b) Compute the approximate value of the standard deviation, σ, of the ages.

Formula 3.4 for computing the population mean of grouped data is $\mu = (\Sigma xf)/N$. Using algebra, we can rewrite this formula as

$$\mu = \Sigma x \cdot \frac{f}{N}$$

Since f/N represents the relative frequency of a class, we have

$$\mu = \Sigma x \cdot \frac{f}{N}$$
(2)
$$= \Sigma[(\text{class mark}) \cdot (\text{relative frequency of class})]$$

This formula permits us to compute the mean of a population when it is grouped in a relative-frequency distribution. We will apply the formula in Exercises 3.107 and 3.108.

3.107 Refer to Exercise 3.103.
a) Construct a relative-frequency distribution for the population of ages.
b) Use Formula (2) to compute μ, and compare your answer to the one obtained in Exercise 3.103(a).

3.108 Refer to Exercise 3.104.
a) Construct a relative-frequency distribution for the number of incorrect answers.
b) Use Formula (2) to compute μ, and compare your answer to the one obtained in Exercise 3.104(a).

3.109 Derive two formulas for computing the standard deviation, σ, of a population when it is grouped in a relative-frequency distribution.

Exercises 3.110 and 3.111 use the results of Exercise 3.109.

3.110 Refer to Exercise 3.108. Use one of the formulas you derived in Exercise 3.109 to compute σ from the relative-frequency distribution you constructed in Exercise 3.108(a). Compare your answer to the one obtained in Exercise 3.104(b).

3.111 Refer to Exercise 3.107. Use one of the formulas you derived in Exercise 3.109 to compute σ from the relative-frequency distribution you constructed in Exercise 3.107(a). Compare your answer to the one obtained in Exercise 3.103(b).

3.112 Consider the three data sets below:

Data Set 1	Data Set 2	Data Set 3
2 4	7 5 5 3	4 7 8 9 7
7 3	9 8 6	4 5 3 4 5

a) Assuming that these data sets are each a sample of a population, compute their standard deviations. [Round your final answers to two decimal places.]
b) Assuming these data sets are each a population, compute their standard deviations. [Round your final answers to two decimal places.]
c) Using your results from parts (a) and (b), make an educated guess about the answer to the following question: Suppose that both s and σ are computed for the same data set. Will they tend to be closer together if the data set is large or if it is small?

3.113 Consider a data set with m data values. If the data set is a sample from a population, then we compute the sample standard deviation, s. If the data set is a population, then we compute the population standard deviation, σ.
a) Derive a formula that gives the precise relationship between the values of s and σ calculated from a data set having m data values.

b) Verify that your formula in part (a) works for each of the three data sets given in Exercise 3.112.

c) Suppose a data set consists of 15 data values. You compute the population standard deviation of the data and obtain $\sigma = 38.6$. Then you realize that the data set is actually a sample of a population and that you should have computed the sample standard deviation, s, instead. Use your result from part (a) to obtain s.

3.8   Computer packages*

In this chapter we have studied a number of different descriptive measures that can be used to summarize a set of data. By employing statistical computer packages, we can avoid performing the required calculations manually. All we have to do is enter the data into the computer and type the appropriate commands—the computer does the rest.

DESCRIBE

A Minitab program called **DESCRIBE** will determine the mean, median, standard deviation, and several other descriptive measures for a data set. We will illustrate the use of this program by applying it to the exam-score data from Example 3.17.

EXAMPLE 3.24   Illustrates the use of DESCRIBE

The exam scores for the students in an introductory statistics class, given in Table 3.21, are repeated here as Table 3.28.

TABLE 3.28
Exam scores for introductory statistics class.

88	67	64	76	86
85	82	39	75	34
90	63	89	89	84
81	96	100	70	96

To employ DESCRIBE, we first enter the data in Table 3.28 into the computer. Then we type the command DESCRIBE followed by the storage location of the data (C3). This and the resulting output are displayed in Printout 3.1.

PRINTOUT 3.1
Minitab output
for DESCRIBE

```
MTB > DESCRIBE C3

            N       MEAN     MEDIAN    TRMEAN     STDEV    SEMEAN
C3         20      77.70     83.00     78.89     17.57      3.93

           MIN       MAX        Q1        Q3
C3       34.00    100.00     67.75     89.00
```

The first entry of the output gives the number of pieces of data, which in this case is 20. The next two entries display the mean (MEAN) and median (MEDIAN) of the data set. In the fourth entry, labelled TRMEAN, we find the 5% trimmed mean (discussed in Exercise 3.15 on page 62). The fifth entry, STDEV, gives the *sample* standard deviation of the data. SEMEAN, shown next, stands for "standard error of the mean.'' We will discuss this in Chapter 7. The seventh and eighth entries of the output display the smallest data value (MIN) and the largest data value (MAX). The final two entries give the first quartile (Q1) and the third quartile (Q3), respectively. [*Note:* The value given in Printout 3.1 for the first quartile is 67.75, which differs from the value of 68.5 that we obtained in Example 3.17. This discrepancy is due to the fact that Minitab uses a slightly different rule for computing quartiles than we do.] ∎

BOXPLOTS

On pages 92–93 we discussed box-and-whisker diagrams. As we learned, these diagrams are used to provide a graphical display of the variation in a data set. Minitab has a program called **BOXPLOTS** that will construct box-and-whisker diagrams for us.

EXAMPLE 3.25 | **Illustrates the use of BOXPLOTS**

Consider again the exam-score data given in Table 3.28. In Example 3.18, we constructed a box-and-whisker diagram for that data. To use Minitab to obtain the box-and-whisker diagram, we enter the data into the computer and then type BOXPLOTS, followed by the storage location of the data (C3). See Printout 3.2.

PRINTOUT 3.2
Minitab output
for BOXPLOTS

```
MTB > BOXPLOTS C3
```

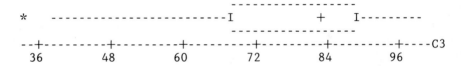

```
                                         -----------------
 *      -------------------------I          +     I---------
                                         -----------------
 --+---------+---------+---------+---------+---------+---------+----C3
   36        48        60        72        84        96
```

You should compare Minitab's box-and-whisker diagram for the exam-score data to the one we constructed by hand in Figure 3.8 on page 93. ∎

◆ **Chapter Review**

FORMULAS

In the following formulas,

n = sample size
$\bar{x}$ = sample mean
s = sample standard deviation
f = class frequency
N = population size
μ = population mean
σ = population standard deviation

Sample mean, 64

$$\bar{x} = \frac{\Sigma x}{n}$$

Range, 68

$$\text{Range} = \text{Largest value} - \text{Smallest value}$$

Sample standard deviation, 72, 73

$$s = \sqrt{\frac{\Sigma(x-\bar{x})^2}{n-1}} \quad \text{or} \quad s = \sqrt{\frac{n(\Sigma x^2) - (\Sigma x)^2}{n(n-1)}}$$

Sample mean (grouped form), 83

$$\bar{x} = \frac{\Sigma xf}{n}$$

(x = class mark)

Sample standard deviation (grouped form), 84

$$s = \sqrt{\frac{\Sigma(x-\bar{x})^2 f}{n-1}} \quad \text{or} \quad s = \sqrt{\frac{n(\Sigma x^2 f) - (\Sigma xf)^2}{n(n-1)}}$$

(x = class mark)

Population mean, 95

$$\mu = \frac{\Sigma x}{N}$$

Population standard deviation, 97

$$\sigma = \sqrt{\frac{\Sigma(x-\mu)^2}{N}} \quad \text{or} \quad \sigma = \sqrt{\frac{\Sigma x^2}{N} - \mu^2}$$

Population mean (grouped form), 99

$$\mu = \frac{\Sigma xf}{N}$$

(x = class mark)

Population standard deviation (grouped form), 99

$$\sigma = \sqrt{\frac{\Sigma(x-\mu)^2 f}{N}} \quad \text{or} \quad \sigma = \sqrt{\frac{\Sigma x^2 f}{N} - \mu^2}$$

(x = class mark)

YOU SHOULD BE ABLE TO . . .

1 use and understand the preceding formulas.

2 explain the purpose of a measure of central tendency.

3 compute and interpret the mean and the median of a data set.

4 choose an appropriate measure of central tendency for a data set.

5 distinguish between a sample mean and a population mean.

6 use and understand summation notation.

7 explain the purpose of a measure of dispersion.

8 compute and interpret the standard deviation of a data set.

9 state and apply Chebychev's theorem.

10 determine and interpret the quartiles, deciles, and percentiles of a data set.

11 construct and interpret a box-and-whisker diagram.

12 distinguish between a sample standard deviation and a population standard deviation.

REVIEW TEST

1 Euromonitor Publications Limited in London provides data on per capita food consumption of major food commodities in various countries. Suppose that samples of 10 Germans and 15 Russians yield the following consumption of fish during last year. (Data are in kilograms.)

Germans		Russians		
17	17	16	21	12
1	9	11	5	23
15	6	19	19	22
10	13	16	23	12
14	11	18	7	17

a) Compute the mean of each data set.
b) Compute the median of each data set.

2 For each part below, decide whether the mean or the median would be a more appropriate measure of central tendency and give a reason for your answer.
a) The U.S. National Center for Health Statistics publishes information on the duration of marriages.

b) The National Education Association publishes statistics on daily attendance in public elementary and secondary schools.

3 Telephone companies conduct surveys to obtain information on the lengths of telephone conversations. Suppose a sample of 12 phone calls yields the following durations to the nearest minute:

4	2	1	2	2	8
6	3	1	3	1	15

a) Compute the sample mean duration, $\bar{x}$.
b) Compute the range of the durations.
c) Compute the sample standard deviation, s, of the durations.

4 Refer to Problem 3.
a) Locate $\bar{x}$ on the graph below and mark off one, two, and three standard deviations to either side of $\bar{x}$.

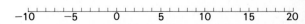

b) Plot the telephone data from Problem 3 on the graph using dots to represent the data values.

c) By Chebychev's theorem, at least _____ % of the data lies within two standard deviations to either side of the mean.

d) What percentage of the telephone data actually lies within two standard deviations to either side of the mean?

e) From parts (c) and (d) we see that Chebychev's theorem does not necessarily give precise estimates. What then makes Chebychev's theorem so valuable?

5 Dr. Thomas Stanley, a marketing professor at Georgia State University, has collected information on millionaires, including their ages, since 1973. Suppose the mean age of a sample of 36 millionaires is $\bar{x} = 58.5$ years, with a standard deviation of $s = 13.4$ years.

a) Complete the following graph:

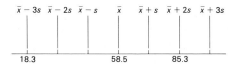

b) Apply Chebychev's theorem with $k = 2$ to make some observations about the ages of the 36 millionaires.

c) Repeat part (b) with $k = 3$.

6 Referring to Problem 5, the actual ages of the 36 millionaires sampled are as follows:

31	64	39	66	68	48	69	71	52
68	45	60	54	66	79	38	48	77
53	52	79	75	67	71	42	39	57
47	74	59	64	42	55	61	79	48

a) Determine the quartiles for the data.

b) Construct a box-and-whisker diagram.

7 An exercise physiologist measured the heart rates of 30 people who had been placed in a long-distance running program. A frequency distribution for these rates is shown at the top of the next column.

a) Compute the mean heart rate, $\bar{x}$.

b) Compute the sample standard deviation, s, of the heart rates.

Rate x	Frequency f	Rate x	Frequency f
67	2	72	3
68	2	73	4
69	2	74	4
70	3	75	3
71	6	77	1

8 The data below give the 1983–1984 enrollment figures for the University of California campuses. Data are in thousands, rounded to the nearest hundred. (Source: *Peterson's Guides.*)

Campus	Enrollment (thousands)
Berkeley	30.0
Davis	19.0
Irvine	11.9
Los Angeles	34.8
Riverside	4.7
San Diego	13.7
San Francisco	3.7
Santa Barbara	15.5
Santa Cruz	6.9

a) Compute the population mean enrollment, μ, per campus.

b) Compute σ.

9 A frequency distribution of the 1984 annual salaries of the state governors in the United States is as follows:

Salary ($thousands)	Frequency
30–under 40	2
40–under 50	6
50–under 60	8
60–under 70	19
70–under 80	10
80–under 90	4
90–under 100	0
100–under 110	1

a) Compute the approximate value of the mean salary, μ.

b) Compute the approximate value of the standard deviation, σ, for this population of salaries.

c) Why are the values that you computed for μ and σ in parts (a) and (b) only approximate?

10 The U.S. Energy Information Administration publishes figures on retail gasoline prices. Data are obtained by sampling 10,000 gasoline service stations from a total of more than 185,000. For the 10,000 stations sampled, suppose the mean and standard deviation of the price per gallon for unleaded regular gasoline are as follows:

$$\text{Mean price} = \$1.27$$
$$\text{Standard deviation} = \$0.08$$

a) Is the mean price given here a sample mean or a population mean? Why?

b) What letter would you use to designate the mean of $1.27?

c) Is the standard deviation given here a sample standard deviation or a population standard deviation? Why?

d) What letter would you use to denote the standard deviation of $0.08?

e) Are the mean and standard deviation given here statistics or parameters? Explain your answer.

◆ F O U R ◆

Up to this point we have concentrated our discussion on *descriptive statistics*—methods for organizing and summarizing data. However, our main purpose in this text is to present the fundamentals of *inferential statistics*—methods of drawing conclusions about a population based on information obtained from a sample of the population. Since inferential statistics involves using information obtained from part of the population (the sample) to make inferences concerning the entire population, we can never be certain that our inferences are correct—we are dealing with uncertainty. Consequently, before we can understand, develop, and apply the methods of statistical inference, we need to become familiar with uncertainty.

The science of uncertainty is called **probability.** Probability enables us to evaluate and control the likelihood that our statistical inference is correct. The next few chapters are devoted to the study of probability.

Probability

CHAPTER OUTLINE

4.1 Introduction; classical probability   Introduces the concept of probability and discusses the classical approach to probability.

4.2 Events   Examines in detail the concept of events and gives some useful shorthand notations.

4.3 Some rules of probability   Presents some of the most important rules that are used in probability.

4.4 Contingency tables; joint and marginal probabilities   Discusses cross-classified data and describes how such data can be analyzed using probability.

4.5 Conditional probability   Introduces methods for computing probabilities given that a specified event has occurred.

4.6 The multiplication rule; independence   Presents some further rules of probability and examines the concept of independent events.

4.1

Introduction; classical probability

Most applications of probability to statistical inference involve large populations. However, the fundamental concepts of probability are most easily illustrated and explained using relatively small populations and games of chance. So, keep in mind that many of the illustrations in this chapter are designed expressly to present the principles of probability in a lucid manner.

Classical probability

We will begin our study of probability by discussing **classical probability.** Classical probability applies when the possible outcomes of an experiment are *equally likely* to occur.

EXAMPLE   4.1   **Introduces classical probability**

The ages of the students in Professor W's introductory statistics class are listed in Table 3.4 on page 60. A grouped-data table for those ages is presented below.

TABLE 4.1

Grouped-data table for ages of the students in Professor W's introductory statistics class.

Age (yrs) x	Frequency f	Relative frequency
17	1	0.025
18	1	0.025
19	8	0.200
20	7	0.175
21	8	0.200
22	5	0.125
23	3	0.075
24	4	0.100
26	1	0.025
35	1	0.025
36	1	0.025
Σ	40	1.000

Suppose that one of the students is selected **at random,** meaning that each student is *equally likely* to be the one selected. What is the probability that the student selected is 20 years old?

SOLUTION

As we see from Table 4.1, seven of the 40 students are 20 years old. Thus, the chances are seven out of 40 for selecting a student who is 20 years old. The probability is therefore

$$\frac{\text{Number of 20-year-olds}}{\text{Total number of students}} = \frac{7}{40}$$

Note that the probability, 7/40, of selecting a student aged 20 is exactly the same as the relative frequency, 0.175, of the students aged 20.   ∎

DEFINITION 4.1 Classical probability

Suppose there are *N equally likely* outcomes for an experiment. Then the probability an event occurs equals the number of ways, *f,* that the event can occur, divided by the total number, *N,* of possible outcomes. That is, the probability equals

$$\frac{f}{N} \quad \begin{array}{l} \leftarrow \quad \text{Number of ways event can occur} \\ \leftarrow \quad \text{Total number of possible outcomes} \end{array}$$

For convenience, we will sometimes refer to the above definition as the ***f/N* rule.**

In Example 4.1, $N = 40$ since there are 40 students in the class. If we are considering the event that the student selected is 20 years old, then $f = 7$, since seven of the students are 20 years old. Therefore, the probability that a student selected at random is 20 years old equals

$$\frac{f}{N} = \frac{7}{40} = 0.175$$

as we noted in Example 4.1.

EXAMPLE   4.2   Illustrates Definition 4.1

For the situation in Example 4.1, find the probability that a student selected at random is younger than 21.

SOLUTION

From Table 4.1, we see that there are 17 $(1 + 1 + 8 + 7)$ students in the class younger than 21. So, $f = 17$ and the probability is

$$\frac{f}{N} = \frac{17}{40} = 0.425$$

In terms of percentages, this means that 42.5% of the students in the class are under 21.   ∎

EXAMPLE   4.3   **Illustrates Definition 4.1**

An income-level distribution for the families in a small city is displayed in Table 4.2.[†]

TABLE 4.2
Frequency distribution
for family incomes.

Income	Frequency
Under $5000	390
$5,000 –$9,999	689
$10,000–$14,999	806
$15,000–$19,999	787
$20,000–$24,999	798
$25,000–$34,999	1268
$35,000–$49,999	1040
$50,000 & over	709
Σ	6487

Find the probability that a randomly selected family makes between $20,000 and $49,999 (inclusive).

SOLUTION
The second column of Table 4.2 shows that there are a total of 6487 families in the city, and thus $N = 6487$. The event in question here is that the randomly selected family makes between $20,000 and $49,999. To find the probability of that event we first need to determine the number of ways, f, that it can occur, that is, the number of families that make between $20,000 and $49,999. We see from the table that the number of such families is $798 + 1268 + 1040 = 3106$, so $f = 3106$. Therefore, the probability that a randomly selected family makes between $20,000 and $49,999 is

$$\frac{f}{N} = \frac{3106}{6487} = 0.479$$

MTB
SPSS

(to three decimal places). In other words, approximately 47.9% of the families make between $20,000 and $49,999. ∎

EXAMPLE   4.4   **Illustrates Definition 4.1**

When a pair of *fair* dice is rolled there are a total of 36 possible *equally likely* outcomes. We have depicted the possibilities in Figure 4.1 on the next page. Find the probability that
    a) the sum of the dice is 11.
    b) doubles are rolled (that is, both dice come up the same number).

SOLUTION
For this experiment $N = 36$.

[†] The data is based on information taken from the *Statistical Abstract of the United States, 1984.*

FIGURE 4.1
Possible outcomes
for rolling
a pair of dice.

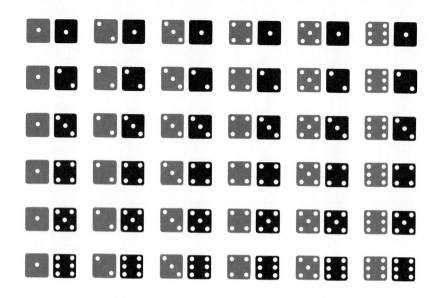

a) The sum of the dice can be 11 in $f = 2$ ways, as is readily seen from Figure 4.1. Thus, the probability is

$$\frac{f}{N} = \frac{2}{36} = 0.056$$

b) There are $f = 6$ ways that doubles can be rolled, and so the probability of rolling doubles is

$$\frac{f}{N} = \frac{6}{36} = 0.167$$

■

Probability and percentages

We pointed out in Example 4.1 that the probability of randomly selecting a student aged 20 is exactly the same as the relative frequency of the students aged 20. Since relative frequencies are just percentages expressed as decimals, we have the following fact.

KEY FACT

When selecting an individual or item at random from a population, the concepts of probability and relative frequency (percentages) are exactly the same.

For example, the fact that 11.9% of the U.S. population is black also means that the probability is 0.119 that a randomly selected U.S. resident will be black.

Note that the probabilities in Examples 4.2 and 4.3 arise from the random selection of an individual or item from a population, whereas those in Example 4.4 do not.

Basic properties of probability

It is important to note some simple but basic properties of probability.

KEY FACT Basic properties of probability

PROPERTY 1 *The probability of an event is always between 0 and 1.*
PROPERTY 2 *The probability of an event that cannot occur is 0. [An event that cannot occur is said to be **impossible**.]*
PROPERTY 3 *The probability of an event that must occur is 1. [An event that must occur is called **certain**.]*

Property 1 indicates that numbers such as 5 or -0.23 could not possibly be probabilities. Thus, if you calculate a probability and get an answer like 5 or -0.23, then you made an error. The next example illustrates Properties 2 and 3.

EXAMPLE 4.5 **Illustrates Properties 2 and 3**

Consider again the age data from Example 4.1. [Refer to Table 4.1 on page 111.]
a) What is the probability that a student selected at random is 25 years old?
b) What is the probability that a student selected at random is younger than 50?

SOLUTION
a) Since *none* of the students in the class are 25 years old, we have $f = 0$. Thus, the probability that a student selected at random is 25 years old equals

$$\frac{f}{N} = \frac{0}{40} = 0$$

This illustrates Property 2.
b) We next determine the probability that a student selected at random is younger than 50. Because *all* of the 40 students in the class are younger than 50, we have $f = 40$. The probability is therefore

$$\frac{f}{N} = \frac{40}{40} = 1$$

This illustrates Property 3. ∎

In this section we have seen how to find probabilities for experiments in which each outcome is equally likely to occur. When that is not the case, it is necessary to resort to other methods in order to determine probabilities. Some of those methods will be discussed later.

Exercises 4.1

In the exercises below, express each probability as a decimal rounded to three places.

4.1 The absentee records over the past year for the employees of Cudahey Masonry, Inc. are presented below in grouped form.

Days missed x	Number of employees f
0	4
1	2
2	14
3	10
4	16
5	18
6	10
7	6
Σ	80

If an employee is selected at random, find the probability that, over the past year, the employee missed
a) three days of work.
b) at most two days of work (that is, two or fewer).
c) between one and five days of work, inclusive.
d) eight days of work.
e) at most eight days of work.

4.2 A frequency distribution of the number of cars owned by each of the 6487 families in Example 4.3 is as follows:

Cars owned	Number of families
0	27
1	1422
2	2865
3	1796
4	324
5	53
Σ	6487

What is the probability that a randomly selected family owns
a) two cars?
b) more than three cars?
c) between one and three cars, inclusive?
d) seven cars?
e) at most seven cars? (that is, seven or fewer)

4.3 According to the Bureau of Labor Statistics, the age distribution of employed persons 16 years old and over is as shown in the following table. The frequencies are given in thousands.

Age (yrs) x	Number of persons f
16–19	6,549
20–24	13,690
25–34	28,149
35–44	20,879
45–54	15,923
55–64	11,414
65 and over	2,923
Σ	99,527

If an employed person is selected at random, find the probability that the person is
a) between 25 and 34 years old.
b) at least 45 years old (that is, 45 years old or older).
c) between 20 and 44 years old.
d) under 20 or over 54.

4.4 The distribution of scientists by field in the U.S. is given below. Frequencies are in thousands. (Source: National Science Foundation.)

Type	Frequency
Life scientists	440
Physical scientists	272
Computer scientists	430
Social scientists	247
Psychologists	142
Mathematical scientists	142
Environmental scientists	116
Σ	1789

What is the probability that a randomly selected scientist is
a) a psychologist?
b) a life or social scientist?
c) not a computer scientist?

4.5 Suppose a pair of fair dice is to be rolled. Calculate the probabilities of the following events. [Refer to Figure 4.1 on page 114.]
a) The sum of the dice is 6.
b) The sum of the dice is even.
c) The sum of the dice is 7 or 11.
d) The sum of the dice is either 2, 3, or 12.

4.6 A balanced dime is tossed three times. The possible outcomes can be represented as

HHH   HTH   THH   TTH
HHT   HTT   THT   TTT

where, for example, HHT means the first two tosses came up heads and the third tails. Since the dime is balanced, each of the eight outcomes above is equally likely. What is the probability that
a) exactly two of the three tosses yield heads?
b) the last two tosses come up tails?
c) all three tosses come up the same?
d) the second toss is a head?

4.7 The U.S. Senate consists of 100 senators, two from each state. If a senator is selected at random to serve as the chair of a subcommittee, what is the probability that the senator selected is from Georgia?

4.8 According to the U.S. Bureau of the Census, housing in the U.S. is occupied as follows (the number of housing units is given in thousands):

Owner-Occupied	Renter-Occupied	Vacant year-round
54,724	29,914	7,037

What is the probability that a housing unit selected at random is
a) owner-occupied?
b) not renter-occupied?
c) vacant year-round?

4.9 The Internal Revenue Service reports that the number (in thousands) of U.S. businesses in various categories is as follows:

Proprietorships	Partnerships	Corporations
12,185	1,461	2,812

What is the probability that a business selected at random is
a) a proprietorship?
b) not a partnership?
c) a corporation?

4.10 Which of the following numbers could not possibly be probabilities?
a) 0.462   b) −0.201   c) 1
d) 5/6   e) 3.5   f) 0

Justify your answer.

4.11 Which of the following numbers could not possibly be probabilities?
a) 0.732   b) 4.75   c) 0.0
d) −3/4   e) 1   f) 4/5

Justify your answer.

4.12 Refer to Exercise 4.2. Which, if any, of the events in parts (a)–(e) are certain? Impossible?

4.13 Refer to Exercise 4.1. Which, if any, of the events in parts (a)–(e) are certain? Impossible?

4.14 Explain what is wrong with the following argument: When two fair dice are rolled, the sum of the dice can be 2, 3, 4, 5, 6, 7, 8, 9, 10, 11, or 12. This gives 11 possibilities. Therefore, the probability that the sum is 2 equals 1/11.

4.15 Explain what is wrong with the following argument: When a balanced coin is tossed twice, the total number of heads obtained can be either 0, 1, or 2. This gives three possibilities. Therefore, the probability of getting 2 heads is 1/3.

4.16 A study of the methods that Americans use to get to work revealed the following data, in hundreds of thousands of workers. (Source: U.S. Bureau of the Census.)

	Type of Worker		
Method		Urban	Rural	Total
	Automobile	450	150	600
	Public transportation	65	5	70
	Total	515	155	670

If a worker is selected at random, what is the probability that he or she
a) is an urban worker?
b) drives an automobile to work?
c) is an urban worker who drives an automobile to work?
d) is an urban worker or drives an automobile to work?
e) is a rural worker who uses public transportation?

4.17 The table on the following page gives a frequency distribution of institutions of higher education in the United States by region and type. (Source: U.S. National Center for Education Statistics.)

Region		Type: Public	Private	Total
Region	Northeast	266	555	821
	Midwest	359	504	863
	South	533	502	1035
	West	313	242	555
	Total	1471	1803	3274

If an institution of higher education is selected at random, what is the probability that
a) it is a public school?
b) it is in the Midwest?
c) it is a public school and in the Midwest?
d) it is a public school or in the Midwest?
e) it is a Northeastern private school?

4.18 Suppose there are N equally likely outcomes for an experiment. Use Definition 4.1 (page 112) to verify that Properties 1–3 given on page 115 hold for such an experiment.

4.2 Events

Before we can continue our study of probability, we need to discuss *events* in greater detail. In the previous section we used the word "event" intuitively. To be more precise, an **event** consists of an outcome or collection of outcomes. Consider Example 4.6, at the top of the next page.

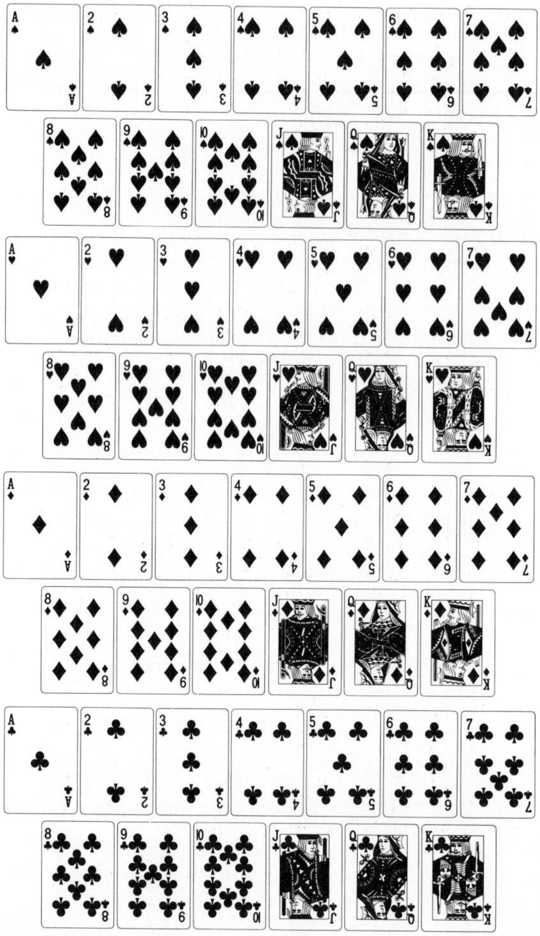

FIGURE 4.2
Deck of cards.

EXAMPLE 4.6 **Illustrates events**

A deck of playing cards consists of 52 cards, as displayed in Figure 4.2 on the previous page. When we perform the experiment of randomly selecting a card from the deck, there are $N = 52$ possible outcomes, namely, the 52 cards pictured in Figure 4.2. There are many events associated with this experiment. Consider these four:

1 The event that the card selected is the king of hearts.
2 The event that the card selected is a king.
3 The event that the card selected is a heart.
4 The event that the card selected is a face card.

List the outcomes making up each of the above events.

SOLUTION
The first event above consists of the single outcome "king of hearts," and is pictured in Figure 4.3.

FIGURE 4.3
The event the king of hearts
is selected.

The second event consists of the four outcomes "king of spades," "king of hearts," "king of diamonds," and "king of clubs." This event is depicted in Figure 4.4.

FIGURE 4.4
The event a king is selected.

There are 13 outcomes comprising the third event; namely, the outcomes "ace of hearts," "two of hearts," . . . , "king of hearts." See Figure 4.5.

FIGURE 4.5
The event a heart
is selected.

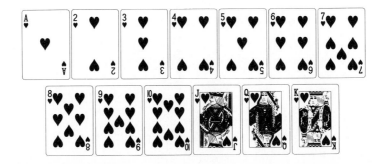

Finally, the fourth event consists of 12 outcomes, the 12 face cards shown in Figure 4.6.

FIGURE 4.6

The event a face card is selected.

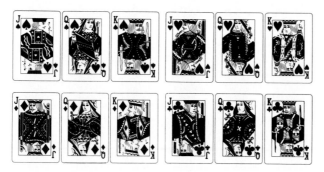

Note that, when the experiment of selecting a card from the deck is performed, an event **occurs** if it includes the card selected. For instance, if the card selected from the deck turns out to be the "king of spades," then the second and fourth events above occur, while the first and third events do not occur.   ■

Notation and graphical displays for events

It is convenient and less cumbersome to employ letters such as *A, B, C, D, . . .* to represent events. For instance, in the card-selection experiment of Example 4.6, we might let

A = event the card selected is the king of hearts
B = event the card selected is a king
C = event the card selected is a heart
D = event the card selected is a face card

Graphical displays of events are also quite useful for explaining and understanding probability. This can be accomplished by the use of **Venn diagrams** (named after the English logician John Venn, 1834–1923). The collection of all possible outcomes is portrayed as a rectangle with the various events drawn as disks inside the rectangle. The simplest situation occurs when only one event is displayed, as in Figure 4.7.

FIGURE 4.7

Venn diagram for event *E*.

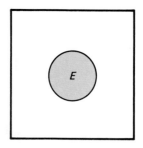

The shaded portion of the figure represents the event *E*.

Relationships among events

To each event E there corresponds another event defined by the condition that "E does not occur." This is called the **complement of E** and is denoted by **(not E).** The event (not E) consists of all outcomes not in E. A Venn diagram makes this idea clearer. See Figure 4.8.

FIGURE 4.8
Venn diagram for (not E).

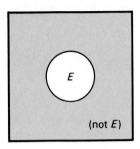

With any two events, A and B, we can associate *two* new events. The first new event is defined by the condition that "*both* event A *and* event B occur" and is denoted by **(A & B).** This event consists of all outcomes common to both event A and event B. A Venn diagram for the event (A & B) is given in Figure 4.9.

FIGURE 4.9
Venn diagram for (A & B).

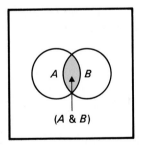

The second new event is defined by the condition that "*either* event A *or* event B *or* both occur;" or, equivalently, that "*at least one* of the events A and B occurs." This event is denoted by **(A or B)** and consists of all outcomes in either event A or event B or both. A Venn diagram for the event (A or B) is displayed in Figure 4.10.

FIGURE 4.10
Venn diagram for (A or B).

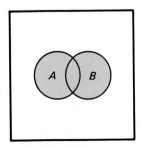

We will now summarize the terms describing relationships among events and illustrate them with some examples.

DEFINITION 4.2 (not *E*), (*A* & *B*), and (*A* or *B*)

Let *E, A,* and *B* denote events. Then the events (not *E*), (*A* & *B*), and (*A* or *B*) are defined as follows:

(not *E*) is the event that "*E* does not occur."
(*A* & *B*) is the event that "both event *A* and event *B* occur."
(*A* or *B*) is the event that "either event *A* or event *B* or both occur."

Note: It is important to realize that the event (*A* & *B*) is the same as the event (*B* & *A*). This is because the event that "both event *A* and event *B* occur" is the same as the event that "both event *B* and event *A* occur." Similarly, the event (*A* or *B*) is the same as the event (*B* or *A*).

EXAMPLE 4.7   Illustrates Definition 4.2

For the experiment of randomly selecting a card from a deck of 52, let

A = event the card selected is the king of hearts
B = event the card selected is a king
C = event the card selected is a heart
D = event the card selected is a face card

The outcomes making up these four events are shown in Figures 4.3–4.6 (pages 119–120). Determine the following events:

a) (not *D*)   c) (*B* or *C*)
b) (*B* & *C*)   d) (*C* & *D*)

SOLUTION

a) (not *D*) is the event that "*D* does not occur"—the event that a face card is *not* selected. Event (not *D*) consists of the 40 cards in the deck that are not face cards. See Figure 4.11.

FIGURE 4.11
The event (not *D*).

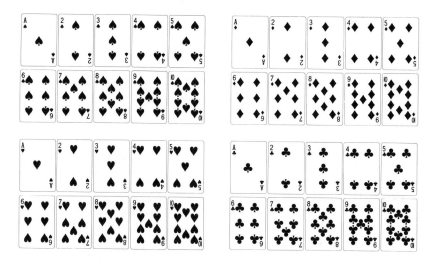

b) (*B* & *C*) is the event that "both event *B* and event *C* occur"—the event that the card selected is *both* a king *and* a heart. This can only happen if the card selected is the king of hearts. In other words, (*B* & *C*) is the event that the card selected is the king of hearts. It consists of the single outcome shown in Figure 4.12.

FIGURE 4.12
The event (*B* & *C*).

[Note that the event (*B* & *C*) is the same as event *A*, so we can write *A* = (*B* & *C*).]

c) (*B* or *C*) is the event that "either event *B* or event *C* or both occur"—the event that the card selected is either a king *or* a heart *or* both. Event (*B* or *C*) consists of 16 outcomes; namely, the four kings and the 12 nonking hearts. See Figure 4.13.

FIGURE 4.13
The event (*B* or *C*).

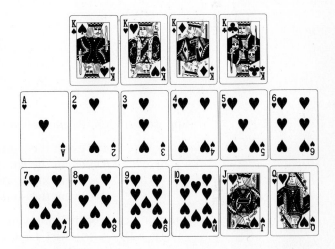

[Note that there are 16 (not 17) ways that the event (*B* or *C*) can occur, since the outcome, "king of hearts," is common to both event *B* and event *C*.]

d) (*C* & *D*) is the event that "both event *C* and event *D* occur"—the event that the card selected is *both* a heart *and* a face card. For this to occur, the card selected must be either the king, queen, or jack of hearts. Thus, event (*C* & *D*) consists of the three outcomes displayed in Figure 4.14.

FIGURE 4.14
The event (*C* & *D*).

These are precisely the outcomes common to both events *C* and *D*. ∎

For the card-selection example above, we listed the outcomes comprising the events in question. Sometimes this is unnecessary, inconvenient, or undesirable. In cases such as the following, we can describe events verbally instead of listing their outcomes.

EXAMPLE   4.8   Illustrates Definition 4.2

A frequency distribution for the ages of the students in Professor W's introductory statistics class is presented in Table 4.3.

TABLE 4.3

Age (yrs) x	Frequency f
17	1
18	1
19	8
20	7
21	8
22	5
23	3
24	4
26	1
35	1
36	1
Σ	40

Suppose a student is selected at random. Consider the following events:

A = event the student selected is under 21
B = event the student selected is over 30
C = event the student selected is in his or her 20s
D = event the student selected is over 18

Determine each of the following events:

a) (not D)   b) (A & D)   c) (A or D)   d) (B or C)

SOLUTION

a) (not D) is the event that "D does not occur"—the event that the student selected is *not* over 18. In other words, (not D) is the event that the student selected is 18 or under. This event is comprised of the two students in the class who are 18 or under (see Table 4.3).

b) (A & D) is the event that "both event A and event D occur"—the event that the student selected is *both* under 21 and over 18. Consequently, (A & D) is the event that the student selected is either 19 or 20. This event is comprised of the 15 students in the class who are 19 or 20.

c) (A or D) is the event that "either event A or event D or both occur"—the event that the student selected is either under 21 or over 18 (or both). But every student is either under 21 or over 18. Consequently, (A or D) contains all students and hence is *certain* to occur.

d) (*B* or *C*) is the event that "either event *B* or event *C* or both occur"—the event that the student selected is either over 30 or in his or her 20s. A glance at Table 4.3 shows that (*B* or *C*) is comprised of the 30 students in the class who are 20 or over. ∎

Mutually exclusive events

Another important concept in probability is that of mutually exclusive events. Two events are said to be **mutually exclusive** if at most one of them can occur when the experiment is performed. This requires that the two events have no outcomes in common.

DEFINITION 4.3 Mutually exclusive events

Two events are said to be **mutually exclusive** if they cannot both occur when the experiment is performed; that is, if *at most one* of the events can occur.

The Venn diagrams in Figures 4.15 and 4.16 portray the difference between events that are mutually exclusive and events that are not mutually exclusive.

FIGURE 4.15
Mutually exclusive events—the events have no outcomes in common.

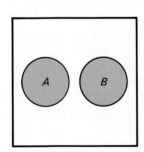

FIGURE 4.16
Events that are *not* mutually exclusive—the events have outcomes in common.

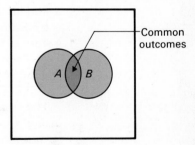

Common outcomes

EXAMPLE 4.9 Illustrates Definition 4.3

For the experiment of selecting a card at random from a deck of 52 playing cards, let

C = event the card selected is a heart
D = event the card selected is a face card
E = event the card selected is an ace

Which of the following pairs of events are mutually exclusive?

a) *C, D*. b) *C, E*. c) *D, E*.

SOLUTION

a) Event *C* and event *D* are *not* mutually exclusive because they have the common outcomes "king of hearts," "queen of hearts," and "jack of hearts." In other words, event *C* and event *D* can *both* occur if the card selected is either the king, queen, or jack of hearts.

b) Event *C* and event *E* are *not* mutually exclusive, since they have the common outcome "ace of hearts." *Both* events can occur if the card selected is the ace of hearts.

c) Event *D* and event *E* *are* mutually exclusive, since they have no common outcomes. They cannot both occur when the experiment is performed, because it is impossible to select a card that is both a face card and an ace.   ■

EXAMPLE 4.10   **Illustrates Definition 4.3**

For the experiment of selecting a student at random from Professor W's introductory statistics course, let

A = event the student selected is under 21
B = event the student selected is over 30
C = event the student selected is in his or her 20s

Which of the following pairs of events are mutually exclusive?
a) *A*, *B*.   b) *B*, *C*.   c) *A*, *C*.

SOLUTION

a) Event *A* and event *B are* mutually exclusive, since it is not possible to select a student who is both under 21 and over 30.

b) Event *B* and event *C are* mutually exclusive, since it is not possible to select a student who is both over 30 and in his or her 20s.

c) Event *A* and event *C* are *not* mutually exclusive. They can both occur if the student selected is 20 years old.   ■

The concept of mutually exclusive events can be extended to more than two events. Consider, for instance, three events, *A*, *B*, and *C*. These events are said to be *mutually exclusive* if at most one of them can occur when the experiment is performed. This requires that no two of them have any outcomes in common. The Venn diagram in Figure 4.17 portrays three mutually exclusive events.

FIGURE 4.17
Three mutually
exclusive events.

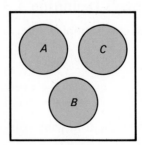

In general, we have the following definition.

DEFINITION 4.4 Mutually exclusive events

Two or more events are said to be **mutually exclusive** if no two of them can occur simultaneously; that is, if *at most one* of the events can occur when the experiment is performed.

EXAMPLE 4.11 Illustrates Definition 4.4

In the card selection experiment, let

D = event the card selected is a face card
E = event the card selected is an ace
F = event the card selected is an "8"
G = event the card selected is a "10" or a jack

a) Are the three events, D, E, and F, mutually exclusive?
b) Are the four events, D, E, F, and G, mutually exclusive?

SOLUTION
a) Events D, E, and F *are* mutually exclusive since no two of them can occur simultaneously.
b) Events D, E, F, and G are *not* mutually exclusive, since event D and event G can both occur if the card selected is a jack. ∎

Exercises 4.2

4.19 When a single die is rolled there are six possible outcomes:

Let

A = event the die comes up even
B = event the die comes up four or more
C = event the die comes up at most two
D = event the die comes up three

List the outcomes comprising each of the events above.

4.20 In a horse race, the odds are as follows:

Horse	Odds (against winning)
#1	8
#2	15
#3	2
#4	3
#5	30
#6	5
#7	10
#8	5

Let

A = event one of the top two favorites wins (The top two favorites are the two horses with the lowest odds, Horse #3 and Horse #4.)

B = event the winning horse has a number above five
C = event the winning horse has a number of at most three
D = event one of the two longshots wins (The two longshots are the two horses with the highest odds.)

List the outcomes comprising each of the events above.

4.21 When a dime is tossed four times, there are 16 possible outcomes:

HHHH	HTHH	THHH	TTHH
HHHT	HTHT	THHT	TTHT
HHTH	HTTH	THTH	TTTH
HHTT	HTTT	THTT	TTTT

Here, for example, HTTH represents the outcome that the first toss is heads, the next two tosses are tails, and the fourth toss is heads. Let

A = event exactly two heads are tossed
B = event the first two tosses are tails
C = event the first toss is a head
D = event all four tosses come up the same

List the outcomes that make up each of these events.

4.22 A committee is formed of five executives: three women and two men. Their first names are

Marie (M)   John (J)
Susan (S)   Bill (B)
Carol (C)

and the letter in parentheses stands for the name. The committee needs to select a chairperson and a secretary. It decides to make the selection randomly by drawing straws. The person getting the longest straw will be appointed chairperson and the one getting the shortest straw will be appointed secretary. We can represent the possible outcomes as follows:

MS	SM	CM	JM	BM
MC	SC	CS	JS	BS
MJ	SJ	CJ	JC	BC
MB	SB	CB	JB	BJ

Here, for example, MS represents the outcome that Marie is appointed chairperson and Susan is appointed secretary. Let

A = event a male is appointed chairperson
B = event Carol is appointed chairperson
C = event Bill is appointed secretary
D = event only females are appointed

List the outcomes comprising each of these events.

4.23 Refer to Exercise 4.19. For each of the following events, list the outcomes that make up the event and describe the event in words.
a) (not A)
b) (A & B)
c) (B or C)

4.24 Refer to Exercise 4.20. For each of the following events, list the outcomes that make up the event and describe the event in words.
a) (not C)
b) (C & D)
c) (A or C)

4.25 Refer to Exercise 4.21. For each of the following events, list the outcomes that make up the event and describe the event in words.
a) (not B)
b) (A & B)
c) (C or D)

4.26 Refer to Exercise 4.22. For each of the following events, list the outcomes that make up the event and describe the event in words.
a) (not A)
b) (B & D)
c) (B or C)

4.27 The absentee records over the past year for the employees of Cudahey Masonry, Inc. are presented below in a frequency distribution.

Days missed	Number of employees
0	4
1	2
2	14
3	10
4	16
5	18
6	10
7	6
Σ	80

For an employee selected at random, let

A = event the employee missed at most three days
B = event the employee missed at least one day
C = event the employee missed between four and six days, inclusive
D = event the employee missed more than six days

Describe each of the following events in words and determine the number of outcomes (employees) that comprise each event.
a) (not A)
b) (A & B)
c) (C or D)

4.28 A frequency distribution for the number of cars owned by each of the families in a small city is presented below.

Cars owned	Number of families
0	27
1	1422
2	2865
3	1796
4	324
5	53
Σ	6487

For a family selected at random, let

A = event the family owns at most three cars
B = event the family owns at least one car
C = event the family owns between two and four cars inclusive
D = event the family owns at least three cars

Describe each of the following events in words and determine the number of outcomes (families) that comprise each event.

a) (not *B*)

b) (*C* & *D*)

c) (*A* or *D*)

4.29 According to the Bureau of Labor Statistics, the age distribution of employed persons 16 years old and over is as shown in the following table. The frequencies are given in thousands.

Age (yrs)	Number of persons (thousands)
16–19	6,549
20–24	13,690
25–34	28,149
35–44	20,879
45–54	15,923
55–64	11,414
65 and over	2,923
Σ	99,527

An employed person is selected at random. Let

A = event the person is under 20

B = event the person is between 20 and 54, inclusive

C = event the person is under 45

D = event the person is at least 55

Describe each of the following events in words and determine the number of outcomes (persons) that comprise each event.

a) (not *C*)

b) (not *B*)

c) (*B* & *C*)

d) (*A* or *D*)

4.30 Each part of this exercise lists events from Exercise 4.20. Indicate whether the events are mutually exclusive.

a) Events *A* and *B*

b) Events *B* and *C*

c) Events *A*, *B*, and *C*

d) Events *A*, *B*, and *D*

e) Events *A*, *B*, *C*, and *D*

4.31 Refer to Exercise 4.19.

a) Are the events *A* and *B* mutually exclusive?

b) Are the events *B* and *C* mutually exclusive?

c) Are the events *A*, *C*, and *D* mutually exclusive?

d) Can you find *three* mutually exclusive events among *A*, *B*, *C*, and *D*? How about four?

4.32 For the following groups of events from Exercise 4.28, determine which are mutually exclusive.

a) Events *A* and *B*

b) Events (not *B*) and *C*

c) Events *A* and *D*

d) Events (not *B*), *C*, and *D*

4.33 For the following groups of events from Exercise 4.29, determine which are mutually exclusive.

a) Events *C* and *D*

b) Events *B* and *C*

c) Events *A*, *B*, and *D*

d) Events *A*, *B*, and *C*

e) Events *A*, *B*, *C*, and *D*

4.34 Draw a Venn diagram portraying *four* mutually exclusive events.

4.35 Draw a Venn diagram portraying four events *A*, *B*, *C*, and *D* with the following property: Events *A*, *B*, and *C* are mutually exclusive, events *A*, *B*, and *D* are mutually exclusive, but no other three of the four events are mutually exclusive.

4.36 Suppose that *A*, *B*, and *C* are three events with the property that they cannot all occur simultaneously. Does this necessarily imply that *A*, *B*, and *C* are mutually exclusive? Justify your answer and illustrate with a Venn diagram.

4.3 Some rules of probability

In this section we will discuss several rules of probability. In preparation, let's look at some additional notation used in probability.

EXAMPLE 4.12 Introduces some additional probability notation

When a fair die is rolled, there are *N* = 6 *equally likely* outcomes. These are pictured in Figure 4.18 on the following page.

FIGURE 4.18

Consider the event that the die comes up even. This event can occur in $f = 3$ ways, namely, if "2," "4," or "6" comes up. Since

$$\frac{f}{N} = \frac{3}{6} = 0.5$$

we see that *the probability that the die comes up even is 0.5.*

Employing probability notation enables us to express the italicized phrase much more concisely. Let

$$A = \text{event the die comes up even}$$

We use the notation $P(A)$ to stand for the probability that event A occurs. Thus, the italicized statement above can be written simply as

$$P(A) = 0.5$$

and is read "the probability of A is 0.5." We should emphasize that the notation A refers to the *event* that the die comes up even, while the notation $P(A)$ refers to the *probability* of that event. ■

DEFINITION 4.5 $P(E)$

If E is an event, then the notation $P(E)$ stands for the probability that the event E occurs, and is read "the probability of E."

The special addition rule

The first rule of probability that we will study is called the **special addition rule.** It states that if two events are mutually exclusive, then the probability that either occurs is equal to the sum of their probabilities.

EXAMPLE 4.13   **Introduces the special addition rule**

The ages of the 40 students in Professor W's introductory statistics class are displayed in a frequency distribution in Table 4.4.

TABLE 4.4

Age (yrs)	Frequency
17	1
18	1
19	8
20	7
21	8
22	5
23	3
24	4
26	1
35	1
36	1
Σ	40

Suppose a student is selected at random. Let

$$E = \text{event the student selected is } 19$$
$$F = \text{event the student selected is } 20$$

The probability that the student selected is 19 is computed as usual:

$$P(E) = \frac{f}{N} = \frac{8}{40} = 0.200$$

Similarly, the probability that the student selected is 20 equals

$$P(F) = \frac{f}{N} = \frac{7}{40} = 0.175$$

In summary

$$P(E) = 0.200 \qquad P(F) = 0.175$$

Next, let us compute the probability that the student selected is either 19 or 20; that is, $P(E \text{ or } F)$. As we see from Table 4.4, a total of $8 + 7 = 15$ students in the class are either 19 or 20. Thus

$$P(E \text{ or } F) = \frac{f}{N} = \frac{8+7}{40} = \frac{15}{40} = 0.375$$

Now here is the important point of this example. In computing $P(E \text{ or } F)$, we can take a slightly different route:

$$P(E \text{ or } F) = \frac{f}{N} = \frac{8+7}{40} = \frac{8}{40} + \frac{7}{40} = 0.200 + 0.175 = 0.375$$

Note that the two numbers in color are, respectively, $P(E)$ and $P(F)$. Consequently, in this case we have

$$P(E \text{ or } F) = P(E) + P(F)$$

This formula is the *special addition rule.* It holds not only for the events E and F given here, but for any two events that are *mutually exclusive.*   ■

FORMULA 4.1 The special addition rule

If event A and event B are *mutually exclusive,* then

$$P(A \text{ or } B) = P(A) + P(B)$$

More generally, if events $A, B, C, \ldots$ are *mutually exclusive,* then

$$P(A \text{ or } B \text{ or } C \text{ or } \cdots) = P(A) + P(B) + P(C) + \cdots$$

EXAMPLE 4.14 **Illustrates Formula 4.1**

A relative-frequency distribution for the size of farms in the U.S. is presented in the first two columns of Table 4.5 on the following page.

TABLE 4.5
Size of farms in the U.S.

Size (acres)	Relative frequency	Event
Under 10	0.087	A
10–49	0.192	B
50–99	0.156	C
100–179	0.173	D
180–259	0.098	E
260–499	0.143	F
500–999	0.087	G
1000–1999	0.040	H
2000 & over	0.026	I

Source: U.S. Bureau of the Census.

Table 4.5 shows, for instance, that 17.3% (0.173) of the farms in the U.S. have between 100 and 179 acres.

In the third column of Table 4.5 we have introduced events that correspond to each size class. For example, if a farm is selected at random, then D is the event that the farm selected has between 100 and 179 acres. The probabilities for the events in the third column are simply the relative frequencies given in the second column. Thus, the probability that a randomly selected farm has between 100 and 179 acres is $P(D) = 0.173$.

Use Table 4.5 and the special addition rule to determine the probability that a randomly selected farm has

a) under 50 acres.

b) between 100 and 499 acres, inclusive.

S O L U T I O N
To find the desired probabilities, we express each event in (a) and (b) in terms of the *mutually exclusive* events A through I and then apply the special addition rule.

a) The event that a randomly selected farm has under 50 acres can be expressed as (A or B), as we see from Table 4.5. Since event A and event B are mutually exclusive (why?), we see by the special addition rule that

$$P(A \text{ or } B) = P(A) + P(B) = 0.087 + 0.192 = 0.279$$

In terms of percentages, this means that 27.9% of the farms in the U.S. have under 50 acres.

b) The event that a randomly selected farm has between 100 and 499 acres can be expressed as (D or E or F). The events D, E, and F are mutually exclusive. Thus by the special addition rule

$$P(D \text{ or } E \text{ or } F) = P(D) + P(E) + P(F)$$
$$= 0.173 + 0.098 + 0.143$$
$$= 0.414$$

In other words, 41.4% of U.S. farms have between 100 and 499 acres.   ∎

The complementation rule

The special addition rule has several consequences. One such consequence is called the **complementation rule.** It states that the probability that an event occurs is equal to one minus the probability that it doesn't occur.

EXAMPLE 4.15

Introduces the complementation rule

The first two columns of Table 4.6 provide a relative-frequency distribution for the number of cars owned by each of the families in a small city.

TABLE 4.6

Cars owned	Relative frequency	Event
0	0.004	A
1	0.219	B
2	0.422	C
3	0.277	D
4	0.050	E
5	0.008	F
Σ	1.000	

What is the probability that a randomly selected family owns *at least* one car?

SOLUTION
Let

$$G = \text{event the family selected owns at least one car}$$

We can express G in terms of the events B through F given in the third column of Table 4.6:

$$G = (B \text{ or } C \text{ or } D \text{ or } E \text{ or } F)$$

Since the events B through F are mutually exclusive, the special addition rule implies that

$$
\begin{aligned}
P(G) &= P(B \text{ or } C \text{ or } D \text{ or } E \text{ or } F) \\
&= P(B) + P(C) + P(D) + P(E) + P(F) \\
&= 0.219 + 0.422 + 0.277 + 0.050 + 0.008 \\
&= 0.996
\end{aligned}
$$

There is, however, an easier way to find $P(G)$. Consider the event that "G does not occur" — (not G). Since G is the event that the family selected owns at least one car, (not G) is the event that the family selected does *not* own at least one car — that is, the family owns 0 cars. From the table we find that

$$P(\text{not } G) = 0.004$$

Now, here is the important point. Either the event G occurs or it doesn't. In other words, either event G occurs or event (not G) occurs. This means that the event (G or (not G)) is *certain.* Consequently,

$$P(G \text{ or } (\text{not } G)) = 1$$

Moreover, event *G* and event (not *G*) are mutually exclusive, so that the special addition rule implies

$$P(G \text{ or } (\text{not } G)) = P(G) + P(\text{not } G)$$

This and the previous equation yield the equation

$$P(G) + P(\text{not } G) = 1$$

Rearranging terms, we get the *complementation rule*

$$P(G) = 1 - P(\text{not } G)$$

Since, as we have seen, $P(\text{not } G) = 0.004$, we can quickly find $P(G)$ using the complementation rule:

$$P(G) = 1 - P(\text{not } G) = 1 - 0.004 = 0.996$$

This, of course, is the same value for $P(G)$ that we obtained earlier. However, the second method for determining $P(G)$ is quicker and easier.   ■

FORMULA 4.2 The complementation rule

For any event *E*,

$$P(E) = 1 - P(\text{not } E)$$

In words, the probability that an event *E* occurs equals one minus the probability that its complement, (not *E*), occurs.

EXAMPLE 4.16   **Illustrates Formula 4.2**

In Example 4.14 we considered the distribution for the size of farms in the U.S. The relative-frequency distribution from Table 4.5 is repeated here as Table 4.7.

TABLE 4.7
Size of farms in the U.S.

Size (acres)	Relative frequency	Event
Under 10	0.087	*A*
10–49	0.192	*B*
50–99	0.156	*C*
100–179	0.173	*D*
180–259	0.098	*E*
260–499	0.143	*F*
500–999	0.087	*G*
1000–1999	0.040	*H*
2000 & over	0.026	*I*

Source: U.S. Bureau of the Census.

Determine the probability that a randomly selected farm has
   a) less than 2000 acres.
   b) at least 50 acres.

S O L U T I O N

a) Let

$$J = \text{event the farm selected has less than 2000 acres}$$

To find $P(J)$ we will use the complementation rule, since it is easier to compute $P(\text{not } J)$. Note that (not J) is the event that the farm selected has 2000 or more acres, which is event I in Table 4.7. So

$$P(\text{not } J) = P(I) = 0.026$$

Applying the complementation rule we find that

$$P(J) = 1 - P(\text{not } J) = 1 - 0.026 = 0.974$$

b) Let

$$K = \text{event the farm selected has at least 50 acres}$$

We will apply the complementation rule to find $P(K)$. Now, (not K) is the event that the farm selected has less than 50 acres. From Table 4.7 we see that (not K) is the same as the event (A or B). Since event A and event B are mutually exclusive, the special addition rule implies that

$$\begin{aligned} P(\text{not } K) = P(A \text{ or } B) &= P(A) + P(B) \\ &= 0.087 + 0.192 = 0.279 \end{aligned}$$

Using this and the complementation rule we conclude that

$$P(K) = 1 - P(\text{not } K) = 1 - 0.279 = 0.721 \qquad \blacksquare$$

The general addition rule

Recall that the *special addition rule* provides a formula for $P(A \text{ or } B)$ in terms of $P(A)$ and $P(B)$; namely,

$$P(A \text{ or } B) = P(A) + P(B)$$

The formula holds provided event A and event B are *mutually exclusive*. What happens if event A and event B are *not* mutually exclusive? The answer is supplied by the *general addition rule,* which we introduce in Example 4.17.

EXAMPLE 4.17 **Introduces the general addition rule**

Consider again the experiment of randomly selecting a card from a deck of 52 playing cards. Let

$$A = \text{event a king is selected}$$
$$B = \text{event a heart is selected}$$

The event A can occur in four ways:

The event *B* can occur in 13 ways:

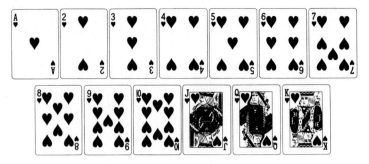

Consequently,

$$P(A) = \frac{f}{N} = \frac{4}{52} = 0.077$$

and

$$P(B) = \frac{f}{N} = \frac{13}{52} = 0.250$$

Now let's consider the event (*A* or *B*):

(*A* or *B*) = event a king or a heart is selected

This event can occur in 16 (not 17) ways:

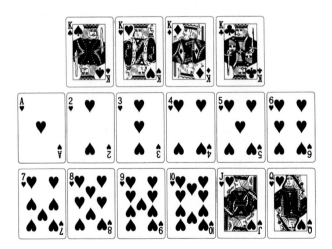

Thus,

$$P(A \text{ or } B) = \frac{f}{N} = \frac{16}{52} = 0.308$$

Note that

$$P(A \text{ or } B) \neq P(A) + P(B)$$

since

$$0.308 \neq 0.077 + 0.250$$

The reason that $P(A \text{ or } B) \neq P(A) + P(B)$ is because event A and event B are *not* mutually exclusive. They have the common outcome "king of hearts." So when we add $P(A)$ to $P(B)$ we count the king of hearts twice, instead of once. The probability of this common outcome is

$$P(A \ \& \ B) = \frac{f}{N} = \frac{1}{52} = 0.019$$

and when we subtract this from $P(A) + P(B)$ we *do* get $P(A \text{ or } B)$:

$$P(A \text{ or } B) = P(A) + P(B) - P(A \ \& \ B)$$

since

$$0.308 = 0.077 + 0.250 - 0.019$$

∎

FORMULA 4.3 The general addition rule

If A and B are any two events, then

$$P(A \text{ or } B) = P(A) + P(B) - P(A \ \& \ B)$$

In words, the probability that either event A or event B occurs equals the probability that event A occurs *plus* the probability that event B occurs *minus* the probability that both occur.

Frequently we have a choice for computing $P(A \text{ or } B)$ either directly or by using the general addition rule. The next example illustrates such a case.

EXAMPLE 4.18   **Illustrates Formula 4.3**

In the card selection experiment above, determine the probability that the card selected is either a spade or a face card
   a) without using the general addition rule.
   b) with the aid of the general addition rule.

SOLUTION
   a) Let

$E =$ event the card selected is either a spade or a face card

The event E consists of 22 cards; namely, the 13 spades plus the other nine face cards that are not spades. See Figure 4.19 on the following page.
   Consequently,

$$P(E) = \frac{f}{N} = \frac{22}{52} = 0.423$$

FIGURE 4.19

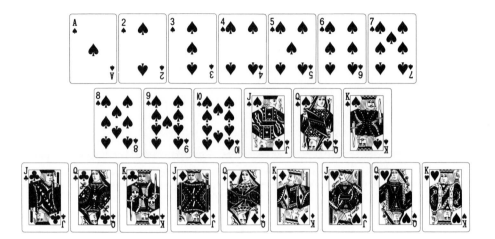

b) To find $P(E)$ using the general addition rule, we first let

C = event the card selected is a spade
D = event the card selected is a face card

and note that $E = (C \text{ or } D)$. Event C consists of the 13 spades and event D consists of the 12 face cards. Also, event $(C \& D)$ consists of the three spades that are face cards—the king, queen, and jack of spades. Applying the general addition rule we get

$$P(E) = P(C \text{ or } D) = P(C) + P(D) - P(C \& D)$$

$$= \frac{13}{52} + \frac{12}{52} - \frac{3}{52} = 0.250 + 0.231 - 0.058$$

$$= 0.423$$

which agrees with the result obtained in part (a).   ■

We have computed the probability of selecting either a spade or a face card in two ways—with and without the aid of the general addition rule. In this case, computing the probability is easier without using the general addition rule. There are many cases, however, in which the use of the general addition rule is the easier or even the only way to compute a probability. Consider the following example.

EXAMPLE 4.19   **Illustrates Formula 4.3**

According to the U.S. Bureau of the Census, 83.7% of the people arrested in 1982 were male, 17.9% were under 18, and 14.2% were males under 18. Suppose a person arrested in 1982 is selected at random. Find the probability that the person selected is either a male or under 18.

S O L U T I O N
Let

$$M = \text{event the person selected is a male}$$
$$E = \text{event the person selected is under 18}$$

From the percent data given above, we have

$$P(M) = 0.837 \qquad P(E) = 0.179 \qquad P(M \,\&\, E) = 0.142$$

The probability that the person selected at random is either a male or under 18 is $P(M \text{ or } E)$. Employing the general addition rule we conclude that

$$
\begin{aligned}
P(M \text{ or } E) &= P(M) + P(E) - P(M \,\&\, E) \\
&= 0.837 + 0.179 - 0.142 \\
&= 0.874
\end{aligned}
$$

In terms of percentages this means that 87.4% of those arrested in 1982 were either male or under 18. ∎

It is important to realize that the general addition rule is consistent with the special addition rule. That is, if two events *are* mutually exclusive, then the general addition rule gives the same result as the special addition rule. To see this, suppose event A and event B are mutually exclusive. Then they cannot both occur when the experiment is performed. This means that the event $(A \,\&\, B)$ is *impossible* and hence has a probability of 0. Applying the general addition rule we get

$$
\begin{aligned}
P(A \text{ or } B) &= P(A) + P(B) - P(A \,\&\, B) \\
&= P(A) + P(B) - 0 \\
&= P(A) + P(B)
\end{aligned}
$$

which is the same result given by the special addition rule. Thus, if you are not sure whether to use the special addition rule or the general addition rule, it is always safe to use the general addition rule.

Finally, the general addition rule we have given here is for two events only. We will examine how it can be extended to apply to three or more events in Exercises 4.52 through 4.54.

Exercises 4.3

4.37 A bowl contains 10 marbles, of which three are red, two are white, and five are blue. If a marble is selected at random from the bowl, let

$$E = \text{event the marble selected is white}$$

Determine the probability that the marble selected is white, and use probability notation to express your answer.

4.38 Suppose that you hold 20 out of a total of 500 tickets sold for a lottery. The grand-prize winner is determined by the random selection of one of the 500 tickets. Let

$$G = \text{event you win the grand prize}$$

Find the probability that you win the grand prize, and express your answer using probability notation.

4.39   An age distribution of the U.S. Senators in the 98th Congress is given in the table below. (Source: U.S. Congress, Joint Committee on Printing.)

Age (yrs)	Number of Senators f
Under 40	7
40–49	28
50–59	39
60–69	20
70–79	3
80 and over	3
Σ	100

Suppose a senator is selected at random. Let

A = event the senator is under 40
B = event the senator is in his or her 40s
C = event the senator is in his or her 50s
S = event the senator is under 60

a) Use the table and the f/N rule to find $P(S)$.
b) Express the event S in terms of the events $A, B,$ and C.
c) Determine $P(A)$, $P(B)$, and $P(C)$.
d) Compute $P(S)$ using the special addition rule and your results from parts (b) and (c). Compare your answer with the one found in part (a).

4.40   For the year 1981, the number of commercial failures by type of industry is as follows. (Source: *Dun & Bradstreet.*)

Industry	Failures
Mining and Manufacturing	2,223
Wholesale trade	1,709
Retail trade	6,882
Construction	3,614
Commercial service	2,366
Σ	16,794

Suppose a failed business is selected at random. Let

A = event it was in wholesale trade
B = event it was in retail trade
T = event it was in either wholesale or retail trade

a) Use the table and the f/N rule to find $P(T)$.
b) Express the event T in terms of the events A and B.
c) Determine $P(A)$ and $P(B)$.
d) Compute $P(T)$ using the special addition rule and your results from parts (b) and (c). Compare your answer with the one found in part (a).

4.41   A relative-frequency distribution for U.S. businesses by receipts received from sales and services in 1980 is shown below. (Source: Internal Revenue Service.)

Receipts	Relative frequency	Event
Under $25,000	0.613	A
$25,000–$49,999	0.108	B
$50,000–$99,999	0.095	C
$100,000–$499,999	0.132	D
$500,000–$999,999	0.025	E
$1,000,000 or more	0.029	F

The table shows, for example, that 61.3% of U.S. businesses had receipts of under $25,000. Now, suppose a business is selected at random. Let

A = event the business selected had receipts under $25,000
B = event the business selected had receipts between $25,000 and $49,999

and so on. [See the third column of the table.] Determine the probability that a randomly selected business had receipts of
a) under $100,000.
b) at least $500,000.
c) between $25,000 and $499,999.
d) Interpret each of your results in parts (a) through (c) in terms of percentages.

4.42   The following is a percentage distribution for the number of years of school completed by U.S. adults aged 25 years old and over. (Source: U.S. Bureau of the Census.)

	Years completed	Percent	Event
Elementary school	0–4	3.1	A
	5–7	5.6	B
	8	7.1	C
High school	9–11	13.3	D
	12	37.9	E
College	13–15	15.3	F
	16 or more	17.7	G

The table shows, for instance, that 37.9% of adults 25 years old and over have completed exactly 12 years of school. Suppose an adult, 25 years old or over, is randomly selected from the population. Let

A = event the person selected has completed between zero and four years of school
B = event the person selected has completed between five and seven years of school

and so forth. [See the third column of the table.] Determine the probability that a randomly selected U.S. adult 25 years old or over

a) has at most an elementary-school education.
b) has at most a high-school education.
c) has completed at least one year of college.
d) Interpret each of your results in parts (a) through (c) in terms of percentages.

In Exercises 4.43 through 4.46, find the designated probabilities with the aid of the *complementation rule.*

4.43 Refer to Exercise 4.39. Compute the probability that a randomly selected senator in the 98th Congress is
a) at least 40 years old.
b) under 70 years old.

4.44 Refer to Exercise 4.40. Determine the probability that a failed business selected at random was not in construction.

4.45 In Exercise 4.41, find the probability that a randomly selected business had receipts of
a) under $1,000,000.
b) at least $50,000.

4.46 In Exercise 4.42, calculate the probability that a randomly selected person has completed
a) less than four years of college.
b) at least five years of school.

4.47 In the game of craps, a pair of fair dice is rolled. There are 36 equally likely outcomes possible. See Figure 4.1 on page 114. Let

A = event the sum of the dice is 7
B = event the sum of the dice is 11
C = event the sum of the dice is 2
D = event the sum of the dice is 3
E = event the sum of the dice is 12
F = event the sum of the dice is 8
G = event doubles are rolled

a) Compute the probabilities of the seven events listed above. [See Figure 4.1.]
b) The player wins on the first roll if the sum of the dice is 7 or 11. Find the probability of this event using the special addition rule and your results from part (a).
c) The player loses on the first roll if the sum of the dice is 2, 3, or 12. Determine the probability of that event using the special addition rule and your results from part (a).
d) Compute the probability that either the sum of the dice is 8 or doubles are rolled (i) without

using the general addition rule, and (ii) with the aid of the general addition rule.

4.48 According to the U.S. Bureau of Justice Statistics, about 56.5% of jail inmates are white, 94.0% are male, and 53.5% are white males. Suppose an inmate is selected at random. Let

W = event the inmate selected is white
M = event the inmate selected is male

a) Find $P(W)$, $P(M)$, and $P(W \& M)$.
b) Determine $P(W \text{ or } M)$ and interpret your results in terms of percentages.
c) What is the probability that a randomly selected inmate is female?

4.49 The U.S. Bureau of the Census reports that of U.S. adults, 52.6% are female, 7.0% are divorced, and 4.2% are divorced females. For a U.S. adult selected at random, let

F = event the person selected is female
D = event the person selected is divorced

a) Find $P(F)$, $P(D)$, and $P(F \& D)$.
b) Determine $P(F \text{ or } D)$ and interpret your results in terms of percentages.
c) What is the probability that a randomly selected adult is male?

4.50 Suppose A and B are events with $P(A) = 1/4$, $P(B) = 1/3$, and $P(A \text{ or } B) = 1/2$.
a) Are A and B mutually exclusive?
b) What is $P(A \& B)$?

4.51 If $P(A) = 1/3$, $P(A \text{ or } B) = 1/2$, and $P(A \& B) = 1/10$, what is $P(B)$?

4.52 The general addition rule for *three* events is

$$P(A \text{ or } B \text{ or } C) = P(A) + P(B) + P(C)$$
$$- P(A \& B) - P(A \& C) - P(B \& C)$$
$$+ P(A \& B \& C)$$

Prove the above rule. *Hint:* Let $D = (B \text{ or } C)$ and apply the general addition rule to $(A \text{ or } D)$. Also, note that $(A \& D) = ((A \& B) \text{ or } (A \& C))$.

4.53 When a balanced dime is tossed three times, there are eight equally likely outcomes possible:

HHH   HTH   THH   TTH
HHT   HTT   THT   TTT

Let

A = event the first toss is a head
B = event the second toss is a tail
C = event the third toss is a head

a) List the outcomes comprising each of the three events above.
b) Find $P(A)$, $P(B)$, and $P(C)$.
c) Describe each of the following events in words and list the outcomes comprising each one: $(A \& B)$, $(A \& C)$, $(B \& C)$, $(A \& B \& C)$.
d) Determine the probability for each event in part (c).
e) Describe the event $(A \text{ or } B \text{ or } C)$ in words and list the outcomes that make it up.
f) Find $P(A \text{ or } B \text{ or } C)$ using your result from part

(e) and the f/N rule.
g) Find $P(A \text{ or } B \text{ or } C)$ using your results from parts (b) and (d), along with the general addition rule for three events given in Exercise 4.52.

4.54   On page 137 we stated the general addition rule for two events and in Exercise 4.52 that for three events.
a) Guess what the general addition rule is for four events.
b) Prove your answer in part (a).

4.4   ## Contingency tables; joint and marginal probabilities

An important application of probability and statistics involves the analysis of data obtained by cross classifying the members of a population or sample according to *two* characteristics. Some examples of cross classifications are annual income vs. educational level, automobile accident frequency vs. age, and political affiliation vs. religion. Frequencies for cross-classified data are most easily displayed using *contingency tables*.

EXAMPLE 4.20   ### Introduces contingency tables

The data displayed in Table 4.8 are adapted from the *Arizona State University Statistical Summary*. The table provides a frequency distribution obtained by cross classifying the faculty according to the two characteristics age and rank.

TABLE 4.8

Contingency table for age vs. rank of faculty members.

		Rank				
		Full Professor R_1	Associate Professor R_2	Assistant Professor R_3	Instructor R_4	**Total**
	Under 30 A_1	2	3	57	6	68
	30–39 A_2	52	170	163	17	402
Age	40–49 A_3	156	125	61	6	348
	50–59 A_4	145	68	36	4	253
	60 & over A_5	75	15	3	0	93
	Total	430	381	320	33	1164

The number 2 in the upper left-hand corner of the table tells us that two faculty members are both under 30 and at the rank of full professor. The number 170 diagonally below the upper left-hand corner shows that 170 faculty members are associate professors in their 30s. The row total in the first row of the table indicates that 68 (2 + 3 + 57 + 6) of the faculty members are under 30. Similarly, the column total in the third column shows that 320 of the faculty members are assistant professors. The number 1164 in the lower right-hand corner of the table gives the total number of faculty. That total can be found by summing *either* the row totals *or* the column totals.

MTB
SPSS

A table such as Table 4.8 that gives a frequency distribution for cross-classified data is called a **contingency table.** The table includes all the different possibilities in the cross-classification. In other words, it accounts for all contingencies. The boxes inside the heavy lines give the frequencies for the various contingencies. These boxes are called the **cells** of the contingency table. There are 20 cells in Table 4.8. ■

Joint and marginal probabilities

We return to the age vs. rank data for the faculty members. That data will be used to introduce the concepts of *joint and marginal probabilities.*

EXAMPLE 4.21 Introduces joint and marginal probabilities

The contingency table for the age vs. rank data in Example 4.20 is repeated below in Table 4.9.

TABLE 4.9
Contingency table for age vs. rank of faculty members.

		Rank				
		Full Professor R_1	Associate Professor R_2	Assistant Professor R_3	Instructor R_4	**Total**
Age	Under 30 A_1	2	3	57	6	68
	30–39 A_2	52	170	163	17	402
	40–49 A_3	156	125	61	6	348
	50–59 A_4	145	68	36	4	253
	60 & over A_5	75	15	3	0	93
	Total	430	381	320	33	1164

Now, suppose a faculty member is selected at random. If you look at Table 4.9, you will note that the rows and columns are labelled with letters. The first row is labelled A_1, which represents the event that the randomly selected faculty member is under 30:

$$A_1 = \text{event the faculty member selected is under 30}$$

Similarly,

$$R_2 = \text{event the faculty member selected is an associate professor}$$

and so forth. Note that the events A_1 through A_5 are mutually exclusive, as are the events R_1 through R_4 (why?).

In addition to considering the events A_1 through A_5 and R_1 through R_4 separately, we can also consider them *jointly*. For example, the event that the randomly selected faculty member is in his or her 40s (event A_3) *and* is also an associate professor (event R_2) is (A_3 & R_2). The event (A_3 & R_2) is represented by the cell in the third row and second column of Table 4.9. As another illustration, we have the joint event

(A_2 & R_4) = event the faculty member selected is an instructor in his or her 30s

This event is represented by the cell in the second row and fourth column of the table. There are 20 different joint events of this type, one for each cell of the table. In fact, it is useful to think of the table as a sort of Venn diagram. This is pictured in Figure 4.20.

FIGURE 4.20

Venn diagram
to accompany
Table 4.9.

	R_1	R_2	R_3	R_4
A_1	(A_1 & R_1)	(A_1 & R_2)	(A_1 & R_3)	(A_1 & R_4)
A_2	(A_2 & R_1)	(A_2 & R_2)	(A_2 & R_3)	(A_2 & R_4)
A_3	(A_3 & R_1)	(A_3 & R_2)	(A_3 & R_3)	(A_3 & R_4)
A_4	(A_4 & R_1)	(A_4 & R_2)	(A_4 & R_3)	(A_4 & R_4)
A_5	(A_5 & R_1)	(A_5 & R_2)	(A_5 & R_3)	(A_5 & R_4)

The Venn diagram makes it clear that the twenty joint events (A_1 & R_1), (A_1 & R_2), . . ., (A_5 & R_4) are mutually exclusive.

Let us now move on to an examination of probabilities. Here we have $N = 1164$, since there are 1164 faculty members altogether. To find, for example, the probability that a randomly selected faculty member is an assistant professor (event R_3), we note from Table 4.9 that $f = 320$, and so

$$P(R_3) = \frac{f}{N} = \frac{320}{1164} = 0.275$$

Similarly, the probability that a randomly selected faculty member is in his or her 50s equals

$$P(A_4) = \frac{f}{N} = \frac{253}{1164} = 0.217$$

[In terms of percentages, these last two probabilities indicate that 27.5% of the faculty are assistant professors and that 21.7% of the faculty are in their 50s.]

We also can easily find probabilities for joint events—so-called **joint probabilities.** For instance, the probability that a randomly selected faculty member is an associate professor in his or her 40s equals

$$P(A_3 \text{ \& } R_2) = \frac{f}{N} = \frac{125}{1164} = 0.107$$

In Table 4.10 we have replaced the joint frequency distribution in Table 4.9 with a **joint probability distribution.** The probabilities in the table are determined in the same way as the three probabilities we just computed.

TABLE 4.10

Joint probability distribution corresponding to Table 4.9.

	Full Professor R_1	Associate Professor R_2	Assistant Professor R_3	Instructor R_4	$P(A_i)$
Under 30 A_1	0.002	0.003	0.049	0.005	0.058
30–39 A_2	0.045	0.146	0.140	0.015	0.345
40–49 A_3	0.134	0.107	0.052	0.005	0.299
50–59 A_4	0.125	0.058	0.031	0.003	0.217
60 & over A_5	0.064	0.013	0.003	0.000	0.080
$P(R_j)$	0.369	0.327	0.275	0.028	1.000

Note that the joint probabilities are given inside the heavy lines of the table. Also observe that the row labelled "Total" and the column labelled "Total" in Table 4.9 have been relabelled as $P(R_j)$ and $P(A_i)$, respectively. This is because the last row of Table 4.10 gives the probabilities of the events R_1 through R_4, and the last column gives the probabilities of the events A_1 through A_5. These probabilities are often called **marginal probabilities** since they are "in the margin" of the joint probability distribution table.

Finally, we should point out that the sum of the joint probabilities in a given row or column must equal the marginal probability in that row or column. (Any observed discrepancy is due to roundoff error.) For example, consider the row A_4 of Table 4.10. The sum of the joint probabilities in that row equals

$$P(A_4 \text{ \& } R_1) + P(A_4 \text{ \& } R_2) + P(A_4 \text{ \& } R_3) + P(A_4 \text{ \& } R_4)$$
$$= 0.125 + 0.058 + 0.031 + 0.003$$
$$= 0.217$$

which is precisely the marginal probability at the end of the row A_4, that is, $P(A_4)$.

The fact that the sum of a row or column of joint probabilities equals the marginal probability in that row or column follows from the special addition rule. [See Exercise 4.68.]

MTB
SPSS

Exercises   4.4

4.55   The following contingency table cross classifies institutions of higher education in the United States by region and type. (Source: U.S. National Center for Education Statistics.)

<table>
<tr><td rowspan="2"></td><td colspan="3" align="center">Type</td></tr>
<tr><td>Public
T_1</td><td>Private
T_2</td><td>Total</td></tr>
<tr><td>Northeast
R_1</td><td>266</td><td>555</td><td>821</td></tr>
<tr><td>Midwest
R_2</td><td>359</td><td>504</td><td>863</td></tr>
<tr><td>South
R_3</td><td>533</td><td>502</td><td>1035</td></tr>
<tr><td>West
R_4</td><td>313</td><td>242</td><td>555</td></tr>
<tr><td>Total</td><td>1471</td><td>1803</td><td>3274</td></tr>
</table>

Region (label on left of table)

a) How many cells does this contingency table have?
b) What is the total number of institutions of higher education in the United States?
c) How many institutions of higher education are in the Midwest?
d) How many are public?
e) How many institutions of higher education are private schools in the South?

4.56   The number of cars and trucks in use by age is presented in the following contingency table. Frequencies are in millions. (Source: Motor Vehicle Manufacturers Association of the United States.)

<table>
<tr><td rowspan="2"></td><td colspan="3" align="center">Type</td></tr>
<tr><td>Car
V_1</td><td>Truck
V_2</td><td>Total</td></tr>
<tr><td>Under 3
A_1</td><td>21.5</td><td>3.2</td><td>24.7</td></tr>
<tr><td>3–5
A_2</td><td>29.9</td><td>2.5</td><td>32.4</td></tr>
<tr><td>6–8
A_3</td><td>22.2</td><td>2.0</td><td>24.2</td></tr>
<tr><td>9–11
A_4</td><td>17.9</td><td>1.8</td><td>19.7</td></tr>
<tr><td>12 & over
A_5</td><td>15.4</td><td>3.6</td><td>19.0</td></tr>
<tr><td>Total</td><td>106.9</td><td>13.1</td><td>120.0</td></tr>
</table>

Age (yrs) (label on left of table)

a) How many cells does this contingency table have?
b) What is the total number of cars and trucks in use?
c) How many vehicles are trucks?
d) How many vehicles are between three and five years old?
e) How many vehicles are nine- to eleven-year-old trucks?

4.57   The following contingency table gives a joint frequency distribution for the civilian labor force in the U.S. based on a cross classification of employment status and educational level. Frequencies are given in thousands. (Source: U.S. Bureau of Labor Statistics.)

<table>
<tr><td rowspan="2"></td><td colspan="3" align="center">Employment status</td></tr>
<tr><td>Employed
E_1</td><td>Unemployed
E_2</td><td>Total</td></tr>
<tr><td>Less than 8
S_1</td><td>3535.5</td><td>612.2</td><td>4147.7</td></tr>
<tr><td>8
S_2</td><td>3240.9</td><td>517.1</td><td>3758.0</td></tr>
<tr><td>9–11
S_3</td><td>12767.0</td><td>2807.4</td><td>15574.4</td></tr>
<tr><td>12
S_4</td><td>40068.9</td><td>4601.5</td><td>44670.4</td></tr>
<tr><td>13–15
S_5</td><td>18266.7</td><td>1340.4</td><td>19607.1</td></tr>
<tr><td>16 or more
S_6</td><td>20230.8</td><td>686.0</td><td>20916.8</td></tr>
<tr><td>Total</td><td>98109.8</td><td>10564.6</td><td>108674.4</td></tr>
</table>

Years of school completed (label on left of table)

a) How many cells does this contingency table have?
b) What is the size of the civilian labor force?
c) How many people in the civilian labor force have completed exactly 12 years of school?
d) How many are unemployed?
e) How many with 16 or more years of school are unemployed?

4.58   The contingency table at the top of the next page results from cross-classifying U.S. hospitals by type and number of beds. (Source: U.S. National Center for Health Statistics.)

Number of beds

Type		6–24 B_1	25–74 B_2	75+ B_3	Total
	General H_1	299	1894	3945	6138
	Psychiatric H_2	17	121	378	516
	Chronic H_3	0	7	40	47
	Tuberculosis H_4	0	1	10	11
	Other H_5		131		
	Total	338	2154	4535	7027

a) Fill in the missing entries.
b) How many cells does this contingency table have?
c) How many hospitals have at least 75 beds?
d) How many psychiatric hospitals are there?
e) How many general hospitals are there with between 25 and 74 beds?

4.59 A contingency table is presented below giving a joint frequency distribution for U.S. farms resulting from a cross classification of acreage and tenure of operator. Frequencies are in thousands of farms. (Source: U.S. Bureau of the Census.)

Tenure of operator

Acreage		Full owner T_1	Part owner T_2	Tenant T_3	Total
	Under 50 A_1	532	74	84	690
	50–179 A_2	563		94	814
	180–499 A_3	262		87	596
	500–999 A_4	57	128		215
	1000+ A_5	36	107	18	161
	Total	1450	713	313	2476

a) Fill in the three empty cells.
b) How many cells does this contingency table have?
c) How many farms have under 50 acres?
d) How many farms are tenant operated?

e) How many farms are operated by part owners and have between 500 and 999 acres?
f) How many farms are not full-owner operated?

4.60 A contingency table for annual income of families by type of family is shown below. Frequencies are in thousands. (Source: U.S. Bureau of the Census.)

Type of family

Income level		Married couple F_1	Husband only F_2	Wife only F_3	Total
	Under $10,000 I_1	5,852	392	4,308	10,552
	$10,000–$19,999 I_2	12,326		2,992	15,925
	$20,000–$49,999 I_3	26,240			29,109
	$50,000 and over I_4	5,213	107	111	5,431
	Total	49,631	1,984	9,402	61,017

a) Fill in the three empty cells.
b) How many cells does this contingency table have?
c) How many families make between $20,000 and $49,999?
d) How many families have only the husband present?
e) How many families have only the wife present and make under $10,000?
f) How many families make at least $20,000?
g) How many married couples make less than $20,000?

4.61 Refer to Exercise 4.55.
a) For a randomly selected institution of higher education, describe the following events in words: (i) T_2 (ii) R_3 (iii) (T_1 & R_4)
b) Compute the probability of each event in (a). Interpret your results in terms of percentages.
c) Construct a joint probability distribution similar to Table 4.10 on page 145.
d) Verify that the sum of each row and column of joint probabilities equals the corresponding marginal probability. (Rounding may cause slight deviations here.)

4.62 In Exercise 4.56, a contingency table is given for the number of cars and trucks in use by age. Suppose a vehicle (car or truck) is selected at random.
  a) Describe each of the following events in words:
     (i) A_3 (ii) V_1 (iii) $(A_3 \& V_1)$
  b) Determine the probability of each event in part (a), and interpret your results in terms of percentages.
  c) Construct a joint probability distribution similar to Table 4.10 on page 145.
  d) Verify that the sum of each row and column of joint probabilities equals (up to rounding error) the corresponding marginal probability.

4.63 The contingency table in Exercise 4.57 cross classifies the civilian labor force according to employment status and educational level. Suppose a person in the civilian labor force is selected at random.
  a) Describe each of the following events in words:
     (i) E_1 (ii) S_5 (iii) $(E_1 \& S_5)$
  b) Compute the probability of each event in part (a).
  c) Compute $P(E_1 \text{ or } S_5)$ by
     (i) using the contingency table and the f/N rule.
     (ii) using the general addition rule and your results from part (b).
  d) Construct a joint probability distribution.

4.64 Refer to Exercise 4.58. Suppose a hospital is selected at random.
  a) Describe each of the following events in words:
     (i) H_2 (ii) B_2 (iii) $(H_2 \& B_2)$ (iv) $(H_4 \& B_1)$
  b) Compute the probability of each event in part (a).
  c) Compute $P(H_2 \text{ or } B_2)$ by
     (i) using the contingency table and the f/N rule.
     (ii) using the general addition rule and your results from part (b).
  d) Construct a joint probability distribution.

4.65 The contingency table of Exercise 4.59 cross classifies U.S. farms by acreage and tenure of operator. Suppose a U.S. farm is selected at random.
  a) Use the letters in the margins of the contingency table to represent each of the following events:
     (i) The farm selected has between 180 and 499 acres.
     (ii) The farm selected is part-owner operated.
     (iii) The farm selected is full-owner operated with at least 1000 acres.

  b) Compute the probability of each event in part (a).
  c) Construct a **joint percentage distribution.** [This is similar to a joint probability distribution, but with percentages replacing probabilities.]

4.66 Refer to Exercise 4.60. Suppose a family is selected at random.
  a) Use the letters in the margins of the contingency table to represent each of the events below.
     (i) The family selected has only the wife present.
     (ii) The family selected makes at least $50,000.
     (iii) The family selected has only the husband present and makes between $10,000 and $19,999.
  b) Determine the probability of each event in part (a).
  c) Construct a joint percentage distribution.

4.67 Explain why the joint events in a contingency table are always mutually exclusive.

4.68 This exercise supplies a proof of the fact that the sum of the joint probabilities in a given row or column equals the marginal probability in that row or column. Consider a joint probability distribution in the following form:

	C_1	$\cdots$	C_n	$P(R_i)$
R_1	$P(R_1 \& C_1)$	$\cdots$	$P(R_1 \& C_n)$	$P(R_1)$
.	.	$\cdots$	.	.
.	.	$\cdots$	.	.
.	.	$\cdots$	.	.
R_m	$P(R_m \& C_1)$	$\cdots$	$P(R_m \& C_n)$	$P(R_m)$
$P(C_j)$	$P(C_1)$	$\cdots$	$P(C_n)$	1

  a) Explain why we can write

$$R_1 = [(R_1 \& C_1) \text{ or } \cdots \text{ or } (R_1 \& C_n)]$$

  b) Why are the events $(R_1 \& C_1), \ldots, (R_1 \& C_n)$ mutually exclusive?
  c) Explain why we can conclude from parts (a) and (b) that

$$P(R_1) = P(R_1 \& C_1) + \cdots + P(R_1 \& C_n)$$

This equation shows that the first row of joint probabilities sums to the marginal probability at the end of that row. A similar argument applies to any other row or column.

4.5 Conditional probability

In this section we will introduce the concept of *conditional probability*. The **conditional probability** of an event is the probability of the event given that another event has occurred. Consider the following example.

EXAMPLE 4.22 **Introduces conditional probability**

When a fair die is rolled, there are $N = 6$ equally likely outcomes. These are displayed in Figure 4.21.

FIGURE 4.21
The six equally likely
outcomes possible when
a balanced die is rolled.

If, for example, we let

$$F = \text{event a "4" is rolled}$$

then the probability that event F occurs is:

$$P(F) = \frac{f}{N} = \frac{1}{6} = 0.167$$

To introduce conditional probability, let us suppose the die is rolled and, although we cannot see the die, we are told that the result is an even number. *Now* what is the probability that event F occurs? That is, if we are *given* that the die comes up even, what is the probability that it is a "4"?

The answer is quite simple, because if we know the die comes up even, there are no longer six equally likely outcomes possible—there are only three, as shown in Figure 4.22.

FIGURE 4.22
The three equally likely
outcomes possible given that
the die comes up even.

Consequently, the probability that event F (a "4") occurs is no longer $1/6 = 0.167$. It is now

$$\frac{f}{N} = \frac{1}{3} = 0.333$$

Thus, the information that the die comes up even affects the probability that event F occurs. The new probability, 0.333, is called a *conditional probability*, since it is computed under the *condition* that the die comes up even.

It is useful to have a notation for conditional probabilities, just as it is for ordinary probabilities. To this end, recall that

$$F = \text{event a "4" is rolled}$$

and let

$$E = \text{event the die comes up even}$$

Then we use the notation

$$P(F|E)$$

to represent the conditional probability that event F (a "4") occurs *given* that event E (even) occurs. The symbol $P(F|E)$ is read "the probability of F *given* E." Just remember that the vertical bar "|" stands for "given." In summary, we have

$$P(F) = \frac{1}{6} = 0.167 \qquad P(F|E) = \frac{1}{3} = 0.333$$

To further illustrate conditional probability using this example, let

$$C = \text{event the die comes up odd}$$
$$D = \text{event the die comes up three or less}$$

Let us compute the conditional probability that event C occurs given that event D occurs; that is, the probability the die comes up odd *given* that it comes up three or less. Given that event D occurs, the possible outcomes are the three shown in Figure 4.23.

FIGURE 4.23
The three equally likely outcomes possible given that the die comes up three or less.

And, under these circumstances, event C (odd) can occur in $f = 2$ ways. Consequently,

$$P(C|D) = \frac{f}{N} = \frac{2}{3} = 0.667$$

Note that the "unconditional" probability that event C occurs is

$$P(C) = \frac{f}{N} = \frac{3}{6} = 0.500$$

since the die can come up odd in three of the six possibilities. So again, additional information concerning the experiment has affected the probability. ■

We now summarize the concepts and notation introduced in the previous example. Following that we will present another example to further illustrate conditional probability.

DEFINITION 4.6 Conditional probability

If A and B are events, then the probability that event B occurs *given* that event A has occurred is called a **conditional probability.** It is denoted by the symbol **$P(B|A)$,** which is read "the probability of B given A."

In Section 4.4 we introduced contingency tables as a method for tabulating cross-classified data. Conditional probability plays a crucial role in the analysis of cross-classified data. We now illustrate the procedure for obtaining conditional probabilities for cross-classified data directly from the contingency table.

EXAMPLE 4.23 Illustrates Definition 4.6

In Example 4.20 we presented a contingency table resulting from cross classifying the faculty at Arizona State University according to age and rank. We repeat that table here as Table 4.11.

TABLE 4.11

Contingency table for age vs. rank of faculty members.

Age		Full Professor R_1	Associate Professor R_2	Assistant Professor R_3	Instructor R_4	Total
	Under 30 A_1	2	3	57	6	68
	30–39 A_2	52	170	163	17	402
	40–49 A_3	156	125	61	6	348
	50–59 A_4	145	68	36	4	253
	60 & over A_5	75	15	3	0	93
	Total	430	381	320	33	1164

(Rank spans columns R_1–R_4)

Suppose a faculty member is selected at random.

a) Find the probability that the faculty member selected is in his or her 50s.

b) Find the (conditional) probability that the faculty member selected is in his or her 50s *given* that the selection is made from among the assistant professors.

S O L U T I O N

a) The probability that a randomly selected faculty member is in his or her 50s is $P(A_4)$. From Table 4.11 we see that $N = 1164$, since there are 1164 faculty members altogether. Also, $f = 253$, since 253 of the faculty members are in their 50s. Therefore,

$$P(A_4) = \frac{f}{N} = \frac{253}{1164} = 0.217$$

b) Here we are to find the probability that a randomly selected faculty member is in his or her 50s (event A_4) *given* that the selection is made from among the assistant professors (that is, given that event R_3 occurs). In other words, we want to determine $P(A_4|R_3)$. To find this probability, we simply restrict our attention to the "assistant professor" column of the contingency table. We have $N = 320$, since there are a total of 320 assistant professors. Also, $f = 36$, since 36 of the assistant professors are in their 50s. Thus,

$$P(A_4|R_3) = \frac{f}{N} = \frac{36}{320} = 0.113$$

In summary then

$$P(A_4) = 0.217$$

while

$$P(A_4|R_3) = 0.113$$

It is necessary to be able to interpret these and similar probability results in terms of percentages. $P(A_4) = 0.217$ indicates that 21.7% of the *faculty* are in their 50s, while $P(A_4|R_3) = 0.113$ indicates that 11.3% of the *assistant professors* are in their 50s.   ∎

The conditional probability rule

In the previous examples we have computed conditional probabilities directly. That is, we first obtained the new set of possible outcomes determined by the given event, and then calculated probabilities from there in the usual manner. For instance, in the die illustration of Example 4.22 on page 149, we computed the conditional probability of a "4" given that the die comes up even. To do this, we first obtained the new set of possible outcomes (in this case "2, 4, 6") and then went on from there.

Sometimes, however, we cannot determine conditional probabilities directly, but must compute them in terms of ordinary (unconditional) probabilities. To see how this can be done, we return to the situation of Example 4.23.

EXAMPLE   4.24   **Introduces the conditional probability rule**

At the top of this page we computed the conditional probability that a randomly selected faculty member is in his or her 50s (event A_4), *given* that the faculty member selected is an assistant professor (event R_3). To accomplish this, we restricted our attention to the "R_3" column of Table 4.11 and obtained

$$P(A_4|R_3) = \frac{36}{320} = 0.113$$

To find this probability using unconditional probabilities, proceed as follows. First note that the number "36" in the numerator of the above fraction is the number of assistant professors in their 50s, and that this corresponds to the event $(R_3 \ \& \ A_4)$—"assistant professor" *and* "in 50s." Next observe that the "320" in the denominator is the total number of assistant professors, and that this corresponds to the event R_3. Thus, the numbers "36" and "320" are precisely those used to compute the *unconditional* probabilities of the events $(R_3 \ \& \ A_4)$ and R_3, respectively:

$$P(R_3 \& A_4) = \frac{36}{1164} = 0.031 \qquad P(R_3) = \frac{320}{1164} = 0.275$$

So, by arithmetic and the previous three probabilities, we see that

$$P(A_4|R_3) = \frac{36}{320} = \frac{36/1164}{320/1164} = \frac{P(R_3 \& A_4)}{P(R_3)}$$

Consequently, the conditional probability $P(A_4|R_3)$ can be computed from the unconditional probabilities $P(R_3 \& A_4)$ and $P(R_3)$ by using the formula

$$P(A_4|R_3) = \frac{P(R_3 \& A_4)}{P(R_3)}$$

This formula holds in general, and we will refer to it as the **conditional probability rule.** ∎

FORMULA 4.4 The conditional probability rule

If A and B are any two events, then

$$P(B|A) = \frac{P(A \& B)}{P(A)}$$

In words, the conditional probability of event B given event A is equal to the joint probability of event A and event B divided by the (marginal) probability of event A.

Note that in computing conditional probabilities using the conditional probability rule, we divide by the probability of the *given* event; that is, the event on the right of the "|."

We should emphasize that for the faculty-member example above, conditional probabilities can be computed either directly or by using the conditional probability rule. However, as the next example illustrates, the conditional probability rule is sometimes the *only* way that conditional probabilities can be determined.

EXAMPLE 4.25 **Illustrates Formula 4.4**

Table 4.12 gives a joint probability distribution for the marital status of U.S. adults by sex.

TABLE 4.12
Joint probability distribution for marital status vs. sex.

| | | Marital status | | | | |
		Single M_1	Married M_2	Widowed M_3	Divorced M_4	$P(S_i)$
Sex	Male S_1	0.116	0.319	0.012	0.028	0.475
	Female S_2	0.093	0.325	0.066	0.041	0.525
	$P(M_j)$	0.209	0.644	0.078	0.069	1.000

Source: U.S. Bureau of the Census.

The table shows, for example, that the probability is 0.475 that a randomly selected adult is a male—$P(S_1) = 0.475$. That is, 47.5% of adults are male. Similarly, the table indicates that the probability is 0.028 that a randomly selected adult is a divorced male—$P(S_1 \& M_4) = 0.028$. That is, 2.8% of adults are divorced males.

Note that, unlike our previous illustrations with contingency tables, we do not have the frequency data here, but only the probability (relative-frequency) data. Because of this, we cannot compute conditional probabilities directly—we *must* use the conditional probability rule.

To illustrate, let us find the conditional probability that a randomly selected adult is divorced, given that the adult is a male: $P(M_4|S_1)$. Using the conditional probability rule along with Table 4.12 we obtain

$$P(M_4|S_1) = \frac{P(S_1 \& M_4)}{P(S_1)} = \frac{0.028}{0.475} = 0.059$$

In other words, 5.9% of adult males are divorced.

A similar problem involves the determination of the conditional probability that a randomly selected adult is a male, given that the adult is divorced: $P(S_1|M_4)$. We have

$$P(S_1|M_4) = \frac{P(M_4 \& S_1)}{P(M_4)} = \frac{0.028}{0.069} = 0.406$$

Thus, 40.6% of divorced adults are males.   ∎

Exercises 4.5

For Exercises 4.69 through 4.74, compute conditional probabilities directly; that is, do not use the conditional probability rule.

4.69   Suppose a card is selected at random from an ordinary deck of 52 playing cards. Let

A = event a face card is selected
B = event a king is selected
C = event a heart is selected

Determine the following probabilities and express your results in words.
a) $P(B)$ b) $P(B|A)$ c) $P(B|C)$ d) $P(B|(\text{not } A))$
e) $P(A)$ f) $P(A|B)$ g) $P(A|C)$ h) $P(A|(\text{not } B))$

4.70   A balanced dime is tossed twice. The four equally likely outcomes are

HH   HT   TH   TT

Let

A = event the first toss is a head
B = event the second toss is a head
C = event at least one toss is heads

Determine the following probabilities and express your results in words.
a) $P(B)$ b) $P(B|A)$ c) $P(B|C)$
d) $P(C)$ e) $P(C|A)$ f) $P(C|(\text{not } B))$

4.71   The absentee records over the past year for the employees of Cudahey Masonry, Inc. are presented below in a frequency distribution.

Days missed	Number of employees
0	4
1	2
2	14
3	10
4	16
5	18
6	10
7	6
Σ	80

a) Find the probability that a randomly selected employee missed exactly three days of work.

b) Find the conditional probability that a randomly selected employee missed exactly three days, given that the employee missed at least one day.

c) Find the conditional probability that a randomly selected employee missed at most three days, given that the employee missed at least one day.

d) Interpret your results in parts (a) through (c) in terms of percentages.

4.72 A frequency distribution is given below for the population of the states in the U.S. and Washington, D.C. (Source: U.S. Bureau of the Census.)

Population size (millions)	Number of states
Under 1	7
1–under 2	7
2–under 3	13
3–under 5	6
5–under 10	6
10 or over	12
Σ	51

Suppose a state is selected at random. Determine the probability that the population of the state selected

a) is between two and three million.

b) is between two and three million, given that it is at least one million.

c) is less than five million, given that it is at least one million.

d) Interpret your results in parts (a) through (c) in terms of percentages.

4.73 The following contingency table cross classifies institutions of higher education in the United States by region and type. (Source: U.S. National Center for Education Statistics.)

	Type		
	Public T_1	Private T_2	Total
Northeast R_1	266	555	821
Midwest R_2	359	504	863
South R_3	533	502	1035
West R_4	313	242	555
Total	1471	1803	3274

Suppose an institution of higher education is selected at random. Determine the probability that the institution selected

a) is in the Northeast.

b) is in the Northeast, given that it is a private school.

c) is a private school, given that it is in the Northeast.

d) Interpret your results in parts (a) through (c) in terms of percentages.

4.74 The number of cars and trucks in use by age is presented in the following contingency table. Frequencies are in millions. (Source: Motor Vehicle Manufacturers Association of the United States.)

	Type		
	Car V_1	Truck V_2	Total
Under 3 A_1	21.5	3.2	24.7
3–5 A_2	29.9	2.5	32.4
6–8 A_3	22.2	2.0	24.2
9–11 A_4	17.9	1.8	19.7
12 & over A_5	15.4	3.6	19.0
Total	106.9	13.1	120.0

Suppose a vehicle is selected at random. What is the probability that the vehicle selected

a) is under three years old?

b) is under three years old, given that it is a car?

c) is a car?

d) is a car, given that it is under three years old?

e) Interpret each of your results in parts (a) through (d) in terms of percentages.

4.75 A contingency table is presented on the following page for acreage by tenure of operator of U.S. farms. Frequencies are in thousands of farms. (Source: U.S. Bureau of the Census.) Suppose a U.S. farm is selected at random.

a) Find $P(T_3)$.

b) Find $P(T_3 \& A_3)$.

c) Compute $P(A_3|T_3)$ directly from the table.

d) Compute $P(A_3|T_3)$ using the conditional probability rule and your results from parts (a) and (b).

e) State your results in (a) through (c) in words.

Tenure of operator

		Full owner T_1	Part owner T_2	Tenant T_3	Total
Acreage	Under 50 A_1	532	74	84	690
	50–179 A_2	563	157	94	814
	180–499 A_3	262	247	87	596
	500–999 A_4	57	128	30	215
	1000+ A_5	36	107	18	161
	Total	1450	713	313	2476

4.76 The contingency table below results from cross classifying U.S. hospitals by type and number of beds. (Source: U.S. National Center for Health Statistics.)

Number of beds

		6–24 B_1	25–74 B_2	75+ B_3	Total
Type	General H_1	299	1894	3945	6138
	Psychiatric H_2	17	121	378	516
	Chronic H_3	0	7	40	47
	Tuberculosis H_4	0	1	10	11
	Other H_5	22	131	162	315
	Total	338	2154	4535	7027

Suppose a U.S. hospital is randomly selected.
a) Find $P(H_1)$.
b) Find $P(H_1 \text{ \& } B_3)$.
c) Compute $P(B_3|H_1)$ directly from the table.
d) Compute $P(B_3|H_1)$ using the conditional probability rule and your results from parts (a) and (b).
e) State each of your results in parts (a) through (c) in words.

4.77 The following is a joint probability (relative-frequency) distribution for the members of the 98th Congress by political party. (Source: U.S. Congress, Joint Committee on Printing.)

	Reps C_1	Senators C_2	$P(P_i)$
Democrats P_1	0.500	0.084	0.584
Republicans P_2	0.313	0.103	0.416
$P(C_j)$	0.813	0.187	1.000

If a member of the 98th Congress is selected at random, what is the probability that the member selected
a) is a senator?
b) is a Republican senator?
c) is a Republican, given that he or she is a senator?
d) is a senator, given that he or she is a Republican?
e) Interpret each of your results in parts (a) through (d) in terms of percentages.

4.78 A joint probability (relative-frequency) distribution is shown below for engineers and scientists in the U.S. by highest degree obtained. (Source: U.S. National Science Foundation.)

Type

		Engineers T_1	Scientists T_2	$P(D_i)$
Highest degree	Bachelor's D_1	0.343	0.289	0.632
	Master's D_2	0.098	0.146	0.244
	Doctorate D_3	0.017	0.091	0.108
	Other D_4	0.013	0.003	0.016
	$P(T_j)$	0.471	0.529	1.000

A person is selected at random from among the engineers and scientists. Determine the probability that the person selected
a) is an engineer.
b) has a doctorate.
c) is an engineer with a doctorate.
d) is an engineer, given that the person has a doctorate.

e) has a doctorate, given that the person is an engineer.

f) Interpret each of your results in parts (a) through (e) in terms of percentages.

4.79 According to the U.S. Bureau of the Census, 10.6% of U.S. families are black and 3.5% are black and have incomes below the poverty level. What percentage of black families have incomes below the poverty level?

4.80 According to the U.S. National Center for Education Statistics, 13.4% of all students (at all levels) attend private institutions. Also, 4.5% of all students attend private colleges. What percentage of students in private schools are in college?

4.81 Refer to the contingency table of Exercise 4.74.
a) Construct a joint probability distribution.

b) Find the probability distribution of age for cars in use; that is, construct a table giving the (conditional) probabilities that a car in use is under 3 years old, 3–5 years old, 6–8 years old, etc.

c) Find the probability distribution of type for vehicles 3–5 years old.

d) The probability distributions in parts (b) and (c) are examples of **conditional probability distributions.** Find two other conditional probability distributions for the type vs. age data of motor vehicles in use.

4.82 Do you think that the conditional probability of one event B given another event A must always be different than the (unconditional) probability of event B? That is, do you think it is always true that $P(B|A) \neq P(B)$? *Hint:* Try to construct an example in which $P(B|A) = P(B)$.

4.6   The multiplication rule; independence

The *conditional probability rule,* given on page 153, is used to compute conditional probabilities in terms of ordinary probabilities. The formula is

$$P(B|A) = \frac{P(A \ \& \ B)}{P(A)}$$

By suitably rewriting this, we can also obtain a formula for computing joint probabilities in terms of marginal and conditional probabilities. To get this formula, we multiply both sides of the above equation by $P(A)$. The result is the following rule of probability.

FORMULA 4.5 The general multiplication rule

If A and B are any two events, then

$$P(A \ \& \ B) = P(A)P(B|A)$$

In words, the probability that both event A and event B occur equals the probability that event A occurs *times* the conditional probability that event B occurs given that event A occurs.

Note that the *conditional probability rule* and the *general multiplication rule* are simply variations of each other. In cases where the joint probabilities and marginal probabilities are known (or easily determined directly), we can use the conditional probability rule to obtain conditional probabilities. On the other hand, in cases where the marginal and conditional probabilities are known (or easily determined directly), we can use the general multiplication rule to obtain joint probabilities.

EXAMPLE   4.26   **Illustrates Formula 4.5**

According to the U.S. Bureau of Labor Statistics, 30.9% of all employed black people are white-collar workers. In addition, 9.2% of all employed workers are black. What is the probability that a randomly selected employed worker is a black white-collar worker?

S O L U T I O N
To determine the desired probability, we first label the relevant events with letters. Let

B = event a randomly selected employed worker is black
W = event a randomly selected employed worker is a white-collar worker

The probability that a randomly selected employed worker is a black white-collar worker can be expressed as $P(B \& W)$.

Now, according to the given data, we have $P(B) = 0.092$, since 9.2% of all employed workers are black. Also $P(W|B) = 0.309$, since 30.9% of all employed black people are white-collar workers. Applying the general multiplication rule, we find that

$$P(B \& W) = P(B)P(W|B) = 0.092 \cdot 0.309 = 0.028$$

Thus, the probability is 0.028 that a randomly selected employed worker is a black white-collar worker. Expressed in percentages, 2.8% of all employed workers are black white-collar workers. ∎

Another application of the general multiplication rule arises in situations when *two or more* individuals or items are selected from a population. We illustrate this in Example 4.27.

EXAMPLE   4.27   **Illustrates Formula 4.5**

In Professor W's introductory statistics class, the number of males and females are as given in the following frequency distribution.

TABLE 4.13
Frequency distribution of males and females in Professor W's introductory statistics class.

Sex	Frequency
Male	17
Female	23
Σ	40

Suppose *two* students are selected at random from the class. Assume that the first student selected is *not* returned to the population (class) before the second student is selected. What is the probability that the first student selected is a female and the second is a male?

S O L U T I O N
Let us employ the following notation:

$$F1 = \text{event the first student selected is female}$$
$$M2 = \text{event the second student selected is male}$$

The problem is to determine the probability that the first student selected is female *and* the second is male, that is, $P(F1 \& M2)$. We apply the general multiplication rule to write

$$P(F1 \& M2) = P(F1)P(M2|F1)$$

As we see from Table 4.13, 23 out of the 40 students are female. Therefore,

$$P(F1) = \frac{f}{N} = \frac{23}{40} = 0.575$$

Now, given that the first student selected is female (that is, given that event $F1$ occurs), there are 39 students remaining in the class, of which 17 are male and 22 are female. Thus, the conditional probability that the second student selected is male, given that the first student selected is female, equals

$$P(M2|F1) = \frac{f}{N} = \frac{17}{39} = 0.436$$

Substituting these two probabilities into the general multiplication rule, we get

$$P(F1 \& M2) = P(F1)P(M2|F1) = 0.575 \cdot 0.436 = 0.251$$

Consequently, the probability is 0.251 that the first student selected is a female and the second is a male.

We can also use a device called a **tree diagram** to calculate probabilities when employing the general multiplication rule. A tree diagram for the present example is displayed in Figure 4.24.

FIGURE 4.24
Tree diagram for student-selection problem.

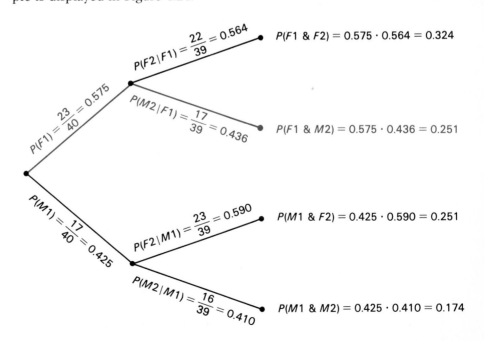

$$P(F2|F1) = \frac{22}{39} = 0.564 \qquad P(F1 \& F2) = 0.575 \cdot 0.564 = 0.324$$

$$P(F1) = \frac{23}{40} = 0.575$$

$$P(M2|F1) = \frac{17}{39} = 0.436 \qquad P(F1 \& M2) = 0.575 \cdot 0.436 = 0.251$$

$$P(M1) = \frac{17}{40} = 0.425$$

$$P(F2|M1) = \frac{23}{39} = 0.590 \qquad P(M1 \& F2) = 0.425 \cdot 0.590 = 0.251$$

$$P(M2|M1) = \frac{16}{39} = 0.410 \qquad P(M1 \& M2) = 0.425 \cdot 0.410 = 0.174$$

The part of the tree diagram representing the determination of the probability $P(F1 \& M2)$, which we just computed, is shown in color. ∎

Note: When two or more members are sampled from a population and those members selected are *not* returned to the population for possible reselection, then the sampling is said to be done **without replacement.**

The general multiplication rule given on page 157 can be extended to more than two events. This will be explored in Exercise 4.99.

Independence

We next discuss the concept of *statistical independence.* Two events are **statistically independent** if the occurrence (or nonoccurrence) of one of the events does not affect the probability of the occurrence of the other event. Consider the following example.

EXAMPLE 4.28 **Introduces the concept of independence**

In the experiment of randomly selecting a card from an ordinary deck of 52 playing cards, let

$$A = \text{event a face card is selected}$$
$$B = \text{event a king is selected}$$
$$C = \text{event a heart is selected}$$

First we compute $P(B|A)$. Given that event A occurs, there are 12 possible outcomes (four kings, four queens, and four jacks), and event B can occur in four of these 12 possibilities. So,

$$P(B|A) = \frac{f}{N} = \frac{4}{12} = 0.333$$

Note that the unconditional probability of event B is

$$P(B) = \frac{f}{N} = \frac{4}{52} = \frac{1}{13} = 0.077$$

Thus, the fact that event A occurs *affects* the probability of the occurrence of event B.

Next, we compute $P(B|C)$. Given that event C occurs, there are 13 possible outcomes (the 13 hearts), and event B can occur in one of these 13 possibilities. Therefore,

$$P(B|C) = \frac{f}{N} = \frac{1}{13} = 0.077$$

This is the same as the unconditional probability of B. That is, in this case, the fact that event C occurs *does not affect* the probability of the occurrence of event B. To express this succinctly, we say that event B is *statistically independent* of event C. It is useful to observe that this independence stems from the following fact: The percentage of kings among the hearts is the same as the percentage of kings among all the cards, namely 7.7%. ∎

With the above example in mind, we now present a formal definition of statistical independence. For brevity, we will use the term *independence* instead of *statistical independence*.

DEFINITION 4.7 Independence

Event *B* is said to be **independent** of event *A* if the occurrence of event *A* does not affect the probability of the occurrence of event *B*. In symbols:

$$P(B|A) = P(B)$$

EXAMPLE 4.29 **Illustrates Definition 4.7**

Table 4.14 gives a contingency table for the number of surgeons in the U. S. cross classified by specialty and base of practice.

TABLE 4.14
Surgeons by specialty and base of practice. Frequencies are given in thousands.

		Base			
		Office B_1	Hospital B_2	Other B_3	Total
Specialty	General Surgery S_1	22.7	10.5	1.5	34.7
	Obstetrics/Gynecology S_2	20.9	5.3	1.0	27.2
	Orthopedic S_3	11.6	2.7	0.4	14.7
	Ophthalmology S_4	11.3	1.5	0.5	13.3
	Total	66.5	20.0	3.4	89.9

Source: American Medical Association.

Suppose a surgeon is selected at random. Is the event that the surgeon selected is office based independent of the event that the surgeon selected is an orthopedist? That is, is event B_1 independent of event S_3?

SOLUTION
To solve this problem we need to compute $P(B_1|S_3)$ and $P(B_1)$. If those two probabilities are equal, then event B_1 is independent of event S_3. Otherwise, event B_1 is not independent of event S_3. From the table we find that

$$P(B_1|S_3) = \frac{11.6}{14.7} = 0.789$$

and

$$P(B_1) = \frac{66.5}{89.9} = 0.740$$

Thus, $P(B_1|S_3) \neq P(B_1)$ and so the event B_1 is *not* independent of the event S_3. That is, the event that the surgeon selected is office based is *not* independent of the event that the surgeon selected is an orthopedist. This lack of independence results from the fact that the percentage of orthopedists who are office based (78.9%) is not the same as the percentage of all surgeons who are office based (74.0%).   ∎

It can be shown that if event B is independent of event A, then it is also true that event A is independent of event B (see Exercise 4.100). Consequently, in such cases we often say that **event A and event B are independent** or that **A and B are independent events.** Finally, if two events are not independent, then they are said to be **dependent.** Hence, the events S_3 and B_1 in Example 4.29 are dependent.

The special multiplication rule

Recall that the *general multiplication rule* states that for *any* two events A and B,

$$P(A \& B) = P(A)P(B|A)$$

If A and B are *independent* events, then this formula can be simplified. Indeed, if event B is independent of event A, then $P(B|A) = P(B)$. Consequently, by substituting $P(B)$ for $P(B|A)$ in the above formula, we get the following formula:

FORMULA 4.6 The special multiplication rule

If event A and event B are *independent,* then

$$P(A \& B) = P(A)P(B)$$

In words, the joint probability is equal to the product of the marginal probabilities.

In the previous examples we have used the condition $P(B|A) = P(B)$ to check whether or not event A and event B are independent. Alternatively, this check can be accomplished with the aid of the special multiplication rule—that is, by determining whether or not

$$P(A \& B) = P(A)P(B)$$

The special multiplication rule is also used to compute joint probabilities when we know or can reasonably assume that two events are independent. Such a situation is illustrated in the next example.

EXAMPLE 4.30   **Illustrates Formula 4.6**

A roulette wheel contains 38 numbers, of which 18 are red, 18 are black, and 2 are green. When the roulette ball is spun it is equally likely to land on any of the 38 numbers. A gambler is playing roulette and decides to bet on red each time.

In two plays at the wheel, what is the probability that the gambler wins both times?

SOLUTION

We need to determine the probability that in two plays at the wheel, the ball lands on red both times. Let

$$R1 = \text{event the ball lands on red the first time}$$
$$R2 = \text{event the ball lands on red the second time}$$

Then we need to find $P(R1 \& R2)$. It is reasonable to suppose that the outcomes on successive plays of the wheel are *independent*. Hence, we can assume that event $R1$ and event $R2$ are independent. Consequently, the special multiplication rule implies that

$$P(R1 \& R2) = P(R1)P(R2)$$

We have

$$P(R1) = \frac{f}{N} = \frac{18}{38} = 0.474$$

The same is true for $P(R2)$. Thus,

$$P(R1 \& R2) = P(R1)P(R2) = 0.474 \cdot 0.474 = 0.225$$

The gambler has a 22.5% chance of the ball landing on red twice in a row. ■

The definition of independence for *three or more* events is more complicated than that for two events. Nevertheless, the special multiplication rule still holds. For example, if A, B, C, and D are independent events, then

$$P(A \& B \& C \& D) = P(A)P(B)P(C)P(D)$$

FORMULA 4.7 The special multiplication rule

If the events A, B, C, . . . are *independent,* then

$$P(A \& B \& C \& \cdots) = P(A)P(B)P(C) \cdots$$

EXAMPLE 4.31   **Illustrates Formula 4.7**

What is the probability that in five plays at the roulette wheel, the ball lands on red all five times?

SOLUTION
Let

$$R1 = \text{event the ball lands on red the first time}$$
$$R2 = \text{event the ball lands on red the second time}$$

and so forth. We need to compute $P(R1 \& R2 \& R3 \& R4 \& R5)$. Since we may assume the results on successive plays at the wheel are independent, the special multiplication rule applies to give

$$P(R1 \ \& \ R2 \ \& \ R3 \ \& \ R4 \ \& \ R5) = P(R1)P(R2)P(R3)P(R4)P(R5)$$

$$= \frac{18}{38} \cdot \frac{18}{38} \cdot \frac{18}{38} \cdot \frac{18}{38} \cdot \frac{18}{38}$$

$$= 0.024 \qquad\qquad \blacksquare$$

Mutually exclusive vs. independent

It is not uncommon for students to confuse the concept of *mutually exclusive events* with that of *independent events*. These terms do *not* mean the same thing. The concept of "mutually exclusive" involves whether or not two events can occur simultaneously, whereas the concept of "independence" involves whether or not the occurrence of one event affects the probability of the occurrence of another. We will make some general statements concerning the relationship between the two ideas in Exercise 4.104. The important point, however, is that the concepts of mutually exclusive events and independent events are quite different.

Exercises 4.6

4.83 According to the U.S. Bureau of the Census, 29.2% of all farm families make at least $25,000 per year. Also, 2.6% of all families are farm families. Use the general multiplication rule to find the probability that a randomly selected family is a farm family making at least $25,000 per year. Interpret your results in terms of percentages.

4.84 The U.S. National Center for Educational Statistics reports that 43.9% of all public elementary schools have between 250 and 499 students. Also, 51.4% of all public schools are elementary schools. Use the general multiplication rule to determine the probability that a randomly selected public school is an elementary school with between 250 and 499 students. Interpret your results in terms of percentages.

4.85 Cards numbered 1, 2, 3, . . . , 10 are placed in a box. The box is shaken and a blindfolded person selects two successive cards without replacement.
a) What is the probability that the first card selected is numbered 6?
b) Given that the first card is numbered 6, what is the probability that the second is numbered 9?
c) What is the probability of selecting a 6 and then a 9?

d) What is the probability that both cards selected are over 5?

4.86 A person has agreed to participate in an ESP experiment. He is asked to pick two random numbers between 1 and 6. The second number must be different from the first. Let

H = event that the first number picked is a 3
K = event that the second number picked exceeds 4

Find
a) $P(H)$ b) $P(K|H)$ c) $P(H \ \& \ K)$

Find the probability that
d) both numbers picked are less than 3.
e) both numbers picked are greater than 3.

4.87 Below is a frequency distribution of the party affiliations of U.S. governors.

Party	Frequency
Democratic	34
Republican	16
Σ	50

Suppose *two* governors are selected at random without replacement. What is the probability that

a) the first selected is a Republican and the second is a Democrat?
b) both governors selected are Republicans?
c) Draw a tree diagram for this problem similar to Figure 4.24 on page 159.
d) What is the probability that both governors selected belong to the same party?

4.88 A frequency distribution for the class of students in a midwest high school is as follows:

Class	Frequency
Freshman	89
Sophomore	127
Junior	118
Senior	93
Σ	427

Suppose *two* students are randomly selected without replacement. Determine the probability that
a) the first student selected is a junior and the second a senior.
b) both students selected are sophomores.
c) Draw a tree diagram for this problem similar to Figure 4.24 on page 159.
d) What is the probability that one of the students selected is a freshman and the other is a sophomore?

4.89 A contingency table for injuries in the U.S. by circumstance and sex is presented below. Frequencies are in millions. (Source: U.S. National Center for Health Statistics.)

Circumstance

Sex	Work C_1	Home C_2	Other C_3	Total
Male S_1	8.0	9.8	17.8	35.6
Female S_2	1.3	11.6	12.9	25.8
Total	9.3	21.4	30.7	61.4

a) Find $P(C_1)$.
b) Find $P(C_1|S_2)$.
c) Are the events C_1 and S_2 independent? Why?
d) Is the event that an injured person is male independent of the event that an injured person was hurt at home? Justify your answer.

4.90 A study of the methods that Americans use to get to work yielded the following contingency table. Frequencies are in millions. (Source: U.S. Bureau of the Census.)

Residence

Method	Urban R_1	Rural R_2	Total
Automobile M_1	45.0	15.0	60.0
Public transportation M_2	6.5	0.5	7.0
Total	51.5	15.5	67.0

a) Find $P(M_1)$.
b) Find $P(M_1|R_2)$.
c) Are the events M_1 and R_2 independent? Why? Explain your results in terms of percentages.
d) Is the event that a worker resides in an urban area independent of the event that the worker uses an automobile to get to work? Justify your answer.

4.91 When a balanced dime is tossed three times, there are eight equally likely outcomes:

HHH HTH THH TTH
HHT HTT THT TTT

Let

A = event the first toss is heads
B = event the third toss is tails
C = event the total number of heads is exactly one

a) Compute $P(A)$, $P(B)$, and $P(C)$.
b) Compute $P(B|A)$.
c) Are the events A and B independent? Why?
d) Compute $P(C|A)$.
e) Are the events A and C independent? Why?

4.92 When a pair of fair dice is rolled, there are 36 equally likely outcomes possible, as shown in Figure 4.1 on page 114. Let

A = event the blue die comes up even
B = event the black die comes up odd
C = event the sum of the dice is 10
D = event the sum of the dice is even

a) Compute $P(A)$, $P(B)$, $P(C)$, and $P(D)$.
b) Compute $P(B|A)$.
c) Are the events A and B independent? Why?
d) Compute $P(C|A)$.
e) Are the events A and C independent? Why?
f) Compute $P(D|A)$.
g) Are the events A and D independent? Why?

4.93   The following is a joint probability distribution for the members of the 98th Congress by political party. (Source: U.S. Congress, Joint Committee on Printing.)

	Reps C_1	Senators C_2	$P(P_i)$
Democrats P_1	0.500	0.084	0.584
Republicans P_2	0.313	0.103	0.416
$P(C_j)$	0.813	0.187	1.000

a) Determine $P(P_1)$, $P(C_2)$, and $P(P_1 \& C_2)$.
b) Are the events P_1 and C_2 independent? [Employ the special multiplication rule to answer this question.]

4.94   A joint probability distribution is shown below for engineers and scientists in the U.S. by highest degree obtained. (Source: U.S. National Science Foundation.)

		Type		
		Engineers T_1	Scientists T_2	$P(D_i)$
	Bachelor's D_1	0.343	0.289	0.632
Highest degree	Master's D_2	0.098	0.146	0.244
	Doctorate D_3	0.017	0.091	0.108
	Other D_4	0.013	0.003	0.016
	$P(T_j)$	0.471	0.529	1.000

a) Determine $P(T_2)$, $P(D_3)$, and $P(T_2 \& D_3)$.
b) Are the events T_2 and D_3 independent? Why?

4.95   Two cards are drawn from an ordinary deck of 52 cards. What is the probability that both cards are aces, if

a) the first card is replaced before the second card is drawn?
b) the first card is not replaced before the second card is drawn?

4.96   In the game of Yahtzee, five fair dice are rolled.
a) What is the probability of rolling all 2s? *Hint:* Use the fact that the outcomes of different dice are independent, and apply the special multiplication rule extended to five events.
b) What is the probability that all the dice come up the same number? *Hint:* You will need to use both the special addition rule and the special multiplication rule.

4.97   Suppose E and F are independent events with $P(E) = 1/3$ and $P(F) = 1/4$. What is
a) $P(E \& F)$?   b) $P(E$ or $F)$?

4.98   A family has two portable computers that run on batteries. There is a 70% chance that a given computer will run for over 60 operating hours without a change of batteries. Determine the probability that
a) both computers run for over 60 hours without a change of batteries.
b) one or the other or both will run for over 60 hours without a change of batteries.

4.99   The general multiplication rule for three events is

$$P(A \& B \& C) = P(A)P(B|A)P[C|(A \& B)]$$

a) If *three* cards are randomly selected without replacement from an ordinary deck of 52, find the probability that all three cards are hearts.
b) State the general multiplication rule for four events.

4.100   Prove that if $P(B|A) = P(B)$, then $P(A|B) = P(A)$. This shows that if event B is independent of event A, then it is also true that event A is independent of event B.

4.101   Three events A, B, and C are said to be *independent* if

$$P(A \& B) = P(A)P(B)$$
$$P(A \& C) = P(A)P(C)$$
$$P(B \& C) = P(B)P(C)$$

and

$$P(A \& B \& C) = P(A)P(B)P(C)$$

What do you think is required for four events to be independent?

4.102 Consider the experiment of rolling a pair of fair dice.

Let

    A = event blue die comes up even
    B = event black die comes up even
    C = event sum of dice is even
    D = event blue die comes up 1, 2, or 3
    E = event blue die comes up 3, 4, or 5
    F = event sum of dice is 5

Use the definition of three events being independent (Ex. 4.101) to solve the following problems.
a) Are A, B, and C independent?
b) Show that $P(D\ \&\ E\ \&\ F) = P(D)P(E)P(F)$ but that D, E, and F are not independent.

4.103 Suppose a fair coin is tossed four times. There are 16 equally likely outcomes:

HHHH	THHH	THHT	THTT
HHHT	HHTT	THTH	TTHT
HHTH	HTHT	TTHH	TTTH
HTHH	HTTH	HTTT	TTTT

Let

    A = event first toss is heads
    B = event second toss is tails
    C = event last two tosses are heads

Use the definition of three independent events given in Exercise 4.101 to show that A, B, and C are independent.

4.104 This exercise explores the relationship between the concepts of *mutually exclusive events* and *independent events*. Consider any two events A and B, neither of which is impossible (that is, assume $P(A) > 0$ and $P(B) > 0$).
a) Show that if A and B are independent events, then they cannot be mutually exclusive. *Hint:* Use the special multiplication rule to show that $P(A\ \&\ B) > 0$.
b) Show that if A and B are mutually exclusive events, then they cannot be independent. *Hint:* Argue that $P(B|A) = 0$ so that $P(B|A) \neq P(B)$.
c) Find an example of two events that are neither mutually exclusive nor independent.

◆ **Chapter Review**

FORMULAS

Classical probability (f/N rule for equally likely outcomes), 112

$$P(E) = \frac{f}{N}$$

(f= number of ways E can occur, N= total number of possible outcomes)

Special addition rule, 131

$$P(A \text{ or } B \text{ or } C \text{ or } \cdots) = P(A) + P(B) + P(C) + \cdots$$

(A, B, C, ... mutually exclusive)

Complementation rule, 134

$$P(E) = 1 - P(\text{not } E)$$

General addition rule, 137

$$P(A \text{ or } B) = P(A) + P(B) - P(A \& B)$$

Conditional probability rule, 153

$$P(B|A) = \frac{P(A \& B)}{P(A)}$$

General multiplication rule, 157

$$P(A \& B) = P(A)P(B|A)$$

Special multiplication rule, 163

$$P(A \& B \& C \& \cdots) = P(A)P(B)P(C) \cdots$$

(A, B, C, ... independent)

YOU SHOULD BE
ABLE TO . . .

1 use and understand the preceding formulas.

2 find and describe (not E), ($A \& B$), and (A or B).

3 determine whether two or more events are mutually exclusive.

4 read and interpret contingency tables.

5 construct a joint probability distribution from a contingency table.

6 compute conditional probabilities both directly and by using the conditional probability rule.

7 determine whether two events are independent.

REVIEW TEST

1 The first two columns of Table 4.15 (at the top of the next page) give a frequency distribution of the adjusted gross incomes from 1982 individual federal income tax returns. Frequencies are in thousands. (Source: U.S. Internal Revenue Service.) Suppose a 1982 income tax return is selected at random.

a) Find $P(A)$, the probability that a randomly se-

lected return shows an adjusted gross income under $10,000.

b) Determine the probability that a randomly selected return shows an adjusted gross income between $30,000 and $99,999.

c) Compute the probabilities of the seven events in the third column of Table 4.15 and record these probabilities in the fourth column.

TABLE 4.15

Adjusted gross income	Number of returns	Event	Probability
Under $10,000	34,081	A	
$10,000–$19,999	24,842	B	
$20,000–$29,999	16,425	C	
$30,000–$39,999	9,863	D	
$40,000–$49,999	4,717	E	
$50,000–$99,999	3,759	F	
$100,000 and over	740	G	
Σ	94,427		

2 Refer to Problem 1. Let

$H=$ event the return selected shows an adjusted gross income between $20,000 and $99,999
$I \ =$ event the return selected shows an adjusted gross income of at most $49,999
$J=$ event the return selected shows an adjusted gross income of at most $99,999
$K=$ event the return selected shows an adjusted gross income of at least $50,000

Describe each of the following events in words and determine the number of outcomes (returns) that comprise each event.
a) (not J)
b) (H & I)
c) (H or K)
d) (H & K)

3 For the following groups of events from Problem 2, determine which are mutually exclusive.
a) The events H and I.
b) The events I and K.
c) The events H and (not J).
d) The events H, (not J), and K.

4 Refer to Problems 1 and 2.
a) Use the second column of Table 4.15 and the f/N rule to compute the probability of each of the four events $H, I, J,$ and K.
b) Express each of the events $H, I, J,$ and K in terms of the mutually exclusive events A through G in the third column of Table 4.15.
c) Compute the probability of each of the four events $H, I, J,$ and K using your results from part (b), the special addition rule, and the fourth column of Table 4.15, which you completed in Problem 1(c).

5 Consider the events (not J), (H & I), (H or K), and (H & K) discussed in Problem 2.
a) Determine the probabilities of the four events above using the f/N rule. [Refer to the second column of Table 4.15.]
b) Compute $P(J)$ using the complementation rule and your result for $P(\text{not } J)$ in part (a).
c) In Problem 4 you found that $P(H) = 0.368$ and $P(K) = 0.048$; and in part (a) of this problem you found that $P(H \ \& \ K) = 0.040$. Using these probabilities along with the general addition rule, determine $P(H \text{ or } K)$. Compare your result with the value for $P(H \text{ or } K)$ from part (a).

6 Table 4.16 gives a contingency table for enrollment in public and private schools by level. Data are in thousands of students. (Source: U.S. National Center for Education Statistics.)

TABLE 4.16

		Type		
		Public T_1	Private T_2	Total
Level	Elementary L_1	26,951	3,600	30,551
	High school L_2	12,215	1,400	13,615
	College L_3	9,612	2,562	12,174
	Total	48,778	7,562	56,340

a) How many cells does this contingency table have?
b) How many students are in high school?
c) How many students attend public schools?
d) How many students attend private colleges?

7 Refer to Problem 6. Suppose a student is selected at random.
a) Describe in words each of the following events:
   (i) L_3 (ii) T_1 (iii) (T_1 & L_3)
b) Determine the probability of each event in part (a), and interpret your results in terms of percentages.
c) Construct a joint probability distribution for Table 4.16.
d) Compute $P(T_1 \text{ or } L_3)$
   (i) using Table 4.16 and the f/N rule.
   (ii) using the general addition rule and your results from part (b).

8 Refer to Problem 6. Suppose a student is selected at random.

a) Find $P(L_3|T_1)$ directly by using Table 4.16 and the f/N rule. Interpret your result in terms of percentages.
b) Find $P(L_3|T_1)$ using the conditional probability rule and your results from Problem 7(b).

9 Refer to Problem 6. Suppose a student is selected at random.

a) Using Table 4.16, compute $P(T_2)$ and $P(T_2|L_2)$.
b) Are the events L_2 and T_2 independent? Why? Explain your results in terms of percentages.
c) Are the events L_2 and T_2 mutually exclusive? Why?
d) Is the event that a student is in elementary school independent of the event that the student attends public school? Justify your answer.

10 The College of Public Programs at Arizona State University awarded the following number of master's degrees in 1982:

Type of degree	Frequency
Master of Arts	3
Master of Public Administration	28
Master of Science	19
Σ	50

Suppose *two* students who received master's degrees are selected at random without replacement. What is the probability that

a) the first student selected received a master of arts and the second student selected received a master of science?
b) both students selected received a master of public administration?
c) Draw a tree diagram for this problem similar to Figure 4.24 on page 159.
d) What is the probability that both students selected received the same degree?

11 According to a manufacturer of electric water heaters, there is a 25% chance that one of the electric water heaters will last more than 10 years. What is the probability that four such electric water heaters all last more than 10 years?

In Chapter 4 we began our study of the fundamentals of probability. We continue that study in this chapter by examining the concepts of *random variables* and *probability distributions*. As you will see, random variables and probability distributions play a key role in the design and analysis of procedures for making statistical inferences.

Discrete random variables

5.1   Discrete random variables; probability distributions

In this section we will examine the concepts of discrete random variables and their probability distributions. We begin with the following illustration.

EXAMPLE   5.1   Introduces random variables

A grouped-data table is presented in Table 5.1 for the number of cars owned by each of the families in a small city.

TABLE 5.1
Grouped-data table for the number of cars owned by each of the families in a small city.

Cars owned x	Number of families f	Relative frequency f/N
0	27	0.004
1	1422	0.219
2	2865	0.422
3	1796	0.277
4	324	0.050
5	53	0.008
Σ	6487	1.000

Table 5.1 shows, for instance, that 1796 out of the 6487 families, or 27.7%, own three cars.

Since the "number of cars owned" varies from family to family it is called a *variable*. When a family is to be selected at random, the "number of cars owned" by the family selected is called a **random variable** since its value depends on chance—namely, upon which family is selected.   ∎

DEFINITION 5.1 Random variable

A **random variable** is a *numerical quantity* whose value depends on chance.

The random variable in Example 5.1 is the "number of cars owned" by a randomly selected family. The possible values of this random variable form a *discrete* data set; namely, the numbers 0, 1, 2, 3, 4, and 5 (see Table 5.1). Consequently, such a random variable is called a *discrete* random variable.

DEFINITION 5.2 Discrete random variable

A **discrete random variable** is a random variable whose possible values form a discrete data set.

A discrete random variable usually involves a count of something, like the number of cars owned by a randomly selected family, the number of people waiting for a haircut in a barber shop, or the number of households in a sample that own a color television set.

It is customary to denote random variables by letters such as x, y, and z. In doing so, we can develop some useful notation. For instance, in Example 5.1 we might let x denote the number of cars owned by a randomly selected family. Then, for instance, we can represent the event that "the family selected owns three cars" by

$$\{x = 3\}$$

read "x equals three." Similarly, the event that "the family selected owns *at least* one car" can be expressed as $\{x \geq 1\}$, read "x is greater than or equal to one."

Probability distributions

We next discuss the concept of a *probability distribution*. Refer again to Example 5.1, and as before, let x denote the number of cars owned by a randomly selected family. Since, in this case, probabilities are just relative frequencies, we can list the probabilities for x by simply referring to Table 5.1. This is done in Table 5.2.

TABLE 5.2
Probability distribution of
the random variable x.

Cars owned x	Probability $P(x)$
0	0.004
1	0.219
2	0.422
3	0.277
4	0.050
5	0.008
Σ	1.000

Just as a table of relative frequencies is called a relative-frequency distribution, a table of probabilities is called a **probability distribution.** Thus, Table 5.2 gives us the **probability distribution of the random variable x**—it tells us the distribution of probabilities for the various values of the random variable x.

Note that the probabilities for x, given in the second column of Table 5.2, add up to 1. This is always true.

KEY FACT

The sum of the probabilities of any discrete random variable is always equal to 1. That is, if x is a discrete random variable, then

$$\Sigma P(x) = 1$$

As a further shorthand notation, we can write the statement "the probability the family selected owns three cars equals 0.277" simply as

$$P(x = 3) = 0.277$$

This is read "the probability x equals three is 0.277." When there is no question about which random variable is under consideration, we will frequently write $P(3)$ instead of $P(x = 3)$.

We should emphasize a point that students often find confusing. The notation $\{x = 3\}$ refers to the *event* that the family selected owns three cars, while the notation $P(x = 3)$ refers to the *probability* of that event.

EXAMPLE 5.2 Illustrates random variable notation and probability distributions

Table 5.3 gives a frequency distribution for the enrollment by grade in public elementary schools in the U.S. ($0 = $ kindergarten, $1 = $ first grade, and so on). Frequencies are in thousands of students.

TABLE 5.3

Frequency distribution for enrollment by grade in U.S. public elementary schools.

Grade y	Enrollment (thous.) f
0	2,700
1	2,951
2	2,786
3	2,812
4	2,926
5	3,131
6	3,180
7	3,184
8	3,062
Σ	26,732

Source: U.S. National Center for Education Statistics.

Suppose a student in elementary school is to be selected at random. Let y denote the grade of the student selected. Then y is a random variable whose possible values are 0, 1, 2, . . . , 8.

a) Use random variable notation to represent the event that the student selected is in the fifth grade.

b) Determine $P(y = 5)$ and express your results in terms of percentages.

c) Find $P(7)$.

d) Find the probability distribution of the random variable y.

SOLUTION

a) The *event* that the student selected is in the fifth grade can be represented as $\{y = 5\}$.

b) $P(y = 5)$ is the *probability* that the student selected is in the fifth grade. Using Table 5.3 and the f/N rule we can obtain that probability:

$$P(y = 5) = \frac{f}{N} = \frac{3,131}{26,732} = 0.117$$

In terms of percentages, this means that 11.7% of the students in elementary school are in the fifth grade.

c) $P(7)$ is the probability that the student selected is in the seventh grade. We have

$$P(7) = \frac{f}{N} = \frac{3,184}{26,732} = 0.119$$

d) The probability distribution of y is obtained by computing $P(y)$ for $y = 0, 1, 2, \ldots, 8$. We have already done this for $y = 5$ and $y = 7$. The other probabilities are computed similarly and are displayed in the second column of Table 5.4.

TABLE 5.4
Probability distribution of the random variable y.

Grade y	Probability $P(y)$	
0	0.101	
1	0.110	
2	0.104	
3	0.105	
4	0.109	
5	0.117	
6	0.119	
7	0.119	
8	0.115	
Σ	0.999	*(rounding error)*

MTB
SPSS

In Table 5.4 the sum of the probabilities is given as 0.999. As we know from the Key Fact on page 174, the sum of the probabilities must be *exactly* 1, but our computation is off by a little since we rounded the probabilities for y to three decimal places.

Once we have the probability distribution of a discrete random variable, it is easy to determine any probability involving that random variable. The basic tool for accomplishing this is the *special addition rule* given on page 131. Consider the following example.

EXAMPLE 5.3 **Illustrates random variable notation and probability distributions**

When a balanced dime is tossed three times, eight equally likely outcomes are possible:

HHH	HTH	THH	TTH
HHT	HTT	THT	TTT

Here, for example, HHT means that the first two tosses are heads and the third is tails. Let x denote the total number of heads obtained in the three tosses. Then x is a random variable whose possible values are 0, 1, 2, and 3.

a) Use random variable notation to represent the event that *exactly* two heads are tossed.

b) Determine $P(x = 2)$.

c) Find the probability distribution of the random variable x.

d) Use random variable notation to represent the event that *at most* two heads are tossed.

e) Find $P(x \leq 2)$.

S O L U T I O N

a) The event that exactly two heads are tossed is $\{x = 2\}$.

b) $P(x = 2)$ is the probability that exactly two heads are tossed. We have

$$P(x = 2) = \frac{f}{N} = \frac{3}{8} = 0.375$$

This is because there are three ways to get a total of two heads and there are eight possible outcomes altogether.

c) The remaining probabilities for x are computed as in part (b). The probability distribution of the random variable x is shown in Table 5.5.

TABLE 5.5

Probability distribution for the number of heads, x, obtained in three tosses of a balanced dime.

Total number of heads x	Probability $P(x)$
0	0.125
1	0.375
2	0.375
3	0.125
Σ	1.000

d) The event that at most two heads are tossed can be represented as $\{x \leq 2\}$, read "x is less than or equal to two."

e) $P(x \leq 2)$ is the probability that at most two heads are tossed. The event that at most two heads are tossed can be expressed as

$$\{x \leq 2\} = (\{x = 0\} \text{ or } \{x = 1\} \text{ or } \{x = 2\})$$

Since the three events on the right are mutually exclusive (why?), we can use the *special addition rule* and Table 5.5 to conclude that

$$P(x \leq 2) = P(x = 0) + P(x = 1) + P(x = 2)$$
$$= 0.125 + 0.375 + 0.375 = 0.875$$

Thus the probability is 0.875 that at most two heads are tossed.   ∎

Random variables and numerical populations

A population is said to be a **numerical population** if its members are numbers. It is important to understand that the word "population" is used here in the *statistical* sense. Consider the following example.

EXAMPLE   5.4   **Illustrates numerical populations**

The number of complete terms served by each of the first 40 U.S. Presidents, as of 1986, is displayed in Table 5.6.

TABLE 5.6
Number of terms completed
by each of the first 40 U.S.
Presidents as of 1986.

2	1	2	2	2	1	2	1
0	0	1	0	0	1	1	1
0	2	1	0	0	1	1	1
1	1	1	2	0	1	1	3
1	2	0	1	1	0	1	1

If we are interested in analyzing the number of complete terms served by each of these Presidents, then what is the population?

S O L U T I O N
The population is *not* the collection of the 40 Presidents per se. It is the collection of the 40 numerical values in Table 5.6 that give the number of terms completed by each President. Hence this is a numerical population. If we were interested in the party affiliations or religions of the Presidents, the population would not be numerical.   ∎

It is natural to associate with any numerical population, a random variable—namely, the value of a randomly selected member of the population. The probability distribution of such a random variable coincides with the percentage (relative-frequency) distribution of the population. To see this, let us return to Example 5.4. A relative-frequency distribution for the number of completed terms can be obtained from Table 5.6. We present that in Table 5.7.

TABLE 5.7
Relative-frequency
distribution for
number of terms
completed.

Terms completed x	Relative frequency f/N
0	0.250
1	0.525
2	0.200
3	0.025
Σ	1.000

Now, suppose one of the first 40 Presidents is to be selected at random, and we let x denote the number of complete terms served by the President chosen. Then x is a random variable. Since, in this case, probabilities are just relative frequencies, Table 5.7 also serves as the probability distribution of the random variable x. We summarize these last points below.

KEY FACT

With any *numerical* population we can associate a random variable—namely, the value of a member selected at random from the population. For this type of random variable, probabilities are the same as relative frequencies. In other words, the probability distribution of such a random variable is the same as the relative-frequency distribution of the population.

Exercises 5.1

5.1   According to the U.S. Bureau of the Census, a frequency distribution for the number of persons per household in the U.S. is as follows (frequencies in millions).

Number of persons x	Number of households f
1	19.4
2	26.5
3	14.6
4	12.9
5	6.1
6	2.5
7†	1.6
Σ	83.6

† Actually this is "7 or more" but we will consider it to be "7" for illustrative purposes.

Let x denote the number of persons in a household selected at random.
a) What are the possible values of the random variable x?
b) Use random variable notation to represent the event that the household selected has exactly five persons.
c) Determine $P(x=5)$ and interpret your results in terms of percentages.
d) Find $P(3)$.
e) Find the probability distribution of the random variable x.

5.2   Below is a frequency distribution for the enrollment by grade in public secondary schools. Frequencies are in thousands. (Source: U.S. National Center for Education Statistics.)

Grade x	Enrollment f
9	3,290
10	3,223
11	3,041
12	2,908
Σ	12,462

Suppose a student in public secondary school is to be selected at random. Let x denote the grade level of the student chosen.
a) What are the possible values of the random variable x?
b) Use random variable notation to represent the event that the student selected is in the tenth grade.
c) Determine $P(x=10)$ and interpret your results in terms of percentages.
d) Find $P(12)$.
e) Find the probability distribution of the random variable x.

5.3   When a pair of fair dice is rolled, 36 equally likely outcomes are possible. [See Figure 4.1 on page 114.] Let y denote the sum of the dice.
a) What are the possible values of the random variable y?
b) Use random variable notation to represent the event that the sum of the dice is 7.
c) Find $P(y=7)$.
d) Determine $P(11)$.
e) Determine the probability distribution of the random variable y. [Leave your probabilities in fraction form.]

5.4   When a pair of fair dice is rolled there are 36 possible equally likely outcomes. [See Figure 4.1 on page 114.] Let x denote the larger number showing on the two dice. For example, if the outcome is

then $x=5$. [If both dice come up the same, then x equals that common value.]
a) What are the possible values of the random variable x?
b) Use random variable notation to represent the event that the larger number is four.
c) Find $P(x=4)$.
d) Determine $P(2)$.
e) Determine the probability distribution of the random variable x. [Leave your probabilities in fraction form.]

5.5   The Prescott National Bank has six tellers available to serve customers. The number of busy tellers varies from time to time, so it is a random variable, which we shall call x. From past records, the probability distribution of x is known. It is given in the following table:

Number busy x	Probability $P(x)$
0	0.03
1	0.05
2	0.08
3	0.15
4	0.21
5	0.26
6	0.22

The table indicates, for example, that the probability is 0.26 that exactly five of the tellers are busy with customers (that is, about 26% of the time exactly five tellers are busy with customers). Use random variable notation to represent each of the following events:

a) Exactly four tellers are busy.
b) At least two tellers are busy.
c) Less than five tellers are busy.
d) At least two but less than five tellers are busy.

Use the special addition rule and the table to determine

e) $P(x = 4)$. g) $P(x < 5)$.
f) $P(x \geq 2)$. h) $P(2 \leq x < 5)$.

5.6 On the basis of past experience, a car salesperson knows that the number of cars she sells per week is a random variable, y, with a probability distribution as shown here.

Cars sold y	Probability $P(y)$
0	0.135
1	0.271
2	0.271
3	0.180
4	0.090
5	0.036
6	0.012
7	0.003
8	0.002

Suppose a week is selected at random. Use random variable notation to represent each of the following events: She sells

a) exactly three cars.
b) at least three cars.
c) less than seven cars.

d) at least three but less than seven cars.

Use the special addition rule and the table to find

e) $P(y = 3)$. g) $P(y < 7)$.
f) $P(y \geq 3)$. h) $P(3 \leq y < 7)$.

5.7 Benny's Barber Shop in Cleveland has five chairs for waiting customers. Previous records indicate that the probability distribution for the number of customers waiting, y, is as follows:

Customers waiting y	Probability $P(y)$
0	0.424
1	0.161
2	0.134
3	0.111
4	0.093
5	0.077

Represent each of the events below using random variable notation.

a) Exactly two customers are waiting.
b) At most four customers are waiting.
c) More than one customer is waiting.
d) Between two and four customers, inclusive, are waiting.

Apply the special addition rule and the table to determine

e) $P(y = 2)$. h) $P(2 \leq y \leq 4)$.
f) $P(y \leq 4)$. i) $P(1 < y \leq 4)$.
g) $P(y > 1)$.

5.8 A small company in Tulsa, Oklahoma has a total of 10 management employees. Six are male and four are female. Three of these 10 employees are to be selected to represent the company at a small business conference sponsored by the SBA (Small Business Administration). If the selection is done randomly, then the number of women selected is a random variable, x, with a probability distribution given by

Women chosen x	Probability $P(x)$
0	0.167
1	0.500
2	0.300
3	0.033

Use random variable notation to express each of the following events: The number of women selected is

a) exactly two. c) at least one.
b) at most two. d) either one or two.

Apply the special addition rule and the table to compute

e) $P(x=2)$. g) $P(x \geq 1)$.
f) $P(x \leq 2)$. h) $P(x=1 \text{ or } 2)$.

5.9 Suppose that z is a random variable and that $P(z > 1.96) = 0.025$. What is $P(z \leq 1.96)$? *Hint:* Use the complementation rule (page 134).

5.10 Suppose that t is a random variable and that $P(t > 2.02) = 0.05$ and $P(t < -2.02) = 0.05$. What is $P(-2.02 \leq t \leq 2.02)$?

5.11 Let z be a random variable and suppose that $P(-1.64 \leq z \leq 1.64) = 0.90$. Also suppose that

$P(z > 1.64) = P(z < -1.64)$. Find $P(z > 1.64)$.

5.12 Assume that x is a random variable and that $P(x > c) = a$, where c and a are numbers, and $0 \leq a \leq 1$. What is $P(x \leq c)$?

5.13 Assume that y is a random variable and that $P(y > c) = a/2$, where c and a are numbers, and $0 \leq a \leq 1$. Also suppose $P(y < -c) = P(y > c)$. Find $P(-c \leq y \leq c)$ in terms of a.

5.14 Suppose that t is a random variable and that $P(-c \leq t \leq c) = 1 - a$. Further assume that $P(t < -c) = P(t > c)$. Find $P(t > c)$ in terms of a.

5.2   The mean and standard deviation of a discrete random variable

In this section we will study the mean and standard deviation of a discrete random variable. As you will see, the mean and standard deviation of a random variable are analogous to the mean and standard deviation of a population. In fact, when a random variable is defined to be the value of a randomly selected member from a numerical population, the mean and standard deviation of the random variable are precisely the same as the mean and standard deviation of the population.

Mean of a discrete random variable

We first introduce the concept of the mean of a discrete random variable with the aid of the following illustrative example.

EXAMPLE 5.5 Introduces the mean of a discrete random variable

CJ^2 Business Services is a small company specializing in word processing. The company employs eight people. A frequency distribution for the weekly salaries of these eight employees is presented in Table 5.8.

TABLE 5.8
Frequency distribution for the weekly salaries of CJ^2 employees.

Salary x	Number of employees f
$240	3
320	2
450	1
600	2
Σ	8

The (population) mean salary of these employees can be obtained using Formula 3.4 on page 99:

$$\mu = \frac{\Sigma xf}{N}$$

This is the formula used to compute the mean of a population when the data are given in grouped form. To apply the formula we append an xf-column to Table 5.8 and proceed as usual.

TABLE 5.9

x	f	xf
240	3	720
320	2	640
450	1	450
600	2	1200
Σ	8	3010

Thus, the mean weekly salary of the employees of CJ^2 is

$$\mu = \frac{\Sigma xf}{N} = \frac{\$3010}{8} = \$376.25$$

Now, we can associate with this population of eight weekly salaries, as we can with any numerical population, a random variable x—namely, the weekly salary of an employee selected at random. Probabilities for the random variable x are just relative frequencies. Thus, for each x-value, the quantities $P(x)$ and f/N are exactly the same. For instance, the relative frequency of the salary $240 is

$$\frac{f}{N} = \frac{3}{8} = 0.375$$

This is the same as the probability, $P(240)$, of randomly selecting an employee whose weekly salary is $240. Because probabilities and relative frequencies are the same here, we can rewrite the formula for μ in terms of probabilities as follows:

$$\mu = \frac{\Sigma xf}{N} = \frac{240 \cdot 3 + 320 \cdot 2 + 450 \cdot 1 + 600 \cdot 2}{8}$$

$$= 240 \cdot \frac{3}{8} + 320 \cdot \frac{2}{8} + 450 \cdot \frac{1}{8} + 600 \cdot \frac{2}{8}$$

$$= \Sigma x \cdot \frac{f}{N} = \Sigma x P(x)$$

Since, in this form, the formula for the mean μ involves the probability distribution of the random variable x, we make the following definition. ∎

DEFINITION 5.3 Mean of a discrete random variable

The **mean† of a discrete random variable** x is

$$\mu_x = \Sigma x P(x)$$

$\dagger$ The terms **expected value** and **expectation** are also commonly used.

Note: The μ has the *subscript x* to emphasize that it is the mean of the random variable x. If the random variable under consideration is called y, then we would write μ_y for its mean and express the formula in Definition 5.3 as $\mu_y = \Sigma y P(y)$.

As we have seen, the computation of descriptive measures is most easily accomplished using tables. The same is true here: it is usually easiest to compute the mean of a discrete random variable using a probability distribution table. We illustrate this in Example 5.6.

EXAMPLE 5.6 Illustrates Definition 5.3

In Example 5.5 we considered the weekly salaries of the eight employees of CJ² Business Services. Suppose an employee is to be selected at random and we let x denote the salary of the randomly chosen employee. Then x is a random variable. Find the mean, μ_x, of the random variable x.

S O L U T I O N

To compute μ_x we must first obtain the probability distribution of x. Since probabilities for x are just relative frequencies, we can obtain the probability distribution of x from the frequency distribution given in Table 5.8 on page 180. See Table 5.10.

TABLE 5.10
Probability distribution of the random variable x—the salary of a randomly selected CJ² employee.

Salary x	Probability $P(x)$	
240	0.375	← 3/8
320	0.250	← 2/8
450	0.125	← 1/8
600	0.250	← 2/8

Now we can compute μ_x by appending an $xP(x)$-column to Table 5.10 and employing the formula

$$\mu_x = \Sigma x P(x)$$

This is done in Table 5.11.

TABLE 5.11
Table for computation of μ_x.

x	$P(x)$	$xP(x)$
240	0.375	90.00
320	0.250	80.00
450	0.125	56.25
600	0.250	150.00
Σ		376.25

MTB
SPSS

Consequently, the mean of the random variable x is

$$\mu_x = \Sigma x P(x) = \$376.25$$

∎

Before we consider the mean of a random variable further, let us look again at Examples 5.5 and 5.6. From Example 5.5 we know that the population mean salary of the employees of CJ² Business Services is

$$\mu = \frac{\Sigma x f}{N} = \frac{\$3010}{8} = \$376.25$$

From Example 5.6 we know that if x denotes the salary of a randomly selected employee of CJ², then the mean of the random variable x is

$$\mu_x = \Sigma x P(x) = \$376.25$$

Thus, as expected, the population mean salary, μ, and the mean, μ_x, of the random variable x are exactly the same. More generally, we can make the following statement:

KEY FACT

When a random variable x is defined to be the value of a randomly selected member from a numerical population, then the mean, μ_x, of the random variable is exactly the same as the mean, μ, of the population.

In Examples 5.5 and 5.6, the numerical population consists of the eight weekly salaries of the employees of CJ² Business Services, and the random variable x is the value of a randomly selected salary. As we have seen, the mean, μ, of this numerical population and the mean, μ_x, of the random variable x are both equal to $\$376.25$.

We also need to emphasize that the definition of the mean of a discrete random variable, given in Definition 5.3, applies to any discrete random variable, not just to those arising from numerical populations. In the next example, we calculate the mean of a discrete random variable that does *not* arise from numerical population data.

EXAMPLE   5.7   Illustrates Definition 5.3

The Prescott National Bank has six tellers available to serve customers. Of course, the number of tellers actually busy with customers varies from time to time—it is a random variable. On the basis of past records, it is known that the probability distribution for this random variable is as given in Table 5.12.

TABLE 5.12
Probability distribution for number of tellers busy with customers.

Number busy x	Probability $P(x)$
0	0.03
1	0.05
2	0.08
3	0.15
4	0.21
5	0.26
6	0.22

[The table indicates, for instance, that about 26% of the time exactly five tellers are busy with customers.] Find the mean, μ_x, of the random variable x. That is, find the mean number of tellers that are busy with customers.

SOLUTION

The mean of the random variable x is obtained in the usual manner by appending an $xP(x)$-column to Table 5.12 and summing that column of numbers. We do this in Table 5.13.

TABLE 5.13
Table for computation of μ_x.

x	$P(x)$	$xP(x)$
0	0.03	0.00
1	0.05	0.05
2	0.08	0.16
3	0.15	0.45
4	0.21	0.84
5	0.26	1.30
6	0.22	1.32
Σ		4.12

Consequently, $\mu_x = \Sigma xP(x) = 4.12$. In other words, the mean number of tellers busy with customers is 4.12. ∎

Interpretation of the mean of a random variable

As you know, the mean of a finite population is just the arithmetic average of the population values. A similar interpretation can be given to the mean of a discrete random variable. For instance, in Example 5.7 the random variable x denotes the number of tellers busy with customers. The mean of this random variable is $\mu_x = 4.12$. Of course, we can never observe a time when there are 4.12 busy tellers. The mean of 4.12 simply indicates that if we observe the number of busy tellers at many different times, then we would expect *an average* of about 4.12 busy tellers per observation. This kind of interpretation holds in all cases.

KEY FACT

The mean, μ_x, of a random variable x can be given the following interpretation: It is the value of x that we would expect to observe *on the average* if the experiment were repeated many times.

Standard deviation of a discrete random variable

We next examine the concept of the standard deviation of a discrete random variable. To begin, let us return to the salary data of Example 5.5.

EXAMPLE   5.8   Introduces the standard deviation of a discrete random variable

A frequency distribution for the weekly salaries of the employees of CJ² Business Services is repeated in Table 5.14.

TABLE 5.14
Frequency distribution
for the weekly salaries
of CJ² employees.

Salary x	Number of employees f
$240	3
320	2
450	1
600	2
Σ	8

The population standard deviation of these salaries can be found by applying Formula 3.5 on page 99:

$$\sigma = \sqrt{\frac{\Sigma(x-\mu)^2 f}{N}}$$

This is the formula used to compute the standard deviation of a population when the data are in grouped form.

As in Example 5.5, we can associate with this population of salaries a random variable x—the salary of a randomly selected employee. Then probabilities for x are just relative frequencies. Consequently, the formula for σ can be rewritten in terms of the $P(x)$-values as follows:

$$\sigma = \sqrt{\frac{\Sigma(x-\mu)^2 f}{N}} = \sqrt{\Sigma(x-\mu)^2 \cdot \frac{f}{N}} = \sqrt{\Sigma(x-\mu)^2 P(x)}$$

The last expression is the definition of the standard deviation of a discrete random variable. ∎

DEFINITION 5.4 Standard deviation of a discrete random variable

The **standard deviation of a discrete random variable** x is

$$\sigma_x = \sqrt{\Sigma(x-\mu_x)^2 P(x)}$$

Recall that the *variance* of a population is σ^2, the square of the standard deviation of the population. Similarly, σ_x^2 is called the **variance of the random variable** x.

Although we motivated Definition 5.4 using a random variable arising from a numerical population, it is important to realize that the definition applies to any discrete random variable. However, when the random variable *does* arise from a numerical population, we have the following analogue of the statement given for the mean on page 183.

KEY FACT

When a random variable x is defined to be the value of a randomly selected member from a numerical population, then the standard deviation, σ_x, of the random variable is exactly the same as the standard deviation, σ, of the population.

We now present an example to show the calculations involved in computing the standard deviation of a discrete random variable.

EXAMPLE   5.9   Illustrates Definition 5.4

Compute the standard deviation, σ_x, of the random variable x, if x is the salary of a randomly selected employee of CJ².

SOLUTION

The probability distribution of the random variable x, displayed in Table 5.10, is repeated below in Table 5.15.

TABLE 5.15

Probability distribution of the random variable x—the salary of a randomly selected CJ² employee.

Salary x	Probability $P(x)$
240	0.375
320	0.250
450	0.125
600	0.250

The mean of the random variable x was found in Example 5.6—$\mu_x = \$376.25$. To apply the formula

$$\sigma_x = \sqrt{\Sigma(x-\mu_x)^2 P(x)}$$

we construct Table 5.16.

TABLE 5.16

Table for computation of σ_x.

x	$P(x)$	$x-\mu_x$	$(x-\mu_x)^2$	$(x-\mu_x)^2 P(x)$
240	0.375	−136.25	18,564.0625	6,961.5234375
320	0.250	−56.25	3,164.0625	791.0156250
450	0.125	73.75	5,439.0625	679.8828125
600	0.250	223.75	50,064.0625	12,516.0156250
Σ				20,948.4375000

The last column of Table 5.16 shows that $\Sigma(x-\mu_x)^2 P(x) = 20{,}948.4375$. Taking the square root we get σ_x:

$$\sigma_x = \sqrt{\Sigma(x-\mu_x)^2 P(x)} = \sqrt{20{,}948.4375} = 144.74$$

(to two decimal places). Consequently, the standard deviation of the random variable x is $\sigma_x = \$144.74$. ∎

MTB
SPSS

Shortcut formula for σ_x

The previous example makes it clear that the computations required to find σ_x are time-consuming and tedious even in simple cases. To deal with this problem we can use a computer or develop a shortcut formula. The latter approach is presented in Formula 5.1.

FORMULA 5.1 Shortcut formula for σ_x

The shortcut formula for computing the standard deviation of a discrete random variable x is

$$\sigma_x = \sqrt{\Sigma x^2 P(x) - \mu_x^2}$$

EXAMPLE 5.10   **Illustrates Formula 5.1**

For purposes of comparison we will use the shortcut formula to compute the standard deviation of the random variable x from Example 5.9. We have already seen that $\mu_x = \$376.25$. To employ the shortcut formula we also need to determine $\Sigma x^2 P(x)$. This is done in Table 5.17.

TABLE 5.17
Table for computation of σ_x using the shortcut formula.

x	$P(x)$	x^2	$x^2 P(x)$
240	0.375	57,600	21,600.0
320	0.250	102,400	25,600.0
450	0.125	202,500	25,312.5
600	0.250	360,000	90,000.0
Σ			162,512.5

Thus, $\Sigma x^2 P(x) = 162{,}512.5$. Applying Formula 5.1 we get

$$\sigma_x = \sqrt{\Sigma x^2 P(x) - \mu_x^2} = \sqrt{162{,}512.5 - (376.25)^2}$$
$$= \sqrt{20{,}948.4375} = 144.74$$

This, of course, is the same result obtained in Example 5.9. However, the computations done here using the shortcut formula are significantly easier.   ∎

Interpretation of the standard deviation of a random variable

Recall that the standard deviation of a population is a measure of the dispersion of the population values. Roughly speaking, it measures how far the population values are from the mean, on the average. A similar interpretation can be given to the standard deviation of a random variable.

KEY FACT

The standard deviation, σ_x, of a random variable x measures the dispersion of the possible values of the random variable relative to its mean—the smaller the standard deviation, the more likely that x will take a value close to its mean.

Exercises 5.2

In Exercises 5.15 through 5.22 you will be given the probability distributions of the random variables considered in Exercises 5.1 through 5.8 of Section 5.1. For each exercise,

a) compute and interpret the mean of the random variable.
b) compute the standard deviation of the random variable using the defining formula (page 185).
c) compute the standard deviation of the random variable using the shortcut formula (page 187).

5.15 The random variable x is the number of persons in a randomly selected U.S. household. Its probability distribution is

Number of persons x	Probability $P(x)$
1	0.232
2	0.317
3	0.175
4	0.154
5	0.073
6	0.030
7	0.019

5.16 The random variable x is the grade level of a secondary-school student selected at random. Its probability distribution is

Grade x	Probability $P(x)$
9	0.264
10	0.259
11	0.244
12	0.233

5.17 The random variable y is the sum of the dice when a pair of fair dice is rolled. Its probability distribution is

Sum of dice y	Probability $P(y)$
2	1/36
3	1/18
4	1/12
5	1/9
6	5/36
7	1/6
8	5/36
9	1/9
10	1/12
11	1/18
12	1/36

[Do your computations using fractions.]

5.18 The random variable x is the larger number showing, when a pair of fair dice is rolled. Its probability distribution is

Larger number x	Probability $P(x)$
1	1/36
2	1/12
3	5/36
4	7/36
5	1/4
6	11/36

[Do your computations using fractions.]

5.19 The random variable x is the number of busy tellers at the Prescott National Bank. Its probability distribution is

Number busy x	Probability $P(x)$
0	0.03
1	0.05
2	0.08
3	0.15
4	0.21
5	0.26
6	0.22

5.20 The random variable y is the number of cars sold per week by a car salesperson. Its probability distribution is as follows:

Cars sold y	Probability $P(y)$
0	0.135
1	0.271
2	0.271
3	0.180
4	0.090
5	0.036
6	0.012
7	0.003
8	0.002

5.21 The random variable y is the number of customers waiting at Benny's Barber Shop. Its probability distribution is as follows:

Customers waiting y	Probability $P(y)$
0	0.424
1	0.161
2	0.134
3	0.111
4	0.093
5	0.077

5.22 The random variable x is the number of women selected to represent a company at a small business conference. Its probability distribution is

Women chosen x	Probability $P(x)$
0	0.167
1	0.500
2	0.300
3	0.033

As mentioned earlier, the *mean* of a random variable is also called the *expected value*. This terminology is especially useful in gambling and decision theory, as illustrated in Exercises 5.23 and 5.24.

5.23 A roulette wheel contains 38 numbers, of which 18 are red, 18 are black, and two are green. When the roulette ball is spun, it is equally likely to land on any of the 38 numbers. Suppose you bet $1 on red. Then, if the ball lands on a red number, you win $1; otherwise, you lose your $1. Let x be the amount you win on your $1 bet. Then x is a random variable with probability distribution

x	$P(x)$
1	0.474
−1	0.526

a) Verify that the probability distribution shown in the table is correct.
b) Find the *expected value* (mean) of the random variable x.
c) On the average, how much will you lose per play? *Hint:* Refer to the interpretation of the mean of a random variable given on page 184.
d) About how much would you expect to lose if you bet $1 on red 100 times? 1000 times?
e) Do you think that roulette is a profitable game to play?

5.24 An investor plans to invest $50,000 in one of four investments. The return on each investment depends on whether next year's economy is strong or weak. The table in the next column summarizes the possible *payoffs* of the four investments. Let v denote the payoff of the certificate of deposit, w that of the investment in the office-complex, x that of the investment in the land, and y that of the investment in the tech school. Then v, w, x, and y are random variables. Assume that next year's economy has a 40% chance of being strong and a 60% chance of being weak.

a) Find the probability distribution of each of the four random variables v, w, x, and y.
b) Determine the expected value of each of the four random variables.
c) Which investment has the best expected payoff? Which has the worst?
d) Which investment would *you* select? Why?

	Next year's economy	
Investment	Strong	Weak
Certificate of deposit	$ 6,000	$ 6,000
Office complex	$15,000	$ 5,000
Land speculation	$33,000	−$17,000
Tech school	$ 5,500	$10,000

5.25 A factory manager collected data on the number of equipment breakdowns per day. From this data she derived the following probability distribution:

Number of breakdowns w	Probability $P(w)$
0	0.80
1	0.15
2	0.05

a) Determine μ_w and σ_w.
b) On the average, how many breakdowns are there per day?
c) About how many breakdowns are expected per year, assuming 250 work days per year?

Some important properties of the mean and standard deviation of a random variable are developed in Exercises 5.26 through 5.29.

5.26 Refer to Exercise 5.25. Assume that the number of breakdowns on different days are *independent* of one another. Let x and y denote the number of breakdowns in each of two consecutive days.

a) Fill in the joint probability distribution table below.

		y		
		0	1	2
x	0			
	1			
	2			

Hint: For the upper-righthand corner, we have by the special multiplication rule and the table in Exercise 5.25 that

$$P(x=0 \ \& \ y=2) = P(x=0)P(y=2)$$
$$= (0.80) \cdot (0.05) = 0.04$$

b) Use your answer from part (a) to find the probability distribution of the random variable $x + y$, the total number of breakdowns in two days. That is, complete the following table:

$x+y$	$P(x+y)$
0	
1	
2	
3	
4	

c) Use your answer from part (b) to find μ_{x+y} and σ_{x+y}.

d) Use part (c) to verify that the following equations are true for this example.

$$\mu_{x+y} = \mu_x + \mu_y$$
$$\sigma_{x+y} = \sqrt{\sigma_x^2 + \sigma_y^2}$$

[The mean and standard deviation of x and y are the same as that of w in Exercise 5.25.]

e) The equations in part (d) hold in general. That is,

If x and y are *any* two random variables then

$$\mu_{x+y} = \mu_x + \mu_y$$

If, in addition, x and y are *independent* then

$$\sigma_{x+y} = \sqrt{\sigma_x^2 + \sigma_y^2}$$

Interpret these two equations in words.

5.27   The factory manager in Exercise 5.25 estimates that each breakdown costs the company $100 in repairs and loss of production. If w is the number of breakdowns in a day, then $100 \cdot w$ is the cost due to breakdowns for that day.

a) Find the probability distribution of the random variable $100 \cdot w$. [Refer to the probability distribution in Exercise 5.25.]

b) Determine the mean daily breakdown cost, $\mu_{100 \cdot w}$, using your answer from part (a).

c) What is the relationship between $\mu_{100 \cdot w}$ and μ_w? [From Exercise 5.25, $\mu_w = 0.25$]

d) Find $\sigma_{100 \cdot w}$ using your answer from part (a).

e) What is the apparent relationship between $\sigma_{100 \cdot w}$ and σ_w? [From Exercise 5.25, $\sigma_w = 0.536$]

f) The results in parts (c) and (e) are true in general. That is,

If w is *any* random variable and c is a constant, then

$$\mu_{c \cdot w} = c\mu_w$$

and

$$\sigma_{c \cdot w} = |c|\sigma_w$$

Interpret these two equations in words.

5.3

Binomial coefficients; Bernoulli trials

A tremendous number of applied problems in probability and statistics involve situations where an experiment is performed several times, and in which there are only *two* possible outcomes on any given trial. For example, consider testing the effectiveness of a drug. Several patients take the drug (the trials) and, for each patient, the drug is either effective or not effective (the two possible outcomes). Or, consider the weekly sales of a car salesperson. The salesperson has several customers during the week (the trials) and, for each customer, the salesperson either makes a sale or does not make a sale (the two possible outcomes). Or, consider taste tests for colas. A number of people taste two different colas (the trials) and, for each person, the preference is either for the first cola or the second cola (the two possible outcomes).

The statistical analysis of repeated trials, each of which has only two possible outcomes, requires knowledge of **Bernoulli trials** and the **binomial distribution.** [The "bi" in "binomial distribution" refers to the fact that there are *two* possible outcomes for each trial.] We will study Bernoulli trials in this section and the binomial distribution in the next section.

To perform the necessary computations in binomial distribution problems, we need an understanding of some additional mathematical concepts—*factorials* and *binomial coefficients*. We will discuss these first.

Factorials

A useful concept in mathematics is that of *factorial*. The factorial of a positive integer is defined as follows:

DEFINITION 5.5 Factorials

For a positive integer n, we define **$n!$** (read "**n factorial**") as

$$n! = n(n-1) \cdots 2 \cdot 1$$

That is, to obtain $n!$ we start with n and keep multiplying by the next smallest integer until we get to 1.

EXAMPLE 5.11 **Illustrates Definition 5.5**

Find 3!, 4!, and 5!.

SOLUTION

$$3! = 3 \cdot 2 \cdot 1 = 6$$
$$4! = 4 \cdot 3 \cdot 2 \cdot 1 = 24$$
$$5! = 5 \cdot 4 \cdot 3 \cdot 2 \cdot 1 = 120$$

■

It is necessary to define 0!. Although you might think it is natural to set this equal to 0, it turns out to be convenient to let it be 1:

$$0! = 1$$

Binomial coefficients

We now define the *binomial coefficients*. You may have already come upon these in algebra when you studied the binomial expansion, that is, the expansion of $(a+b)^n$. We do not require that you remember this, but if you would like a quick refresher, see Exercise 5.42.

DEFINITION 5.6 Binomial coefficients

If n is a positive integer and x is a nonnegative integer less than or equal to n, we define the **binomial coefficient** $\binom{n}{x}$ as

$$\binom{n}{x} = \frac{n!}{x! \, (n-x)!}$$

EXAMPLE 5.12 **Illustrates Definition 5.6**

Determine the value of the following binomial coefficients:

a) $\binom{3}{1}$

b) $\binom{5}{2}$

c) $\binom{7}{3}$

d) $\binom{4}{4}$

S O L U T I O N

a) $\binom{3}{1} = \dfrac{3!}{1!\,(3-1)!} = \dfrac{3!}{1!2!} = \dfrac{3 \cdot 2 \cdot 1}{(1)(2 \cdot 1)} = \dfrac{6}{1 \cdot 2} = 3$

b) $\binom{5}{2} = \dfrac{5!}{2!(5-2)!} = \dfrac{5!}{2!3!} = \dfrac{5 \cdot 4 \cdot 3 \cdot 2 \cdot 1}{(2 \cdot 1)(3 \cdot 2 \cdot 1)} = \dfrac{120}{2 \cdot 6} = 10$

c) $\binom{7}{3} = \dfrac{7!}{3!(7-3)!} = \dfrac{7!}{3!4!} = \dfrac{7 \cdot 6 \cdot 5 \cdot 4 \cdot 3 \cdot 2 \cdot 1}{(3 \cdot 2 \cdot 1)(4 \cdot 3 \cdot 2 \cdot 1)} = \dfrac{5040}{6 \cdot 24} = 35$

d) $\binom{4}{4} = \dfrac{4!}{4!(4-4)!} = \dfrac{4!}{4!0!} = \dfrac{4 \cdot 3 \cdot 2 \cdot 1}{(4 \cdot 3 \cdot 2 \cdot 1)(1)} = \dfrac{24}{24 \cdot 1} = 1$

■

Bernoulli trials; assignment of probabilities

Bernoulli trials, named after the Swiss mathematician James Bernoulli (1654–1705), are repeated independent trials of an experiment whose outcomes are classified in one of two categories. Let us look at an example.

EXAMPLE 5.13 **Introduces Bernoulli trials**

A roulette wheel consists of 38 numbers, of which 18 are red, 18 are black, and 2 are green. When the roulette ball is spun, it is equally likely to land on any of the 38 numbers.

A gambler is playing roulette and decides to bet on "red" each time. On any given play, there are 38 possible outcomes (the 38 different numbers). However, as far as the gambler is concerned, it is only important whether or not the ball lands on a red number. From this point of view, the possible outcomes on any given play can be classified in one of two categories; namely, either the ball lands on a red number or it doesn't.

There is some standard terminology that is used with Bernoulli trials. We introduce that terminology by relating it to the present example. Each repetition of the basic experiment is called a **trial**. In this case, each play at the wheel constitutes a trial. The two possible outcomes of each trial are designated by *s* and *f,* for **success** and **failure.** Here we let

s = ball lands on a red number
f = ball does not land on a red number

The probability of a success on a given trial, denoted by the letter **p,** is called the **success probability.** In this case,

$$p = P(s) = \frac{18}{38} = 0.474$$

since 18 of the 38 numbers on the wheel are red. The probability of a failure on a given trial is just $1 - p$, which here is $1 - 0.474 = 0.526$.   ∎

With the above example in mind, we now make the following definition.

DEFINITION 5.7 Bernoulli trials

Repeated trials of an experiment are called **Bernoulli trials** if

1 The result of each trial is classified as either the occurrence (success) or non-occurrence (failure) of a specified event.
2 The trials are independent of one another.
3 The probability, p, of a success remains the same from trial to trial.

In Example 5.13, a trial is a single play at the wheel, the specified event is "the ball lands on a red number," and the success probability is $p = 0.474$.

A crucial problem is to determine the probabilities of the possible outcomes in a sequence of Bernoulli trials. To see how this is done, we return to the roulette example.

EXAMPLE   5.14   **Illustrates the calculation of probabilities in Bernoulli trials**

A gambler is playing roulette and is betting on "red" each time. On any given play at the wheel (trial), we will consider the result a success, s, if the ball lands on a red number and a failure, f, if it does not. Determine the probabilities of the possible outcomes in *three* plays at the wheel—a sequence of three Bernoulli trials.

SOLUTION
To begin, we list the possible outcomes in three plays at the wheel. They are

$$sss \quad sfs \quad fss \quad ffs$$
$$ssf \quad sff \quad fsf \quad fff$$

For example, *ssf* means that each of the first two trials results in success (red) and the third results in a failure (not red).

As we see from the list, there are eight possible outcomes in three plays at the wheel. However, these eight possible outcomes are *not* equally likely. To determine the probabilities of these outcomes, we proceed as follows: First of all, as we have seen, the probability of a success on any given trial is

$$p = P(s) = \frac{18}{38} = 0.474$$

and the probability of a failure is $P(f) = 1 - p = 0.526$.

Since the trials are independent (Assumption 2), we can apply the *special multiplication rule* to obtain the probabilities of the eight possible outcomes. For example, the probability of the outcome *ssf* is

$$P(ssf) = P(s)P(s)P(f) = (0.474)(0.474)(0.526) = 0.118$$

(to three decimals) and that for *fsf* is

$$P(fsf) = P(f)P(s)P(f) = (0.526)(0.474)(0.526) = 0.131$$

The remaining probabilities are computed in a similar manner. See Table 5.18.

TABLE 5.18
Outcomes and probabilities
for roulette experiment.

Outcome	Probability
sss	0.106
ssf	0.118
sfs	0.118
sff	0.131
fss	0.118
fsf	0.131
ffs	0.131
fff	0.146

A *tree diagram* is useful to organize and summarize the outcomes and their probabilities. This is given in Figure 5.1.

FIGURE 5.1   Tree diagram corresponding to Table 5.18.

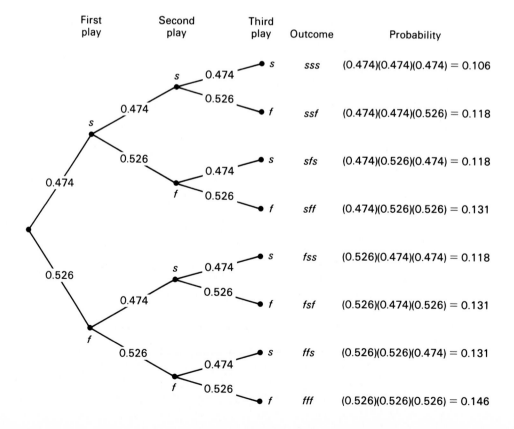

Note that outcomes with the same number of successes have the same probability. For example, there are three outcomes with exactly two successes: *ssf, sfs,* and *fss.* All of these outcomes have the same probability—0.118. This is because each probability is obtained by multiplying two success probabilities of 0.474 and one failure probability of 0.526.   ∎

EXAMPLE 5.15   **Illustrates Bernoulli trials and the calculation of probabilities**

A drug is known to be 80% effective in curing a certain disease. *Four* people with the disease are to be given the drug. Determine the probabilities of the possible "cure-noncure" results.

S O L U T I O N

We can consider the experiment a sequence of *four* Bernoulli trials. If we let a success correspond to the event that a patient is cured, then the success probability is $p = 0.8$ (80%).

There are 16 possible outcomes. These outcomes, along with their probabilities, are presented in Table 5.19. The probabilities are obtained by once again applying the special multiplication rule. For example, *sssf* represents the outcome that the first three patients are cured and the fourth is not. The probability of that outcome is

$$P(sssf) = P(s)P(s)P(s)P(f) = (0.8)(0.8)(0.8)(0.2) = 0.1024$$

TABLE 5.19

Outcomes and probabilities for drug experiment.

Outcome	Probability
ssss	$(0.8)(0.8)(0.8)(0.8) = 0.4096$
sssf	$(0.8)(0.8)(0.8)(0.2) = 0.1024$
ssfs	$(0.8)(0.8)(0.2)(0.8) = 0.1024$
ssff	$(0.8)(0.8)(0.2)(0.2) = 0.0256$
sfss	$(0.8)(0.2)(0.8)(0.8) = 0.1024$
sfsf	$(0.8)(0.2)(0.8)(0.2) = 0.0256$
sffs	$(0.8)(0.2)(0.2)(0.8) = 0.0256$
sfff	$(0.8)(0.2)(0.2)(0.2) = 0.0064$
fsss	$(0.2)(0.8)(0.8)(0.8) = 0.1024$
fssf	$(0.2)(0.8)(0.8)(0.2) = 0.0256$
fsfs	$(0.2)(0.8)(0.2)(0.8) = 0.0256$
fsff	$(0.2)(0.8)(0.2)(0.2) = 0.0064$
ffss	$(0.2)(0.2)(0.8)(0.8) = 0.0256$
ffsf	$(0.2)(0.2)(0.8)(0.2) = 0.0064$
fffs	$(0.2)(0.2)(0.2)(0.8) = 0.0064$
ffff	$(0.2)(0.2)(0.2)(0.2) = 0.0016$

Note again that outcomes with the same number of successes have the same probability. For instance, there are four outcomes with exactly three successes: *sssf, ssfs, sfss,* and *fsss.* All of these outcomes have the same probability—0.1024, since each probability is obtained by multiplying three success probabilities of 0.8 and one failure probability of 0.2.   ∎

Bernoulli trials and sampling from two-category populations

Many statistical studies are concerned with the percentage of a population that has a specified attribute. For instance, we might be interested in the percentage of U.S. households that own a color television set. Here the population consists of all U.S. households, and the specified attribute is "owns a color television set." Or we might want to know the percentage of American businesses that are minority owned. In this case, the population is comprised of all American businesses, and the specified attribute is "minority owned."

A population in which each member is classified as either having or not having a specified attribute is called a **two-category population.** In most applications, the population under consideration will be large (or infinite). Consequently, we usually cannot find the *exact* percentage of a two-category population that has the specified attribute. We must instead *estimate* the percentage from sample data, that is, from the percentage of a sample of the population that has the specified attribute.

If sampling is done *with replacement* (that is, members selected are returned to the population for possible reselection), then the sampling process constitutes Bernoulli trials. Indeed, each selection of a member from the population corresponds to a single trial. A success occurs on a given trial if the member selected in that trial has the specified attribute; otherwise, a failure occurs. The trials are independent of one another since the sampling is done with replacement. The success probability, *p,* remains the same from trial to trial—it always equals the percentage of the population that has the specified attribute.

In reality, however, sampling is usually done *without replacement;* that is, members selected are not returned to the population—no member can be selected more than once. Under these circumstances, the sampling process does *not* constitute Bernoulli trials because the trials are not independent and the success probability varies from trial to trial.

To illustrate our discussion, we will consider a small population; specifically, the students in Professor W's introductory statistics class. Let us take the specified attribute to be "female." From Table 4.13 on page 158, we see that the class consists of 17 males and 23 females. Thus the percentage of the population with the specified attribute is $23/40 = 0.575$, or 57.5%.

As we said, if sampling is done with replacement, then the sampling process constitutes Bernoulli trials. Here a success occurs on a given trial if the student selected is a female; the trials are independent, since the sampling is done with replacement; and the success probability, *p,* always equals 0.575.

On the other hand, suppose the sampling is done without replacement. Then the sampling process does not constitute Bernoulli trials, because the trials are not independent and the success probability varies from trial to trial. For instance, the probability of a success (female) on the first trial (selection) is $23/40 = 0.575$. However, given that the first trial results in a success, the probability of a success on the second trial is $22/39 = 0.564$ (why?).

Consequently, we see that sampling without replacement from a finite two-category population does not constitute Bernoulli trials. Nonetheless, if the sample size is small relative to the size of the population, then for all practical purposes, the sampling process can be *considered* to constitute Bernoulli trials.[†] The reason for this is as follows: If the sample size is small relative to the population size, then there is very little difference between sampling without replacement and sampling with replacement; and, in the latter case, we do indeed have Bernoulli trials. [See Exercise 5.43.] We summarize the above discussion in the following Key Fact.

KEY FACT

Sampling without replacement from a two-category population can be regarded as Bernoulli trials provided that the sample size is small relative to the size of the population.

Exercises **5.3**

5.28 Find 1!, 2!, and 6!.

5.29 Compute 7!, 8!, and 9!.

5.30 Evaluate the following binomial coefficients.

a) $\binom{4}{1}$ b) $\binom{6}{2}$

c) $\binom{8}{3}$ d) $\binom{9}{6}$

5.31 Determine the value of each of the following binomial coefficients.

a) $\binom{5}{2}$ b) $\binom{7}{4}$

c) $\binom{10}{3}$ d) $\binom{12}{5}$

5.32 Determine the value of each of the following binomial coefficients.

a) $\binom{3}{2}$ b) $\binom{6}{0}$

c) $\binom{6}{6}$ d) $\binom{7}{3}$

5.33 Evaluate the following binomial coefficients.

a) $\binom{5}{3}$ b) $\binom{10}{0}$

c) $\binom{10}{10}$ d) $\binom{9}{5}$

5.34 In an ESP experiment, a person in one room randomly selects one of ten cards numbered 1 through 10, and a person in another room tries to guess the number. This experiment is to be repeated *three* times, with the card chosen being replaced before the next selection. Assuming that the person guessing lacks ESP, the probability that he guesses correctly on any given trial is $1/10 = 0.1$. Suppose we consider a success, *s*, on a given trial to be a correct guess. So, for example, *ssf* represents a correct guess on the first two trials and an incorrect guess on the third.

a) What is the success probability, p?
b) Construct a table similar to Table 5.19 on page 195 for the three guesses.
c) Construct a tree diagram for this problem similar to Figure 5.1 on page 194.
d) List the outcomes with exactly one success.
e) What is the probability of each outcome in part (d)? Why are these probabilities all the same?

5.35 Based on data from the *Statistical Abstract of the United States,* the probability that a newborn baby will be a girl is about 0.487. Suppose we consider a success, *s,* in a given birth to be "a girl."

a) What is the success probability, p?
b) Construct a table similar to Table 5.19 on page 195 for the next *three* births.[Display the probabilities to three decimal places.]

[†] A rule of thumb is that Bernoulli trials may be assumed whenever the sample size is *at most* 5% of the population size.

c) Draw a tree diagram for this problem similar to Figure 5.1 on page 194.

d) List the outcomes in which exactly two of the three babies born are girls.

e) What is the probability of each outcome in part (d)? Why are these probabilities all the same?

5.36 According to a survey done by the Opinion Research Corporation, 60% of all clerical workers in the U.S. "like their jobs very much." Suppose *four* randomly selected clerical workers are to be asked whether they like their jobs very much. Let us consider an affirmative response by a given clerical worker to be a success, *s*.

a) What is the success probability, *p*?

b) Construct a table similar to Table 5.19 on page 195 for the possible responses of the four clerical workers. [Display the probabilities to four decimal places.]

c) Draw a tree diagram for this problem similar to Figure 5.1 on page 194.

d) List the outcomes in which exactly two of the four clerical workers respond affirmatively.

e) What is the probability of each outcome in part (d)? Why are these probabilities all the same?

5.37 The National Institute of Mental Health reports that 20% of adult Americans suffer from at least one psychiatric disorder. Suppose *four* randomly selected adult Americans are to be examined for psychiatric disorders.

a) If we let a success, *s*, correspond to an adult American having a psychiatric disorder, then what is the success probability, *p*? [*Note:* The use of the word "success" in Bernoulli trials need not conform to the ordinarily positive connotation of the word.]

b) Construct a table similar to Table 5.19 on page 195 for the four people examined. [Display the probabilities to four decimal places.]

c) Draw a tree diagram for this problem similar to Figure 5.1 on page 194.

d) List the outcomes in which exactly three of the four people examined have a psychiatric disorder.

e) What is the probability of each outcome in part (d)? Why are these probabilities all the same?

5.38 A 1980 telephone company study in Phoenix revealed that the mean duration of residential phone calls is 3.8 minutes and that the probability is about 0.25 that a randomly selected phone call will last

longer than the mean duration of 3.8 minutes. For a given phone call, suppose we consider a success, *s*, to be that the call lasts *at most* 3.8 minutes.

a) What is the success probability, *p*?

b) Construct a table similar to Table 5.19 on page 195 for the possible "success-failure" results of *three* randomly selected calls. [Display the probabilities to three decimal places.]

c) List the outcomes in which exactly two of the three calls last at most 3.8 minutes.

d) What is the probability of each outcome in part (c)? Why are these probabilities all the same?

5.39 According to a study done by the Chicago Title Insurance Company, the probability is 0.768 that a home-buyer will purchase an existing (resale) home. Let a success correspond to the purchase of a *new* (non-resale) home.

a) What is the success probability, *p*?

b) Construct a table similar to Table 5.19 on page 195 for the possible "success-failure" results of *four* randomly selected home purchases. [Display the probabilities to three decimal places.]

c) List the outcomes in which exactly two out of the four purchases are new homes.

d) What is the probability of each outcome in part (c)? Why are these probabilities all the same?

5.40 Show that for any positive integer n, $n! = n(n-1)!$.

5.41 Establish the following equalities:

a) $\dbinom{n}{0} = 1$ b) $\dbinom{n}{n} = 1$ c) $\dbinom{n}{x} = \dbinom{n}{n-x}$

5.42 The **binomial theorem** states that

$$(a+b)^n = \binom{n}{0}a^n b^0 + \binom{n}{1}a^{n-1}b^1 + \cdots + \binom{n}{n}a^0 b^n$$

a) Verify the binomial theorem for $n = 2$ by computing $(a+b)^2 = (a+b)(a+b)$ and showing the result is identical to

$$\binom{2}{0}a^2 b^0 + \binom{2}{1}a^1 b^1 + \binom{2}{2}a^0 b^2$$

b) Repeat part (a) with $n = 3$.

5.43 When doing this exercise, refer to the discussion on pages 196–197 of the relationship between Bernoulli trials and sampling without replacement from a finite, two-category population. Consider the following frequency distribution for students in Professor W's introductory statistics class.

Sex	Frequency
Male	17
Female	23
Σ	40

a) Suppose *two* students are to be selected at random. Find the probability that both students selected are male, if the selection is done
  (i) with replacement. [Use the special multiplication rule.]
  (ii) without replacement. [Use the general multiplication rule.]
b) Compare your two answers in part (a).

Suppose Professor W's class had 10 times the students, but in the same proportions. That is, suppose the frequency distribution were

Sex	Frequency
Male	170
Female	230
Σ	400

c) Repeat parts (a) and (b) above with this second frequency distribution.
d) In which case is there less difference between sampling without and with replacement? Can you explain why this is so?

5.4 The binomial distribution

Now that we have studied binomial coefficients and Bernoulli trials, we are in a position to examine the binomial distribution. The **binomial distribution** is the probability distribution for the *number of successes* in a sequence of Bernoulli trials. We introduce this by returning to the drug experiment of Example 5.15.

EXAMPLE 5.16 Introduces the binomial distribution

A drug is known to be 80% effective in curing a certain disease; that is, the probability is 0.80 that a person with the disease will be cured by the drug. If *four* people with the disease are to be given the drug, find the probability that
  a) exactly two are cured.
  b) exactly three are cured.

SOLUTION
To begin, recall that we can consider the experiment of giving four people the drug a sequence of four Bernoulli trials. If we let a success, *s*, correspond to the event that a patient is cured, then the success probability is

$$p = 0.8$$

and the failure probability is $1 - p = 0.2$. The sixteen possible "cure-noncure" results given in Table 5.19 are repeated here in Table 5.20, on the following page. Using this table, we can solve the two problems posed in parts (a) and (b).
  a) The problem here is to determine the probability that exactly two of the four patients are cured. As we see from Table 5.20, the event that exactly two of the four patients are cured consists of the *six* outcomes *ssff, sfsf, sffs, fssf, fsfs,* and *ffss.* Moreover, these six outcomes have the same probability—0.0256. [This is because each probability is obtained by multiplying two success probabilities of

0.8 and two failure probabilities of 0.2.] Consequently, by the special addition rule, we have

$$P(\text{Exactly two are cured})$$
$$= P(ssff) + P(sfsf) + P(sffs) + P(fssf) + P(fsfs) + P(ffss)$$
$$= 0.0256 + 0.0256 + 0.0256 + 0.0256 + 0.0256 + 0.0256$$
$$= 6 \cdot (0.0256) = 0.1536$$

Thus, the probability that exactly two are cured is 0.1536.

b) For this part we need to find the probability that exactly three of the four patients are cured. From Table 5.20, we observe that the event exactly three of the four patients are cured consists of the *four* outcomes *sssf, ssfs, sfss,* and *fsss*. These outcomes all have the same probability—0.1024. [This is because each probability is obtained by multiplying three success probabilities of 0.8 and one failure probability of 0.2.] Thus by the special addition rule,

$$P(\text{Exactly three are cured})$$
$$= P(sssf) + P(ssfs) + P(sfss) + P(fsss)$$
$$= 0.1024 + 0.1024 + 0.1024 + 0.1024$$
$$= 4 \cdot (0.1024) = 0.4096$$

Consequently, the probability that exactly three are cured is 0.4096.

TABLE 5.20

Outcomes and probabilities for drug experiment.

Outcome	Probability
ssss	$(0.8)(0.8)(0.8)(0.8) = 0.4096$
sssf	$(0.8)(0.8)(0.8)(0.2) = 0.1024$
ssfs	$(0.8)(0.8)(0.2)(0.8) = 0.1024$
ssff	$(0.8)(0.8)(0.2)(0.2) = 0.0256$
sfss	$(0.8)(0.2)(0.8)(0.8) = 0.1024$
sfsf	$(0.8)(0.2)(0.8)(0.2) = 0.0256$
sffs	$(0.8)(0.2)(0.2)(0.8) = 0.0256$
sfff	$(0.8)(0.2)(0.2)(0.2) = 0.0064$
fsss	$(0.2)(0.8)(0.8)(0.8) = 0.1024$
fssf	$(0.2)(0.8)(0.8)(0.2) = 0.0256$
fsfs	$(0.2)(0.8)(0.2)(0.8) = 0.0256$
fsff	$(0.2)(0.8)(0.2)(0.2) = 0.0064$
ffss	$(0.2)(0.2)(0.8)(0.8) = 0.0256$
ffsf	$(0.2)(0.2)(0.8)(0.2) = 0.0064$
fffs	$(0.2)(0.2)(0.2)(0.8) = 0.0064$
ffff	$(0.2)(0.2)(0.2)(0.2) = 0.0016$

∎

As we have seen, Table 5.20 gives the sixteen possible "success-failure" (cure-noncure) outcomes and their probabilities. However, in this and many other examples, it is not really the "success-failure" outcomes that are of interest. Rather it is the total *number of successes*. If we denote this total number of successes (cures) by x, then x is a random variable—its value depends on chance, namely upon how many patients are cured.

The event that "exactly two of the four patients are cured" can be written simply as $\{x = 2\}$. In part (a) of Example 5.16, we determined the probability of that event to be 0.1536. We can express this concisely as $P(x = 2) = 0.1536$ or, more briefly, $P(2) = 0.1536$. This is recorded in the third row of Table 5.21. In part (b) we found that the probability is 0.4096 that exactly three of the four patients are cured. Thus, $P(3) = 0.4096$. This is recorded in the fourth row of Table 5.21.

Using similar reasoning, we can determine the remaining probabilities for x and hence obtain the probability distribution of the random variable x (Table 5.21).

TABLE 5.21

Probability distribution of the number of people cured out of the four given the drug.

Number cured x	Probability $P(x)$
0	0.0016
1	0.0256
2	0.1536
3	0.4096
4	0.4096

The binomial probability formula

In Table 5.21, we displayed the probability distribution of the random variable x—the number of people, out of four, cured by a drug. Although the work done to get that probability distribution is quite involved, it is rather simple compared to most situations. This is because, in practice, the number of trials is generally much larger than four. For instance, if 20 people instead of four were given the drug, then there would be over a million possible outcomes as opposed to 16. In such a case, the tabulation method used in Example 5.16 (Table 5.20) would be totally impractical. Fortunately, there is a *formula* for binomial probabilities. The first step in developing this formula is the following:

KEY FACT

If n Bernoulli trials are performed, then the number of ways to get exactly x successes is equal to the *binomial coefficient* $\binom{n}{x}$. That is, the event of obtaining exactly x successes in n trials consists of $\binom{n}{x}$ outcomes.

We will not stop to prove this fact, but let us quickly check to see that it makes sense for Example 5.16. In part (a) of that example, we found from Table 5.20 that there are *six* outcomes in which exactly two of the four patients are cured: *ssff, sfsf, sffs, fssf, fsfs,* and *ffss.* In other words, the event of exactly two successes ($x = 2$) in four trials ($n = 4$) consists of *six* outcomes. Using binomial coefficients, we can determine that fact without resorting to a direct listing. We have

$$\begin{bmatrix} \text{Number of outcomes} \\ \text{comprising the event} \\ \text{of exactly two cures} \end{bmatrix} = \binom{4}{2} = \frac{4!}{2!(4-2)!} = \frac{4!}{2!2!} = \frac{24}{2 \cdot 2} = 6$$

The same approach holds for part (b) of Example 5.16:

$$\begin{bmatrix} \text{Number of outcomes} \\ \text{comprising the event} \\ \text{of exactly three cures} \end{bmatrix} = \binom{4}{3} = \frac{4!}{3!(4-3)!} = \frac{4!}{3!1!} = \frac{24}{6 \cdot 1} = 4$$

We can now develop a probability formula for the number of successes in Bernoulli trials. We indicate briefly how that formula is derived by referring to Example 5.16. [You may also want to consult Table 5.20 on page 200.] For instance, to determine the probability, $P(x = 3)$, of exactly three cures, we reason as follows: Any *particular* outcome with exactly three cures (for example, *ssfs*) has probability

$$\overset{\text{three cures}}{\underset{\underset{\substack{\text{probability} \\ \text{of a cure}}}{\uparrow}}{(0.8)^3}} \cdot \overset{\text{one noncure}}{\underset{\underset{\substack{\text{probability} \\ \text{of a noncure}}}{\uparrow}}{(0.2)^1}} = 0.512 \cdot 0.2 = 0.1024$$

obtained by multiplying three success probabilities of 0.8 and one failure probability of 0.2. Also, the number of outcomes with exactly three cures is

$$\overset{\text{number of trials}}{\underset{\underset{\text{number of cures}}{\uparrow}}{\binom{4}{3}}} = \frac{4!}{3!(4-3)!} = \frac{24}{6 \cdot 1} = 4$$

Consequently, by the special addition rule, the probability of exactly three cures is

$$P(x = 3) = \binom{4}{3} \cdot (0.8)^3 (0.2)^1 = 4 \cdot (0.1024) = 0.4096$$

Of course, this is the same result obtained in part (b) of Example 5.16. However, this time we determined the probability quickly and easily—no tabulation, no listing. More importantly, the reasoning applies to *any* sequence of Bernoulli trials and leads to the following fundamental formula.

FORMULA 5.2   Binomial probability formula

Suppose n Bernoulli trials are performed, with the probability of success on any given trial being p. Let x denote the total number of successes in the n trials. Then the probability distribution of the random variable x is given by the formula

$$P(x) = \binom{n}{x} p^x (1 - p)^{n-x}$$

The random variable x is said to have the **binomial distribution** *with parameters n and p.*

To determine and apply the binomial probability formula in specific problems, it is useful to have a well-organized strategy. Such a strategy is presented in the following step-by-step procedure.

PROCEDURE 5.1 **To find a binomial probability formula.**

STEP 1 *Identify a success.*
STEP 2 *Determine p, the success probability.*
STEP 3 *Determine n, the number of trials.*
STEP 4 *The binomial probability formula for the number of successes, x, is*

$$P(x) = \binom{n}{x} p^x (1-p)^{n-x}$$

Once you practice the procedure a few times, you will find that binomial probability problems are all quite similar and are relatively easy to solve. Now to some examples.

EXAMPLE 5.17 **Illustrates Procedure 5.1**

According to a 1980 telephone company study in Phoenix, the probability is about 0.25 that a randomly selected phone call will last longer than the mean duration of 3.8 minutes. What is the probability that, out of *three* randomly selected calls,
   a) exactly two last longer than 3.8 minutes?
   b) none last longer than 3.8 minutes?

SOLUTION
To find the indicated probabilities, we first apply Procedure 5.1.

STEP 1 *Identify a success.*

Here we will take a success to be any particular call lasting longer than 3.8 minutes.

STEP 2 *Determine p, the success probability.*

This is the probability that any particular call lasts longer than 3.8 minutes, which is 0.25. So, $p = 0.25$.

STEP 3 *Determine n, the number of trials.*

In this case, the number of trials is the number of calls selected, which is three. Thus, $n = 3$.

STEP 4 *The binomial probability formula for the number of successes, x, is*

$$P(x) = \binom{n}{x} p^x (1-p)^{n-x}$$

Since $n = 3$ and $p = 0.25$, the formula becomes

$$P(x) = \binom{3}{x}(0.25)^x(0.75)^{3-x}$$

Having completed Procedure 5.1, we can now answer the questions posed in (a) and (b).

a) For this part, we want the probability that exactly two calls last longer than 3.8 minutes, that is, the probability of exactly two successes. Applying the formula with $x = 2$ we get

$$P(2) = \binom{3}{2}(0.25)^2(0.75)^{3-2}$$

$$= \frac{3!}{2!(3-2)!}(0.25)^2(0.75)^1 = 0.141$$

The probability is 0.141 that exactly two out of the three randomly selected calls will last longer than 3.8 minutes—about a 14.1% chance.

b) The probability that *none* of the three calls last longer than 3.8 minutes is

$$P(0) = \binom{3}{0}(0.25)^0(0.75)^{3-0}$$

$$= \frac{3!}{0!(3-0)!}(0.25)^0(0.75)^3 = 0.422$$

∎

EXAMPLE 5.18   **Illustrates Procedure 5.1**

A new-car salesperson knows from past experience that, on the average, she will make a sale to about 20% of her customers. What is the probability that, in five randomly selected presentations, she makes a sale to

    a) exactly three customers?
    b) at most one customer?
    c) at least one customer?
    d) Determine the probability distribution for the number of sales.

S O L U T I O N
To answer the questions, we first apply Procedure 5.1.

STEP 1 *Identify a success.*

In this problem a success is a sale to a customer.

STEP 2 *Determine p, the success probability.*

This is the probability that the salesperson makes a sale to any particular customer, which is given as 20%. So, $p = 0.2$.

STEP 3 *Determine n, the number of trials.*

In this case, the number of trials is just the number of customers for which a sale is attempted, namely, five. Thus, $n = 5$.

STEP 4 *The binomial probability formula for the number of successes, x, is*

$$P(x) = \binom{n}{x}p^x(1-p)^{n-x}$$

Since $n = 5$ and $p = 0.2$, the formula becomes

$$P(x) = \binom{5}{x}(0.2)^x(0.8)^{5-x}$$

Now that we have applied Procedure 5.1, it is relatively easy to solve the problems posed in parts (a) through (d).

a) Here we want the probability of exactly three sales, that is, the probability of exactly three successes. Applying the formula with $x = 3$ gives

$$P(3) = \binom{5}{3}(0.2)^3(0.8)^{5-3}$$

$$= \frac{5!}{3!(5-3)!}(0.2)^3(0.8)^2 = 0.051$$

The probability that the salesperson makes exactly three sales in five attempts is 0.051 or 5.1%.

b) The probability of at most one sale is

$$P(x \le 1) = P(0) + P(1)$$

$$= \binom{5}{0}(0.2)^0(0.8)^{5-0} + \binom{5}{1}(0.2)^1(0.8)^{5-1}$$

$$= 0.328 + 0.410 = 0.738$$

c) The probability of at least one sale is $P(x \ge 1)$. This can be computed using the fact that

$$P(x \ge 1) = P(1) + P(2) + P(3) + P(4) + P(5)$$

and applying the binomial probability formula to calculate each of the five terms on the right. However, it is easier to use the complementation rule:

$$P(x \ge 1) = 1 - P(x < 1) = 1 - P(x = 0)$$

$$= 1 - P(0) = 1 - \binom{5}{0}(0.2)^0(0.8)^{5-0}$$

$$= 1 - 0.328 = 0.672$$

Consequently, the salesperson has a 67.2% chance of making at least one sale in five attempts.

d) Finally, we are to determine the probability distribution of the number of sales, x, in five attempts. Thus, we need to compute $P(x)$ for $x = 0, 1, 2, 3, 4$, and 5 using the formula

$$P(x) = \binom{5}{x}(0.2)^x(0.8)^{5-x}$$

This has already been done for $x = 0$, 1, and 3 in parts (a) and (b). For $x = 5$ we have

$$P(5) = \binom{5}{5}(0.2)^5(0.8)^{5-5} = (0.2)^5 = 0.000$$

to three decimal places. [The exact value of $P(5)$ is 0.00032.] Similar computations yield $P(2) = 0.205$ and $P(4) = 0.006$. The probability distribution for the number of sales, x, is as shown in Table 5.22.

TABLE 5.22
Probability distribution of the random variable x—the number of sales in five attempts.

Number of sales x	Probability $P(x)$
0	0.328
1	0.410
2	0.205
3	0.051
4	0.006
5	0.000

MTB
SPSS

■

Binomial probability tables

Because of the importance of the binomial distribution, tables of binomial probabilities have been extensively compiled. Such a table, Table I, can be found in the appendix. In this table, the number of trials, n, is given in the far left column; the number of successes, x, in the next column to the right; and the success probability, p, along the top.

EXAMPLE 5.19 **Illustrates the use of binomial probability tables**

Use Table I to find the probability computed in part (b) of Example 5.18. That is, determine the probability that a new-car salesperson will make at most one sale in five attempts, assuming a success probability of 0.2.

SOLUTION
We have $n = 5$ and $p = 0.2$. The binomial distribution with those two parameters is displayed on the first page of Table I. Using this, it is a simple matter to compute the probability of at most one sale:

$$P(x \le 1) = P(0) + P(1) = 0.328 + 0.410 = 0.738$$

Of course, this is the same answer we obtained in part (b) of Example 5.18 using the binomial probability formula. However, by using Table I, we have significantly reduced the computations required.   ■

As we have just observed, binomial probability tables eliminate most of the computations necessary in dealing with the binomial distribution. But such tables are of limited usefulness because they contain only a relatively small number of different values for n and p. For example, our table has only eleven different values for p and stops at $n = 20$. Consequently, if we want to determine a binomial probability whose n or p parameter is not included in the table, we must either use the binomial probability formula or employ a computer program. The latter method is discussed in Section 5.6.

Other discrete probability distributions

The binomial distribution is the most important and widely used discrete probability distribution. However, many other discrete probability distributions occur regularly in applications. Two significant ones are the *hypergeometric distribution* and the *Poisson distribution*. We will explore the hypergeometric distribution in Exercise 5.59.

Exercises 5.4

5.44 This exercise uses the results of Exercise 5.34 on page 197. [If you have not done that exercise, you should do it before proceeding.] In an ESP experiment, a person in one room randomly selects one of ten cards numbered 1 through 10, and a person in another room tries to guess the number. This experiment is to be repeated *three* times, with the card chosen being replaced before the next selection. Assuming that the person guessing lacks ESP, the probability that he guesses correctly on any given trial is $1/10 = 0.1$. Suppose we consider a success, s, on a given trial to be a correct guess.
   a) Use your results from Exercise 5.34 to determine the probability of exactly one correct guess in three tries.
   b) Find the probability in part (a) by using the binomial probability formula. [Employ Procedure 5.1 on page 203.]

5.45 This exercise uses the results of Exercise 5.37 on page 198. [If you have not done that exercise, you should do it before proceeding with this problem.] According to the National Institute of Mental Health, 20% of adult Americans suffer from a psychiatric disorder. Suppose *four* randomly selected adult Americans are to be examined for psychiatric disorders.
   a) Use your results from Exercise 5.37 to determine the probability that exactly three of the four people examined have a psychiatric disorder.
   b) Find the probability in part (a) by using the binomial probability formula. [Employ Procedure 5.1 on page 203.]

Use Procedure 5.1 on page 203 to solve Exercises 5.46 through 5.49. Express each probability as a decimal rounded to three places.

5.46 According to the *Daily Racing Form,* the probability is about 0.67 that the favorite in a horse race will finish in the money (1st, 2nd, or 3rd). In the next *five* races, what is the probability that the favorite finishes in the money
   a) exactly twice?
   b) exactly four times?
   c) at least four times?
   d) Determine the probability distribution of the random variable x—the number of times the favorite finishes in the money in the next five races.

5.47 Based on data in the *Statistical Abstract of the United States,* the probability that a newborn baby will be a girl is about 0.487. In the next *three* births, what is the probability that
   a) exactly one is a girl?
   b) at most one is a girl?
   c) at least one is a girl?
   d) Determine the probability distribution of the random variable x—the number of girls born in the next three births.

5.48 According to a survey done by the Opinion Research Corporation, 60% of all clerical workers in the U.S. "like their jobs very much." Suppose *four* randomly selected clerical workers are to be asked whether or not they like their jobs very much. Find the probability that the number of affirmative responses is
   a) exactly two.

b) at most two.

c) at least one.

d) Determine the probability distribution of the random variable x—the number of affirmative responses.

5.49 The A.C. Nielsen Company reports that about 85.5% of U.S. households have a color television set. If *six* households are randomly selected, what is the probability that the number of those households that have a color television set is

a) exactly four?

b) at least four?

c) at most five?

d) Determine the probability distribution of the random variable x—the number of households out of six that have a color television set.

For Exercises 5.50 through 5.53 determine the indicated probabilities using Procedure 5.1. If possible, use Table I to check your results.

5.50 According to the U.S. Bureau of the Census, 25% of U.S. children (under 18 years old) are not living with both parents. If 10 U.S. children are selected at random, find the probability that the number not living with both parents is

a) exactly two.

b) at most two.

5.51 The U.S. Energy Information Administration reports that 36.7% of households in the U.S. use automatic dishwashers. Suppose 12 households are randomly selected. What is the probability that

a) exactly three of these households use automatic dishwashers?

b) at most three use automatic dishwashers?

c) at least one uses an automatic dishwasher?

5.52 The *Statistical Abstract of the United States* reports that 71% of U.S. adults have completed high school. Suppose *eight* U.S. adults are selected at random. What is the probability that

a) exactly six have completed high school?

b) at least six have completed high school?

c) at least three have completed high school?

5.53 Surveys indicate that in 80% of Swedish couples both partners work. If *nine* Swedish couples are selected at random, what is the probability that the number of working couples is

a) exactly seven?

b) at least seven?

c) at most nine?

In Exercises 5.54 and 5.55 use Table I to determine the indicated probabilities.

5.54 Studies show that 60% of American families use physical aggression to resolve conflict. If 20 families are selected at random, find the probability that the number that use physical aggression to resolve conflict is

a) exactly 10.

b) between 10 and 15, inclusive.

c) over 75% of those surveyed.

5.55 According to the American Bankers Association, only one in 10 people are dissatisfied with their local bank. If 15 people are randomly selected, what is the probability that the number dissatisfied with their local bank is

a) exactly two?

b) at most two?

c) at least two?

5.56 A sales representative for a tire manufacturer claims that the company's steel-belted radials get at least 35,000 miles. A tire dealer decides to check this claim by testing eight of the tires. If 75% or more of the eight tires he tests get at least 35,000 miles, he will purchase tires from the sales representative. If, in fact, 90% of the steel-belted radials produced by the manufacturer get at least 35,000 miles, what is the probability that the tire dealer will purchase tires from the sales representative?

5.57 From past experience, the owner of a restaurant knows that, on the average, 4% of the parties making reservations never show up. How many reservations can the owner take and still be at least 80% sure that all parties making a reservation will show up?

5.58 Suppose x has the binomial distribution with parameters n and p. Show that

$$P(x \geq 1) = 1 - (1 - p)^n$$

5.59 When sampling *without replacement* from a finite, two-category population, the exact probability distribution for the number of "successes" (that is, the number of members selected having the specified attribute) is called the **hypergeometric distribution.** Its probability formula is

$$P(x) = \frac{\binom{Np}{x}\binom{N(1 - p)}{n - x}}{\binom{N}{n}}$$

Here N is the population size and n is the sample size. Also p is the proportion of "successes" in the population; that is, the percentage of the population that has the specified attribute.

To illustrate, suppose a customer purchases *four* fuses from a shipment of 250, of which 94% are not defective. Let a "success" correspond to a fuse that is not defective.

a) What is N?
b) What is n?
c) What is p?
d) Use the hypergeometric probability formula to find the probability distribution of the random variable x—the number of fuses the customer gets that are not defective.

At the end of Section 5.3 (page 197) we pointed out that sampling without replacement from a two-category population can be treated as a sequence of Bernoulli trials provided the sample size is small relative to the population size. Therefore, in such circumstances, we can use the binomial distribution as a reasonable approximation to the hypergeometric distribution. In particular, we can use the binomial probability formula

$$P(x) = \binom{n}{x} p^x (1-p)^{n-x}$$

with $n = 4$ and $p = 0.94$ to *approximate* the probability distribution of the number of fuses the customer gets that are not defective.

e) Find the probability distribution of the binomial distribution with parameters $n = 4$ and $p = 0.94$.
f) Compare the hypergeometric distribution in part (d) with the binomial distribution in part (e).

5.5 The mean and standard deviation of a binomial distribution

In many of the applications of the binomial distribution it will be necessary for us to know the *mean* and *standard deviation*. To compute these, we can use the formulas

$$\mu_x = \Sigma x P(x) \quad \text{and} \quad \sigma_x = \sqrt{\Sigma x^2 P(x) - \mu_x^2}$$

given on pages 181 and 187 in Section 5.2. Let us look at an example.

EXAMPLE 5.20 Illustrates the calculation of the mean and standard deviation of a binomial distribution

According to the U.S. National Center for Health Statistics, about 60% of all eye operations are performed on females. Suppose *three* people who have had eye operations are selected at random. Let x denote the number out of the three that are female. Then x is a random variable having the *binomial distribution* with parameters $n = 3$ and $p = 0.6$ (why?). Find the mean and standard deviation of the random variable x.

SOLUTION
To apply the formulas for μ_x and σ_x, we first need to determine the probability distribution of the random variable x. Since x has the binomial distribution with parameters $n = 3$ and $p = 0.6$, its probabilities can be computed using the formula

$$P(x) = \binom{3}{x}(0.6)^x (0.4)^{3-x}$$

Applying this formula, we obtained the probability distribution of x. See the first two columns of Table 5.23.

TABLE 5.23

x	$P(x)$	$xP(x)$	x^2	$x^2P(x)$
0	0.064	0.000	0	0.000
1	0.288	0.288	1	0.288
2	0.432	0.864	4	1.728
3	0.216	0.648	9	1.944
Σ		1.800		3.960

From the third column of the table, we conclude that the mean of the random variable x is

$$\mu_x = \Sigma xP(x) = 1.8$$

Using that along with the fifth column of the table, we can compute the standard deviation of x:

$$\sigma_x = \sqrt{\Sigma x^2 P(x) - \mu_x^2} = \sqrt{3.96 - (1.8)^2}$$

$$= \sqrt{0.72} = 0.85 \qquad \blacksquare$$

We just found that, if three eye-operation patients are selected at random and x denotes the number that are female, then the mean of the random variable x is $\mu_x = 1.8$. This indicates that, *on the average,* about 1.8 out of every three eye-operation patients are female. You might have guessed that result before doing any computations if you had used the following reasoning: "60% of all eye-operation patients are female. Thus, out of three eye-operation patients, we would expect about 60% of them to be female. 60% of three is 1.8. Consequently, on the average, we would expect about 1.8 out of every three eye-operation patients to be female." This is the same result obtained for μ_x in Example 5.20. However, here we did almost no calculations. We simply multiplied the number of trials, $n = 3$, by the success probability, $p = 0.6$.

It can be shown that this type of reasoning will always work. That is, the mean of a random variable with a binomial distribution is equal to the number of trials, n, times the success probability, p.

FORMULA 5.3 Mean of a binomial distribution

If x has the *binomial distribution* with parameters n and p, then the mean of the random variable x is

$$\mu_x = np$$

This formula is derived mathematically by substituting the binomial probability formula

$$P(x) = \binom{n}{x}p^x(1 - p)^{n-x}$$

into the formula

$$\mu_x = \Sigma xP(x)$$

and simplifying. The details are somewhat tricky, but the idea is not.

The formula for the standard deviation of a binomially distributed random variable is not intuitively clear like the formula for the mean. Nonetheless it is derived in basically the same way: substitute the formula for $P(x)$ into $\sigma_x = \sqrt{\Sigma x^2 P(x) - \mu_x^2}$ and simplify. The result of doing this is given in Formula 5.4.

FORMULA 5.4 Standard deviation of a binomial distribution

If x has the *binomial distribution* with parameters n and p, then the standard deviation of the random variable x is

$$\sigma_x = \sqrt{np(1-p)}$$

EXAMPLE 5.21 **Illustrates Formulas 5.3 and 5.4**

In Example 5.20 we used the general formulas[†]

$$\mu_x = \Sigma x P(x) \quad \text{and} \quad \sigma_x = \sqrt{\Sigma x^2 P(x) - \mu_x^2}$$

to find the mean and standard deviation of a random variable x—the number of eye operations out of three that are performed on females. Alternatively, since that random variable has the *binomial distribution* (with parameters $n = 3$ and $p = 0.6$), we can instead use the special formulas

$$\mu_x = np \quad \text{and} \quad \sigma_x = \sqrt{np(1-p)}$$

applicable *only* to binomial distributions. Employing the special formulas we get

$$\mu_x = np = 3 \cdot (0.6) = 1.8$$

and

$$\sigma_x = \sqrt{np(1-p)} = \sqrt{3 \cdot (0.6)(0.4)} = \sqrt{0.72} = 0.85$$

The values we just obtained for μ_x and σ_x are, of course, the same as those found in Example 5.20. However, the calculations required here are significantly easier. ∎

As we have seen, the special formulas (Formulas 5.3 and 5.4) are real time-savers in computing the mean and standard deviation of a binomially distributed random variable, even when the number of trials, *n,* is small. In most practical applications n is quite large and, in those cases, the special formulas are indispensable.

EXAMPLE 5.22 **Illustrates Formulas 5.3 and 5.4**

Based on detailed records, a small airline in the southwest has determined that the no-show rate for reservations is about 16%. In other words, the probability is

---

[†] The formulas are *general* in the sense that they can be used to find the mean and standard deviation of *any* discrete random variable.

0.16 that a party making a reservation will not show up. Suppose the next flight has 42 parties with advance reservations. Find the mean and standard deviation of the number of no-shows.

SOLUTION

Let x denote the number of no-shows out of the 42 advance reservations. We need to compute the mean and standard deviation of the random variable x. The random variable x has the binomial distribution with parameters $n = 42$ and $p = 0.16$. Thus,

$$\mu_x = np = 42 \cdot (0.16) = 6.72$$

and

$$\sigma_x = \sqrt{np(1-p)} = \sqrt{42 \cdot (0.16)(0.84)} = 2.38$$

In particular, the airline can expect about 6.72 no-shows on the next flight. ∎

Exercises 5.5

5.60 In an ESP experiment, a person in one room randomly selects one of ten cards numbered 1 through 10, and a person in another room tries to guess the number. This experiment is to be repeated three times, with the card chosen being replaced before the next selection. Assuming that the person guessing lacks ESP, the probability that he guesses correctly on a given trial is $1/10 = 0.1$. Let x denote the number of correct guesses out of the three tries.

a) Determine the mean and standard deviation of the random variable x using the general formulas

$$\mu_x = \Sigma x P(x) \qquad \sigma_x = \sqrt{\Sigma x^2 P(x) - \mu_x^2}$$

[First obtain the probability distribution of x by employing the binomial probability formula $P(x) = \binom{3}{x}(0.1)^x(0.9)^{3-x}$; next construct a table similar to Table 5.23 on page 210; and then apply the general formulas for μ_x and σ_x.]

b) Determine the mean and standard deviation of x using the special formulas

$$\mu_x = np \qquad \sigma_x = \sqrt{np(1-p)}$$

c) Compare the amount of work required in parts (a) and (b).

5.61 According to the National Institute of Mental Health, 20% of all adult Americans suffer from a psychiatric disorder. Suppose *four* adult Americans are to be examined for psychiatric disorders, and let

x denote the number out of the four that have a psychiatric disorder.

a) Determine the mean and standard deviation of the random variable x using the general formulas

$$\mu_x = \Sigma x P(x) \qquad \sigma_x = \sqrt{\Sigma x^2 P(x) - \mu_x^2}$$

[First obtain the probability distribution of x by employing the binomial probability formula $P(x) = \binom{4}{x}(0.2)^x(0.8)^{4-x}$; next construct a table similar to Table 5.23 on page 210; and then apply the general formulas for μ_x and σ_x.]

b) Determine the mean and standard deviation of x using the special formulas

$$\mu_x = np \qquad \sigma_x = \sqrt{np(1-p)}$$

c) Compare the amount of work required in parts (a) and (b).

In Exercises 5.62 through 5.71, use the special formulas

$$\mu_x = np \qquad \sigma_x = \sqrt{np(1-p)}$$

to determine the mean and standard deviation of the given binomially distributed random variable. Interpret your result for the mean in words.

5.62 According to the *Daily Racing Form,* the probability is about 0.67 that the favorite in a horse race will finish in the money (1st, 2nd, or 3rd). Find the mean and standard deviation of the number of favorites showing in the next five races.

5.63 The A.C. Nielsen Company reports that about 85.5% of U.S. households have a color T.V. Find the mean and standard deviation of the number of households in a sample of six that have a color T.V.

5.64 According to the U.S. Bureau of the Census, 25% of U.S. children under 18 years old are not living with both parents. Find the mean and standard deviation of the number of children under 18 out of a sample of 10 that are not living with both parents.

5.65 The U.S. Energy Information Administration reports that 36.7% of U.S. households use automatic dishwashers. If 12 households are selected at random, find the mean and standard deviation of the number that use an automatic dishwasher.

5.66 The *Statistical Abstract of the United States* reports that 71% of U.S. adults have completed high school. If eight U.S. adults are selected at random, find the mean and standard deviation of the number that have completed high school.

5.67 Surveys indicate that in 80% of Swedish couples both partners work. If nine Swedish couples are selected at random, find the mean and standard deviation of the number that both work.

5.68 Studies show that 60% of American families use physical aggression to resolve conflict. If 20 families are selected at random, find the mean and standard deviation of the number using physical aggression to resolve conflict.

5.69 According to the American Bankers Association, only one in 10 people are dissatisfied with their local bank. If 15 people are selected at random, find the mean and standard deviation of the number dissatisfied with their local bank.

5.70 An air conditioner contractor knows from past experience that about 40% of the evaporative coolers he sells are fiberglass (as opposed to metal). Find the mean and standard deviation of the number of fiberglass units sold out of the next 20 evaporative-cooler sales.

5.71 A student takes a multiple-choice test with 50 questions. Each question has four possible answers. If the student guesses at each question, find the mean and standard deviation of the number of questions answered correctly.

5.72 In a sequence of n Bernoulli trials, let x denote the number of successes, and $\bar{p}$ denote the **proportion of successes**; that is, $\bar{p} = x/n$. Use the special formulas for μ_x and σ_x along with the results of part (f) of Exercise 5.27 (page 190) to show that
a) $\mu_{\bar{p}} = p$
b) $\sigma_{\bar{p}} = \sqrt{p(1-p)/n}$
[We will use these formulas extensively in later chapters.]

5.6

Computer packages*

We learned in Section 5.4 that if x is a random variable having a binomial distribution with parameters n and p, then probabilities for x can be computed using the binomial probability formula

$$P(x) = \binom{n}{x} p^x (1-p)^{n-x}$$

We also learned that tables, such as Table I, are available that allow us to simply look up binomial probabilities. Those tables eliminate the computations necessitated by the binomial probability formula, but are limited due to the relatively few values of n and p that are included.

Computer packages gives us the best of both worlds. They eliminate the computations while providing an extensive choice of parameters. [In fact, there is usually no restriction on p.] To obtain a binomial distribution, the user only needs to enter the parameters n and p. Everything else is done by the computer. In what follows, we will discuss Minitab's program for binomial probabilities.

PDF; BINOMIAL

Minitab's program for computing binomial probabilities is called **PDF; BINOMIAL.** We will illustrate the use of that program in the next example.

EXAMPLE 5.23 **Illustrates the use of PDF; BINOMIAL**

In Example 5.18 on page 204, we considered the number of sales by a new-car salesperson in five randomly selected presentations. As we saw, the number of sales, x, has the binomial distribution with parameters $n = 5$ and $p = 0.2$. The probability distribution of x can be determined by applying the binomial probability formula

$$P(x) = \binom{5}{x}(0.2)^x(0.8)^{5-x}$$

with $x = 0, 1, 2, 3, 4$, and 5. Doing that, we obtained Table 5.24.

TABLE 5.24
Probability distribution of the random variable x—the number of sales in five attempts.

Number of sales x	Probability $P(x)$
0	0.328
1	0.410
2	0.205
3	0.051
4	0.006
5	0.000

Using Minitab, we can get the binomial distribution in Table 5.24 without performing any calculations. To do so, we first type the command PDF. Then, on the next line, we type the subcommand BINOMIAL followed by the values of the parameters (N = 5, P = 0.2). These commands and the resulting output are shown in Printout 5.1.

PRINTOUT 5.1
Minitab output for PDF; BINOMIAL.

```
MTB > PDF;
SUBC > BINOMIAL N=5, P=0.2.

BINOMIAL WITH N =    5  P = 0.200000
       K          P( X = K)
       0            0.3277
       1            0.4096
       2            0.2048
       3            0.0512
       4            0.0064
       5            0.0003
```

Printout 5.1 is Minitab's version of Table 5.24.

We can also use Minitab to obtain individual binomial probabilities. For instance, in part (a) of Example 5.18 we computed the probability, $P(3)$, of exactly three sales in five attempts. To get this using Minitab, we proceed as above except that we type a "3" after the command PDF. See Printout 5.2.

PRINTOUT 5.2
Minitab output for PDF 3;
BINOMIAL.

```
MTB > PDF 3;
SUBC > BINOMIAL N=5, P=0.2.
          K            P( X  = K)
          3               0.0512
```

Thus the probability of exactly three sales in five attempts is 0.0512.   ∎

◆ Chapter Review

KEY TERMS

Bernoulli trials, 193
binomial coefficients, 191, 201
binomial distribution, 202
binomial probability formula, 202
discrete random variable, 173
factorial, 191
failure, 192
mean of a random variable, 181
numerical population, 176

PDF; BINOMIAL*, 214
probability distribution, 173
random variable, 172
standard deviation of a random
   variable, 185
success, 192
success probability, 192
trial, 192
two-category population, 196

FORMULAS

Mean of a random variable x, 181

$$\mu_x = \Sigma x P(x)$$

Standard deviation of a random variable x, 185, 187

$$\sigma_x = \sqrt{\Sigma (x - \mu_x)^2 P(x)} \quad \text{or} \quad \sigma_x = \sqrt{\Sigma x^2 P(x) - \mu_x^2}$$

Factorial, 191

$$n! = n(n-1) \cdots 2 \cdot 1$$

Binomial coefficients, 191

$$\binom{n}{x} = \frac{n!}{x!(n-x)!}$$

Binomial probability formula, 202

$$P(x) = \binom{n}{x} p^x (1-p)^{n-x}$$

(n = number of trials, p = success probability)

Mean and standard deviation of a binomially distributed random variable, 210, 211

$$\mu_x = np \qquad \sigma_x = \sqrt{np(1-p)}$$

(n = number of trials, p = success probability)

YOU SHOULD BE ABLE TO . . .

1 use and understand the preceding formulas.

2 determine probability distributions of discrete random variables.

3 describe events using random variable notation, when appropriate.

4 find and interpret the mean and standard deviation of a discrete random variable.

5 define and apply the concept of Bernoulli trials.

6 assign probabilities to the outcomes in a sequence of Bernoulli trials.

7 apply Procedure 5.1 on page 203 to obtain binomial probabilities.

8 use the binomial probability table (Table I in the appendix).

9 compute the mean and standard deviation of a binomially distributed random variable using the special formulas, Formulas 5.3 and 5.4.

REVIEW TEST

1 Here is a frequency distribution for the number of undergraduate students at Arizona State University by class level. [1 = Freshman, 2 = Sophomore, 3 = Junior, and 4 = Senior.]

Class level x	Number of students f
1	5,745
2	6,240
3	7,486
4	10,063
Σ	29,534

Suppose an undergraduate at ASU is selected at random. Let x denote the class level of the student chosen.
a) What are the possible values of the random variable x?
b) Use random variable notation to represent the event that the student selected is a junior (class level 3).
c) Determine $P(x = 3)$ and interpret your results in terms of percentages.
d) Find $P(2)$.
e) Determine the probability distribution of the random variable x.

2 An accounting office has six incoming telephone lines. The probability distribution of the number of busy lines, y, is:

Number busy y	Probability $P(y)$
0	0.052
1	0.154
2	0.232
3	0.240
4	0.174
5	0.105
6	0.043

Use random variable notation to express each of the following events. The number of busy lines is
a) exactly four.
b) at least four.
c) between two and four, inclusive.
d) at least one.
Apply the special addition rule and the above table to determine
e) $P(y = 4)$.
f) $P(y \geq 4)$.
g) $P(2 \leq y \leq 4)$.
h) $P(y \geq 1)$.

3 Refer to the probability distribution in Problem 2.
a) Find the mean of the random variable y.
b) On the average, how many lines are busy?
c) Compute the standard deviation of y using the defining formula.
d) Compute the standard deviation of y using the shortcut formula.

4 Determine 0!, 3!, 4!, and 7!.

5 Find the value of each of the following binomial coefficients:

a) $\binom{8}{3}$ b) $\binom{8}{5}$ c) $\binom{6}{6}$

d) $\binom{10}{2}$ e) $\binom{40}{4}$ f) $\binom{100}{0}$

6 The game of craps is played by rolling a pair of fair dice. A first roll of a sum of seven or eleven wins, and a first roll of two, three, or twelve loses. To win with any other first sum, that sum must be repeated before a seven is thrown. It can be shown that the probability a player wins is 0.493. Suppose we consider a win by the player to be a success, *s*.
a) What is the success probability, *p*?
b) Construct a table giving the possible "win-lose" results and their probabilities for *three* games of craps. Express each probability as a decimal rounded to three places.
c) Draw a tree diagram.
d) List the outcomes in which the player wins exactly two out of the three times.

e) What is the probability of each outcome in part (d)? Why are all these probabilities the same?

7 The nation of Surinam is located on the northern coast of South America. Studies indicate that 80% of the population is literate. Suppose *four* people from Surinam are selected at random. Use the binomial probability formula to find the probability that the number of literate people in the sample is
a) exactly three.
b) at least three.

8 Solve Problem 7 by employing Table I.

9 Refer to Problem 7. Determine the mean and standard deviation of the number of literate people in a sample of four.

10 Strictly speaking, why doesn't the sampling in Problem 7 constitute a sequence of four Bernoulli trials? Why is it permissible to use the binomial distribution?

In this chapter we will introduce what is considered to be the most important probability distribution—the **normal distribution.** The normal distribution is often referred to as the "bell–shaped curve."

In our study of statistics, the normal distribution arises in two significant places. First of all, it has been discovered that many physical measurements involving things such as length, weight, volume, and time have distributions that are bell-shaped. Consequently, it is often appropriate to use the normal distribution as the distribution of a population or random variable. Our second use of the normal distribution pertains to inferential statistics. For example, we will employ the normal distribution in Chapters 8 and 9 when making inferences about a population mean. Additional applications of the normal distribution to statistical inference will be encountered throughout the text.

The normal distribution

CHAPTER OUTLINE

6.1

Introduction; the standard normal curve

In Chapter 5 we examined discrete random variables and discrete probability distributions. The possible values of a discrete random variable form a discrete data set. As we mentioned, a discrete random variable usually involves a *count* of something, such as the number of television sets owned by a randomly selected family.

In contrast to discrete random variables, there are **continuous random variables.** The possible values of a continuous random variable form a continuous data set, usually some interval of real numbers. Generally, a continuous random variable involves a *measurement* of something, like the gas mileage of a car, the height of a person, or the length of time it takes a runner to complete a marathon.

A **continuous probability distribution** is a probability distribution that gives the probabilities for some continuous random variable. The most important continuous probability distribution is the **normal distribution**—the so-called "bell-shaped curve." This chapter is devoted to introducing the normal distribution. In future chapters, we will consider some other widely used continuous probability distributions such as Student's *t*-distribution and Fisher's *F*-distribution.

The standard normal curve

We observe a tremendous variety of populations and random variables in the world around us. Although many populations and random variables are intrinsically different, a large number of them share an important characteristic. Namely, that the probabilities associated with them are equal, at least approximately, to areas under an appropriate "bell-shaped" curve—that is, a curve similar to the one shown at the top of the following page. In this section and the next, we will learn how to find areas under such curves, which are called **normal curves.**

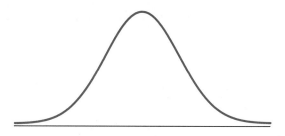

There are many, in fact infinitely many, normal curves. But fortunately there is a way to find areas under any normal curve once we know how to find areas under one specific normal curve—the **standard normal curve.** This curve is shown in Figure 6.1.[†] For convenience, we sometimes refer to the standard normal curve as the **z-curve.**

FIGURE 6.1
The standard normal
curve (*z*-curve).

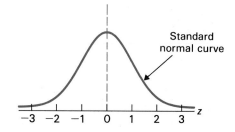

Some important characteristics of the standard normal curve are given below.

KEY FACT Basic properties of the standard normal curve

PROPERTY 1 *The area under the standard normal curve is equal to 1.*[††]
PROPERTY 2 *The standard normal curve extends indefinitely in both directions, approaching the horizontal axis as it does so.*
PROPERTY 3 *The standard normal curve is* **symmetric** *about 0. That is, the part of the curve to the left of the dashed line in Figure 6.1 is the mirror image of the part of the curve to the right of it.*
PROPERTY 4 *Most of the area under the standard normal curve lies between −3 and 3.*

Using the standard normal table

Because of the importance of finding areas under the standard normal curve, tables of those areas have been constructed. We present such a table in Table II,

---

† The equation of the standard normal curve is

$$y = \frac{1}{\sqrt{2\pi}} e^{-z^2/2}$$

where $e \approx 2.718$ and $\pi \approx 3.142$.

†† This is not uniquely a property of the standard normal curve. In fact, the total area under any curve representing a continuous probability distribution is equal to 1.

which can be found in the inside front cover of the book. A typical four-decimal-place number in the *body* of Table II gives the area under the standard normal curve between 0 and the specified value of z.

EXAMPLE   6.1   Illustrates the use of the standard normal table (Table II)

Find the area under the standard normal curve between 0 and $z = 1.83$.

S O L U T I O N
We use Table II. First go down the left-hand column, labelled z, to "1.8." Then go across this row until you are under the "0.03" in the top row. The number in the body of the table there, which is 0.4664, is the area under the standard normal curve between $z = 0$ and $z = 1.83$. See Figure 6.2.   ∎

FIGURE 6.2
Area under the standard
normal curve between
$z = 0$ and $z = 1.83$.

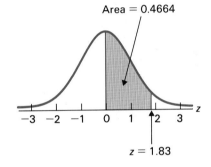

Area = 0.4664

$z = 1.83$

KEY FACT

A number in the body of Table II gives the area under the standard normal curve between 0 and the specified value of z.

To find the area under the standard normal curve between 0 and a *negative* value of z, we apply the fact that the standard normal curve is *symmetric* about 0 (Property 3 on page 220).

**EXAMPLE   6.2   Illustrates how to find the area between 0 and
a negative value of z**

Determine the area under the standard normal curve between $z = -1.45$ and $z = 0$.

S O L U T I O N
We first find the area between $z = 0$ and $z = 1.45$. As in Example 6.1, we go down the column labelled z to "1.4" and then across that row until we are under the "0.05" in the top row. The number there, 0.4265, is the area under the standard normal curve between $z = 0$ and $z = 1.45$. By the *symmetry* of the curve, 0.4265 is also the area between $z = -1.45$ and $z = 0$. We picture the procedure just given in Figure 6.3.   ∎

FIGURE 6.3   Finding the area between $z = -1.45$ and $z = 0$.

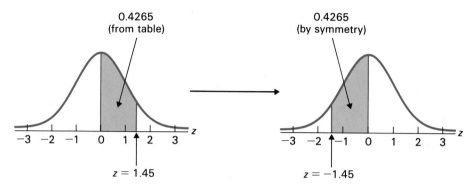

Note: Although z-values can be negative, areas must be positive. In fact, an area under any curve representing a continuous probability distribution, including the standard normal curve, must always be between 0 and 1, inclusive.

Next we will learn how to obtain the area under the standard normal curve to the right of a positive value of z. To accomplish that, we use Properties 1 and 3 on page 220.

EXAMPLE   6.3   Illustrates how to find the area to the right of a positive value of z

Find the area under the standard normal curve to the right of $z = 1.25$.

S O L U T I O N
First of all, since the total area under the curve is 1 (Property 1) and the curve is symmetric about 0 (Property 3), it follows that the area under the curve to the right of $z = 0$ is 0.5. Next, we see from Table II that the area *between* $z = 0$ and $z = 1.25$ is 0.3944. Because the total area to the right of $z = 0$ is 0.5, the area to the right of $z = 1.25$ must be

$$0.5000 - 0.3944 = 0.1056$$

See Figure 6.4.

FIGURE 6.4   Finding the area to the right of $z = 1.25$.

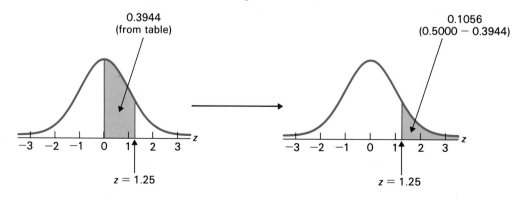

Just one more step will also give us the area under the standard normal curve to the *left* of $z = -1.25$. Applying the symmetry property to the second diagram in Figure 6.4 we obtain Figure 6.5. ■

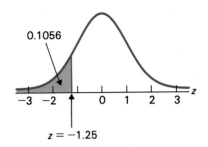

FIGURE 6.5
Area to the
left of $z = -1.25$.

In the following example we will see how to determine the area under the standard normal curve between two positive values of z. A simple subtraction procedure does the trick.

EXAMPLE 6.4 **Illustrates how to find the area between two positive z-values**

Find the area under the standard normal curve between $z = 0.46$ and $z = 1.75$.

SOLUTION
By Table II, the area between $z = 0$ and $z = 0.46$ is 0.1772 and that between $z = 0$ and $z = 1.75$ is 0.4599. See Figure 6.6.

FIGURE 6.6

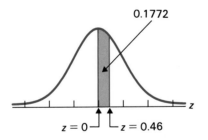
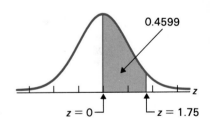

The area between $z = 0.46$ and $z = 1.75$ is the area between $z = 0$ and $z = 1.75$ *less* the area between $z = 0$ and $z = 0.46$. So we subtract:

$$\text{Area} = 0.4599 - 0.1772 = 0.2827$$

See Figure 6.7.

FIGURE 6.7
Area between $z = 0.46$
and $z = 1.75$.

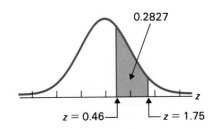

Using the same procedure we can find the area under the standard normal curve between two *negative z*-values. ■

Our next illustration indicates the method for finding the area under the standard normal curve to the left of a positive value of z.

EXAMPLE   6.5   **Illustrates how to find the area to the left of a positive z-value**

Obtain the area under the standard normal curve to the left of $z = 1.96$.

SOLUTION
We begin by determining the area between $z = 0$ and $z = 1.96$. From Table II this area is 0.4750. Since the area to the left of $z = 0$ is 0.5000 (why?), we conclude that the area to the *left* of $z = 1.96$ is

$$0.4750 + 0.5000 = 0.9750$$

This procedure is depicted in Figure 6.8. ■

FIGURE 6.8
Area to the left of $z = 1.96$ is the total shaded area:
$0.4750 + 0.5000 = 0.9750$

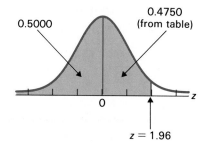

The example below shows how we can compute the area under the standard normal curve between a negative value of z and a positive value of z. This is done by adding two areas from Table II.

EXAMPLE   6.6   **Illustrates how to find the area between a negative z-value and a positive z-value**

Determine the area under the standard normal curve between $z = -1.56$ and $z = 2.12$.

SOLUTION
To obtain the desired area, we simply add the area between $z = -1.56$ and $z = 0$ to the area between $z = 0$ and $z = 2.12$. The area between $z = -1.56$ and $z = 0$ is the same as the area between $z = 0$ and $z = 1.56$, which is 0.4406. The area between $z = 0$ and $z = 2.12$ is 0.4830. Consequently, the area under the standard

normal curve between $z = -1.56$ and $z = 2.12$ is

$$0.4406 + 0.4830 = 0.9236$$

Figure 6.9 summarizes the discussion.   ■

FIGURE 6.9
Area under the standard
normal curve between
$z = -1.56$ and $z = 2.12$ is
the total shaded area:
$0.4406 + 0.4830 = 0.9236$

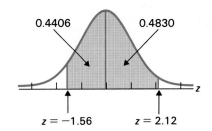

We have not covered every possible normal-curve-area problem. However, based on what you have learned in Examples 6.1 through 6.6, you should be able to use Table II and Properties 1 and 3 on page 220 to determine any required area under the standard normal curve.

MTB
SPSS

Some important areas under the standard normal curve

Property 4 on page 220 states that most of the area under the standard normal curve lies between $z = -3$ and $z = 3$. Using Table II we can find out just how much area that is. The area between $z = 0$ and $z = 3$ is 0.4987; so is the area between $z = -3$ and $z = 0$. Thus, the area under the standard normal curve between $z = -3$ and $z = 3$ is $2 \cdot (0.4987) = 0.9974$. Since the total area under the standard normal curve is equal to 1 (Property 1), we conclude that 99.74% of the area lies between $z = -3$ and $z = 3$. See Figure 6.10.

FIGURE 6.10
Area between $z = -3$
and $z = 3$.

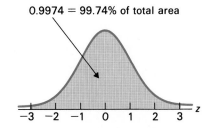

By applying similar arguments, we can get the useful information given in Table 6.1. A visual display of that information is depicted in Figure 6.11.

TABLE 6.1
Some important areas under
the standard normal curve.

z	Area under curve between $-z$ and z	Percentage of total area
1	0.6826	68.26%
2	0.9544	95.44%
3	0.9974	99.74%

FIGURE 6.11 Percentage of area under the standard normal curve that lies between (a) $z = -1$ and $z = 1$, (b) $z = -2$ and $z = 2$, (c) $z = -3$ and $z = 3$.

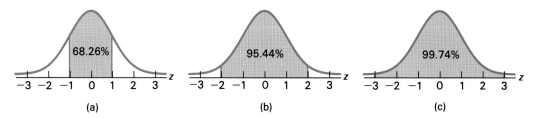

A note concerning Table II, the standard-normal table

The last value given in Table II is for $z = 3.90$. According to the table, the area between $z = 0$ and $z = 3.90$ is 0.5000. This does not mean that the area under the standard normal curve between $z = 0$ and $z = 3.90$ is exactly 0.5, but only that it is 0.5 *to four decimal places.*[†] Since the area to the right of $z = 0$ is exactly 0.5 and the curve extends indefinitely to the right without ever touching the axis, the area between 0 and any value of z will be less than 0.5 no matter how far to the right we go.

Exercises 6.1

Use Table II to find the areas under the standard normal curve specified in Exercises 6.1 through 6.16. Sketch a standard normal curve and shade the area of interest in each problem.

6.1 Find the area under the standard normal curve *between* $z = 0$ and
a) $z = 1.$
b) $z = 2.$
c) $z = 3.$
d) $z = 1.28.$
e) $z = 1.64.$
f) $z = 1.96.$

6.2 Find the area under the standard normal curve *between* $z = 0$ and
a) $z = 2.33.$
b) $z = 2.58.$
c) $z = 3.08.$
d) $z = 0.5.$
e) $z = 1.5.$
f) $z = 2.5.$

6.3 Determine the area under the standard normal curve *between* $z = 0$ and
a) $z = -0.46.$
b) $z = -2.12.$
c) $z = -1.1.$
d) $z = -3.62.$
e) $z = -0.05.$
f) $z = -1.74.$

6.4 Determine the area under the standard normal curve *between* $z = 0$ and
a) $z = -2.35.$
b) $z = -0.12.$
c) $z = -1.7.$
d) $z = -3.04.$
e) $z = -2.$
f) $z = -1.65.$

6.5 Find the area under the standard normal curve to the *right* of
a) $z = 1.64.$
b) $z = 1.96.$
c) $z = 2.33.$
d) $z = 2.58.$

6.6 Find the area under the standard normal curve to the *right* of
a) $z = 1.$
b) $z = 2.$
c) $z = 3.$
d) $z = 1.28.$

6.7 Determine the area under the standard normal curve *between*
a) $z = 1$ and $z = 2.$
b) $z = 2$ and $z = 3.$
c) $z = -1$ and $z = -0.5.$
d) $z = -1.5$ and $z = -1.$

6.8 Determine the area under the standard normal curve *between*
a) $z = 1.5$ and $z = 2.$
b) $z = 2$ and $z = 2.5.$
c) $z = -3$ and $z = -1.$
d) $z = -2$ and $z = -1.$

6.9 Find the area under the standard normal curve to the *left* of
a) $z = 1.78.$
b) $z = 0.94.$
c) $z = 2.4.$
d) $z = 3.1.$

[†] The area to nine decimal places is 0.499951904.

6.10 Find the area under the standard normal curve to the *left* of
a) $z = 2.03$
b) $z = 0.04$.
c) $z = 1.95$.
d) $z = 1.31$.

6.11 Obtain the area under the standard normal curve *between*
a) $z = -1$ and $z = 2$.
b) $z = -2$ and $z = 3$.
c) $z = -1.75$ and $z = 0.58$.
d) $z = -0.75$ and $z = 2.34$.

6.12 Obtain the area under the standard normal curve *between*
a) $z = -2$ and $z = 1$.
b) $z = -3$ and $z = 1$.
c) $z = -1.83$ and $z = 2.12$.
d) $z = -0.63$ and $z = 1.77$.

6.13 Find the area under the standard normal curve to the *left* of
a) $z = -1$.
b) $z = -2$.
c) $z = -1.28$.
d) $z = -1.64$.

6.14 Find the area under the standard normal curve to the *left* of
a) $z = -3$.
b) $z = -1.96$.
c) $z = -2.33$.
d) $z = -2.58$.

6.15 Find the area under the standard normal curve that is
a) to the *right* of $z = -1.38$.
b) either to the *left* of $z = -2.12$ or to the *right* of $z = 1.67$.
c) either to the *left* of $z = 0.63$ or to the *right* of $z = 1.54$.

6.16 Find the area under the standard normal curve that is
a) to the *right* of $z = -1.96$.
b) either to the *left* of $z = -1$ or to the *right* of $z = 2$.
c) either to the *left* of $z = 1$ or to the *right* of $z = 2.5$.

In Exercises 6.17 and 6.18 use Table II to determine the shaded area under the standard normal curve.

6.17 a)

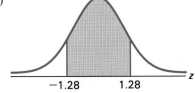

b)

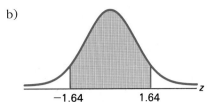

c)

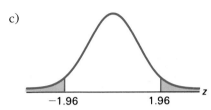

d)

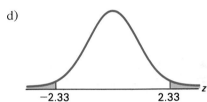

6.18 a)

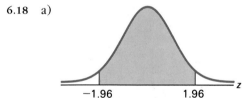

b)

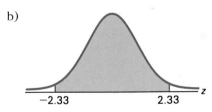

c)

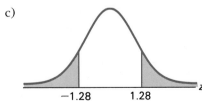

d)

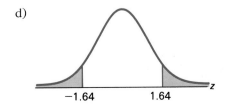

6.19   Verify that the information in Table 6.1 on page 225 is correct. Draw a standard normal curve for each of the three entries and shade the appropriate area.

6.20   Below is a standard normal curve. The total area under the curve is divided into eight regions.

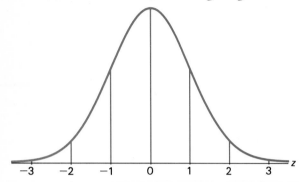

a) Determine the area of each of the eight regions.
b) Complete the following table:

Region	Area	Percentage of total area
$-\infty$ to -3	0.0013	0.13
-3 to -2		
-2 to -1		
-1 to 0		
0 to 1	0.3413	34.13
1 to 2		
2 to 3		
3 to ∞		
Σ	1.0000	100.00

6.2   Areas under normal curves

Recall that it is important to be able to determine areas under normal curves, since these areas correspond to probabilities for many populations and random variables. In the previous section we learned how to find areas under the *standard* normal curve. With that knowledge we can obtain areas under *any* normal curve, as we shall discover in this section.

Each normal curve can be identified by two **parameters,** usually denoted by μ and σ. The parameter μ tells us where the normal curve is *centered,* and the parameter σ indicates its *spread* (shape).[†] Two normal curves with the same μ will be centered at the same place, and two normal curves with the same σ will have the same shape. Below are three normal curves. The normal curve on the left has parameters $\mu = -2$ and $\sigma = 1$; the one in the center has parameters $\mu = 3$ and $\sigma = 1/2$; and the one on the right has parameters $\mu = 8$ and $\sigma = 2$.

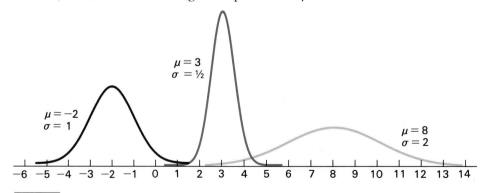

[†] The equation of the normal curve with parameters μ and σ is

$$y = \frac{1}{\sqrt{2\pi}\sigma} e^{-(x-\mu)^2/2\sigma^2}$$

Note that (1) the normal curves are centered at μ and (2) the larger the value of σ, the more the curve is spread out. Some of the more important properties of normal curves are as follows:

KEY FACT Basic properties of normal curves

PROPERTY 1 *The area under any normal curve is 1.*
PROPERTY 2 *A normal curve extends indefinitely in both directions, approaching the horizontal axis as it does so.*
PROPERTY 3 *A normal curve with parameters μ and σ is **symmetric** about μ. That is, the part of the curve to the left of μ is the mirror image of the part of the curve to the right of μ.*
PROPERTY 4 *Most of the area under the normal curve with parameters μ and σ lies between $\mu - 3\sigma$ and $\mu + 3\sigma$.*

Sketching normal curves

Although it is not necessary to draw normal curves exactly, it is useful to be able to sketch them. Using the properties of normal curves, it is easy to make a quick sketch. We illustrate how to do this in the following example.

EXAMPLE   6.7   Illustrates sketching normal curves

Sketch the normal curve with parameters
  a) $\mu = 5$ and $\sigma = 2$.
  b) $\mu = 0$ and $\sigma = 1$.

SOLUTION
  a) By Property 3, the normal curve with parameters $\mu = 5$ and $\sigma = 2$ is symmetric about $\mu = 5$. Also, by Property 4, most of the area under that normal curve lies between

$$\mu - 3\sigma = 5 - 3 \cdot 2 = -1$$

and

$$\mu + 3\sigma = 5 + 3 \cdot 2 = 11$$

So, we sketch this curve as shown in Figure 6.12.

FIGURE 6.12
Sketch of normal
curve with
parameters $\mu = 5$
and $\sigma = 2$.

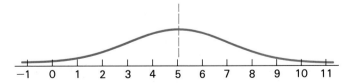

  b) Next we sketch the normal curve with parameters $\mu = 0$ and $\sigma = 1$. By Property 3, we know that this normal curve is symmetric about $\mu = 0$ and, by Property 4, most of the area under this normal curve lies between

$$\mu - 3\sigma = 0 - 3 \cdot 1 = -3$$

and

$$\mu + 3\sigma = 0 + 3 \cdot 1 = 3$$

Thus, we can sketch this curve as pictured in Figure 6.13.   ■

FIGURE 6.13
Sketch of normal curve with
parameters $\mu = 0$ and $\sigma = 1$.

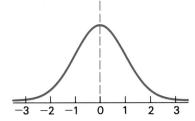

The curve in Figure 6.13 should look familiar. It is the standard normal curve. (See Figure 6.1 on page 220.) In other words:

KEY FACT

The normal curve with parameters $\mu = 0$ and $\sigma = 1$ is the *standard normal curve.*

Finding areas under normal curves

We will now proceed to the main business of this section—how to find areas under normal curves. The basic idea is to relate any given normal curve to the standard normal curve.

EXAMPLE   6.8   Illustrates how to find areas under a normal curve

Determine the area under the normal curve with parameters $\mu = 5$ and $\sigma = 2$ that lies
   a) to the right of 7.
   b) between 3 and 8.

S O L U T I O N
   a) We begin by sketching the normal curve with parameters $\mu = 5$ and $\sigma = 2$ and shading the area to the right of 7. See Figure 6.14.

FIGURE 6.14   Normal curve with parameters $\mu = 5$ and $\sigma = 2$, with area to the right of 7 shaded.

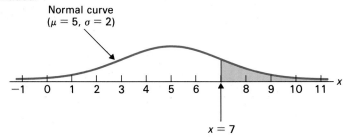

Note that we have labelled the horizontal axis in Figure 6.14 with an "x." This will help us to keep things straight, as you will see.

It can be shown mathematically (see Exercise 6.40) that the area under the normal curve with parameters $\mu = 5$ and $\sigma = 2$ that lies to the right of $x = 7$ is equal to the area under the *standard* normal curve that lies to the right of

$$z = \frac{x - \mu}{\sigma} = \frac{7 - 5}{2} = 1$$

This latter area is pictured in Figure 6.15.

FIGURE 6.15

Standard normal curve, with area to the right of 1 shaded.

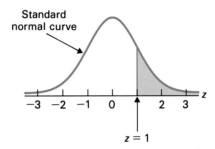

In other words, the shaded areas in Figures 6.14 and 6.15 are equal. By Table II, the area shaded in Figure 6.15 is

$$0.5000 - 0.3413 = 0.1587$$

Thus, the area shaded in Figure 6.14 is also 0.1587. That is, the area under the normal curve with parameters $\mu = 5$ and $\sigma = 2$ that lies to the right of 7 equals 0.1587.

b) Here we need to determine the area under the normal curve with parameters $\mu = 5$ and $\sigma = 2$ that lies between 3 and 8. See Figure 6.16.

FIGURE 6.16 Normal curve with parameters $\mu = 5$ and $\sigma = 2$, with area between 3 and 8 shaded.

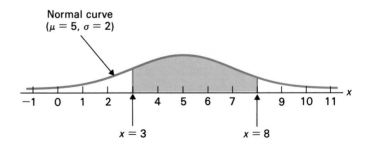

We obtain this area as follows: The area under the normal curve with parameters $\mu = 5$ and $\sigma = 2$ that lies between $x = 3$ and $x = 8$ is equal to the area under the *standard* normal curve that lies between

$$z = \frac{x - \mu}{\sigma} = \frac{3 - 5}{2} = -1$$

and

$$z = \frac{x - \mu}{\sigma} = \frac{8 - 5}{2} = 1.5$$

This latter area is shown in Figure 6.17.

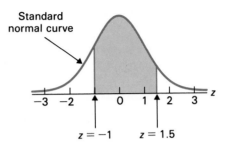

FIGURE 6.17
Standard normal
curve, with area
between −1 and
1.5 shaded.

By Table II, the shaded area in Figure 6.17 equals

$$0.3413 + 0.4332 = 0.7745$$

Consequently, the shaded area in Figure 6.16 also equals 0.7745.   ∎

z-scores

As indicated in Example 6.8, finding areas under normal curves entails converting x-values to z-values by subtracting μ and then dividing by σ:

$$z = \frac{x - \mu}{\sigma}$$

This process is often referred to as **standardizing.** The value z is called the **standard score** or **z-score** for the value x.

We now summarize the fundamental fact that permits us to determine areas under any normal curve by using the Table II values of areas under the standard normal curve.

KEY FACT

The area under the normal curve with parameters μ and σ that lies between $x = a$ and $x = b$ is equal to the area under the *standard* normal curve that lies between

$$z = \frac{a - \mu}{\sigma} \quad \text{and} \quad z = \frac{b - \mu}{\sigma}$$

See Figure 6.18.

Areas under the normal curve with parameters μ and σ that lie to the right (or left) of a given x-value are found similarly; namely, by converting to z-scores and then finding the area under the *standard* normal curve that lies to the right (or left) of the z-score.

FIGURE 6.18

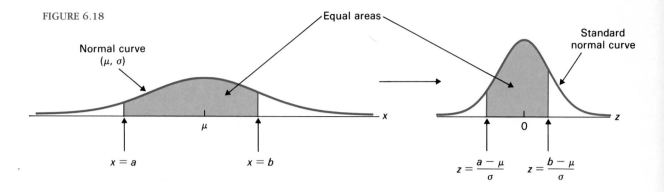

FIGURE 6.18

A streamlined procedure for finding normal curve areas

In Example 6.8, pages 230–232, we drew *two* diagrams to solve each area problem—one for the normal curve with parameters $\mu = 5$ and $\sigma = 2$ and one for the standard normal curve. Generally, the "two-diagram approach" for solving normal curve area problems is as depicted in Figure 6.18. Although this approach is quite useful for explaining and understanding the necessary ideas, it is somewhat cumbersome and lengthy. Here we present a streamlined procedure that is faster and simpler to use.

EXAMPLE   6.9   Illustrates a streamlined procedure for finding normal curve areas

Determine the area under the normal curve with parameters $\mu = 3$ and $\sigma = 4$ that lies between -6 and 9.

SOLUTION

STEP 1 *Sketch the normal curve with parameters μ and σ.*

Here $\mu = 3$ and $\sigma = 4$. We have sketched this normal curve in Figure 6.19. Note that the tick marks are $\sigma = 4$ units apart.

FIGURE 6.19

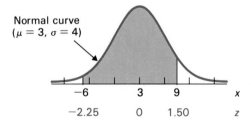

z-score computations:

$$x = -6 \longrightarrow z = \frac{-6 - 3}{4} = -2.25$$

$$x = 9 \longrightarrow z = \frac{9 - 3}{4} = 1.50$$

Area between 0 and z:

0.4878

0.4332

Shaded area $= 0.4878 + 0.4332 = 0.9210$

STEP 2 *Indicate on the graph the area to be determined.*

See the shaded area in Figure 6.19.

STEP 3 *Compute the required z-scores, and mark them on the graph beneath the x-values.*

$$x = -6 \quad \longrightarrow \quad z = \frac{-6 - 3}{4} = -2.25$$

$$x = 9 \quad \longrightarrow \quad z = \frac{9 - 3}{4} = 1.50$$

We have marked the z-scores beneath the x-values in Figure 6.19.

STEP 4 *Use Table II to obtain the desired area.*

The area under the standard normal curve between $z = -2.25$ and $z = 0$ is 0.4878, and the area under the standard normal curve between $z = 0$ and $z = 1.50$ is 0.4332. Consequently, the desired area (the area shaded in Figure 6.19) equals

$$0.4878 + 0.4332 = 0.9210$$

MTB
SPSS

This entire four-step procedure can be accomplished quickly and easily by performing all the steps in a "picture" as in Figure 6.19.   ■

The step-by-step procedure, illustrated in Example 6.9, is given as Procedure 6.1. Remember that this procedure can be carried out most efficiently by organizing your work in the manner shown in Figure 6.19.

PROCEDURE 6.1 To determine areas under the normal curve with parameters μ and σ.

STEP 1 *Sketch the normal curve with parameters μ and σ.*
STEP 2 *Indicate on the graph the area to be determined.*
STEP 3 *Compute the required z-scores, and mark them on the graph beneath the x-values.*
STEP 4 *Use Table II to obtain the desired area.*

EXAMPLE 6.10 **Illustrates Procedure 6.1**

Determine the area under the normal curve with parameters $\mu = 100$ and $\sigma = 16$ that lies to the right of 120.

S O L U T I O N
The results of applying Procedure 6.1 are shown in Figure 6.20 at the top of the following page.

Thus, the desired area is 0.1056.   ■

FIGURE 6.20

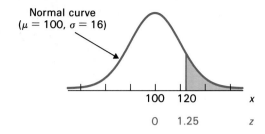

z-score computation:

$$x = 120 \longrightarrow z = \frac{120 - 100}{16} = 1.25$$

Area between 0 and z:

0.3944

Shaded area $= 0.5000 - 0.3944 = 0.1056$

A note concerning rounding

As you can see, the z-values in Table II are given to two decimal places. Thus, when computing z-scores, always round the resulting z-values to two decimal places.

Exercises 6.2

6.21   Which normal curve has a wider spread, the one with parameters $\mu = 1$ and $\sigma = 2$ or the one with parameters $\mu = 2$ and $\sigma = 1$?

6.22   Is the following statement true or false? "The normal curve with parameters $\mu = -4$ and $\sigma = 3$ and the normal curve with parameters $\mu = 6$ and $\sigma = 3$ have the same shape." Explain your answer.

6.23   Answer true or false and give a reason for your answer. "The value of the parameter μ has no effect on the shape of a normal curve."

6.24   Answer true or false and give a reason for your answer. "The normal curve with parameters $\mu = -4$ and $\sigma = 3$ and the normal curve with parameters $\mu = 6$ and $\sigma = 3$ are centered at the same place."

6.25   Apply the procedure illustrated in Example 6.7 on page 229 to sketch the normal curve with parameters
a) $\mu = 3$ and $\sigma = 3$.
b) $\mu = 1$ and $\sigma = 3$.
c) $\mu = 3$ and $\sigma = 1$.

6.26   Use the procedure illustrated in Example 6.7 on page 229 to sketch the normal curve with parameters
a) $\mu = -2$ and $\sigma = 2$.

b) $\mu = -2$ and $\sigma = 1/2$.
c) $\mu = 0$ and $\sigma = 2$.

In Exercises 6.27 through 6.30, use the *two-diagram* procedure (portrayed in Figure 6.18 on page 233) to find the designated areas.

6.27   Find the area under the normal curve with parameters $\mu = 1$ and $\sigma = 2.5$ that lies
a) to the right of 0.
b) to the left of -1.5.
c) between -2 and 2.

6.28   Find the area under the normal curve with parameters $\mu = -1.5$ and $\sigma = 1$ that lies
a) between 0 and 1.4.
b) to the left of -1.5.
c) to the right of 1.

6.29   Find the area under the normal curve with parameters $\mu = 2$ and $\sigma = 1/2$ that lies
a) to the left of 2.87.
b) between 1 and 1.5.
c) to the right of 2.75.

6.30   Find the area under the normal curve with parameters $\mu = 3$ and $\sigma = 0.75$ that lies
a) to the left of 2.5.
b) between 2 and 4.
c) to the right of 1.5.

In Exercises 6.31 through 6.36, use Procedure 6.1 (page 234) to determine the indicated areas.

6.31 Find the area under the normal curve with parameters $\mu = 74$ and $\sigma = 2$ that lies
a) between 71 and 78.
b) to the right of 76.5.

6.32 Obtain the area under the normal curve with parameters $\mu = 7.3$ and $\sigma = 2$ that lies
a) between 4.3 and 11.8.
b) to the left of 9.94.

6.33 Determine the area under the normal curve with parameters $\mu = 335$ and $\sigma = 10$ that lies
a) to the left of 348.5.
b) between 340 and 350.

6.34 Find the area under the normal curve with parameters $\mu = 64.4$ and $\sigma = 2.4$ that lies
a) to the right of 70.
b) between 68 and 71.

6.35 Find the area under the normal curve with parameters $\mu = 40.9$ and $\sigma = 7.1$ that lies
a) to the left of 35.
b) between 25 and 30.

6.36 Determine the area under the normal curve with parameters $\mu = 15.6$ and $\sigma = 5.1$ that lies
a) to the left of 5.
b) between 3.42 and 10.

6.37 Find the indicated area under the normal curve with the given parameters. [Use Procedure 6.1 on page 234.]
a) $\mu = 0$, $\sigma = 1$; area between -3 and 3.
b) $\mu = 4$, $\sigma = 2$; area between -2 and 10.
c) $\mu = \mu$, $\sigma = \sigma$; area between $\mu - 3\sigma$ and $\mu + 3\sigma$.

6.38 a) Complete the following table for the normal curve with parameters $\mu = 4$ and $\sigma = 3$. Draw three pictures illustrating your results.

Given x-values	Corresponding z-scores	Area between x-values
1 & 7	−1 & 1	0.6826
−2 & 10	−2 & 2	
−5 & 13		

b) Complete the following table for the normal curve with parameters μ and σ, and draw three pictures illustrating your results.

Given x-values	Corresponding z-scores	Area between x-values
$\mu - \sigma$ & $\mu + \sigma$	−1 & 1	0.6826
$\mu - 2\sigma$ & $\mu + 2\sigma$		
$\mu - 3\sigma$ & $\mu + 3\sigma$		

6.39 a) Below is a sketch of the normal curve with parameters $\mu = 4$ and $\sigma = 3$. The total area under the curve is divided into eight regions. We determined the area of the region between 7 and 10 and recorded that area on the picture in color. Find and record the z-scores and areas for the remaining regions.

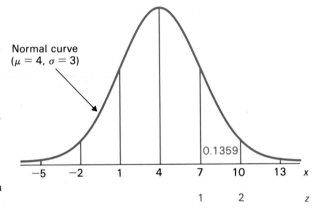

Normal curve
($\mu = 4$, $\sigma = 3$)

0.1359

b) Repeat part (a) for the normal curve with parameters μ and σ.

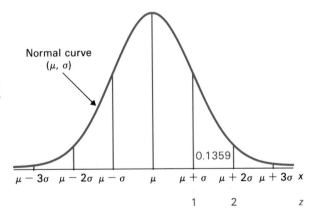

Normal curve
(μ, σ)

0.1359

6.40 This exercise requires elementary calculus and justifies the procedure described on page 232.
a) Show that the area under the normal curve with

parameters μ and σ that lies between $x = a$ and $x = b$ equals

$$\int_a^b \frac{1}{\sqrt{2\pi}\sigma} e^{-(x-\mu)^2/2\sigma^2} \, dx$$

[See the footnote on page 228.]

b) Making the substitution $z = (x - \mu)/\sigma$, show that the integral in part (a) equals

$$\int_{(a-\mu)/\sigma}^{(b-\mu)/\sigma} \frac{1}{\sqrt{2\pi}} e^{-z^2/2} \, dz$$

c) What area does the integral in part (b) equal? [See the footnote on page 220.]

6.3 Normally distributed populations

As we mentioned in Section 6.1, many populations have probability distributions that can be represented by normal curves. This means that percentages for the population are equal, at least approximately, to areas under a suitable normal curve. The following example should help make this idea clear.

EXAMPLE 6.11 **Illustrates a normally distributed population**

A midwestern college has an enrollment of 3264 female students. Records show that the *population* mean height of these students is 64.4 inches, with a standard deviation of 2.4 inches:

$$\mu = 64.4 \text{ inches} \qquad \sigma = 2.4 \text{ inches}$$

A frequency and relative-frequency distribution for this population of heights is given in Table 6.2.

TABLE 6.2
Heights of the 3264
female students.

Height (inches)	Frequency f	Relative frequency
56–under 57	3	0.0009
57–under 58	6	0.0018
58–under 59	26	0.0080
59–under 60	74	0.0227
60–under 61	147	0.0450
61–under 62	247	0.0757
62–under 63	382	0.1170
63–under 64	483	0.1480
64–under 65	559	0.1713
65–under 66	514	0.1575
66–under 67	359	0.1100
67–under 68	240	0.0735
68–under 69	122	0.0374
69–under 70	65	0.0199
70–under 71	24	0.0074
71–under 72	7	0.0021
72–under 73	5	0.0015
73–under 74	1	0.0003
Σ	3264	1.0000

Table 6.2 shows, for instance, that 7.35% (0.0735) of the female students are at least 67 inches tall but less than 68 inches tall. The relative-frequency histogram for these heights is displayed in Figure 6.21.

FIGURE 6.21   Relative-frequency histogram for heights of female students.

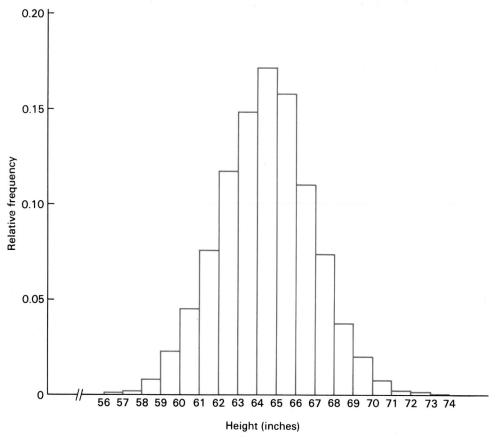

You can see that the histogram in Figure 6.21 is bell-shaped. Because of this, we will be able to approximate percentages (relative frequencies, probabilities) for the population by using areas under a suitable normal curve. As you might guess, the appropriate normal curve is the one with parameters μ and σ, where μ is the mean of the population and σ is its standard deviation. Here $\mu = 64.4$ and $\sigma = 2.4$ (see page 237). In Figure 6.22 we have superimposed the normal curve with parameters $\mu = 64.4$ and $\sigma = 2.4$ upon the histogram in Figure 6.21.

Now, let us see just how we can approximate percentages for this population of heights by using areas under the normal curve with parameters $\mu = 64.4$ and $\sigma = 2.4$. To be specific, let us consider the percentage of female students who are between 67 and 68 inches tall (that is, in the "67–under 68" class). According to Table 6.2 on page 237, the *exact* percentage is 7.35%, or 0.0735. Note that 0.0735 also equals the area of the shaded rectangle in Figure 6.22. This is because the height of the rectangle is 0.0735 and the width of the rectangle is 1.

FIGURE 6.22   Relative-frequency histogram for heights with superimposed normal curve.

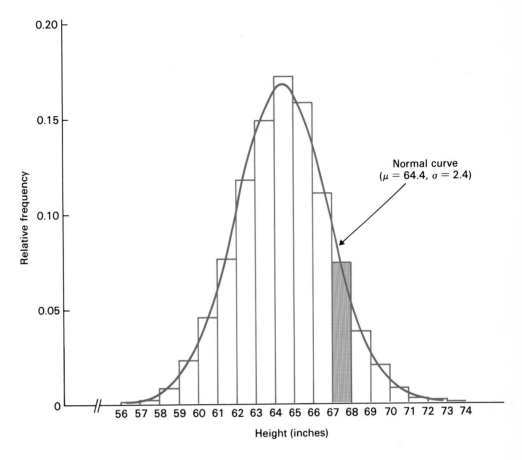

We now examine the parts of the histogram and normal curve in Figure 6.22 between 67 and 68. This is shown in Figure 6.23.

FIGURE 6.23

We have shaded the area under the normal curve between 67 and 68. As you can see from the picture, the area under the normal curve is very close to the area of the rectangle. But we have also noted that the area of the rectangle equals the percentage of students who are between 67 and 68 inches tall. Consequently, *we can approximate the percentage of students who are between 67 and 68 inches tall by the area under the normal curve between 67 and 68.*

Although the main point of this example has now been made, it is interesting to actually compare the numerical values under consideration here. We already know that the percentage of students between 67 and 68 inches tall is exactly 7.35%. To determine the area under the normal curve between 67 and 68, we proceed as explained in Section 6.2. See Figure 6.24.

FIGURE 6.24
Determination of the area under the normal curve with parameters $\mu = 64.4$ and $\sigma = 2.4$ that lies between 67 and 68.

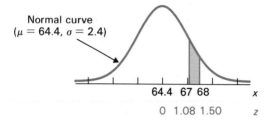

z-score computations:

$$x = 67 \longrightarrow z = \frac{67 - 64.4}{2.4} = 1.08$$

Area between 0 and z:

0.3599

$$x = 68 \longrightarrow z = \frac{68 - 64.4}{2.4} = 1.50$$

0.4332

Shaded area = 0.4332 − 0.3599 = 0.0733

So the area under the normal curve between 67 and 68 is 0.0733 or 7.33%, whereas the exact percentage of female students between 67 and 68 inches tall is 7.35%. The approximation using the area under the normal curve gives excellent results. ∎

The important point of Example 6.11 is that for certain populations, percentages are approximately equal to areas under a suitable normal curve. Not all populations have that property but, if a population does, then we say that it is **normally distributed.**

DEFINITION 6.1 Normally distributed population

A population is said to be (approximately) **normally distributed** if percentages for the population are (approximately) equal to areas under a normal curve. If such a population has mean μ and standard deviation σ, then the normal curve used is the one with parameters μ and σ.

It is a good idea to keep in mind what it means *qualitatively* for a population to be normally distributed; namely, that the histogram of the population is bell-shaped.

There are many instances where it is *known* that a population is normally distributed on the basis of past experience, previous statistical studies, or theoretical considerations. In these cases it is simple to determine any desired percentage. We illustrate such a case in the next example.

EXAMPLE 6.12 **Illustrates how to determine percentages for a normally distributed population**

In a 1905 study, R. Pearl determined that brain weights of Swedish men are approximately *normally distributed* with a mean weight of $\mu = 1.40$ kg (kilograms) and a standard deviation of $\sigma = 0.11$ kg.[†] What percentage of Swedish men have brain weights of

a) between 1.50 and 1.70 kg?
b) less than 1.20 kg?

SOLUTION
Since the brain weights are normally distributed, percentages are equal to areas under a normal curve; namely, the normal curve whose parameters are the same as the mean and standard deviation of the population—in this case $\mu = 1.40$ and $\sigma = 0.11$.

a) Here we want the percentage of Swedish men with brain weights between 1.50 and 1.70 kg. To obtain this, we need to determine the area under the normal curve with parameters $\mu = 1.40$ and $\sigma = 0.11$ that lies between 1.50 and 1.70. That area is computed in Figure 6.25.

FIGURE 6.25
Determination of the area under the normal curve with parameters $\mu = 1.40$ and $\sigma = 0.11$ that lies between 1.50 and 1.70.

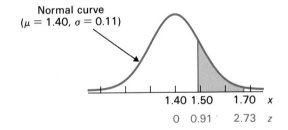

Normal curve
($\mu = 1.40$, $\sigma = 0.11$)

1.40 1.50   1.70 x

0   0.91   2.73 z

z-score computations: Area between 0 and z:

$x = 1.5 \longrightarrow z = \dfrac{1.50 - 1.40}{0.11} = 0.91$ 0.3186

$x = 1.7 \longrightarrow z = \dfrac{1.70 - 1.40}{0.11} = 2.73$ 0.4968

Shaded area = 0.4968 − 0.3186 = 0.1782

[†] Pearl, R. 1905. Biometrical studies on man. I. Variation and correlation in brain weight. *Biometrica,* vol. 4, pp. 13–104.

Thus, we see that 17.82% (0.1782) of Swedish men have brain weights between 1.50 and 1.70 kg.

b) To determine the percentage of Swedish men with brain weights less than 1.20 kg, we need to find the area under the normal curve with parameters $\mu = 1.40$ and $\sigma = 0.11$ that lies to the left of 1.20. This is done in Figure 6.26.

FIGURE 6.26
Determination of the area under the normal curve with parameters $\mu = 1.40$ and $\sigma = 0.11$ that lies to the left of 1.20.

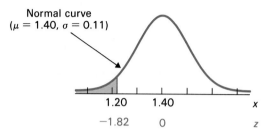

Normal curve
($\mu = 1.40$, $\sigma = 0.11$)

1.20   1.40 x
−1.82   0 z

z-score computation: Area between 0 and z:

$x = 1.2 \longrightarrow z = \dfrac{1.20 - 1.40}{0.11} = -1.82$ 0.4656

Shaded area $= 0.5000 - 0.4656 = 0.0344$

Consequently, only about 3.44% of Swedish men have brain weights less than 1.20 kg.   ■

Interpretation and application of z-scores for normally distributed populations

We introduced z-scores as a method of finding areas under normal curves. But what does the z-score of a *population value* represent? To answer this question, we consider the following example.

EXAMPLE 6.13   **Illustrates the meaning of the z-score of a population value**

Recall from Example 6.12 that the brain weights of Swedish men are normally distributed with a mean of $\mu = 1.40$ kg and a standard deviation of $\sigma = 0.11$ kg. Let us compute the z-score for a particular brain weight, say 1.62 kg. The z-score for that population value is

$$z = \frac{x - \mu}{\sigma} = \frac{1.62 - 1.40}{0.11} = \frac{0.22}{0.11} = 2$$

This indicates that a brain weight of 1.62 kg is *two standard deviations* away from the mean of $\mu = 1.40$ kg. To understand this clearly, reason as follows. A brain weight of 1.62 kg is

$$1.62 - \mu = 1.62 - 1.40 = 0.22 \text{ kg}$$

away from the mean. Since $\sigma = 0.11$ kg, 0.22 kg away from the mean is the same as

$$\frac{0.22}{0.11} = 2 \text{ standard deviations}$$

away from the mean. See Figure 6.27.

FIGURE 6.27

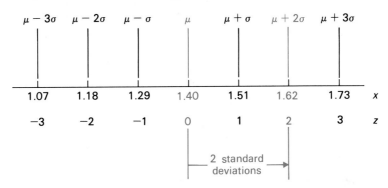

As another example, consider a brain weight of 1.235 kg. The z-score for that weight is

$$z = \frac{x - \mu}{\sigma} = \frac{1.235 - 1.40}{0.11} = \frac{-0.165}{0.11} = -1.5$$

Thus, a brain weight of 1.235 kg is -1.5 standard deviations away from the mean. The minus sign in -1.5 indicates that the brain weight of 1.235 kg is below (to the left of) the mean. ∎

The ideas discussed in the above example can be summarized as follows.

KEY FACT

The z-score of a population value gives the number of standard deviations that the value is away from the mean of the population. A negative z-score indicates that the population value is below (to the left of) the mean, while a positive z-score indicates that the population value is above (to the right of) the mean.

The interpretation of a z-score as the number of standard deviations a population value is away from the mean holds for *any* population, not just for normally distributed populations. However, in case the population *is* normally distributed, the above interpretation of z-scores is particularly useful. Let us look at an example.

EXAMPLE 6.14   **Illustrates an application of z-scores for normally distributed populations**

Consider once more the population of brain weights of Swedish men. Recall that this population is *normally distributed* with a mean of $\mu = 1.40$ kg and a standard

deviation of $\sigma = 0.11$ kg. What percentage of Swedish men have brain weights
a) within one standard deviation to either side of the mean?
b) within two standard deviations to either side of the mean?
c) within three standard deviations to either side of the mean?

SOLUTION

a) The z-score for a brain weight that is one standard deviation to the left of
the mean is $z = -1$ and that for a brain weight that is one standard deviation to
the right of the mean is $z = 1$. Consequently, the percentage of Swedish men that
have brain weights within one standard deviation to either side of the mean is
just the area under the standard normal curve that lies between $z = -1$ and $z = 1$,
which is 0.6826, or 68.26%. [To obtain the area 0.6826, we can either use Table II
in the usual way or simply refer to Table 6.1 on page 225.]

b) The z-scores for brain weights that are two standard deviations to the left
and right of the mean are, respectively, $z = -2$ and $z = 2$. Therefore, the percentage of Swedish men whose brain weights are within two standard deviations to
either side of the mean is just the area under the standard normal curve that lies
between $z = -2$ and $z = 2$, which is 0.9544 or 95.44%.

c) The solution to this part follows the same pattern as parts (a) and (b), and is
left as an exercise.   ■

The arguments employed in Example 6.14 apply to *any* normally distributed
population. We never used the information that the population consisted of brain
weights or that $\mu = 1.40$ kg and $\sigma = 0.11$ kg. All we used is the fact that the population is *normally distributed*. Thus, we have the following fact:

KEY FACT

For any *normally distributed* population:

PROPERTY 1 *About 68.26% of the population values lie within* one *standard deviation to either side of the mean.*

PROPERTY 2 *About 95.44% of the population values lie within* two *standard deviations to either side of the mean.*

PROPERTY 3 *About 99.74% of the population values lie within* three *standard deviations to either side of the mean.*

FIGURE 6.28   Percentage of population values in a *normally distributed* population that
lie within (a) one, (b) two, and (c) three standard deviations to either side of the
mean.

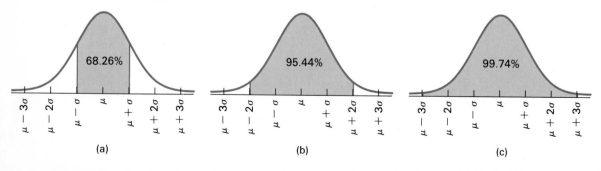

(a)   (b)   (c)

Using these properties, we can quickly and easily obtain a good "picture" of any normally distributed population. To see how, we consider the following example.

EXAMPLE   6.15   Illustrates the previous Key Fact

Intelligence quotients (IQs) are known to be normally distributed with a mean of $\mu = 100$ and a standard deviation of $\sigma = 16$. Apply the previous Key Fact to get some information about IQs.

S O L U T I O N
1 From Property 1, about 68.26% of the population have IQs within one standard deviation to either side of the mean. Now, one standard deviation to the left of the mean corresponds to an IQ of

$$100 - 1 \cdot 16 = 84$$
$$\quad \underset{\mu}{\uparrow} \qquad \underset{\sigma}{\uparrow}$$

and one standard deviation to the right of the mean corresponds to an IQ of

$$100 + 1 \cdot 16 = 116$$

Thus, about 68.26% of the population have IQs between 84 and 116.
2 From Property 2, we conclude that about 95.44% of the population have IQs between 68 ($= 100 - 2 \cdot 16$) and 132 ($= 100 + 2 \cdot 16$).
3 From Property 3, we see that about 99.74% (almost all) of the population have IQs between 52 ($= 100 - 3 \cdot 16$) and 148 ($= 100 + 3 \cdot 16$).

The results we have just obtained are summarized graphically in Figure 6.29. ■

FIGURE 6.29

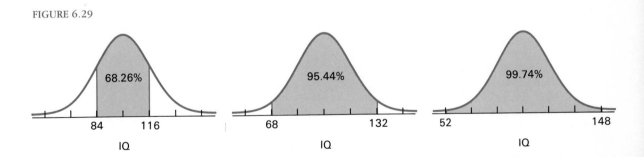

Properties 1–3 on page 244 allow us to determine useful information about a normally distributed population quickly and easily, as illustrated in Example 6.15. We should point out, however, that similar facts are obtainable for any number of standard deviations away from the mean. For instance, we can use Table II to conclude that about 86.64% of the values in a normally distributed population lie within 1.5 standard deviations to either side of the mean.

Exercises 6.3

A relative-frequency distribution for the heights of the 3264 female students attending a midwestern college is given in Table 6.2 on page 237. The mean of that population of heights is $\mu = 64.4$ inches with a standard deviation of $\sigma = 2.4$ inches. In each part of Exercises 6.41 and 6.42, obtain the exact percentage from Table 6.2. Then determine the corresponding area under the normal curve with parameters $\mu = 64.4$ and $\sigma = 2.4$, and compare your results.

6.41 The percentage of students with heights
    a) between 62 and 63 inches.
    b) between 65 and 70 inches.

6.42 The percentage of students with heights
    a) between 71 and 72 inches.
    b) between 61 and 65 inches.

6.43 The annual wages, excluding board, of U.S. farm laborers in 1926 were normally distributed with a mean of $\mu = \$586$ and a standard deviation of $\sigma = \$97$. What percentage of U.S. farm laborers had an annual wage of
    a) between $500 and $700?
    b) at least $400?

6.44 The lengths of adult yellow-bellied sapsuckers are normally distributed with a mean of $\mu = 8.5$ inches and a standard deviation of $\sigma = 0.17$ inch. What percentage of yellow-bellied sapsuckers are
    a) between 8.3 and 9.0 inches long?
    b) less than 8.8 inches long?

6.45 According to the U.S. National Center for Health Statistics, males who are six feet tall and between 18 and 24 years of age have a mean weight of $\mu = 175$ pounds. If the weights are normally distributed with a standard deviation of $\sigma = 14$ pounds, find the percentage of such males that weigh
    a) between 190 and 210 pounds.
    b) less than 150 pounds.

6.46 An issue of *Scientific American* reveals that the batting averages of major league baseball players are approximately normally distributed with a mean of $\mu = 0.270$ and a standard deviation of $\sigma = 0.015$. Determine the percentage of major league baseball players with batting averages
    a) between 0.225 and 0.250.
    b) at least 0.300.

6.47 Refer to Exercise 6.43. How many standard deviations away from the mean was a 1926 U.S. farm laborer's annual wage of
    a) $780?   b) $416.25?   c) $586?
    d) Draw a graph for this problem similar to Figure 6.27 on page 243.

6.48 Refer to Exercise 6.44. How many standard deviations away from the mean is the length of a yellow-bellied sapsucker
    a) 8.7125 inches long?
    b) 8.16 inches long?
    c) 8.5 inches long?
    d) Draw a graph for this problem similar to Figure 6.27 on page 243.

6.49 In Exercise 6.45 we pointed out that U.S. males who are six feet tall and between 18 and 24 years of age have a mean weight of $\mu = 175$ pounds. If the weights are normally distributed with $\sigma = 14$ pounds, what percentage of such males have weights
    a) within one standard deviation to either side of the mean?
    b) within two standard deviations to either side of the mean?
    c) within three standard deviations to either side of the mean?
    d) within 1.5 standard deviations to either side of the mean?

6.50 Refer to Exercise 6.46. What percentage of major league baseball players have batting averages
    a) within one standard deviation to either side of the mean?
    b) within two standard deviations to either side of the mean?
    c) within three standard deviations to either side of the mean?
    d) within 2.5 standard deviations to either side of the mean?

6.51 According to the U.S. Department of Agriculture, the mean weekly food cost for a couple with two children 6–11 years old is $\mu = \$95.40$. Presuming that these weekly food costs are normally distributed with a standard deviation of $\sigma = \$17.20$, fill in the following blanks:
    a) About 68.26% of such couples have weekly food costs between $_____ and $_____.

b) About 95.44% of such couples have weekly food costs between $____ and $____.

c) About 99.74% of such couples have weekly food costs between $____ and $____.

d) Draw graphs similar to the ones in Figure 6.29 on page 245 to illustrate your results.

6.52 The A.C. Nielsen Company reports that the mean TV-viewing time per week by children 2–5 years old is $\mu = 27.15$ hours. Assuming that the weekly TV-viewing times of these children are normally distributed with a standard deviation of $\sigma = 6.23$ hours, fill in the following blanks:

a) About 68.26% of all such children watch between ____ and ____ hours of TV per week.

b) About 95.44% of all such children watch between ____ and ____ hours of TV per week.

c) About 99.74% of all such children watch between ____ and ____ hours of TV per week.

d) Draw graphs similar to the ones in Figure 6.29 on page 245 to illustrate your results.

6.53 For a normally distributed population, fill in the following blanks:

a) About ____ % of the population values lie within 1.96 standard deviations to either side of the mean.

b) About ____ % of the population values lie within 1.64 standard deviations to either side of the mean.

6.54 For a normally distributed population, fill in the following blanks:

a) About 99% of the population values lie within ____ standard deviations to either side of the mean.

b) About 80% of the population values lie within ____ standard deviations to either side of the mean.

6.4   Normally distributed random variables

In the previous section we discussed normally distributed populations. These are populations for which percentages are equal, at least approximately, to areas under a normal curve (the normal curve whose parameters μ and σ are the same as the mean and standard deviation of the population).

As you know, we can associate with any numerical population a random variable x—namely, the value of a randomly selected member from the population. Probabilities for such a random variable are the same as percentages for the population. Consequently, if the population is normally distributed, then probabilities for the random variable x are equal, at least approximately, to areas under a normal curve. We illustrate this discussion in Example 6.16.

EXAMPLE 6.16   **Introduces normally distributed random variables**

The ages of the employees of an aerospace company are approximately *normally distributed* with a mean of $\mu = 40.9$ years and a standard deviation of $\sigma = 7.1$ years. If we let x be the age of a randomly selected employee, then x is a random variable. The mean and standard deviation of x are just the mean and standard deviation of the population of ages. So, $\mu_x = 40.9$ years and $\sigma_x = 7.1$ years.

Additionally, probabilities for the random variable x are the same as relative frequencies (percentages) for the population. Since, in this case, the population is normally distributed, percentages for the population are equal to areas under the normal curve with parameters $\mu = 40.9$ and $\sigma = 7.1$. Consequently, probabilities for the random variable x are also equal to areas under that curve.

For instance, let us determine $P(35 \leq x \leq 50)$, that is, the probability that a randomly selected employee is between 35 and 50 years old. To accomplish this, we must find the area under the normal curve with parameters $\mu = 40.9$ and $\sigma = 7.1$ that lies between 35 and 50. See Figure 6.30.

FIGURE 6.30
Determination of the
area under the nor-
mal curve with para-
meters $\mu = 40.9$
and $\sigma = 7.1$ that lies
between 35 and 50.

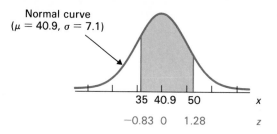

Normal curve
$(\mu = 40.9, \sigma = 7.1)$

35 40.9 50 x

−0.83 0   1.28 z

z-score computations: Area between 0 and z:

$$x = 35 \longrightarrow z = \frac{35 - 40.9}{7.1} = -0.83 \qquad 0.2967$$

$$x = 50 \longrightarrow z = \frac{50 - 40.9}{7.1} = 1.28 \qquad 0.3997$$

Shaded area $= 0.2967 + 0.3997 = 0.6964$

Consequently, $P(35 \leq x \leq 50) = 0.6964$—there is about a 70% chance that a randomly selected employee will be between 35 and 50 years old.   ■

DEFINITION 6.2 Normally distributed random variable

A random variable is said to be (approximately) **normally distributed** if probabilities for the random variable are (approximately) equal to areas under a normal curve. If such a random variable has mean μ_x and standard deviation σ_x, then
the normal curve used is the one with parameters μ_x and σ_x.

We should emphasize that Definition 6.2 applies to *any* random variable, not
just to those associated with numerical populations.

EXAMPLE 6.17   **Illustrates how to determine probabilities for a normally
distributed random variable**

A local bottling plant fills both bottles and cans of soda pop for distribution in the
surrounding area. While the advertised content is 354 ml, the filling process is
actually set to a mean content of 356 ml. As a matter of fact, the amount of soda
put in a bottle or can is *normally distributed* with a mean of 356 ml and a standard deviation of 1.63 ml. What is the probability that a randomly selected bottle
will contain less than the advertised amount of 354 ml?

S O L U T I O N
Let x denote the content, in ml, of a randomly selected bottle of soda. Then, x is
a normally distributed random variable with $\mu_x = 356$ ml and $\sigma_x = 1.63$ ml.
Therefore, probabilities for x are equal to areas under the normal curve with parameters $\mu = 356$ and $\sigma = 1.63$. In particular, the probability that a randomly selected bottle contains less than 354 ml, $P(x < 354)$, is equal to the area under the
normal curve with parameters $\mu = 356$ and $\sigma = 1.63$ that lies to the left of 354.
This area is found in Figure 6.31.

FIGURE 6.31
Determination of the area under the normal curve with parameters $\mu = 356$ and $\sigma = 1.63$ that lies to the left of 354.

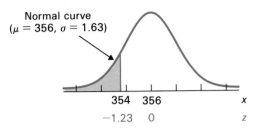

Normal curve
($\mu = 356$, $\sigma = 1.63$)

354   356 x
−1.23   0 z

z-score computation:

$x = 354 \longrightarrow z = \dfrac{354 - 356}{1.63} = -1.23$

Area between 0 and z:

0.3907

Shaded area $= 0.5000 - 0.3907 = 0.1093$

MTB
SPSS

Thus, $P(x < 354) = 0.1093$. In other words, there is about an 11% chance that a randomly selected bottle will contain less than the advertised amount of 354 ml. The relative-frequency interpretation is that about 11% of the bottles will contain less than the advertised amount of 354 ml. ■

The Key Fact on page 244 gives some useful statements concerning percentages for normally distributed populations. Here are the corresponding probability statements for normally distributed random variables:

KEY FACT

For any *normally distributed* random variable x:

PROPERTY 1 *The probability is 0.6826 that x will take a value within* one *standard deviation to either side of its mean.*
PROPERTY 2 *The probability is 0.9544 that x will take a value within* two *standard deviations to either side of its mean.*
PROPERTY 3 *The probability is 0.9974 that x will take a value within* three *standard deviations to either side of its mean.*

It is important to be able to interpret Properties 1–3 and similar statements in terms of percentages. For instance, Property 2 can be interpreted as follows: "About 95.44% of the time, a normally distributed random variable will take a value within *two* standard deviations to either side of its mean." Similarly, Property 3 indicates that a normally distributed random variable will almost always (about 99.74% of the time) take a value within *three* standard deviations to either side of its mean. We will apply the previous Key Fact in Example 6.18.

EXAMPLE 6.18   **Illustrates the previous Key Fact**

In Example 6.17 we examined the filling of soda-pop bottles by a bottling plant. The amount of soda put in a bottle is normally distributed with a mean of 356 ml and a standard deviation of 1.63 ml. If we let x denote the content of a randomly selected bottle of soda, then x is a normally distributed random variable with mean $\mu_x = 356$ ml and standard deviation $\sigma_x = 1.63$ ml. Apply the previous Key Fact to make some pertinent statements about the contents of the bottles.

SOLUTION

1 By Property 1, the probability is 0.6826 that x will take a value within *one* standard deviation of its mean. Now, one standard deviation to the left of the mean corresponds to

$$356 - 1 \cdot 1.63 = 354.37 \text{ ml}$$

$$\uparrow \qquad \uparrow$$
$$\mu_x \qquad \sigma_x$$

and one standard deviation to the right of the mean corresponds to

$$356 + 1 \cdot 1.63 = 357.63 \text{ ml}$$

Thus, the probability is 0.6826 that x will take a value between 354.37 and 357.63 ml. In other words, about 68.26% of the bottles filled will contain between 354.37 and 357.63 ml of soda pop.

2 From Property 2, we conclude that the probability is 0.9544 that x will take a value between 352.74 ($= 356 - 2 \cdot 1.63$) and 359.26 ($= 356 + 2 \cdot 1.63$). That is, about 95.44% of the bottles filled will contain between 352.74 and 359.26 ml of soda pop.

3 Applying Property 3, we deduce that the probability is 0.9974 that x will take a value between 351.11 ($= 356 - 3 \cdot 1.63$) and 360.89 ($= 356 + 3 \cdot 1.63$). That is, about 99.74% of the bottles filled will contain between 351.11 and 360.89 ml of soda pop. ∎

MTB
SPSS

An interpretation of the standard deviation for a normally distributed random variable

In Chapter 5, page 187, we gave the following interpretation to the standard deviation, σ_x, of a random variable x: It is a measure of the dispersion of the possible values of the random variable relative to its mean—the smaller the value of σ_x, the more likely that x will take a value near its mean. Using the Key Fact on page 249, we can make the interpretation of σ_x even more explicit when the random variable is *normally distributed.*

For example, by Property 2 we know that there is a 95.44% chance that a normally distributed random variable will take a value within two standard deviations (that is, within $\pm 2 \cdot \sigma_x$) of its mean. If σ_x is small, then two standard deviations to either side of the mean will also be small, thus making it likely (a 95.44% chance) that x will take a value close to its mean. If σ_x is large, then two standard deviations to either side of the mean will also be large and, consequently, we cannot expect x to take a value close to its mean, although it may. A graphical summary of the discussion in this paragraph is presented in Figure 6.32.

FIGURE 6.32

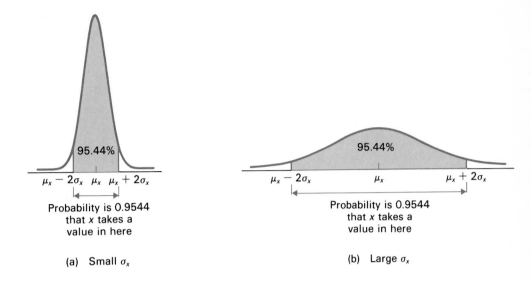

95.44%

$\mu_x - 2\sigma_x \quad \mu_x \quad \mu_x + 2\sigma_x$

Probability is 0.9544
that x takes a
value in here

(a) Small σ_x

95.44%

$\mu_x - 2\sigma_x \qquad \mu_x \qquad \mu_x + 2\sigma_x$

Probability is 0.9544
that x takes a
value in here

(b) Large σ_x

Standardizing a random variable

In Section 6.2 we introduced z-scores (standard scores) as a method for finding areas under normal curves. Then, in Section 6.3, we found that the z-score of a population value gives the number of standard deviations that the value is away from the population mean. It is also useful to consider the process of **standardizing a random variable.** We introduce this concept in Example 6.19.

EXAMPLE 6.19   ### Introduces standardized random variables

A one-man barber shop has an average weekly gross income of $640, with a standard deviation of $50. Let x be the shop's gross income during a randomly selected week. Then x is a random variable with mean $\mu_x = 640$ and standard deviation $\sigma_x = 50$. We can get a new random variable, which we call z, by *standardizing* the random variable x—that is, by subtracting from x its mean and then dividing by its standard deviation:

$$z = \frac{x - \mu_x}{\sigma_x}$$

Since in this case $\mu_x = 640$ and $\sigma_x = 50$, the standardized version of x is

$$z = \frac{x - 640}{50}$$

We emphasize that, since x is a random variable, so is z.

If we select a week at random, then the random variable x tells us that week's gross. What does the *standardized* random variable z tell us? It tells us how many standard deviations that week's gross is away from the mean. For instance, if a randomly selected week has a gross of $740, then $x = 740$ and consequently

$$z = \frac{x - 640}{50} = \frac{740 - 640}{50} = 2$$

Thus, that week's gross is two standard deviations away from, and to the right of, the mean.   ∎

DEFINITION 6.3 Standardized version of a random variable

Let x be a random variable. Then we define the **standardized version** of x to be the random variable

$$z = \frac{x - \mu_x}{\sigma_x}$$

The standardized random variable z gives the number of standard deviations that an observed value of x is away from its mean.

There is one additional point that we need to examine in this section. Recall that the normal curve with parameters $\mu = 0$ and $\sigma = 1$ is the *standard normal curve*. Suppose a normally distributed random variable has mean 0 and standard deviation 1. Then we say that it has the **standard normal distribution,** because probabilities for such a random variable are equal to areas under the standard normal curve.

It is important to note that we can form the standardized version of any random variable x regardless of whether x is normally distributed. If x is not normally distributed, then neither is z. But if x *is* normally distributed, then so is z. In fact, z then has the standard normal distribution.

KEY FACT

Suppose x is a *normally distributed* random variable with mean μ_x and standard deviation σ_x. Then the standardized random variable

$$z = \frac{x - \mu_x}{\sigma_x}$$

has the *standard normal distribution.* That is, z is a normally distributed random variable with mean $\mu_z = 0$ and standard deviation $\sigma_z = 1$. Thus, probabilities for the standardized random variable z are equal to areas under the *standard normal curve.*

EXAMPLE 6.20   **Illustrates Definition 6.3**

Consider the bottle-filling process discussed in Examples 6.17 and 6.18. The amount of soda put in a bottle is *normally distributed* with a mean of 356 ml and a standard deviation of 1.63 ml. Let x be the content of a randomly selected bottle of soda.

a) Determine the standardized version of the random variable x.
b) What is the probability distribution of the standardized random variable?

S O L U T I O N

a) The random variable x has mean $\mu_x = 356$ and standard deviation $\sigma_x = 1.63$. Thus, the standardized version of x is the random variable

$$z = \frac{x - 356}{1.63}$$

b) Since the random variable x is normally distributed, the standardized random variable

$$z = \frac{x - 356}{1.63}$$

has the standard normal distribution. That is, probabilities for that random variable are equal to areas under the standard normal curve. ∎

Exercises 6.4

6.55 According to the National Education Association, the mean salary for secondary school teachers is $21.1 thousand. Assume the salaries are *normally distributed* with a standard deviation of $3.2 thousand. Let x be the salary, in thousands of dollars, of a randomly selected secondary school teacher. Find
a) $P(x < 18)$.
b) $P(25 \leq x \leq 30)$.
c) Interpret each of your results in parts (a) and (b) in words.

6.56 *The World Almanac* reports that the mean travel time to work in New York is 29 minutes. Let x be the time, in minutes, that it takes a randomly selected worker to get to work on a randomly selected day. If the travel times are *normally distributed* with a standard deviation of 9.3 minutes, find
a) $P(x < 45)$.
b) $P(20 \leq x \leq 30)$.
c) Interpret each of your results in parts (a) and (b) in words.

6.57 The lifetime of a brand of flashlight battery is normally distributed with a mean of 30 hours and a standard deviation of 5.6 hours. Let x be the life of a randomly selected flashlight battery of this brand. Determine
a) $P(x > 20)$.
b) $P(15 \leq x \leq 45)$.
c) Interpret each of your results in parts (a) and (b) in terms of percentages.

6.58 A manufacturer of timepieces claims that the weekly error, in seconds, of the watches she makes has a normal distribution with a mean of 0 and a standard deviation of 1. Let x denote the amount of time, in seconds, that one of these watches is off by at the end of a randomly selected week. Find
a) $P(x < -1)$.
b) $P(x < -2 \text{ or } x > 2)$.
c) Interpret each of your results in parts (a) and (b) in words.

6.59 According to *Runner's World* magazine, the times of the finishers in the New York City 10-km run are normally distributed with a mean of 61 minutes and a standard deviation of 9 minutes. Let x be the time of a randomly selected finisher. Find
a) $P(x > 75)$.
b) $P(x < 50 \text{ or } x > 70)$.
c) Interpret each of your results in parts (a) and (b) in words.

6.60 The length of the western rattlesnake is normally distributed with a mean of 42 inches and a standard deviation of 2.04 inches. Let x be the length of one of these snakes selected at random. Determine
a) $P(x > 45)$.
b) $P(35 \leq x \leq 40)$.
c) Interpret each of your results in parts (a) and (b) in terms of percentages.

6.61 A manufacturer needs to produce bolts approximately 10 mm in diameter to fit into a circular hole

10.4 mm in diameter. In reality, the diameters of the bolts produced are normally distributed with a mean of 10 mm and a standard deviation of 0.1 mm. Let x be the diameter of a randomly selected bolt. Apply the Key Fact on page 249 to make some pertinent statements concerning the diameters of the bolts produced. [See Example 6.18 on page 249 for a model.]

6.62   An electronics company gives a general aptitude test to all prospective employees. The test is designed to take about one hour to complete. Experience indicates that the completion times are normally distributed with a mean of 62.4 minutes and a standard deviation of 7.2 minutes. Let x denote the time it takes a randomly selected applicant to complete the test. Apply the Key Fact on page 249 to make some pertinent statements concerning the completion times for the aptitude test.

6.63   Refer to Exercise 6.57. Use the Key Fact on page 249 to obtain some relevant information regarding the lifetimes of flashlight batteries of the given brand.

6.64   Refer to Exercise 6.58. Use the Key Fact on page 249 to obtain some relevant information regarding the weekly error of one of the manufacturer's watches.

6.65   The U.S. Bureau of Prisons reports that, for prisoners released from federal institutions for the first time, the mean time served is 16.3 months. Take the standard deviation for the times served to be 17.9 months. Let x be the time served by a randomly selected prisoner released for the first time from a federal institution.

a) Find the standardized version, z, of the random variable x.
b) How many standard deviations away from the mean is a time served of
   (i) 20.3 months?   (ii) 64.7 months?
   (iii) 4.2 months?
c) Is it necessary to assume the times served are normally distributed to answer parts (a) and (b)? Why?

6.66   According to the Health Insurance Association of America, the average daily room charge for a semi-private hospital room is $178. Assume a standard deviation of $25. Let x be the daily semiprivate room charge of a randomly selected hospital.
a) Find the standardized version, z, of the random variable x.
b) How many standard deviations away from the mean is a daily room charge of
   (i) $231?   (ii) $97?   (iii) $144?
c) Is it necessary to assume that the daily semi-private room charges are normally distributed in order to answer parts (a) and (b)? Why?

6.67   Refer to Exercise 6.59.
a) Determine the standardized version, z, of the random variable x.
b) What does the random variable z represent?
c) What is the probability distribution of the random variable z?

6.68   Refer to Exercise 6.60.
a) Determine the standardized version, z, of the random variable x.
b) What does the random variable z represent?
c) What is the probability distribution of the random variable z?

6.5   The normal approximation to the binomial distribution*

One of the earliest uses of the normal distribution was to approximate the binomial distribution.[†] Specifically, we shall see that under certain conditions on n and p, we can use areas under normal curves to approximate binomial probabilities. Before beginning, we will briefly review the binomial distribution (see Section 5.4 for details).

Suppose n independent success-failure experiments (Bernoulli trials) are performed, with the probability of success on any given trial being p. Let x denote

---

[†] The mathematical theory for doing this is credited to Abraham DeMoivre (1667–1754) and Pierre Laplace (1749–1827).

the total number of successes in the n trials. Then the probability distribution of the random variable x is given by the binomial probability formula

$$P(x) = \binom{n}{x} p^x (1-p)^{n-x}$$

Moreover, we say that x has the *binomial distribution* with parameters n and p.

You might be wondering why we would use normal curve areas to *approximate* binomial probabilities when we can obtain them *exactly* by employing the binomial probability formula. The following example should help explain why.

EXAMPLE 6.21   **Illustrates the need to approximate binomial probabilities**

Insurance companies use mortality tables to compute life insurance premiums, retirement pensions, annuity payments, and other related items. Mortality tables enable actuaries to determine the probability that a person at any given age will live a specified number of years.

The probability is about 0.8 (an 80% chance) that a person aged 70 will be alive at age 75. If 500 people aged 70 are selected at random, determine the probability that

a) exactly 390 of them will be alive at age 75.

b) between 375 and 425 of them, inclusive, will be alive at age 75.

SOLUTION

Let x denote the number of people out of the 500 that will be alive at age 75. Then x has the *binomial distribution* with parameters $n = 500$ (the 500 people) and $p = 0.8$ (the probability a person aged 70 will be alive at age 75). Thus, probabilities for x can be determined exactly by using the binomial probability formula

$$P(x) = \binom{500}{x} (0.8)^x (0.2)^{500-x}$$

Let us apply this formula to the problems posed in (a) and (b).

a) Here we want the probability that exactly 390 of the 500 people will still be alive at age 75—$P(x = 390)$. So, the "answer" is

$$P(390) = \binom{500}{390} (0.8)^{390} (0.2)^{110}$$

To actually get the value of the expression on the right is not so simple, however, even with a calculator. We have to be quite careful in doing the computations to avoid such things as roundoff error and getting numbers so large or so small that they are outside the range of the calculator. Fortunately, as you will soon see, we can avoid the computations altogether by using normal curve areas in a very simple way.

b) In this part we need to determine the probability that between 375 and 425, inclusive, of the 500 people will be alive at age 75—$P(375 \leq x \leq 425)$. So, the "answer" is

$$P(375 \le x \le 425) = P(375) + P(376) + \cdots + P(425)$$

$$= \binom{500}{375}(0.8)^{375}(0.2)^{125} + \binom{500}{376}(0.8)^{376}(0.2)^{124} + \cdots + \binom{500}{425}(0.8)^{425}(0.2)^{75}$$

We see that we have the same computational problems as in part (a), except that here we must evaluate 51 difficult expressions instead of one. [There are 51 summands in the sum.] Thus, although in theory we can use the binomial probability formula to get the answer, doing so is another matter. Surprising as it might seem, there is a way to use normal curve areas to get the approximate answer, and *it is easy!* We will return to this problem momentarily and find the values of the probabilities in parts (a) and (b).   ■

The previous example should make it clear to you why we often need to approximate binomial probabilities. For even though we have a formula for computing binomial probabilities exactly, the formula is simply not practical when we are dealing with large values of n.

Under certain conditions on n and p, the histogram of a binomial distribution is bell-shaped. In such cases, we can use normal curve areas to approximate binomial probabilities.

The following example illustrates a bell-shaped binomial distribution. It is simple to calculate probabilities exactly in this example by using the binomial probability formula. However, for purposes of illustration, we will show how to use normal curve areas to approximate the binomial probabilities.

EXAMPLE   6.22   **Introduces how to approximate binomial probabilities using areas under a normal curve**

A student is taking a true-false exam with 10 questions. If the student *guesses* at all 10 questions, what is the probability that he gets either seven or eight correct?

S O L U T I O N
Let x be the number of correct guesses by the student. There are 10 questions, so here $n = 10$. Since the student guesses at each question, the success probability is $p = 0.5$. So x has the binomial distribution with parameters $n = 10$ and $p = 0.5$. That is, probabilities for the random variable x are given by the binomial probability formula

$$P(x) = \binom{10}{x}(0.5)^x(1 - 0.5)^{10-x}$$

Applying that formula we obtain the probability distribution of x. See Table 6.3.

The problem is to find the probability that the student gets either seven or eight questions correct—$P(x = 7$ or $8)$. From Table 6.3 we find that the *exact* probability is

$$P(x = 7 \text{ or } 8) = P(7) + P(8) = 0.1172 + 0.0439 = 0.1611$$

TABLE 6.3   Probability distribution for the number of correct guesses, x, by the student.

Number correct x	Probability $P(x)$
0	0.0010
1	0.0098
2	0.0439
3	0.1172
4	0.2051
5	0.2461
6	0.2051
7	0.1172
8	0.0439
9	0.0098
10	0.0010

Let us now see how we can use areas under a normal curve to determine the approximate value of $P(x = 7 \text{ or } 8)$. First of all, as you might expect, the normal curve used is the one whose parameters are the same as the mean and standard deviation of the random variable x. Since x has the binomial distribution with parameters $n = 10$ and $p = 0.5$, we have from Formulas 5.3 and 5.4 (pages 210 and 211) that

$$\mu_x = np = 10 \cdot (0.5) = 5$$

and

$$\sigma_x = \sqrt{np(1-p)} = \sqrt{10 \cdot (0.5)(1-0.5)} = 1.58$$

Thus, the normal curve that is used here is the one with parameters $\mu = 5$ and $\sigma = 1.58$.

In Figure 6.33 we have drawn a histogram for the random variable x. The heights of the bars are simply the probabilities for x given in Table 6.3. Note that the histogram is bell-shaped. Also shown in the figure is the normal curve with parameters $\mu = 5$ and $\sigma = 1.58$.

FIGURE 6.33
Histogram for x
with superimposed
normal curve.

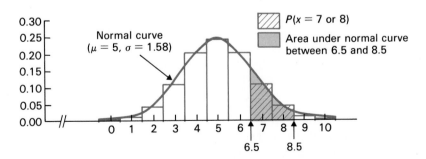

The probability $P(x = 7 \text{ or } 8)$ is exactly equal to the combined area of the corresponding rectangles of the histogram—the cross-hatched area in the figure. By examining Figure 6.33 carefully, we see that the combined area of the rectangles can be *approximated* by the area under the normal curve between 6.5 and 8.5—

the shaded area in the figure. [It should be clear from the picture why we consider the area under the normal curve between 6.5 and 8.5. This is referred to as the **correction for continuity.** Keep this in mind when we present the general procedure for approximating binomial probabilities using normal curve areas.]

Consequently, we see, at least qualitatively, that the probability $P(x = 7 \text{ or } 8)$ is approximately equal to the area under the normal curve between 6.5 and 8.5. To compare these values quantitatively, we must find the normal curve area. This is done in the usual way, as shown in Figure 6.34.

FIGURE 6.34
Determination of the area under the normal curve with parameters $\mu = 5$ and $\sigma = 1.58$ that lies between 6.5 and 8.5.

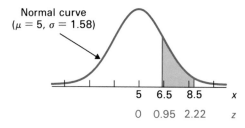

z-score computations:

$x = 6.5 \longrightarrow z = \dfrac{6.5 - 5}{1.58} = 0.95$ 0.3289

$x = 8.5 \longrightarrow z = \dfrac{8.5 - 5}{1.58} = 2.22$ 0.4868

Area between 0 and z:

Shaded area $= 0.4868 - 0.3289 = 0.1579$

Thus, the area under the normal curve that lies between 6.5 and 8.5 is 0.1579, while the exact value of $P(x = 7 \text{ or } 8)$, computed on page 256, is 0.1611. As we expected by looking at Figure 6.33, the normal curve area is an excellent approximation of the exact probability. ∎

It is not always appropriate to use the normal approximation to the binomial distribution, since the histograms for some binomial distributions are not sufficiently bell-shaped. The customary rule of thumb for use of the normal approximation is that *both np and n(1 − p) are at least 5.*

By examining carefully what we did in Example 6.22, we can write down a general procedure for using areas under normal curves to approximate binomial probabilities.

PROCEDURE 6.2 To approximate binomial probabilities using normal curve areas.

STEP 1 *Determine n, the number of trials, and p, the success probability.*
STEP 2 *Check that both np and n(1 − p) are at least 5. If they are not, the normal approximation should not be used.*
STEP 3 *Find μ_x and σ_x using the formulas*

$$\mu_x = np \quad \text{and} \quad \sigma_x = \sqrt{np(1-p)}$$

STEP 4 *Make the correction for continuity, and find the appropriate area under the normal curve with parameters $\mu = np$ and $\sigma = \sqrt{np(1-p)}$.*

Note: Step 4 requires us to make the *correction for continuity*. As illustrated in Example 6.22 this means the following: When using normal curve areas to approximate the probability that a binomial random variable takes a value between two integers, inclusive, we must subtract 0.5 from the smaller integer and add 0.5 to the larger integer before finding the area under the normal curve.

We will now apply Procedure 6.2 to solve the problems presented at the beginning of this section in Example 6.21.

EXAMPLE 6.23   **Illustrates Procedure 6.2**

The probability is 0.80 that a person aged 70 will be alive at age 75. Suppose 500 people aged 70 are selected at random. Determine the probability that
  a) exactly 390 of them will be alive at age 75.
  b) between 375 and 425 of them, inclusive, will be alive at age 75.

S O L U T I O N
We will obtain the approximate values of the probabilities using Procedure 6.2.

STEP 1 *Determine n, the number of trials, and p, the success probability.*

Here $n = 500$ and $p = 0.8$.

STEP 2 *Check that both np and n(1 − p) are at least 5.*

$$np = 500 \cdot (0.8) = 400$$
$$n(1 - p) = 500 \cdot (0.2) = 100$$

STEP 3 *Find μ_x and σ_x using the formulas*

$$\mu_x = np \text{ and } \sigma_x = \sqrt{np(1 - p)}$$

We have

$$\mu_x = np = 500 \cdot (0.8) = 400$$

and

$$\sigma_x = \sqrt{np(1 - p)} = \sqrt{500 \cdot (0.8)(0.2)} = 8.94$$

STEP 4 *Make the correction for continuity, and find the appropriate area under the normal curve with parameters $\mu = 400$ and $\sigma = 8.94$.*

a) Here we want $P(x = 390)$. Making the correction for continuity, we need to find the area under the normal curve with parameters $\mu = 400$ and $\sigma = 8.94$ that lies between 389.5 and 390.5. [We subtracted 0.5 from 390 and added 0.5 to 390.] See Figure 6.35 at the top of the next page.

Therefore, $P(x = 390) = 0.0236$ (approximately). In other words, the probability is about 0.0236 that exactly 390 of the 500 people will be alive at age 75.

b) For this part we want the probability that between 375 and 425 of the 500 people will be alive at age 75—$P(375 \le x \le 425)$. Making the correction for continuity, we need to find the area under the normal curve with parameters $\mu = 400$ and $\sigma = 8.94$ that lies between 374.5 and 425.5. [We subtracted 0.5 from 375 and added 0.5 to 425.] See Figure 6.36.

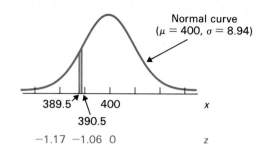

FIGURE 6.35
Determination of
the area under the
normal curve with
parameters $\mu = 400$
and $\sigma = 8.94$ that
lies between 389.5
and 390.5.

z-score computations: Area between 0 and z:

$x = 389.5 \longrightarrow z = \dfrac{389.5 - 400}{8.94} = -1.17$ 0.3790

$x = 390.5 \longrightarrow z = \dfrac{390.5 - 400}{8.94} = -1.06$ 0.3554

Shaded area $= 0.3790 - 0.3554 = 0.0236$

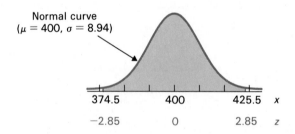

FIGURE 6.36
Determination of
the area under the
normal curve with
parameters $\mu = 400$
and $\sigma = 8.94$ that
lies between 374.5
and 425.5.

z-score computations: Area between 0 and z:

$x = 374.5 \longrightarrow z = \dfrac{374.5 - 400}{8.94} = -2.85$ 0.4978

$x = 425.5 \longrightarrow z = \dfrac{425.5 - 400}{8.94} = 2.85$ 0.4978

Shaded area $= 0.4978 + 0.4978 = 0.9956$

Consequently, $P(375 \leq x \leq 425) = 0.9956$ (approximately). The probability
that between 375 and 425 of the 500 people will be alive at age 75 is about
0.9956.   ■

We need to emphasize that the continuity correction is only necessary when
using the normal distribution to approximate probabilities for a *discrete* probability distribution, such as the binomial distribution. Otherwise no continuity
correction is required.

Exercises 6.5

6.69   Refer to Example 6.22 on page 256.
   a) Use Table 6.3 to find the exact probability that the student guesses correctly
      (i) on four or five questions—$P(x = 4$ or $5)$.
      (ii) on between three and seven questions— $P(3 \leq x \leq 7)$.
   b) Apply Procedure 6.2 on page 258 to approximate the probabilities in part (a) using normal curve areas. Compare your answers.

6.70   Refer to Example 6.22 on page 256.
   a) Use Table 6.3 to find the exact probability that the student guesses correctly
      (i) on at most five questions—$P(0 \leq x \leq 5)$.
      (ii) on at least six questions—$P(6 \leq x \leq 10)$.
   b) Apply Procedure 6.2 on page 258 to approximate the probabilities in part (a) using normal curve areas. Compare your answers.

6.71   If, in Example 6.22, the true-false exam had 30 questions instead of 10, which normal curve would you use to approximate probabilities for the number of correct guesses?

6.72   If, in Example 6.22, the true-false exam had 25 questions instead of 10, which normal curve would you use to approximate probabilities for the number of correct guesses?

In each of Exercises 6.73 through 6.80, apply Procedure 6.2 on page 258 to approximate the indicated binomial probabilities.

6.73   According to the *Daily Racing Form* the probability is about 0.67 that the favorite in a horse race will finish in the money (1st, 2nd, or 3rd). Find the probability that in the next 200 races the favorite will finish in the money
   a) exactly 140 times.
   b) between 120 and 130 times, inclusive.
   c) at least 150 times.

6.74   A small airline in the southwest has determined that the no-show rate for reservations is about 16%. In other words, the probability is 0.16 that a party making a reservation will not show up. The next flight has 42 parties with advance reservations. What is the probability that
   a) exactly five parties do not show up?
   b) between nine and 12, inclusive, do not show up?

   c) at least one does not show up?
   d) at most two do not show up?

6.75   The U.S. National Center for Health Statistics reports that about 38.3% of all injuries in the U.S. occur at home. Out of 500 randomly selected injuries, what is the probability that the number occurring at home is
   a) exactly 200?
   b) between 180 and 210, inclusive?
   c) at most 225?

6.76   The infant mortality rate in India is 139 per 1,000 live births, or 13.9%. Determine the probability that out of 1,000 randomly selected live births there are
   a) exactly 139 infant deaths.
   b) between 120 and 150 infant deaths, inclusive.
   c) at most 130 infant deaths.

6.77   A mail-order firm receives orders on about 12% of its mailings. If 750 brochures are sent, what is the probability that
   a) exactly 100 result in orders?
   b) either less than 80 or more than 105 result in orders?
   c) at least 14% result in orders?

6.78   Data from the *Statistical Abstract of the United States* indicates that there is about a 51.3% chance that a baby born in the U.S. will be male. What is the probability that out of the next 10,000 births at least half will be male?

6.79   A roulette wheel consists of 38 numbers, of which 18 are red, 18 are black, and two are green. When the roulette ball is spun, it is equally likely to land on each of the 38 numbers. A gambler is playing roulette and bets $10 on "red" each time. If the ball lands on a red number the gambler wins $10 from the house; otherwise she loses her $10. What is the probability that the gambler is ahead after
   a) 100 bets?
   b) 1000 bets?
   c) 5000 bets?
   Hint: The gambler is ahead after a series of bets if and only if she has won more than half of the bets.

6.80   A brand of flashlight battery has normally distributed lifetimes with a mean life of 30 hours and a standard deviation of 5 hours. A supermarket purchases 500 of these batteries from the manufacturer. What is the probability that at least 80% of them last longer than 25 hours?

♦ **Chapter Review**

FORMULAS

z-score for an x-value, 232

$$z = \frac{x - \mu}{\sigma}$$

(μ and σ are the parameters for a normal curve *or* the mean and standard deviation of a population)

Standardized version of a random variable x, 252

$$z = \frac{x - \mu_x}{\sigma_x}$$

($\mu_x =$ mean of x, $\sigma_x =$ standard deviation of x)

YOU SHOULD BE
ABLE TO . . .

1 use and understand the preceding formulas.

2 find areas under the standard normal curve using Table II.

3 explain the meaning of the parameters μ and σ for a normal curve.

4 sketch a normal curve.

5 find areas under any normal curve.

6 determine percentages for a normally distributed population.

7 determine probabilities for a normally distributed random variable.

8* know when and how to use normal curve areas to approximate binomial probabilities.

REVIEW TEST

1 What are the two primary reasons for studying normal curves?

2 Determine and sketch the area under the standard normal curve that lies
   a) between $z = 0$ and $z = 2.47$.
   b) between $z = -1.85$ and $z = 0$.
   c) to the right of $z = 0.61$.
   d) to the left of $z = -3.02$.

   e) between $z = 1.11$ and $z = 2.75$.
   f) between $z = -2.06$ and $z = -0.54$.
   g) to the left of $z = 1.59$.
   h) to the right of $z = -1.34$.
   i) between $z = -2$ and $z = 1.5$.
   j) either to the left of $z = 1$ or to the right of $z = 3$.

3 Sketch the normal curve with parameters
   a) $\mu = -1$ and $\sigma = 2$.

b) $\mu = 3$ and $\sigma = 2$.

c) $\mu = -1$ and $\sigma = 0.5$.

4   Consider the normal curves with the following parameters:

$$\mu = 1.5, \sigma = 3$$
$$\mu = 1.5, \sigma = 6.2$$
$$\mu = -2.7, \sigma = 3$$
$$\mu = 0, \sigma = 1$$

a) Which curve has the largest spread?

b) Which curves are centered at the same location?

c) Which curves have the same shape?

d) Which curve is centered furthest to the left?

e) Which curve is the standard normal curve?

5   Find the area under the normal curve with parameters $\mu = -1$ and $\sigma = 2.5$ that lies

a) between 2 and 6.

b) to the right of -5.6.

c) to the right of 3.2.

6   Each year thousands of college seniors take the *Graduate Record Examination*. The scores are transformed so as to have a mean of $\mu = 500$ and a standard deviation of $\sigma = 100$. Furthermore, the scores are known to be approximately *normally distributed*. Determine the percentage of students that score

a) between 350 and 625.

b) at least 375.

c) above 750.

7   Refer to Problem 6. How many standard deviations away from the mean is a score of

a) 645?   b) 320?   c) 500?

8   Refer to Problem 6. Fill in the following blanks:

a) About 68.26% of students score between _____ and _____.

b) About 95.44% of students score between _____ and _____.

c) About 99.74% of students score between _____ and _____.

9   According to a 1973 study by R. R. Paul[†], the mean gestation period of the Morgan horse is 339.6 days.

_____

[†] Paul, R. R. 1973. Foaling date. *The Morgan Horse* 33:40.

The gestation periods are approximately normally distributed with a standard deviation of 13.3 days. Let x denote the gestation period of a randomly selected Morgan horse.

a) What is the mean of the random variable x? What is its standard deviation?

b) Find $P(320 \leq x \leq 330)$.

c) Determine $P(x > 370)$.

d) Interpret your results in parts (b) and (c) in words.

10   Refer to Problem 9. Fill in the following blanks:

a) The probability is 0.6826 that a Morgan horse will have a gestation period between _____ and _____ days.

b) The probability is 0.9544 that a Morgan horse will have a gestation period between _____ and _____ days.

c) The probability is 0.9974 that a Morgan horse will have a gestation period between _____ and _____ days.

11   Refer to Problem 9.

a) Find the standardized version, z, of the random variable x.

b) What does the random variable z represent?

c) How many standard deviations away from the mean is a gestation period of
    (i)   325 days?   (ii) 368 days?
    (iii) 339.6 days?

d) What is the probability distribution of the random variable z?

e) Is it necessary to assume that the gestation periods are normally distributed in answering part (b)? What about part (d)?

12[*]   Acute rotavirus diarrhea is the leading cause of death among children under age five, killing an estimated 4.5 million annually in developing countries. Scientists from Finland and Belgium claim that a new oral vaccine is 80% effective against rotavirus diarrhea. Assuming that the claim is correct, find the probability that, out of 1500 cases, the vaccine will be effective in

a) exactly 1225 cases.

b) at least 1175 cases.

c) between 1150 and 1250 cases, inclusive.

In the previous chapters we have studied descriptive statistics, probability, random variables, and the normal distribution. In this chapter we will begin to learn how these seemingly diverse topics can be combined to lay the groundwork for inferential statistics.

We first discuss the necessity for sampling, indicate some common sampling techniques, and explain why it is necessary to incorporate probability distributions into the design of statistical inferences. Following that, we introduce the *sampling distribution of the mean*. This concept sets the stage for one of the most important statistical inference procedures—using the mean, $\overline{x}$, of a sample from a population to make inferences about the population's mean, μ.

The sampling distribution
of the mean

CHAPTER OUTLINE

7.1   Sampling; random samples

Recall that *inferential statistics* consists of methods of drawing conclusions about a population based on information obtained from a sample of the population. We have already discussed some reasons why using a sample to obtain information about a population is often preferable to conducting a *census* (collecting data for every member of the population). Sampling is less costly, it can be done more quickly than a census, and it is often more accurate than a census.

You might find it curious to see "more accurate" on the list of reasons for sampling. It is often presumed that sampling results are inevitably less accurate than census results. The reasoning here is that because a census examines the entire population and sampling surveys only a portion of the population, it follows that a census is more accurate. However, this is not necessarily the case. Indeed, since sampling is conducted on a smaller scale, response errors and processing errors can be controlled more easily than they can when surveying an entire population.

Once a researcher decides that sampling is appropriate, the next question to consider is *how* to select the sample. That is, what procedure should be employed to obtain the sample from the population? In answering this question, it is necessary to keep in mind the use that will be made of the sample results; namely, that the sample results will be applied to make inferences concerning the entire population.

Clearly then, it is important for the sample to be **representative.** That is, the sample should reflect as closely as possible the relevant characteristic(s) of the population under consideration. For instance, it would not make much sense to use the mean weight of a sample of football players to make an inference about the mean weight of all U.S. adult males. Nor would it be reasonable to try to estimate the mean income of California residents by sampling the incomes of residents of wealthy Beverly Hills.

To see what can happen if a sample is *not* representative, consider the presidential election of 1936. Before the election, the *Literary Digest* magazine conducted an opinion poll of the voting population. Its survey team asked a sample of the voting population whether they would vote for Roosevelt (the Democratic candidate) or for Landon (the Republican candidate). Based on the results of the survey, the magazine predicted an easy win for Landon. The actual election results, of course, were that Roosevelt won by the greatest landslide in the history of the presidency! What happened? Some people think the error occurred because the magazine had obtained its sample from among people who owned a car or had a telephone. At the time, this group included only the more well-to-do people and, historically, such people tend to vote Republican. Whatever the reason, the sample taken by the *Literary Digest* was obviously *not* representative.

Our goal in sampling, then, is to obtain a representative sample. Only then can we expect to make reasonable inferences about a population based on information obtained from a sample. Many different sampling techniques have been developed to cover the diversity of circumstances in which a sample is required. However, all of these sampling procedures have one thing in common—to obtain a representative sample.

Random sampling

The statistical inference methods that you will study in this book are designed for use with one particular sampling procedure—**simple random sampling,** or more briefly, **random sampling.**

DEFINITION 7.1 Random sampling and random samples

A **random sampling** procedure is one in which each possible sample of a given size is equally likely to be the one selected. A sample obtained by random sampling is called a **random sample.**

EXAMPLE   7.1 **Illustrates random sampling**

Suppose that the *population* under consideration consists of the annual salaries of the five top state officials of Oklahoma. Clearly, for such a small population (the population of five annual salaries) there is really no need to sample in order to acquire information. For instance, if we were interested in the population mean salary, μ, we would undoubtedly just obtain the salaries of the five officials and compute μ exactly. However, because this population is so small, it is easy to use it to illustrate the concept of random sampling. The population of salaries is presented in the second column of Table 7.1. Values are in thousands of dollars.

TABLE 7.1

Top five Oklahoma state officials and their annual salaries.

Official	Salary ($thousands)
Governor (G)	70
Lieutenant Governor (L)	40
Secretary of State (S)	37
Attorney General (A)	55
Treasurer (T)	50

Source: *The World Almanac, 1985.*

For convenience, we will use letters (placed parenthetically after each official in Table 7.1) to represent the officials.

Let us consider samples of size $n = 2$. There are 10 possible samples of size two.[†] They are listed in the second column of Table 7.2.

TABLE 7.2
Possible samples of two salaries from the population of five salaries.

Officials selected	Sample obtained
G, L	70, 40
G, S	70, 37
G, A	70, 55
G, T	70, 50
L, S	40, 37
L, A	40, 55
L, T	40, 50
S, A	37, 55
S, T	37, 50
A, T	55, 50

One method we could use to obtain a random sample of size $n = 2$ is the following: Write each of the letters G, L, S, A, and T (corresponding to the five officials) on a separate piece of paper. Next place the five slips of paper into a box, shake the box and then, while blindfolded, pick two of the slips of paper. This procedure would ensure that we are taking a *random sample.* That is, it would guarantee that each of the 10 possible samples of two salaries has the same probability of being the one selected—a probability of 1/10.

If we consider samples of size $n = 4$, there are five possibilities, as indicated in Table 7.3.

TABLE 7.3
Possible samples of four salaries from the population of five salaries.

Officials selected	Sample obtained
G, L, S, A	70, 40, 37, 55
G, L, S, T	70, 40, 37, 50
G, L, A, T	70, 40, 55, 50
G, S, A, T	70, 37, 55, 50
L, S, A, T	40, 37, 55, 50

MTB
SPSS

In this case, a random sampling procedure, such as picking four slips of paper out of the box, gives each of the five possible samples in the second column of Table 7.3 a probability of 1/5 of being the one selected. ∎

The method of picking slips of paper out of a box to get a random sample is not very practical, especially if the population being sampled is large. There are several standard procedures for obtaining random samples. One such procedure involves the use of a **table of random numbers.** Another employs **random-**

[†] The number of samples of size n from a population of size N is given by the binomial coefficient $\binom{N}{n}$. In this case, $n = 2$ and $N = 5$ so that $\binom{N}{n} = \binom{5}{2} = 10$.

number generators, which are available on many calculators and most computers.

Finally, as we mentioned earlier in this section, the statistical inference procedures covered in this book are intended for use with (simple) random samples. Consequently, when we refer to a sample of a population, we will mean a *random sample,* unless we explicitly state otherwise.

Exercises 7.1

7.1   Explain why the following sample is *not* representative. A sample of incomes from 30 dentists in Seattle is taken in order to estimate the median income of all Seattle residents.

7.2   A survey of the political opinions of 150 voters in the retirement community of Sun City, Arizona is used as a sample of the political opinions of all Arizona voters. Is the sample representative? Justify your answer.

7.3   The annual salaries of the five top Oklahoma state officials are repeated below.

Official	Salary ($thousands)
Governor (G)	70
Lieutenant Governor (L)	40
Secretary of State (S)	37
Attorney General (A)	55
Treasurer (T)	50

a) List the 10 possible samples of size $n = 3$ that can be obtained from the population of five salaries. [Construct a table similar to Table 7.3 on page 267.]
b) In a random-sampling procedure for obtaining three salaries, what is the probability of selecting the first sample on your list in part (a)? the second sample? the tenth sample?

7.4   As of the 1984 election, the members of the U.S. House of Representatives from South Carolina were Thomas Hartnett (TH), Floyd Spence (FS), Butler Derrick (BD), Carroll Campbell, Jr. (CC), John Spratt (JS), and Robert Tallon, Jr. (RT).
a) List the 15 possible samples of $n = 2$ representatives that can be selected from the six. Use initials for brevity.
b) Describe a procedure for taking a random sample of two representatives from the six.
c) In a random-sampling procedure to obtain two representatives, what is the probability of selecting TH and FS? BD and RT?

7.5   Refer to Exercise 7.4.
a) List the 15 possible samples of $n = 4$ representatives that can be selected from the six.
b) Describe a procedure for taking a random sample of four representatives from the six.
c) In a random-sampling procedure to obtain four representatives, what is the probability of selecting TH, CC, JS, and RT? FS, BD, CC, and RT?

7.6   Refer to Exercise 7.4.
a) List the 20 possible samples of $n = 3$ representatives that can be selected from the six.
b) Describe a procedure for taking a random sample of three representatives from the six.
c) In a random-sampling procedure to obtain three representatives, what is the probability of selecting TH, FS, and BD? BD, JS, and RT?

7.2   Sampling error; the need for sampling distributions

Now that we have seen why it is important to sample and we have discussed some methods for obtaining a random sample, we need to deal with the following problem. A sample from a population provides us with only a portion of the entire population data. Consequently, it is unreasonable to expect that the sample will yield perfectly accurate information about the population. Thus, we should anticipate that a certain amount of error will result simply because we are sampling—so-called **sampling error.**

EXAMPLE   7.2   **Illustrates sampling error and the need for sampling distributions**

The U.S. Bureau of the Census publishes annual figures on the mean income of households in the United States. Actually, the Census Bureau reports the mean income of a *sample* of 72,000 households out of a total of more than 83 million households. For example, in 1982 the mean income of U.S. households was reported to be $24,309. This is really the *sample mean* income, $\bar{x}$, of the 72,000 households surveyed and not the true (population) mean income, μ, of all U.S. households.

We certainly cannot expect the mean income, $\bar{x}$, of the 72,000 households sampled by the Census Bureau to be exactly the same as the unknown mean income, μ, of all U.S. households—some sampling error is to be anticipated. But how accurate are such estimates likely to be? Is it likely, for example, that the sample mean household income reported by the Census Bureau is within $1,000 of the actual population mean household income?

MTB
SPSS

To answer these and similar questions, we need to know the distribution of all possible sample means that could be obtained by sampling 72,000 households. This type of distribution is called a **sampling distribution.** ∎

In the previous example we introduced the idea of sampling error—the error resulting from the use of a sample to estimate a population quantity. The example dealt specifically with the sampling error arising from using a sample mean, $\bar{x}$, to estimate a population mean, μ. We mentioned that in order to answer questions regarding the accuracy of such estimates, we need to know the distribution of all possible $\bar{x}$-values that could be obtained. That distribution is called the **sampling distribution of the mean.** In this and the next two sections we will study the sampling distribution of the mean. Later we will examine other sampling distributions, such as the sampling distribution of a proportion.

It is difficult to introduce the sampling distribution of the mean with an example that is both realistic and concrete. This is because, even for moderately large populations, the number of possible samples is enormous, thus prohibiting an actual listing of the possibilities. Consequently, we will use an unrealistically small population to introduce the sampling distribution of the mean. Keep in mind, however, that in real-life applications the populations are much larger.

EXAMPLE   7.3   **Introduces the sampling distribution of the mean**

Let us take as our *population* of interest the heights of the five starting players on a men's basketball team. The heights, in inches, are displayed in Table 7.4.

TABLE 7.4
Population of heights.

76	78	79	81	86

The mean of this population is

$$\mu = \frac{\Sigma x}{N} = \frac{76 + 78 + 79 + 81 + 86}{5} = 80$$

Suppose we decide to use the sample mean, $\bar{x}$, of a random sample of two of these heights to estimate the population mean, μ. [We emphasize that this is not realistic. For a population of this size, we would not need to sample, but would simply take a census. The purpose of this example is completely illustrative.]

We are interested in finding the distribution of the possible $\bar{x}$-values for random samples of size $n = 2$. The population here is so small that we can list the possibilities directly. In fact, there are 10 possible samples of size $n = 2$. These 10 samples, along with their means, are displayed in Table 7.5. We have also drawn a dotplot for the $\bar{x}$-values in Figure 7.1 in which the location of each $\bar{x}$-value is shown by a dot.

TABLE 7.5

Possible samples, with their means, for samples of two heights from the population of five heights.

Sample	$\bar{x}$
76, 78	77.0
76, 79	77.5
76, 81	78.5
76, 86	81.0
78, 79	78.5
78, 81	79.5
78, 86	82.0
79, 81	80.0
79, 86	82.5
81, 86	83.5

FIGURE 7.1

Dotplot of $\bar{x}$-values for samples of size $n = 2$.

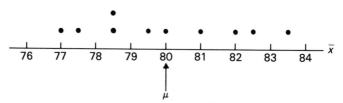

From the second column of Table 7.5, we see that the value of $\bar{x}$ varies from sample to sample. For instance, the first sample has a mean of $\bar{x} = 77.0$, while the second one has a mean of $\bar{x} = 77.5$. Thus, we see that, when sampling from a population, the sample mean, $\bar{x}$, is a *random variable* because its value depends on chance—namely, on which sample is obtained.

Since we are considering *random samples,* each of the 10 possible samples shown in the first column of Table 7.5 has the same probability of being the one selected—a probability of 1/10, or 0.1. Using that fact along with the list of $\bar{x}$-values in the second column of Table 7.5, we can obtain the probability distribution (sampling distribution) of the random variable $\bar{x}$. This is given in Table 7.6.

TABLE 7.6
Probability distribution
(sampling distribution) of $\bar{x}$
for random samples of size
$n = 2$.

Sample mean $\bar{x}$	Probability $P(\bar{x})$	
77.0	0.1	
77.5	0.1	
78.5	0.2	*(Two of the 10 samples have $\bar{x} = 78.5$)*
79.5	0.1	
80.0	0.1	
81.0	0.1	
82.0	0.1	
82.5	0.1	
83.5	0.1	

We can make some simple, but significant, observations from Table 7.6 and Figure 7.1. We see that it is unlikely for the sample mean, $\bar{x}$, of the sample selected to be exactly the same as the population mean $\mu = 80$. As a matter of fact, just one sample out of the 10 has $\bar{x} = 80$, namely, the eighth sample in Table 7.5 on page 270. Thus, in this case, the chances are only 1/10 (10%) that $\bar{x}$ will equal μ—it is quite likely that there will be some sampling error.

Recall that the purpose of determining sampling distributions is to enable us to answer questions regarding the accuracy of sampling results. To illustrate how such questions can be answered once we have obtained the sampling distribution, we will apply Table 7.6 to find the probability that the mean, $\bar{x}$, of the sample selected will be within 1 inch of the true population mean of $\mu = 80$ inches. That is, we will determine $P(79 \leq \bar{x} \leq 81)$. From Table 7.6 we find that

$$P(79 \leq \bar{x} \leq 81) = P(\bar{x} = 79.5, 80.0, \text{ or } 81.0)$$
$$= P(79.5) + P(80.0) + P(81.0)$$
$$= 0.1 + 0.1 + 0.1 = 0.3$$

If we take a random sample of two heights, there is a 30% chance that the mean of the sample selected will be within 1 inch of the true population mean. ∎

Below we summarize one of the important facts discussed in Example 7.3.

KEY FACT

When sampling from a population, the sample mean, $\bar{x}$, is a random variable, because its value depends on chance—namely, on which sample is obtained.

EXAMPLE 7.4 **Further illustrates the sampling distribution of the mean**

Let us return to the population of five heights considered in Example 7.3:

76   78   79   81   86

Up to now, we have considered samples of size $n = 2$. If we consider samples of another size, say $n = 4$, then we obtain a new and different random variable $\bar{x}$.[†]

[†] Some texts use subscripts on $\bar{x}$ in order to emphasize that each sample size gives rise to a different random variable $\bar{x}$. Thus, for instance, $\bar{x}_2$ and $\bar{x}_4$ are used to denote the sample means for samples of sizes two and four, respectively.

There are five possible samples of size $n = 4$. These five samples are listed in Table 7.7 along with their means. A dotplot of the possible $\bar{x}$-values is shown in Figure 7.2.

TABLE 7.7

Possible samples, with their means, for samples of four heights from the population of five heights.

Sample	$\bar{x}$
76, 78, 79, 81	78.50
76, 78, 79, 86	79.75
76, 78, 81, 86	80.25
76, 79, 81, 86	80.50
78, 79, 81, 86	81.00

FIGURE 7.2

Dotplot of $\bar{x}$-values for samples of size $n = 4$.

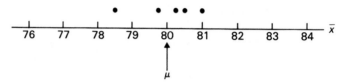

Since we are dealing with random samples, each of the five possible samples of size $n = 4$ has probability $1/5 = 0.2$ of being the one selected. From these facts we can now write down the probability distribution of the random variable $\bar{x}$ for samples of size $n = 4$. See Table 7.8.

TABLE 7.8

Probability distribution (sampling distribution) of $\bar{x}$ for random samples of size $n = 4$.

Sample mean $\bar{x}$	Probability $P(\bar{x})$
78.50	0.2
79.75	0.2
80.25	0.2
80.50	0.2
81.00	0.2

Let us use Table 7.8 to find the probability that the mean, $\bar{x}$, of the sample selected will be within 1 inch of the true population mean of $\mu = 80$:

$$P(79 \leq \bar{x} \leq 81) = P(\bar{x} = 79.75, 80.25, 80.50, \text{ or } 81.00)$$
$$= P(79.75) + P(80.25) + P(80.50) + P(81.00)$$
$$= 0.2 + 0.2 + 0.2 + 0.2 = 0.8$$

If we take a random sample of four heights, there is an 80% chance that the mean of the sample selected will be within 1 inch of the true population mean. ∎

Our examples have shown that, when sampling from a population, the sample mean, $\bar{x}$, is a random variable. As mentioned earlier, the probability distribution of $\bar{x}$ is given a special name—*the sampling distribution of the mean.*

DEFINITION 7.2 The sampling distribution of the mean

The probability distribution of the random variable $\bar{x}$ is called the **sampling distribution of the mean.**

Let us emphasize again that to each sample size there corresponds a different random variable $\overline{x}$ and hence a different sampling distribution of the mean.

Sample size and sampling error

In Figures 7.1 and 7.2 we drew dotplots of the possible $\overline{x}$-values for samples of size $n = 2$ and $n = 4$, respectively. Those two dotplots, as well as one for samples of size $n = 3$, are displayed together in Figure 7.3.

FIGURE 7.3
Dotplots of possible $\overline{x}$-values for sample sizes $n = 2$, $n = 3$, and $n = 4$.

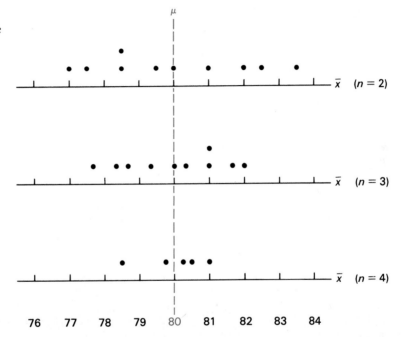

As you can see, the possible $\overline{x}$-values tend to cluster closer around the true mean, μ, as the sample size increases. For example, when $n = 2$, three out of 10, or 30%, of the possible $\overline{x}$-values lie within 1 inch of μ; when $n = 3$, five out of 10, or 50%, of the possible $\overline{x}$-values lie within 1 inch of μ; and when $n = 4$, four out of five, or 80%, of the possible $\overline{x}$-values lie within 1 inch of μ. More generally, we can make the following qualitative statement concerning the sampling distribution of the mean.

KEY FACT

The larger the sample size, n, the greater the tendency for the possible $\overline{x}$-values to cluster closely around the population mean μ, indicating that the sampling error in estimating μ by $\overline{x}$ *tends* to be smaller.

Concluding remarks

We have used a population of five heights to illustrate and explain the importance of the sampling distribution of the mean (that is, the probability distribution of $\bar{x}$). For that small population, it is easy to obtain the sampling distribution of the mean by listing all of the possible samples.

However, as we have noted, the populations we deal with in statistical inference are usually quite large. For such populations, it is not feasible to obtain the sampling distribution of the mean by a direct listing. A more serious practical problem is that, in reality, we do not even know the population values—if we did, there would be no need to sample! Realistically then, we could not list the possible samples to determine the sampling distribution of the mean even if we were willing to "put in the time." So what can we do in the usual case of a *large and unknown* population? Fortunately, there exist mathematical relationships that allow us to determine, at least approximately, the sampling distribution of the mean. We will discuss those relationships in the next two sections.

Exercises 7.2

The following exercises are intended solely to give concrete illustrations of the sampling distribution of the mean. For that reason the populations considered are unrealistically small.

7.7   The annual salaries of the governors of the Great Lakes states are given below in thousands of dollars, rounded to the nearest thousand.

> 76   58   64   82   60

a) Compute the population mean of these five salaries.
b) List the possible samples of size $n = 2$ along with their means. [Construct a table similar to Table 7.5 on page 270.]
c) Draw a dotplot for the possible $\bar{x}$-values like the one in Figure 7.1.
d) Determine the probability distribution (sampling distribution) of $\bar{x}$ for random samples of size $n = 2$. [Construct a table similar to Table 7.6 on page 271.]
e) Find the probability that the sample mean, $\bar{x}$, will equal the population mean μ. That is, determine $P(\bar{x} = \mu)$.
f) What is the probability that the mean of the sample selected will be within 4 (that is, $4,000) of the population mean? Interpret your results in terms of percentages.

7.8   Repeat parts (b)–(f) of Exercise 7.7 for samples of size $n = 3$.

7.9   Repeat parts (b)–(f) of Exercise 7.7 for samples of size $n = 4$.

7.10   This exercise requires that you have done Exercises 7.7 through 7.9.
a) Draw a graph similar to Figure 7.3 on page 273 for sample sizes of $n = 2$, $n = 3$, and $n = 4$ from the population of the five governors' salaries.
b) What does your graph in part (a) illustrate about the impact that increasing sample size has on sampling error?

7.11   Below are the lengths, in centimeters, of a population of six bullfrogs in a small pond:

> 19   14   15   9   16   17

a) Calculate the mean length, μ, of the population of bullfrogs.
b) List the possible samples of size $n = 2$ along with their means. [There are 15 possible samples of size $n = 2$.]
c) Draw a dotplot for the possible $\bar{x}$-values.
d) Determine the sampling distribution of $\bar{x}$ for random samples of size $n = 2$. [Leave your probabilities in fraction form.]
e) Find the probability that the sample mean, $\bar{x}$, of a random sample of size $n = 2$ will equal the population mean, μ.
f) What is the probability that the mean of the sample selected will be within 1 cm of the population mean? Interpret your results in terms of percentages.

7.12   Repeat parts (b)–(f) of Exercise 7.11 for samples of size $n = 3$. [There are 20 possible samples of size $n = 3$.]

7.13   Repeat parts (b)–(f) of Exercise 7.11 for samples of size $n = 4$. [There are 15 possible samples of size $n = 4$.]

7.14   Repeat parts (b)–(f) of Exercise 7.11 for samples of size $n = 5$. [There are six possible samples of size $n = 5$.]

7.15   Repeat parts (b)–(f) of Exercise 7.11 for samples of size $n = 6$. What is the relationship between the only possible sample here and the population?

7.16   Repeat parts (b)–(f) of Exercise 7.11 for samples of size $n = 1$.

7.17   What do the dotplots in parts (c) of Exercises 7.11 through 7.16 illustrate about the impact that increasing sample size has on sampling error?

7.18   Suppose a random sample is to be taken from a finite population of size N. Assume the sampling is without replacement. If the sample size is the same as the population size (that is, $n = N$),
   a) how many possible samples are there?
   b) what are the possible $\bar{x}$-values?
   c) what is the relationship between the only possible sample and the population?

7.19   Suppose a random sample of size $n = 1$ is to be taken from a finite population of size N.
   a) How many possible samples are there?
   b) What is the relationship between the set of all possible $\bar{x}$-values and the population values?
   c) What is the difference between taking a random sample of size $n = 1$ from a population and selecting a member at random from the population?

7.3   The mean and standard deviation of $\bar{x}$

In Section 7.2 we introduced the sampling distribution of the mean (that is, the probability distribution of $\bar{x}$). This sampling distribution is used for making statistical inferences about a population mean based on the mean of a sample from the population.

As we said, it is generally not possible to know the probability distribution of $\bar{x}$ exactly. Fortunately, however, in a large number of circumstances, we can *approximate* the probability distribution of $\bar{x}$ by a *normal distribution.* That is, we can approximate probabilities for $\bar{x}$ by areas under a normal curve. In Chapter 6 we learned that in order to obtain probabilities for a random variable using normal curve areas, we must know the mean and standard deviation of that random variable, because those are also the parameters for the appropriate normal curve. Consequently, this section examines the mean and standard deviation of the random variable $\bar{x}$.

Before we proceed, it is important to recall the notation used for the mean and standard deviation of a random variable. The mean of a random variable is denoted by a μ *subscripted* with the letter representing the random variable. Thus the mean of x is written as μ_x, the mean of y as μ_y, and so on. In particular then, the mean of $\bar{x}$ is denoted by $\mu_{\bar{x}}$. Similarly, the standard deviation of the random variable $\bar{x}$ is denoted by $\sigma_{\bar{x}}$.

The mean of $\bar{x}$

There is a very simple relationship between the mean of the random variable $\bar{x}$ and the mean of the population being sampled. Namely, the mean of $\bar{x}$ is the *same* as the population mean. In symbols,

$$\mu_{\bar{x}} = \mu$$

In other words, for a given sample size, the mean of all possible sample means ($\bar{x}$-values) is the same as the population mean. It is important to note that this holds regardless of the sample size under consideration.

To illustrate the relationship $\mu_{\bar{x}} = \mu$ it is best to use a small population, such as the one considered in Examples 7.3 and 7.4.

EXAMPLE   7.5   Illustrates the relation $\mu_{\bar{x}} = \mu$

The heights, in inches, of the five starting players on a men's basketball team are

$$76 \quad 78 \quad 79 \quad 81 \quad 86$$

Take these five heights to be the population under consideration.

a) Determine the population mean, μ.

b) For samples of size $n = 2$, compute the mean, $\mu_{\bar{x}}$, of the random variable $\bar{x}$ and verify that the relation $\mu_{\bar{x}} = \mu$ holds.

c) Repeat part (b) for samples of size $n = 4$.

S O L U T I O N

a) To determine the population mean, μ, we apply Definition 3.6 (page 95):

$$\mu = \frac{\Sigma x}{N} = \frac{76 + 78 + 79 + 81 + 86}{5} = 80$$

Thus, the population mean height of the five players is $\mu = 80$ inches.

b) The probability distribution of $\bar{x}$ for samples of size $n = 2$, determined in Example 7.3, is repeated in the first two columns of Table 7.9. To compute $\mu_{\bar{x}}$ we proceed as usual: We append an $\bar{x}P(\bar{x})$-column to the probability distribution of $\bar{x}$ and apply Definition 5.3 (page 181). See Table 7.9.

TABLE 7.9
Table for computation of $\mu_{\bar{x}}$
for samples of size $n = 2$.

$\bar{x}$	$P(\bar{x})$	$\bar{x}P(\bar{x})$
77.0	0.1	7.70
77.5	0.1	7.75
78.5	0.2	15.70
79.5	0.1	7.95
80.0	0.1	8.00
81.0	0.1	8.10
82.0	0.1	8.20
82.5	0.1	8.25
83.5	0.1	8.35
Σ		80.00

From the third column of Table 7.9, we find that

$$\mu_{\bar{x}} = \Sigma \bar{x}P(\bar{x}) = 80$$

By part (a), $\mu = 80$. So, we see that $\mu_{\bar{x}} = \mu$.

c) The probability distribution of $\bar{x}$ for samples of size $n = 4$, determined in Example 7.4, is repeated in the first two columns of Table 7.10. The third column displays the $\bar{x}P(\bar{x})$-values.

TABLE 7.10
Table for computation of $\mu_{\bar{x}}$
for samples of size $n = 4$.

$\bar{x}$	$P(\bar{x})$	$\bar{x}P(\bar{x})$
78.50	0.2	15.70
79.75	0.2	15.95
80.25	0.2	16.05
80.50	0.2	16.10
81.00	0.2	16.20
Σ		80.00

From the third column of Table 7.10, we find that $\mu_{\bar{x}} = 80$, which again is the same as μ.   ■

Below we give a formal statement of the relationship between the mean of $\bar{x}$ and the population mean.

FORMULA 7.1   Mean of the random variable $\bar{x}$

The mean of the random variable $\bar{x}$ is the same as the mean of the population, regardless of sample size. In other words,

$$\mu_{\bar{x}} = \mu$$

The standard deviation of $\bar{x}$

Next we will consider the standard deviation of the random variable $\bar{x}$. Let us return once more to our population of heights in order to examine the standard deviation, $\sigma_{\bar{x}}$, of the random variable $\bar{x}$ and to uncover any apparent relationships it has with the population standard deviation, σ.

EXAMPLE   7.6   Examines $\sigma_{\bar{x}}$ and its relation to σ

The population of heights of the five starting players on a men's basketball team is

76   78   79   81   86

a) Determine the population standard deviation, σ, by applying the shortcut formula

$$\sigma = \sqrt{\frac{\Sigma x^2}{N} - \mu^2}$$

(Definition 3.7, page 97).

b) For samples of size $n = 2$, compute the standard deviation, $\sigma_{\bar{x}}$, of the random variable $\bar{x}$ by using the shortcut formula

$$\sigma_{\bar{x}} = \sqrt{\Sigma \bar{x}^2 P(\bar{x}) - \mu_{\bar{x}}^2}$$

(Formula 5.1, page 187). Indicate any apparent relationship between $\sigma_{\bar{x}}$ and σ.

c) Repeat part (b) for samples of sizes $n = 3$ and $n = 4$.

d) Summarize and discuss the results obtained in parts (a)–(c).

S O L U T I O N

a) Recalling that $\mu = 80$, we have

$$\sigma = \sqrt{\frac{\Sigma x^2}{N} - \mu^2} = \sqrt{\frac{76^2 + 78^2 + 79^2 + 81^2 + 86^2}{5} - 80^2}$$

$$= \sqrt{\frac{32{,}058}{5} - 6400} = \sqrt{11.6} = 3.41$$

Thus, the standard deviation of the population of heights is $\sigma = 3.41$ inches.

b) The probability distribution of $\bar{x}$ for samples of size $n = 2$ is repeated in the first two columns of Table 7.11. The third column gives the $\bar{x}^2 P(\bar{x})$-values. These are required to compute $\sigma_{\bar{x}}$ using the shortcut formula.

TABLE 7.11
Table for computation of $\sigma_{\bar{x}}$
for samples of size $n = 2$.

$\bar{x}$	$P(\bar{x})$	$\bar{x}^2 P(\bar{x})$
77.0	0.1	592.900
77.5	0.1	600.625
78.5	0.2	1232.450
79.5	0.1	632.025
80.0	0.1	640.000
81.0	0.1	656.100
82.0	0.1	672.400
82.5	0.1	680.625
83.5	0.1	697.225
Σ		6404.350

Recalling that $\mu_{\bar{x}} = \mu = 80$, we find from the third column of Table 7.11 that

$$\sigma_{\bar{x}} = \sqrt{\Sigma \bar{x}^2 P(\bar{x}) - \mu_{\bar{x}}^2} = \sqrt{6404.350 - 80^2}$$

$$= \sqrt{4.35} = 2.09$$

Thus, for samples of size $n = 2$, the standard deviation of $\bar{x}$ is $\sigma_{\bar{x}} = 2.09$. Note that this is *not* the same as the population standard deviation, which is $\sigma = 3.41$, and that $\sigma_{\bar{x}}$ is *smaller* than σ.

c) Using the same procedure as in part (b), we can compute $\sigma_{\bar{x}}$ for both samples of size $n = 3$ and $n = 4$. For samples of size $n = 3$, $\sigma_{\bar{x}} = 1.39$. For samples of size $n = 4$, $\sigma_{\bar{x}} = 0.85$.

d) Table 7.12 summarizes our results for $\sigma_{\bar{x}}$.

TABLE 7.12
The standard deviation of $\bar{x}$
for samples of sizes $n = 2$,
$n = 3$, and $n = 4$.

Sample size n	Standard deviation of $\bar{x}$ $\sigma_{\bar{x}}$
2	2.09
3	1.39
4	0.85

The table suggests that *the standard deviation of $\bar{x}$ gets smaller as the sample size gets larger.* We could have predicted this phenomenon by looking at the dotplots of the $\bar{x}$-values in Figure 7.3 on page 273 and by recalling that the standard deviation of a random variable measures the spread of the possible values of the random variable. Figure 7.3 indicates graphically that the spread, and hence standard deviation, of the possible $\bar{x}$-values decreases with increasing sample size. Table 7.12 indicates the same thing numerically. ∎

The previous example suggests that as the sample size increases, $\sigma_{\bar{x}}$ gets smaller. Our question now is whether there is a formula that relates the standard deviation of $\bar{x}$ to the standard deviation of the population. The answer is yes! In fact, two different formulas express the precise relationship between the standard deviation of $\bar{x}$ and the standard deviation of the population. One formula is appropriate when the sampling is being done *without* replacement from a finite population, as is the case in the previous example. The formula is

$$\sigma_{\bar{x}} = \sqrt{\frac{N-n}{N-1}} \cdot \frac{\sigma}{\sqrt{n}}$$

where, as usual, n = sample size and N = population size.

The other formula applies when the sampling is being done *with* replacement from a finite population or when the sampling is being done from an infinite population. In this case the formula is

$$\sigma_{\bar{x}} = \frac{\sigma}{\sqrt{n}}$$

If the sample size is small relative to the size of the population, then there is very little difference in the values given by the two formulas (see Exercise 7.37 on page 282). In most applications of statistical inference the sample size *is* small relative to the size of the population. Therefore, we will use the second formula exclusively in this book since it is simpler than the first.

FORMULA 7.2 Standard deviation of the random variable $\bar{x}$

For random samples of size n, the standard deviation of the random variable $\bar{x}$ is equal, at least approximately, to the standard deviation of the population divided by the square root of the sample size. That is,

$$\sigma_{\bar{x}} = \frac{\sigma}{\sqrt{n}}$$

Applying the formulas

We have seen that there are simple formulas relating the mean and standard deviation of the random variable $\bar{x}$ to the mean and standard deviation of the population being sampled. Namely,

$$\mu_{\bar{x}} = \mu \quad \text{and} \quad \sigma_{\bar{x}} = \sigma/\sqrt{n}$$

Let us apply these two formulas in an example.

EXAMPLE 7.7   Illustrates Formulas 7.1 and 7.2

According to the U.S. Energy Information Administration, the mean livable square footage for single-family detached homes is $\mu = 1742$ sq. ft. The standard deviation is $\sigma = 568$ sq. ft.

a) Suppose 25 single-family detached homes are to be selected at random. Let $\bar{x}$ denote the mean livable square footage of the homes selected. Find the mean, $\mu_{\bar{x}}$, and the standard deviation, $\sigma_{\bar{x}}$, of the random variable $\bar{x}$.

b) Repeat part (a) for a sample of size 500.

S O L U T I O N
a) We have

$$\mu_{\bar{x}} = \mu = 1742$$

and, since $n = 25$,

$$\sigma_{\bar{x}} = \frac{\sigma}{\sqrt{n}} = \frac{568}{\sqrt{25}} = 113.60$$

The mean of all possible sample means for samples of size $n = 25$ is 1742, and the standard deviation is 113.60.

b) We have

$$\mu_{\bar{x}} = \mu = 1742$$

and since here $n = 500$,

$$\sigma_{\bar{x}} = \frac{\sigma}{\sqrt{n}} = \frac{568}{\sqrt{500}} = 25.40$$

■

Sample size and sampling error (revisited)

On page 273 of Section 7.2, we pointed out that the sampling error made in estimating a population mean μ by a sample mean $\bar{x}$ tends to be smaller with increasing sample size. We can now make that statement a bit more precise.

Recall from Section 5.2 (page 187) that the standard deviation of a random variable measures the dispersion of the possible values of the random variable relative to its mean—the smaller the standard deviation, the more likely that the random variable will take a value near its mean. In particular, the smaller the value of $\sigma_{\bar{x}}$, the more likely that $\bar{x}$ will take a value near its mean, $\mu_{\bar{x}}$. But the mean of $\bar{x}$ is the same as the population mean ($\mu_{\bar{x}} = \mu$). Thus, the smaller the value of $\sigma_{\bar{x}}$, the more likely that $\bar{x}$ will take a value near μ. That is, the smaller the value of $\sigma_{\bar{x}}$, the greater the tendency for small sampling error.

KEY FACT

The standard deviation $\sigma_{\bar{x}}$ indicates the amount of sampling error to be expected in estimating the mean, μ, of a population by the mean, $\bar{x}$, of a sample from the population. The smaller the value of $\sigma_{\bar{x}}$, the smaller the sampling error tends to be. Hence, $\sigma_{\bar{x}}$ is often referred to as the **standard error of the mean.**

Finally, we have seen that the standard deviation of $\bar{x}$ is equal to the standard deviation of the population divided by the square root of the sample size ($\sigma_{\bar{x}} = \sigma/\sqrt{n}$). Since n appears in the denominator, the larger the sample size, the smaller the value of $\sigma_{\bar{x}}$. Combining this fact with the previous Key Fact, we can make the following fundamental statement:

KEY FACT

When estimating a population mean by a sample mean, the larger the sample size, the greater the likelihood of a small sampling error. In other words, larger sample sizes tend to produce more accurate results.

Exercises 7.3

7.20 Why do we need to know the mean, $\mu_{\bar{x}}$, and standard deviation, $\sigma_{\bar{x}}$, of the random variable $\bar{x}$ if we are going to use normal curve areas to approximate probabilities for $\bar{x}$?

7.21 The annual salaries for the governors of the Great Lakes states are given below in thousands of dollars, rounded to the nearest thousand.

$$76 \quad 58 \quad 64 \quad 82 \quad 60$$

a) Compute the population mean, μ, of these five salaries.
b) Consider samples of size $n = 2$. Find the mean, $\mu_{\bar{x}}$, of the random variable $\bar{x}$ by applying the formula $\mu_{\bar{x}} = \Sigma \bar{x} P(\bar{x})$. [The probability distribution of $\bar{x}$ for samples of size $n = 2$ was found in part (d) of Exercise 7.7.]
c) Find $\mu_{\bar{x}}$ without doing any computations.

7.22 Repeat parts (b) and (c) of Exercise 7.21 for samples of size $n = 3$. [The probability distribution of $\bar{x}$ for samples of size $n = 3$ was found in part (d) of Exercise 7.8.]

7.23 Repeat parts (b) and (c) of Exercise 7.21 for samples of size $n = 4$. [The probability distribution of $\bar{x}$ for samples of size $n = 4$ was found in part (d) of Exercise 7.9.]

In Exercises 7.24 through 7.28, find the mean and standard deviation of $\bar{x}$ for the indicated sample sizes by employing the formulas

$$\mu_{\bar{x}} = \mu \text{ and } \sigma_{\bar{x}} = \frac{\sigma}{\sqrt{n}}$$

7.24 According to the U. S. Bureau of the Census, the mean price of new mobile homes is $\mu = \$19,700$.

The standard deviation is $\sigma = \$4200$.
a) Suppose a random sample of 50 new mobile homes is to be selected. Let $\bar{x}$ be the mean price of the homes obtained. Find $\mu_{\bar{x}}$ and $\sigma_{\bar{x}}$.
b) Repeat part (a) for a sample size of $n = 100$.

7.25 The U.S. National Center for Health Statistics reports that the average stay of female patients in short-stay hospitals is $\mu = 6.9$ days. The standard deviation is $\sigma = 4.3$ days.
a) Suppose 75 female patients are to be selected at random. Let $\bar{x}$ denote the mean hospital stay of the 75 patients chosen. Find $\mu_{\bar{x}}$ and $\sigma_{\bar{x}}$.
b) Repeat part (a) for a random sample of size 500.

7.26 The National Council of the Churches of Christ in the United States of America reports that the mean number of members per local church is $\mu = 408$. Assume the standard deviation is $\sigma = 116.7$.
a) Suppose 100 local churches are to be selected at random. Let $\bar{x}$ denote the mean number of members for the 100 churches chosen. Determine the mean and standard deviation of $\bar{x}$.
b) Repeat part (a) with $n = 200$.

7.27 The average age of cars in use in the U.S. is $\mu = 7.4$ years, with a standard deviation of $\sigma = 2.6$ years.
a) Let $\bar{x}$ denote the mean age of a random sample of $n = 50$ cars. Determine the mean and standard deviation of the random variable $\bar{x}$.
b) Repeat part (a) with $n = 200$.

7.28 According to the U.S. Bureau of Labor Statistics, the average weekly earnings of workers in the trucking industry is $\mu = \$401$. Suppose that n such workers are to be selected at random. Let $\bar{x}$ denote the mean weekly salary of the workers chosen. Assuming a standard deviation of $\sigma = \$63.00$, find the mean and standard deviation of $\bar{x}$ if

a) $n = 100$ b) $n = 200$ c) $n = 400$

7.29 Suppose you are using a sample mean, $\bar{x}$, to estimate a population mean μ. If you want to increase the likelihood of small sampling error, what must you do?

For Exercises 7.30 through 7.37, recall the following fact from page 279. If sampling is done *without* replacement from a finite population of size N, then the precise relationship between the standard deviation of $\bar{x}$ and the population standard deviation is

(1)
$$\sigma_{\bar{x}} = \sqrt{\frac{N-n}{N-1}} \cdot \frac{\sigma}{\sqrt{n}}$$

where n is the sample size. The term $\sqrt{\dfrac{N-n}{N-1}}$ is called the **finite population correction factor.** If the sample size is small relative to the population size, then we can ignore the finite population correction factor and use the simpler formula

(2)
$$\sigma_{\bar{x}} = \frac{\sigma}{\sqrt{n}}$$

The rule of thumb is that the finite population correction factor can be ignored provided that the sample size is no larger than 5% of the size of the population; that is, $n \leq 0.05N$.

7.30 Refer to Example 7.6 on page 277. We computed the standard deviation of the basketball players' heights to be $\sigma = 3.41$ inches. We also applied the formula $\sigma_{\bar{x}} = \sqrt{\Sigma \bar{x}^2 P(\bar{x}) - \mu_{\bar{x}}^2}$ to the probability distribution of $\bar{x}$ to determine the standard deviation of $\bar{x}$ for samples of sizes $n = 2$, $n = 3$, and $n = 4$. The results are summarized in Table 7.12 on page 278. Since the sampling is without replacement from a finite population, we can also use Formula (1) to compute $\sigma_{\bar{x}}$.

a) Use Formula (1) to compute $\sigma_{\bar{x}}$ for samples of size

(i) $n = 2$ (ii) $n = 3$ (iii) $n = 4$

and compare your results to those in Table 7.12.

b) Use the simpler formula, Formula (2), to compute $\sigma_{\bar{x}}$ for samples of size

(i) $n = 2$ (ii) $n = 3$ (iii) $n = 4$

and compare your results to the true values of $\sigma_{\bar{x}}$ in Table 7.12. Why does Formula (2) yield such poor results?

c) What percentage of the population size is a sample of size

(i) $n = 2$?   (ii) $n = 3$?   (iii) $n = 4$?

7.31 Refer to Exercise 7.21.

a) Compute the standard deviation, σ, of the governors' salaries.

b) Consider samples of size $n = 2$. Find the standard deviation, $\sigma_{\bar{x}}$, of the random variable $\bar{x}$ by applying the formula $\sigma_{\bar{x}} = \sqrt{\Sigma \bar{x}^2 P(\bar{x}) - \mu_{\bar{x}}^2}$ to the probability distribution of $\bar{x}$.

c) Which is the appropriate relationship here— Formula (1) or Formula (2)?

d) Use Formula (1) to compute $\sigma_{\bar{x}}$ and compare your result with that from part (b).

e) Use Formula (2) to compute $\sigma_{\bar{x}}$ and compare your result with that from part (b). Why does Formula (2) yield such a poor approximation of the true value of $\sigma_{\bar{x}}$?

7.32 Refer to Exercise 7.22. Repeat parts (b)–(e) of Exercise 7.31 for samples of size $n = 3$. [$\sigma = 9.38$]

7.33 Refer to Exercise 7.23. Repeat parts (b)–(e) of Exercise 7.31 for samples of size $n = 4$. [$\sigma = 9.38$]

7.34 Repeat parts (b)–(e) of Exercise 7.31 for samples of size $n = 5$.

7.35 Repeat parts (b)–(d) of Exercise 7.31 for samples of size $n = 1$.

7.36 Suppose a sample of size n is to be taken from a population of size N.

a) Which formula, Formula (1) or Formula (2), should be used to compute $\sigma_{\bar{x}}$, if the sampling is done without replacement? with replacement?

b) Assume $n = 1$. Compute $\sigma_{\bar{x}}$ using

(i) Formula (1)   (ii) Formula (2)

Why do both formulas give the same result? Explain in words why $\sigma_{\bar{x}} = \sigma$.

c) Assume $n = N$ and that the sampling is done without replacement. Compute $\sigma_{\bar{x}}$ using Formula (1). Could you have guessed the answer without doing any computations? Explain.

7.37 a) Prove that if $n \leq 0.05N$, then

$$0.97 \leq \sqrt{\frac{N-n}{N-1}} \leq 1$$

b) Use part (a) to explain why there is very little difference in the values given by Formulas (1) and (2) when the sample size is no larger than 5% of the population size.

7.38 Exercises 5.26 and 5.27 indicate that if x, y, and w are random variables and c is a positive constant, then

$$\mu_{x+y} = \mu_x + \mu_y$$
$$\mu_{c \bullet w} = c\mu_w$$
$$\sigma_{c \bullet w} = c\sigma_w$$

In addition, if x and y are *independent,* then

$$\sigma_{x+y} = \sqrt{\sigma_x^2 + \sigma_y^2}$$

These four formulas can be used to derive some of the results given in this section. Suppose a random sample of size n is to be taken from a population with mean μ and standard deviation σ. Let x_1 denote the value of the first randomly selected member, x_2 the value of the second, and so on. Then each of the random variables x_1, x_2, ..., x_n has mean μ and standard deviation σ:

$$\mu_{x_i} = \mu, \quad \sigma_{x_i} = \sigma \quad (i = 1, 2, \ldots, n)$$

The sample mean is

$$\bar{x} = \frac{\Sigma x}{n} = \frac{1}{n}(x_1 + x_2 + \cdots + x_n)$$

Use the formulas given at the beginning of this exercise to prove that

a) $\mu_{\bar{x}} = \mu$

and that, if the sampling is done *with* replacement, then

b) $\sigma_{\bar{x}} = \dfrac{\sigma}{\sqrt{n}}$

7.4 The sampling distribution of the mean

Recall that the probability distribution of the random variable $\bar{x}$ is called the *sampling distribution of the mean.* Also recall that we need to know that probability distribution in order to make statistical inferences about the mean of a population based on the mean of a sample from the population. In the previous section we took a first step in describing the sampling distribution of the mean. We discovered that it is easy to express the mean, $\mu_{\bar{x}}$, and standard deviation, $\sigma_{\bar{x}}$, of the random variable $\bar{x}$ in terms of the mean, μ, and standard deviation, σ, of the population being sampled:

$$\mu_{\bar{x}} = \mu$$

and

$$\sigma_{\bar{x}} = \frac{\sigma}{\sqrt{n}} \qquad (n = \text{sample size})$$

In this section, we will examine the second (and final) step for determining the sampling distribution of the mean. It is helpful to distinguish between the case where the population being sampled is normally distributed and the case where it may not be so. We first consider normally distributed populations.

Normally distributed populations

Suppose the population under consideration is *normally distributed.* Recall that qualitatively this means that the histogram of the population is bell-shaped. More precisely, a population with mean μ and standard deviation σ is normally distributed if percentages for the population are equal to areas under the normal curve with parameters μ and σ.

Although it is by no means obvious, when sampling is done from a normally distributed population, the random variable $\bar{x}$ is also normally distributed. In other words, probabilities for the random variable $\bar{x}$ are equal to areas under the normal curve with parameters $\mu_{\bar{x}}$ and $\sigma_{\bar{x}}$. Putting this together with the facts that $\mu_{\bar{x}} = \mu$ and $\sigma_{\bar{x}} = \sigma/\sqrt{n}$, we can now state the following fundamental result concerning the sampling distribution of the mean for normally distributed populations.

KEY FACT The sampling distribution of the mean for normal populations[†]

Suppose a random sample of size n is to be taken from a *normally distributed* population with mean μ and standard deviation σ. Then the random variable $\bar{x}$ is also *normally distributed* and has mean $\mu_{\bar{x}} = \mu$ and standard deviation $\sigma_{\bar{x}} = \sigma/\sqrt{n}$. In other words, if the population being sampled is normally distributed, then probabilities for $\bar{x}$ are equal to areas under the normal curve with parameters μ and $\sigma/\sqrt{n}$.

An intuitive interpretation of this key fact is as follows: If for a fixed sample size n, we compute the $\bar{x}$-value for each possible sample of size n, then the histogram for all of those $\bar{x}$-values will be bell-shaped.

EXAMPLE   7.8   **Illustrates the sampling distribution of the mean for normally distributed populations**

The ages of farm operators in the U.S. are *normally distributed* with a mean of $\mu = 50$ years and a standard deviation of $\sigma = 8$ years. The normal curve for this population of ages is displayed in Figure 7.4.

FIGURE 7.4
Normal curve for ages
of farm operators.

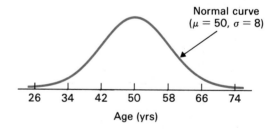

Find the sampling distribution of the mean for samples of size
  a) $n = 3$.
  b) $n = 10$.

SOLUTION
Since the population of ages is normally distributed, the random variable $\bar{x}$ is also normally distributed and has mean $\mu_{\bar{x}} = \mu = 50$ and standard deviation $\sigma_{\bar{x}} = \sigma/\sqrt{n} = 8/\sqrt{n}$.
  a) For samples of size $n = 3$, the mean and standard deviation of $\bar{x}$ are

$$\mu_{\bar{x}} = \mu = 50$$

[†] We will use the phrase "normal population" as an abbreviation of "normally distributed population."

$$\sigma_{\bar{x}} = \frac{\sigma}{\sqrt{n}} = \frac{8}{\sqrt{3}} = 4.62$$

Thus, for samples of size $n = 3$, the sampling distribution of the mean (that is, the probability distribution of $\bar{x}$) is a normal distribution with $\mu_{\bar{x}} = 50$ and $\sigma_{\bar{x}} = 4.62$. The normal curve for $\bar{x}$ is shown in Figure 7.5.

FIGURE 7.5
Sampling distribution of the mean—sample size $n = 3$.

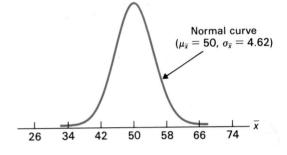

Normal curve
($\mu_{\bar{x}} = 50$, $\sigma_{\bar{x}} = 4.62$)

26   34   42   50   58   66   74 $\bar{x}$

b) For samples of size $n = 10$,

$$\mu_{\bar{x}} = \mu = 50$$

$$\sigma_{\bar{x}} = \frac{\sigma}{\sqrt{n}} = \frac{8}{\sqrt{10}} = 2.53$$

Consequently, for samples of size $n = 10$, the sampling distribution of the mean is a normal distribution with $\mu_{\bar{x}} = 50$ and $\sigma_{\bar{x}} = 2.53$. See Figure 7.6. ∎

FIGURE 7.6
Sampling distribution of the mean—sample size $n = 10$.

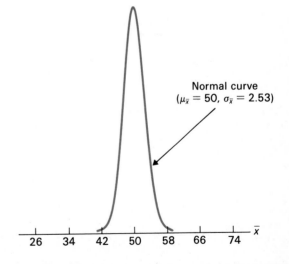

Normal curve
($\mu_{\bar{x}} = 50$, $\sigma_{\bar{x}} = 2.53$)

26   34   42   50   58   66   74 $\bar{x}$

We have drawn the normal curves in Figures 7.5 and 7.6 to scale so that you can compare them visually and observe two important things that you already know. First, both curves are centered at the value of the population mean of $\mu = 50$, since $\mu_{\bar{x}} = \mu = 50$ regardless of the sample size n. Second, the spread of the second curve ($n = 10$) is less extensive than that of the first curve ($n = 3$). This is because the standard deviation, $\sigma_{\bar{x}}$, of the random variable $\bar{x}$ decreases with increasing sample size—$\sigma_{\bar{x}} = \sigma/\sqrt{n}$. The two curves illustrate vividly that the sampling error made in estimating a population mean, μ, by a sample mean, $\bar{x}$, tends to be smaller for larger sample sizes.

Let us now illustrate the computation of probabilities for the random variable $\bar{x}$ when sampling is done from a normally distributed population. Since, under those circumstances, $\bar{x}$ is normally distributed, its probabilities are equal to areas under the normal curve with parameters $\mu_{\bar{x}} = \mu$ and $\sigma_{\bar{x}} = \sigma/\sqrt{n}$.

MTB
SPSS

EXAMPLE   7.9   Illustrates probability computations for $\bar{x}$

The ages of farm operators in the U.S. are *normally distributed* with mean $\mu = 50$ years and standard deviation $\sigma = 8$ years. If $n = 250$ farm operators are to be selected at random, what is the probability that their mean age, $\bar{x}$, will be within one year of the mean age of $\mu = 50$ years of all farm operators? That is, determine the probability that $\bar{x}$ will be between 49 and 51 years.

SOLUTION

Since the population being sampled is normally distributed, the random variable $\bar{x}$ is also *normally distributed*. In addition,

$$\mu_{\bar{x}} = \mu = 50$$

$$\sigma_{\bar{x}} = \frac{\sigma}{\sqrt{n}} = \frac{8}{\sqrt{250}} = 0.51$$

Therefore, probabilities for $\bar{x}$ are equal to areas under the normal curve with parameters $\mu_{\bar{x}} = 50$ and $\sigma_{\bar{x}} = 0.51$. The problem here is to find $P(49 \le \bar{x} \le 51)$— the probability that the mean age of the 250 randomly selected farm operators will be between 49 and 51 years. So, we need to find the area under the normal curve with parameters 50 and 0.51 that lies between 49 and 51. That normal curve area is found in the usual way as depicted in Figure 7.7.

FIGURE 7.7
Determination of the area under the normal curve with parameters $\mu_{\bar{x}} = 50$ and $\sigma_{\bar{x}} = 0.51$ that lies between 49 and 51.

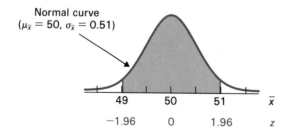

z-score computations:

$$\bar{x} = 49 \longrightarrow z = \frac{49 - 50}{0.51} = -1.96 \qquad 0.4750$$

$$\bar{x} = 51 \longrightarrow z = \frac{51 - 50}{0.51} = 1.96 \qquad 0.4750$$

Area between 0 and z:

Shaded area $= 0.4750 + 0.4750 = 0.9500$

Consequently, $P(49 \leq \bar{x} \leq 51) = 0.95$. We can interpret this result as follows: Suppose 250 randomly selected farm operators are asked their age. Then there is a 95% chance that the mean age, $\bar{x}$, of the 250 farm operators will be within one year of the mean age, $\mu = 50$, of *all* farm operators. We can already begin to see the power of sampling.   ■

General populations; the central limit theorem

We have seen that if the population being sampled is normally distributed, then the random variable $\bar{x}$ is also normally distributed. Remarkably, the random variable $\bar{x}$ is still approximately normally distributed, even if the population being sampled is not normally distributed, *provided only* that the sample size is relatively large. This extraordinary fact is called the **central limit theorem** and is one of the most important results in statistics.

KEY FACT The central limit theorem

For a relatively large sample size, the random variable $\bar{x}$ is approximately *normally distributed,* regardless of how the population is distributed. The approximation becomes better and better with increasing sample size.

Let us consider the significance of the central limit theorem as it pertains to making inferences concerning a population mean, μ, based on the mean, $\bar{x}$, of a sample from the population. As we have noted, such inferences require the ability to determine probabilities for $\bar{x}$. If the population being sampled is normally distributed, we can do this using normal curve areas. More often than not, however, either the population is not normally distributed or we simply do not know its distribution. Nevertheless, because of the central limit theorem, we can still use normal curve areas to approximate probabilities for $\bar{x}$ *no matter what the population looks like,* as long as the sample size is relatively large.

In general, the more "non-normal" the population, the larger the sample size must be in order to use normal curve areas to determine probabilities for $\bar{x}$. As a rule of thumb, we will consider sample sizes of thirty or more ($n \geq 30$) large enough. Thus, we have the following statement regarding the sampling distribution of the mean.

KEY FACT The sampling distribution of the mean for general populations

Suppose a random sample of size $n \geq 30$ is to be taken from a population with mean μ and standard deviation σ. Then regardless of the distribution of the population, the random variable $\bar{x}$ is approximately *normally distributed* and has mean $\mu_{\bar{x}} = \mu$ and standard deviation $\sigma_{\bar{x}} = \sigma/\sqrt{n}$. That is, probabilities for $\bar{x}$ are approximately equal to areas under the normal curve with parameters μ and $\sigma/\sqrt{n}$.

EXAMPLE 7.10 **Illustrates the sampling distribution of the mean**

According to the Chicago Title Insurance Company, the average monthly mortgage payment for recent home buyers is $\mu = \$732$. The standard deviation is $\sigma = \$421$. Suppose a random sample of $n = 125$ recent home buyers is to be selected. What is the probability that their mean monthly mortgage payment will exceed the true mean of $\mu = \$732$ by more than $50?

SOLUTION

The problem here is to find the probability that $\bar{x}$ will exceed $\mu = \$732$ by more than $50—$P(\bar{x} > 782)$. The random variable $\bar{x}$ has mean

$$\mu_{\bar{x}} = \mu = 732$$

and standard deviation

$$\sigma_{\bar{x}} = \frac{\sigma}{\sqrt{n}} = \frac{421}{\sqrt{125}} = 37.66$$

Moreover, since the sample size is $n = 125$, we certainly have $n \geq 30$. Consequently, by the central limit theorem, $\bar{x}$ is approximately normally distributed. This is true regardless of whether or not the population of monthly mortgage payments for recent home buyers is normally distributed.

Therefore, $P(\bar{x} > 782)$ is approximately equal to the area under the normal curve with parameters 732 and 37.66 that lies to the right of 782. We compute this area in Figure 7.8.

FIGURE 7.8
Determination of the area under the normal curve with parameters $\mu_{\bar{x}} = 732$ and $\sigma_{\bar{x}} = 37.66$ that lies to the right of 782.

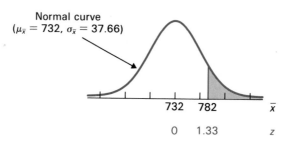

z-score computation:

$$\bar{x} = 782 \longrightarrow z = \frac{782 - 732}{37.66} = 1.33$$

Area between 0 and z:

0.4082

Shaded area $= 0.5000 - 0.4082 = 0.0918$

So, $P(\bar{x} > 782) = 0.0918$. There is less than a 10% chance that the mean monthly mortgage payment of the 125 randomly selected home buyers will exceed the true (population) mean by more than $50. ∎

Some visual displays

In the previous pages of this section we have examined the sampling distribution of the mean. We learned that if the population being sampled is normally distributed, then so is the random variable $\bar{x}$, regardless of the sample size. This fact is portrayed graphically in Figure 7.9.

FIGURE 7.9 Normal population with sampling distributions for $n = 2$, $n = 10$, and $n = 30$.

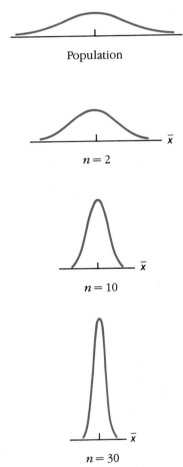

Additionally, the *central limit theorem* indicates that, regardless of the distribution of the population, the random variable $\bar{x}$ is approximately normally distributed, provided that the sample size is relatively large. Figures 7.10 and 7.11 on page 290 give a visual display of the central limit theorem for two non-normal populations. In both of these cases, we see that for samples of size $n = 2$, the random variable $\bar{x}$ is definitely not normally distributed. For $n = 10$, the random variable $\bar{x}$ is already somewhat normally distributed; and, for $n = 30$, the random variable $\bar{x}$ is certainly very close to being normally distributed.

FIGURE 7.10
Skewed population
with sampling
distributions for
$n = 2$, $n = 10$, and
$n = 30$.

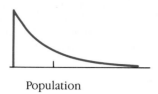

Population

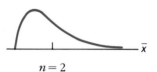

$n = 2$

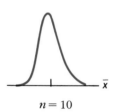

$n = 10$

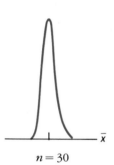

$n = 30$

FIGURE 7.11
Uniform population
with sampling
distributions for
$n = 2$, $n = 10$, and
$n = 30$.

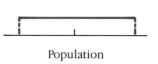

Population

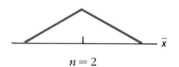

$n = 2$

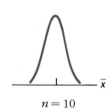

$n = 10$

$n = 30$

MTB
SPSS

Standardized form

In future chapters it will be necessary for us to consider the *standardized* version of the random variable $\bar{x}$, that is,

$$z = \frac{\bar{x} - \mu_{\bar{x}}}{\sigma_{\bar{x}}}$$

Since $\mu_{\bar{x}} = \mu$ and $\sigma_{\bar{x}} = \sigma/\sqrt{n}$, this equation can be written as

$$z = \frac{\bar{x} - \mu}{\sigma/\sqrt{n}}$$

We learned in Section 6.4 (page 252) that if we standardize a *normally distributed* random variable, the resulting random variable has the *standard normal distribution*. Therefore, we can summarize the results of Section 7.3 and the present section as follows:

KEY FACT The sampling distribution of the mean—standardized form

Suppose a random sample of size n is to be taken from a population with mean μ and standard deviation σ. Then the standardized random variable

$$z = \frac{\bar{x} - \mu}{\sigma/\sqrt{n}}$$

1 has exactly the *standard normal distribution* if the population itself is normally distributed, regardless of sample size;
2 has approximately the *standard normal distribution* if $n \geq 30$, regardless of how the population is distributed.

In other words, if *either* the population is normally distributed *or* the sample size is at least 30, then probabilities for the random variable

$$z = \frac{\bar{x} - \mu}{\sigma/\sqrt{n}}$$

are equal, at least approximately, to areas under the *standard normal curve.*

EXAMPLE 7.11 **Illustrates the standardized form of the sampling distribution of the mean**

The mean weekly earnings for workers in the trucking industry is $\mu = \$401$, as reported by the U.S. Bureau of Labor Statistics. The standard deviation is $\sigma = \$63$. Suppose a sample of $n = 81$ workers in the trucking industry is to be selected at random. Let $\bar{x}$ denote the mean weekly earnings of the workers obtained.
a) Determine the standardized version, z, of the random variable $\bar{x}$.
b) What is the probability distribution of the standardized version, z, of $\bar{x}$?
c) Find the probability that the standardized version of $\bar{x}$ will take a value between -2 and 2.

SOLUTION
a) The standardized version of $\bar{x}$ is

$$z = \frac{\bar{x} - \mu_{\bar{x}}}{\sigma_{\bar{x}}} = \frac{\bar{x} - \mu}{\sigma/\sqrt{n}}$$

and since $\mu = 401$, $\sigma = 63$, and $n = 81$, we have

$$z = \frac{\bar{x} - 401}{63/\sqrt{81}} = \frac{\bar{x} - 401}{7}$$

b) Since the sample size here, $n = 81$, exceeds 30, the standardized version of $\bar{x}$,

$$z = \frac{\bar{x} - 401}{7}$$

has *approximately* the standard normal distribution. [Note that this is true regardless of the distribution of the population of weekly earnings, since $n \geq 30$. However, if that population is normally distributed, then $z = (\bar{x} - 401)/7$ has *exactly* the standard normal distribution.]

c) Here we want to find

$$P\left(-2 \leq \frac{\bar{x} - 401}{7} \leq 2\right)$$

Since

$$z = \frac{\bar{x} - 401}{7}$$

has approximately the standard normal distribution, the required probability is approximately equal to the area under the *standard* normal curve between -2 and 2. From Table II we see that this area is $0.4772 + 0.4772 = 0.9544$. Thus,

$$P\left(-2 \leq \frac{\bar{x} - 401}{7} \leq 2\right) = 0.9544$$

approximately.   ■

Exercises 7.4

7.39 Suppose a random sample of size n is to be taken from a normally distributed population with mean μ and standard deviation σ.
a) What is the probability distribution of the random variable $\bar{x}$?
b) Does your answer to part (a) depend on the sample size n? Explain.
c) What is the mean and standard deviation of $\bar{x}$?

7.40 The length of the western rattlesnake is normally distributed with a mean of $\mu = 42$ inches and a standard deviation of $\sigma = 2.04$ inches.
a) Sketch the normal curve for this population.
b) Find the sampling distribution of the mean for samples of size $n = 4$. Draw the normal curve for $\bar{x}$.
c) Repeat part (b) for samples of size $n = 8$.

7.41 According to the U.S. National Center for Health Statistics, the mean weight of males who are six feet tall is $\mu = 175$ lbs. Assume the weights are normally distributed, with a standard deviation of $\sigma = 14$ lbs.
a) Sketch the normal curve for this population of weights.
b) Find the sampling distribution of the mean for samples of size $n = 2$, and sketch the associated normal curve.
c) Repeat part (b) for samples of size $n = 9$.

7.42 Refer to Exercise 7.40. Suppose a random sample of $n = 16$ snakes is to be selected.

a) What is the probability that the mean length, $\bar{x}$, of the snakes obtained will be within one inch of the true (population) mean of $\mu = 42$ inches; that is, between 41 and 43 inches?
b) Interpret your result in part (a) in terms of sampling error.
c) What percentage of the possible sample means, for samples of size $n = 16$, have values that lie within one inch of the true mean of $\mu = 42$ inches?
d) Repeat part (a) for samples of size $n = 50$.

7.43 Refer to Exercise 7.41. Suppose a random sample of $n = 9$ six-foot tall males is to be selected.
a) What is the probability that the mean weight, $\bar{x}$, of the males obtained will be within five pounds of the true (population) mean weight of $\mu = 175$ lbs; that is, between 170 and 180 lbs?
b) Interpret your result in part (a) in terms of sampling error.
c) What percentage of all possible $\bar{x}$-values, for samples of size $n = 9$, lie between 170 and 180?
d) Repeat part (a) for random samples of size $n = 30$.

7.44 The Arizona Regional Multiple Listing Service publishes several books for the real estate industry. One such book deals with the region termed the "Residential Southwest," composed of Mesa, Tempe, and Chandler, Arizona. This book indicates that the mean sale price of homes in that region is

$\mu =$ \$77,500. Suppose you take a random sample of $n = 250$ homes that have recently sold.

a) Assuming a standard deviation of $\sigma =$ \$57,000, what is the probability that the mean sale price of the 250 homes you obtain will be within \$5000 of the population mean sale price of $\mu =$ \$77,500?

b) Do you need to assume the population of sale prices is normally distributed in order to answer part (a)? Explain.

c) Sketch the normal curve for $\bar{x}$ for samples of size $n = 250$.

d) Repeat part (a) for a sample size of $n = 150$.

7.45 The U.S. Bureau of the Census reports that the mean annual alimony income received by women is $\mu =$ \$3000. Assume a standard deviation of $\sigma =$ \$7500. Suppose a random sample of 100 women receiving alimony is to be selected.

a) What is the probability that the mean annual alimony received by the 100 women obtained will be within \$500 of the population mean? That is, find $P(2500 \leq \bar{x} \leq 3500)$.

b) Must you assume that the population of annual alimony payments is normally distributed in order to answer part (a)? Explain.

c) Sketch the normal curve for $\bar{x}$ for samples of size $n = 100$.

d) Repeat part (a) for a random sample of size $n = 1000$.

e) Why is it necessary to take such a large sample in order to be assured of relatively small sampling error?

7.46 An air conditioning contractor is preparing to offer service contracts on the brand of compressor used in all of the units her company installs. Before she can work out the details, it is necessary for her to have an idea of how long these compressors last on the average. The contractor anticipated this need and has kept detailed records on the lifetimes of a random sample of 250 compressors. She plans to use the sample mean lifetime, $\bar{x}$, of these 250 compressors as her estimate for the population mean lifetime, μ, of all such compressors. Suppose that this brand of compressor actually has a mean lifetime of $\mu = 62$ months and a standard deviation of $\sigma = 40$ months. What is the probability that the contractor's estimate will be within five months of the true mean of $\mu = 62$ months?

7.47 An economist employed by the U.S. Department of Agriculture needs to estimate the mean weekly food costs for couples with two children 6–11 years old. The economist plans to randomly sample 500 such families and use the mean weekly food cost, $\bar{x}$, of those families for her estimate of the true mean weekly food cost μ. If, in fact, $\mu =$ \$95.40 and $\sigma =$ \$17.20, then what is the probability that the economist's estimate will be within \$1.00 of the actual mean?

7.48 Look at the graphs in Figures 7.9–7.11 (pages 289 and 290).

a) Why are the four graphs in Figure 7.9 all centered at the same place?

b) Why does the spread of the graphs diminish with increasing sample size? How does this fact affect the sampling error in estimating a population mean μ by a sample mean $\bar{x}$?

c) Why are all the graphs in Figure 7.9 bell-shaped?

d) Why do the graphs in Figures 7.10 and 7.11 become bell-shaped with increasing sample size?

7.49 IQs are normally distributed with a mean of $\mu = 100$ and a standard deviation of $\sigma = 16$. Suppose nine people are to be selected at random. Let $\bar{x}$ denote the mean IQ of the people selected.

a) Determine the standardized version, z, of the random variable $\bar{x}$.

b) What is the probability distribution of the standardized version, z, of $\bar{x}$?

c) Does your answer to (b) depend on the fact that IQs are normally distributed? Explain.

d) Find the probability that the standardized version of $\bar{x}$ will take a value between -1.64 and 1.64.

7.50 The U.S. Energy Information Administration reports that the mean monthly fuel expenditure per household vehicle is $\mu =$ \$58.80. The standard deviation is $\sigma =$ \$30.40. Suppose $n = 50$ household vehicles are to be selected at random, one for each of 50 randomly selected months. Let $\bar{x}$ denote the mean monthly fuel expenditure per vehicle selected.

a) Determine the standardized version, z, of the random variable $\bar{x}$.

b) What is the probability distribution of the standardized version of $\bar{x}$?

c) Does your answer to (b) depend on the fact that the sample size is at least 30? Explain.

d) Find the probability that the standardized version of $\bar{x}$ will take a value less than 1.96.

7.51 A brand of water-softener salt comes in packages marked "net weight 40 lbs." The company claims that, indeed, the bags contain an average of $\mu = 40$ lbs of salt and that the standard deviation in weights is $\sigma = 1.5$ lbs. Furthermore, it is known that the weights are normally distributed.

a) What is the probability that the weight of a randomly selected bag of water-softener salt will be 39 lbs or less, if the company's claim is true?

b) What is the probability that the mean weight of 10 randomly selected bags of water-softener salt will be 39 lbs or less, if the company's claim is true?

c) If you bought a bag of water-softener salt and it weighed 39 lbs, would you consider this to be evidence that the company's claim is incorrect? Explain your answer.

d) If you bought 10 bags of water-softener salt and their mean weight was 39 lbs, would you consider *this* to be evidence that the company's claim is incorrect? Explain your answer.

7.52 According to the Salt River Project, a supplier of electricity to the greater Phoenix area, the mean annual electric bill in 1984 was $852.31. Assume that, for a given year, annual electric bills are normally distributed with $\sigma = \$204$. At the end of 1986, an independent consumer agency wanted to see if the mean annual electric bill had increased over the 1984 figure of $852.31.

a) What is the probability that the mean of a random sample of $n = 25$ annual electric bills for 1986 will be $875 or greater, if in fact the 1986 mean annual electric bill for all customers is still $852.31?

b) If the consumer agency takes a random sample of $n = 25$ annual electric bills for 1986 and finds that $\bar{x} = \$875$, does this provide substantial evidence that the 1986 mean is more than the 1984 mean of $852.31? Why or why not? [Refer to your answer from part (a).]

c) Repeat parts (a) and (b) for $n = 250$ instead of $n = 25$.

◆ Chapter Review

FORMULAS

Mean of the random variable $\bar{x}$, 277

$$\mu_{\bar{x}} = \mu$$

(μ = mean of the population)

Standard deviation of the random variable $\bar{x}$, 279

$$\sigma_{\bar{x}} = \frac{\sigma}{\sqrt{n}}$$

(σ = standard deviation of the population, n = sample size)

YOU SHOULD BE ABLE TO . . .

1 use and understand the preceding formulas.

2 explain what is meant by a representative sample and by a random sample.

3 define sampling error and indicate the need for sampling distributions.

4 find the mean and standard deviation of the random

variable $\bar{x}$, given the mean and standard deviation of the population and the sample size.

5 state and apply the central limit theorem.

6 determine the sampling distribution of the mean when sampling is being done from a normally dis-

tributed population or when the sample size is at least 30.

7 find probabilities for $\bar{x}$ when sampling from a normally distributed population.

8 find probabilities for $\bar{x}$ when sampling from a general population and the sample size is at least 30.

REVIEW TEST

1 An economist wishes to estimate the mean income of the parents of college students. To accomplish this, he surveys a sample of 250 students at Yale. Is this a representative sample? Explain.

2 Classify each sampling procedure below as "random" or "not random."
   a) A college student is hired to interview a random sample of all 200,000 voters in her town. She stays on campus and interviews 100 students in the cafeteria.
   b) A pollster wants to interview a random sample of 20 gas-station managers in Baltimore. He pastes a list of all such managers on his wall, closes his eyes, and tosses a dart at the list 20 times. He interviews the people whose names the dart hits.

3 In 1980, the U.S. Internal Revenue Service sampled 171,700 tax returns to obtain estimates of various parameters. The mean income tax per taxable return was $\bar{x} = \$3387$.
   a) Explain the meaning of sampling error in this context.
   b) If the true (population) mean income tax per taxable return in 1980 was actually $\mu = \$3475$, how much sampling error was made in estimating μ by $\bar{x}$?
   c) If the IRS had sampled 250,000 returns instead of 171,700, would the sampling error necessarily have been smaller? Explain.
   d) In future surveys, how can the IRS increase the likelihood of a small sampling error?

4 According to the U.S. Department of Agriculture, the mean yield of cotton by farms in the U.S. is $\mu = 506$ pounds per acre. The standard deviation is $\sigma = 237$ pounds.
   a) Suppose a random sample of 25 one-acre plots of cotton is to be selected. Let $\bar{x}$ denote the mean yield of the 25 plots obtained. Find the

mean and standard deviation of the random variable $\bar{x}$.
   b) Repeat part (a) for a sample size of $n = 200$.
   c) Suppose a sample of 500 one-acre plots of cotton is to be selected. Without doing any computations, answer the following question: Will the value of $\sigma_{\bar{x}}$ here be larger, smaller, or the same as the value of $\sigma_{\bar{x}}$ in part (b)? Explain.

5 A population, which may or may not be normally distributed, has mean $\mu = 40$ and standard deviation $\sigma = 10$. Suppose a sample of size $n = 100$ is to be selected at random. Decide whether each of the following statements is true or false or whether it is not possible to tell. Give a reason for each of your answers.
   a) There is about a 68.26% chance that the mean of the sample will be between 30 and 50.
   b) About 68.26% of the population values lie between 30 and 50.
   c) There is about a 68.26% chance that the mean of the sample will be between 39 and 41.

6 Repeat Problem 5 under the assumption that the population is normally distributed.

7 The monthly rents for studio apartments in a large city are *normally distributed* with a mean of $\mu = \$285$ and a standard deviation of $\sigma = \$45$.
   a) Sketch the normal curve for this population of monthly rents.
   b) Find the sampling distribution of the mean for samples of size $n = 3$. Draw the normal curve for $\bar{x}$.
   c) Repeat part (b) for samples of size $n = 9$.

8 Refer to Problem 7. Suppose a random sample of $n = 3$ studio apartments is to be selected.
   a) What is the probability that the mean monthly rent, $\bar{x}$, of the apartments selected will be within $10 of the population mean of $\mu = \$285$; that is, between $275 and $295?

b) Interpret your results in part (a) in terms of sampling error.

c) What percentage of the possible sample means, for samples of size $n = 3$, have values within $10 of the population mean of $\mu = \$285$?

d) Repeat part (a) for samples of size $n = 75$.

9 The figure below shows the curve for a normally distributed population (in color). Superimposed on the same graph are the curves for the sampling distribution of the mean for two different sample sizes.

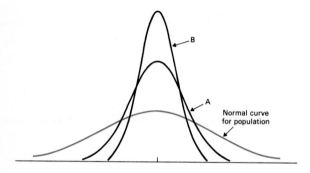

a) Why are all three curves centered at the same place?

b) Which curve corresponds to the larger sample size? Explain.

c) Why is the spread for each curve different?

d) Which of the two sampling-distribution curves corresponds to the sample size that will tend to produce less sampling error? Give a reason for your answer.

10 The American Council of Life Insurance reports that the mean life insurance in force per covered family is $\mu = \$57,300$. Assume a standard deviation of $\sigma = \$30,900$. Suppose $n = 500$ covered families are to be randomly selected.

a) What is the probability that the mean life insurance in force of the 500 families selected will be within $2000 of the population mean of $\mu = \$57,300$?

b) Must you assume that the population of life-insurance amounts is normally distributed in order to answer part (a)? What if the sample size is $n = 20$ instead of $n = 500$?

c) Repeat part (a) for a sample size of $n = 1000$.

11 As reported by the U.S. Social Security Administration, the average monthly benefit to retired workers is $\mu = \$419$. Assume a standard deviation of $\sigma = \$73$. Suppose $n = 125$ retired workers are to be selected at random. Let $\bar{x}$ denote the mean monthly benefit received by the retired workers obtained.

a) Determine the standardized version, z, of the random variable $\bar{x}$.

b) What is the probability distribution of the standardized version, z, of $\bar{x}$?

c) Find the probability that the standardized version of $\bar{x}$ will take a value either greater than 2.33 or less than -2.33.

d) Could you answer part (b) if the sample size were $n = 5$ instead of $n = 125$? Explain.

12 A paint manufacturer in Pittsburgh claims that his penetrating stain will last for an average of $\mu = 5$ years. Assume that paint life is normally distributed with a standard deviation of $\sigma = 0.5$ years.

a) If you painted one house with the paint and the paint lasted 4.5 years, would you consider this to be substantial evidence against the manufacturer's claim? *Hint:* Compute $P(\bar{x} \leq 4.5)$ for a sample of size $n = 1$, assuming the manufacturer's claim is correct.

b) If you painted 10 houses with the manufacturer's paint and the paint lasted an average of 4.5 years for the 10 houses, would you consider *this* to be substantial evidence against the manufacturer's claim?

c) Repeat part (b) if the paint lasted an average of 4.9 years for the 10 homes painted.

We now begin our study of *inferential statistics*. In the present chapter, we will examine methods for estimating the mean of a population by sampling. As you might suspect, the statistic used to estimate a population mean, μ, is the sample mean, $\bar{x}$. Because of sampling error, we cannot expect $\bar{x}$ to be precisely equal to μ. Thus, it is important to provide information about the accuracy of the estimate. This leads to the discussion of *confidence intervals*, which is the main topic of the chapter.

Besides learning to estimate the mean of a population, we will also learn how to estimate the proportion (percentage) of a population that has a specified attribute. For example, we will see how it is possible to estimate the percentage of all U.S. adults who drink alcoholic beverages from the percentage of a sample who do.

Estimating means and proportions

C H A P T E R   O U T L I N E

8.1   Estimating a population mean

An important problem in statistical inference is to obtain information about the mean, μ, of a population. For instance, we might be interested in estimating

1 the mean tar content, μ, of a certain brand of cigarette, or
2 the mean life, μ, of a newly developed steel-belted radial tire, or
3 the mean gas mileage, μ, of a new model car, or
4 the mean annual income, μ, of liberal arts graduates.

If the population is small, we can ordinarily compute μ exactly. In most cases, however, it will be impractical, impossible, or extremely expensive to determine the precise value of μ. Nonetheless, we can usually obtain sufficiently accurate information about μ by just taking a *sample* from the population. Let us look at an example.

EXAMPLE   8.1   Illustrates the use of a sample mean to estimate a population mean

A manufacturer of tobacco products intends to begin marketing a new brand of cigarette. Among other things, the manufacturer needs information regarding the mean tar content, μ, of this new brand. Obviously he cannot test all such cigarettes, so he decides to take a random sample of $n = 50$ of them and test the 50 cigarettes obtained. The results of the tar-content test are displayed in Table 8.1.

TABLE 8.1

Tar content, in milligrams, of the sample of 50 cigarettes tested.

11.70	11.02	11.24	11.12	12.23
10.32	10.33	10.89	11.88	10.72
10.86	11.05	11.23	9.67	10.88
11.36	10.65	11.33	12.00	10.71
10.90	10.74	11.42	10.03	10.35
10.31	11.85	10.88	10.97	10.77
11.06	11.68	10.82	10.16	9.89
10.66	10.94	11.14	10.28	10.35
10.87	11.14	10.79	10.65	11.07
11.43	10.98	10.92	11.20	11.49

Using the sample data, estimate the mean tar content, μ, of all cigarettes of this brand.

SOLUTION

The *sample mean* tar content, $\bar{x}$, of the 50 cigarettes tested can be computed from Table 8.1 in the usual way:

$$\bar{x} = \frac{\Sigma x}{n} = \frac{546.93}{50} = 10.94 \text{ mg}$$

If we had to estimate the population mean tar content, μ, of all cigarettes of this brand based on this sample of 50 cigarettes, we would probably estimate that μ is about 10.94 mg. In other words, we would use the sample mean tar content of $\bar{x} = 10.94$ mg as our estimate for the population mean tar content, μ, of all such cigarettes. An estimate of this type is called a **point estimate** for μ since it consists of a single number, or point. ∎

The term *point estimate* applies to the use of a statistic to estimate any parameter, not just a population mean.

DEFINITION 8.1 Point estimate

A **point estimate** of a parameter is the value of a statistic used to estimate the parameter.

As we learned in Chapter 7, it is unreasonable to expect that a sample mean, $\bar{x}$, will be exactly equal to the population mean, μ—some *sampling error* is to be anticipated. Thus, in addition to reporting a point estimate for μ, we need to provide information that gives us some idea of the accuracy of the estimate. This can be done using a **confidence-interval estimate** for μ. With a confidence-interval estimate for μ, we use a sample mean to construct an interval of numbers and indicate how confident we are that the value of μ lies in that interval. You will see shortly how to construct such confidence intervals.

We summarize the discussion in the above paragraph in Definition 8.2. Again, the terminology applies to any parameter, not just to a population mean.

DEFINITION 8.2 Confidence-interval estimate

A **confidence-interval estimate** of a parameter consists of an interval of numbers obtained from a point estimate, together with a percentage specifying how confident we are that the parameter lies in that interval.

Before proceeding, it will be helpful to us to review the following fact from Section 7.4. It is this result that enables us to make statements about accuracy when a sample mean, $\bar{x}$, is used to estimate a population mean μ.

KEY FACT The sampling distribution of the mean—large samples

Suppose a random sample of size $n \geq 30$ is to be taken from a population with mean μ and standard deviation σ. Then the random variable $\bar{x}$ is approximately *normally distributed* and has mean $\mu_{\bar{x}} = \mu$ and standard deviation $\sigma_{\bar{x}} = \sigma/\sqrt{n}$. In other words, probabilities for $\bar{x}$ are approximately equal to areas under the normal curve with parameters μ and $\sigma/\sqrt{n}$.

EXAMPLE 8.2 Illustrates a use of the preceding Key Fact

The U.S. Bureau of the Census publishes annual mean sale price figures on new mobile homes. The data are obtained from sampling, not from a census. Consequently, the figures published are point estimates and, therefore, sampling error is to be expected. Using the preceding Key Fact, we can get an idea of how accurate such estimates are likely to be.

To illustrate, suppose $n = 40$ new mobile homes are to be selected at random and that their mean sale price, $\bar{x}$, is to be used as an estimate of the mean sale price, μ, of all new mobile homes. What is the probability that the mean sale price, $\bar{x}$, of the 40 new mobile homes selected will be within \$1000 of the population mean sale price, μ, of all new mobile homes?

SOLUTION

For simplicity, we will assume as known that the population standard deviation of the sale prices is $\sigma = \$4200$.[†] The problem is to find the probability that the mean sale price, $\bar{x}$, of the 40 new mobile homes selected will be within \$1000 of the population mean sale price, μ. That is, we need to determine

$$P(\mu - 1000 \leq \bar{x} \leq \mu + 1000)$$

The sample size is $n = 40$, so $n \geq 30$. Consequently, by the preceding Key Fact, the random variable $\bar{x}$ is approximately normally distributed. Specifically, probabilities for $\bar{x}$ are approximately equal to areas under the normal curve with parameters $\mu_{\bar{x}} = \mu$ (which we do not know) and $\sigma_{\bar{x}} = \sigma/\sqrt{n} = 4200/\sqrt{40} = 664.08$.

Thus, to obtain $P(\mu - 1000 \leq \bar{x} \leq \mu + 1000)$ we need only find the area under the normal curve with parameters μ and 664.08 that lies between $\mu - 1000$ and $\mu + 1000$. The difference between this problem and the problems we have

---

[†] We might know this from previous studies of prices of new mobile homes or from a preliminary study of the prices. The more usual case, where σ is unknown, will be covered in Section 8.3.

solved previously is that here we do not know the values of $\mu - 1000$ and $\mu + 1000$. Nevertheless, we can still determine the area in the usual way—by standardizing. This is done in Figure 8.1.

FIGURE 8.1
Determination of
the area under the
normal curve with
parameters $\mu_{\bar{x}} = \mu$
and $\sigma_{\bar{x}} = 664.08$ that
lies between
$\mu - 1000$ and
$\mu + 1000$.

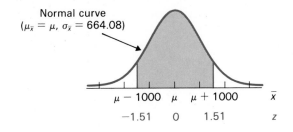

Normal curve
$(\mu_{\bar{x}} = \mu, \sigma_{\bar{x}} = 664.08)$

z-score computations:

Area between 0 and z:

$\bar{x} = \mu - 1000 \longrightarrow z = \dfrac{(\mu - 1000) - \mu}{664.08} = -1.51$ 0.4345

$\bar{x} = \mu + 1000 \longrightarrow z = \dfrac{(\mu + 1000) - \mu}{664.08} = 1.51$ 0.4345

Shaded area $= 0.4345 + 0.4345 = 0.8690$

Consequently,

$$P(\mu - 1000 \leq \bar{x} \leq \mu + 1000) = 0.8690$$

In other words, there is about an 87% chance that the mean sale price, $\bar{x}$, of the 40 new mobile homes selected will be within \$1000 of the true mean sale price, μ, of all new mobile homes. ∎

We will now continue with Example 8.2, since some additional discussion is appropriate. We found that if $n = 40$ new mobile homes are to be selected at random, then the probability is 0.8690, or about 0.87, that their mean sale price, $\bar{x}$, will be within \$1000 of the mean sale price, μ, of all new mobile homes. In symbols:

$$P(\mu - 1000 \leq \bar{x} \leq \mu + 1000) = 0.87$$

However, to say that "$\bar{x}$ will be within \$1000 of μ" is the same as saying that "μ will be within \$1000 of $\bar{x}$." So, we can rewrite the previous equation as

$$P(\bar{x} - 1000 \leq \mu \leq \bar{x} + 1000) = 0.87$$

In other words,

> *If 40 new mobile homes are to be randomly selected, then the probability is 0.87 that the interval from $\bar{x} - 1000$ to $\bar{x} + 1000$ will contain the population mean sale price, μ, of all new mobile homes.*

In terms of percentages, this means the following: Suppose for each possible sample of size $n = 40$, we compute the sample mean sale price, $\bar{x}$, and construct the interval

$$\bar{x} - 1000 \quad \text{to} \quad \bar{x} + 1000$$

Then about 87% of all such intervals will contain the population mean sale price, μ. Or, equivalently, if we repeatedly sampled 40 new mobile homes and constructed the interval from $\bar{x} - 1000$ to $\bar{x} + 1000$, then about 87 out of every 100 intervals would contain μ.

It is very important to realize that in these, and similar statements, the random variable is $\bar{x}$, not μ. The population mean, μ, is a fixed number, although it may be unknown. The value of the sample mean, $\bar{x}$, depends on chance—namely, on which sample is obtained.

EXAMPLE   8.3   Introduces confidence intervals

Let us now examine what can be said once the sample of 40 new mobile homes is obtained and their mean sale price, $\bar{x}$, is computed. Suppose the sale prices of the 40 randomly selected mobile homes are as given in Table 8.2.

TABLE 8.2
Sale prices, to the nearest dollar, of the 40 randomly selected mobile homes.

$21,349	19,540	13,256	29,861
15,746	22,960	19,427	29,682
21,861	12,254	29,663	23,419
20,992	20,488	24,154	26,008
11,371	17,958	24,153	25,417
21,295	23,952	10,753	13,240
18,030	22,236	17,112	23,797
20,718	24,449	20,300	21,609
19,630	6,111	22,398	15,550
18,746	14,766	22,140	22,608

The sample mean sale price of these 40 mobile homes is

$$\bar{x} = \frac{\Sigma x}{n} = \frac{\$808,999}{40} = \$20,225$$

(to the nearest dollar). Hence,

$$\bar{x} - 1000 = \$19,225 \quad \text{and} \quad \bar{x} + 1000 = \$21,225$$

We know from the italicized statement on page 301 that about 87% of such intervals will contain μ. Therefore, we can be 87% *confident* that the mean sale price, μ, of all new mobile homes lies somewhere between $19,225 and $21,225. Consequently, the interval from $19,225 to $21,225 is called an 87% **confidence interval** for μ. The percent confidence, in this case 87%, is called the **confidence level.**   ∎

We need to emphasize that the confidence interval actually obtained depends on the value of $\bar{x}$, which in turn depends on the sample obtained. For example, suppose the sale prices of the sample of 40 mobile homes selected were as given in Table 8.3, instead of as in Table 8.2.

TABLE 8.3

$12,554	19,298	21,567	19,996
16,359	16,089	19,921	13,783
24,042	20,612	20,440	14,434
8,585	27,920	24,696	19,659
16,332	17,811	15,928	20,993
22,716	16,107	17,400	22,501
25,295	19,048	12,379	26,429
14,108	21,047	23,076	21,652
18,627	24,947	19,495	17,636
16,698	16,520	12,845	20,085

Then we would have

$$\bar{x} = \frac{\Sigma x}{n} = \frac{\$759,630}{40} = \$18,991$$

so that

$$\bar{x} - 1000 = \$17,991 \quad \text{and} \quad \bar{x} + 1000 = \$19,991$$

Hence, in this case, the 87% confidence interval for μ would be the interval from $17,991 to $19,991. We could be 87% confident that the mean sale price, μ, of all new mobile homes lies somewhere between $17,991 and $19,991.

It is crucial to remember that the 87% confidence interval that is actually obtained from the sample of 40 new mobile homes may or may not contain the population mean μ—but, we can be 87% confident that it does.

In the final example of this section, we want to stress the interpretation of confidence intervals and the fact that a confidence interval may or may not contain the true value of the population mean, μ.

EXAMPLE 8.4 **Illustrates the interpretation of confidence intervals**

In Examples 8.2 and 8.3 we considered the problem of using the mean sale price, $\bar{x}$, of a random sample of $n = 40$ new mobile homes to estimate the mean sale price, μ, of all new mobile homes. Let us briefly review our results. We found that

$$P(\mu - 1000 \leq \bar{x} \leq \mu + 1000) = 0.87$$

or equivalently

$$P(\bar{x} - 1000 \leq \mu \leq \bar{x} + 1000) = 0.87$$

Consequently, if 40 new mobile homes are to be randomly selected, then the probability is about 0.87 that the interval from $\bar{x} - 1000$ to $\bar{x} + 1000$ will contain μ. So, once the 40 new mobile homes are actually selected and their mean sale price, $\bar{x}$, is computed, an 87% confidence interval for μ can be constructed—namely, the interval from $\bar{x} - 1000$ to $\bar{x} + 1000$.

In order to illustrate that the population mean sale price, μ, of all new mobile homes may or may not lie in the 87% confidence interval obtained, we *simulated* the sampling of 40 new mobile homes twenty times using $\mu = \$20,000$. [Of course, in reality, we do not know μ. We are assuming this value for μ in order to illustrate a point.] For each of the twenty samples of size $n = 40$ we computed $\bar{x}$ and then the corresponding 87% confidence interval from $\bar{x} - 1000$ to $\bar{x} + 1000$. We then observed whether the true mean of $\mu = \$20,000$ was in the confidence interval. The results are summarized in Table 8.4. For each sample, we have drawn a graph (on the right). The dot represents the point estimate, $\bar{x}$, and the horizontal line represents the corresponding confidence-interval estimate. As you can see, the population mean, μ, lies in the confidence interval when and only when the horizontal line crosses the dashed line.

TABLE 8.4

Sample	$\bar{x}$	87% CI $\bar{x} - 1000$ to $\bar{x} + 1000$	Is μ in the CI?
1	20,145	19,145 to 21,145	Yes
2	20,041	19,041 to 21,041	Yes
3	19,608	18,608 to 20,608	Yes
4	20,049	19,049 to 21,049	Yes
5	20,082	19,082 to 21,082	Yes
6	20,248	19,248 to 21,248	Yes
7	21,452	20,452 to 22,452	No
8	19,366	18,366 to 20,366	Yes
9	19,434	18,434 to 20,434	Yes
10	19,563	18,563 to 20,563	Yes
11	19,500	18,500 to 20,500	Yes
12	20,045	19,045 to 21,045	Yes
13	19,576	18,576 to 20,576	Yes
14	20,220	19,220 to 21,220	Yes
15	19,143	18,143 to 20,143	Yes
16	19,487	18,487 to 20,487	Yes
17	19,657	18,657 to 20,657	Yes
18	18,773	17,773 to 19,773	No
19	20,787	19,787 to 21,787	Yes
20	18,884	17,884 to 19,884	No

$\mu = 20K$

17.5K 18.0K 18.5K 19.0K 19.5K   20.5K 21.0K 21.5K 22.0K 22.5K

From Table 8.4 we see that in 17 out of the 20 samples of size $n = 40$, μ is in the 87% confidence interval. In other words, in 85% (17/20) of the 20 samples, μ is in the 87% confidence interval. If instead of 20 samples of size $n = 40$, we simulated, say, 1000 samples of size $n = 40$, then we would find that the percentage of those 1000 samples for which μ is in the 87% confidence interval would be very close indeed to 87%. Thus, we can be 87% confident that any computed 87% confidence interval will actually contain μ.   ■

Exercises   8.1

Note: In the following exercises we will assume that σ is known. The more realistic case in which σ is unknown will be discussed in Section 8.3.

8.1   The U.S. National Center for Health Statistics collects and publishes natality (birth) statistics. Suppose a random sample of $n = 35$ babies is to be selected and that their mean birth weight, $\bar{x}$, is to be used as an estimate of the mean birth weight, μ, of all newborns.

a) What is the probability that $\bar{x}$ will be within 0.5 lb of the true mean weight, μ? [Assume $\sigma = 1.9$ lb, and refer to the argument used in Example 8.2 on pages 300–301.]

b) If the 35 babies selected have the birth weights below, compute $\bar{x}$.

7.4	6.0	8.6	4.5	2.0
7.9	4.0	2.6	5.9	7.3
7.3	7.0	6.3	8.1	7.1
7.3	6.6	5.2	9.8	8.0
10.9	6.3	3.8	5.0	8.0
10.7	9.7	6.0	6.8	10.3
7.6	6.5	7.1	5.8	6.9

[*Note:* The sum of the data is 240.3 lb.]

c) Use your answers from parts (a) and (b) to fill in the blanks: An _____% confidence interval for μ is the interval from _____ lb to _____ lb.

d) Interpret your result for part (c) in words.

e) Does the true mean birth weight, μ, lie in your confidence interval from part (c)? Explain.

8.2 An educational psychologist at a major university wants to estimate the mean IQ of the students. To this end, $n = 30$ students are to be randomly selected and given an IQ test. The mean IQ, $\bar{x}$, of the students selected is to be used to obtain an estimate of the mean IQ, μ, of all students attending the university.

a) What is the probability that $\bar{x}$ will be within 3 points of μ? [Assume $\sigma = 12$.]

b) Suppose the IQ scores of the students selected are as follows:

107	99	101	93	99
103	134	132	103	109
104	103	101	128	113
106	126	103	131	106
119	102	98	116	108
103	111	119	112	105

Compute $\bar{x}$. [*Note:* $\Sigma x = 3294$.]

c) Use your answers from parts (a) and (b) to fill in the blanks: An _____% confidence interval for μ is the interval from _____ to _____.

d) Interpret your results for part (c) in words.

e) Does the true mean IQ, μ, of all students at the university lie in the confidence interval in part (c)? Explain.

8.3 The R.R. Bowker Company of New York collects and publishes data on books and periodicals. Suppose we want to estimate the mean retail price, μ, of all science books. To accomplish this, we plan to take a random sample of $n = 100$ such books and use the mean retail price, $\bar{x}$, of the books selected to construct a confidence interval for μ.

a) What is the probability that the mean retail price, $\bar{x}$, of the 100 science books obtained will be within \$5.00 of the true mean retail price, μ? [Assume $\sigma = \$27.40$.]

b) Suppose that the mean retail price of the 100 science books selected is $\bar{x} = \$46.84$. Use your answer from part (a) to fill in the blanks: A _____% confidence interval for μ is from \$_____ to \$_____.

c) Interpret your results for part (b) in words.

d) Suppose that for each possible sample of 100 science books, we computed the mean retail price, $\bar{x}$, and determined the interval $\bar{x} - 5$ to $\bar{x} + 5$. What percentage of such intervals would contain the true mean retail price, μ, of all science books?

8.4 The A.C. Nielsen Company gathers information on the average number of TV sets per household. Suppose a random sample of $n = 50$ households is to be selected. Let $\bar{x}$ denote the mean number of TV sets per household for the 50 households obtained.

a) Find the probability that $\bar{x}$ will be within 0.25 of the mean number, μ, of TV sets per household for all U.S. households. [Assume $\sigma = 1.06$.]

b) Suppose that the mean number of TV sets for the 50 randomly selected households is $\bar{x} = 1.82$. Fill in the blanks using your answer from part (a): A _____% confidence interval for μ is from _____ to _____ sets per household.

c) Interpret your results from part (b) in words.

d) Suppose the mean number of TV sets per household, $\bar{x}$, is computed for all possible samples of $n = 50$ households. What percentage of the intervals $\bar{x} - 0.25$ to $\bar{x} + 0.25$ will contain the population mean number of TV sets per household?

8.5 Refer to Exercise 8.3. Why were you able to use areas under a normal curve to answer part (a)?

8.6 Refer to Exercise 8.4. Why were you able to use areas under a normal curve to answer part (a)?

8.7 State why a confidence interval for μ is more appropriate than a point estimate for μ.

8.8 On page 301, we made the following statement: To say that "$\bar{x}$ will be within \$1000 of μ" is the same as saying that "μ will be within \$1000 of $\bar{x}$." Mathematically, this means that the inequalities

$$\mu - 1000 \leq \bar{x} \leq \mu + 1000$$

and

$$\bar{x} - 1000 \leq \mu \leq \bar{x} + 1000$$

are equivalent. Verify that by proving the following more general result: The inequalities

$$\mu - E \leq \bar{x} \leq \mu + E$$

and

$$\bar{x} - E \leq \mu \leq \bar{x} + E$$

are equivalent.

8.2   Specifying the confidence level

In the previous section we learned how to find confidence intervals for a population mean, μ, based on the mean, $\bar{x}$, of a sample from the population. In our examples, we specified the sample size and the form of the confidence interval and then computed the confidence level from these specifications. Look again, for instance, at Examples 8.2 and 8.3. There we discussed the problem of finding a confidence interval for the mean sale price, μ, of all new mobile homes. The sample size was set at $n = 40$, and the form of the confidence interval was designated as $\bar{x} - 1000$ to $\bar{x} + 1000$. From these stipulations we computed the confidence level to be 87% (see the equation shown in color on page 301).

However, it is often desirable to specify the confidence level in advance. For example, we might want a 95% confidence interval for the mean sale price, μ, of all new mobile homes, instead of an 87% confidence interval. Before we examine how to find a confidence interval with a specified confidence level, we must learn how to use the standard normal table, Table II, "in reverse." That is, we must learn how to find the z-value for a specified area.

Finding the z-value for a specified area

Up to now we have used Table II to find areas under the standard normal curve between two specified values of z. For example, we have used the table to find the area between $z = 0$ and $z = 1$, which is 0.3413. Now we will learn how to use Table II to find the z-value(s) corresponding to a specified *area* under the standard normal curve. Consider Example 8.5.

EXAMPLE 8.5   **Illustrates how to find the z-value for a specified area**

Find the z-value for which the area under the standard normal curve to the *right* of that value is 0.025. See Figure 8.2.

FIGURE 8.2

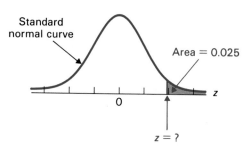

SOLUTION

To find the z-value in question, first recall that the area under the standard normal curve to the right of $z = 0$ equals 0.5. Therefore, the area between $z = 0$ and the desired value of z must be

$$0.500 - 0.025 = 0.475$$

as shown in Figure 8.3.

FIGURE 8.3

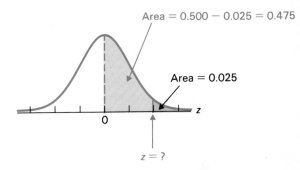

Consequently, we need to use Table II to obtain the z-value corresponding to the area 0.475. For ease of reference we have reproduced a portion of Table II in Table 8.5.

TABLE 8.5
Areas under the standard normal curve.

z				Second decimal place in z					
	0.00	0.01	0.02	0.03	0.04	0.05	0.06	0.07	0.08	0.09
·									
·									
·									
1.0	0.3413	0.3438	0.3461	0.3485	0.3508	0.3531	0.3554	0.3577	0.3599	0.3621
1.1	0.3643	0.3665	0.3686	0.3708	0.3729	0.3749	0.3770	0.3790	0.3810	0.3830
1.2	0.3849	0.3869	0.3888	0.3907	0.3925	0.3944	0.3962	0.3980	0.3997	0.4015
1.3	0.4032	0.4049	0.4066	0.4082	0.4099	0.4115	0.4131	0.4147	0.4162	0.4177
1.4	0.4192	0.4207	0.4222	0.4236	0.4251	0.4265	0.4279	0.4292	0.4306	0.4319
1.5	0.4332	0.4345	0.4357	0.4370	0.4382	0.4394	0.4406	0.4418	0.4429	0.4441
1.6	0.4452	0.4463	0.4474	0.4484	0.4495	0.4505	0.4515	0.4525	0.4535	0.4545
1.7	0.4554	0.4564	0.4573	0.4582	0.4591	0.4599	0.4608	0.4616	0.4625	0.4633
1.8	0.4641	0.4649	0.4656	0.4664	0.4671	0.4678	0.4686	0.4693	0.4699	0.4706
1.9	0.4713	0.4719	0.4726	0.4732	0.4738	0.4744	0.4750	0.4756	0.4761	0.4767
·									
·									
·									

We search the *body* of the table for the area 0.475 and upon finding it, note that the corresponding z-value is 1.96. Thus, the required z-value is $z = 1.96$. See Figure 8.4.

FIGURE 8.4

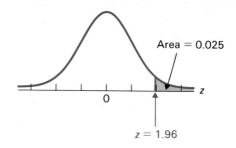

Area = 0.025

z

0

z = 1.96

We will use the notation $z_{0.025}$ to denote the z-value with area 0.025 to its *right*. So, as we have just seen, $z_{0.025} = 1.96$.   ∎

As the previous example indicates, drawing a picture (or pictures) really helps to organize the solving of problems in which we must use Table II in reverse. The next two examples provide additional illustrations.

EXAMPLE   8.6   Illustrates how to find the z-value for a specified area

Use Table II to find
a) $z_{0.10}$
b) $z_{0.05}$

SOLUTION
a) $z_{0.10}$ is the z-value with area 0.10 to its right. See Figure 8.5.

FIGURE 8.5

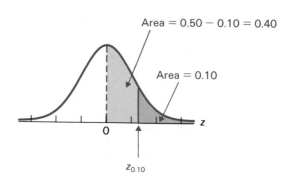

Area = 0.50 − 0.10 = 0.40

Area = 0.10

z

0

$z_{0.10}$

To find the value of $z_{0.10}$, we first note that the area under the curve between 0 and $z_{0.10}$ is $0.50 - 0.10 = 0.40$, as indicated in Figure 8.5. Thus, we search the body of Table II for the area 0.40. We find no such area in the table, and so we use the area closest to 0.40, which is 0.3997. The z-value corresponding to that area is $z = 1.28$. Consequently,

$$z_{0.10} = 1.28$$

approximately.
b) $z_{0.05}$ is the z-value with area 0.05 to its right. See Figure 8.6.

FIGURE 8.6

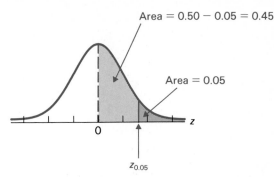

To find $z_{0.05}$, first observe that the area under the curve between 0 and $z_{0.05}$ is $0.50 - 0.05 = 0.45$, as shown in Figure 8.6. Consequently, we search the body of Table II for an area of 0.45. Again we find no such area in the table, and this time there are *two* areas closest to 0.45, namely, 0.4495 and 0.4505. Since the z-values corresponding to these areas are $z = 1.64$ and $z = 1.65$, we take $z_{0.05}$ to be halfway between these. Thus,

$$z_{0.05} = 1.645$$

∎

EXAMPLE   8.7   Illustrates how to find the z-value for a specified area

Determine the *two* z-values that divide the area under the standard normal curve into a middle 0.95 area and two outside 0.025 areas. See Figure 8.7.

FIGURE 8.7

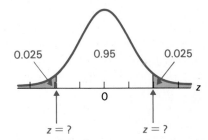

SOLUTION

As we see from Figure 8.7, the area of the shaded region on the right-hand side is 0.025. This means that the z-value on the right is $z_{0.025}$. In Example 8.5, we found that $z_{0.025} = 1.96$. Thus, the z-value on the right is 1.96. Since the standard normal curve is *symmetric* about 0, it follows that the z-value on the left is -1.96. Therefore we have the picture shown in Figure 8.8.

∎

FIGURE 8.8

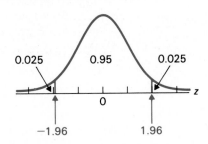

Finding confidence intervals with a specified confidence level

Now that we know how to use Table II to find the z-value(s) corresponding to a specified area under the standard normal curve, we can discuss the method of obtaining confidence intervals when the confidence level is specified in advance. We first recall the following *standardized form* for describing the sampling distribution of the mean, discussed in Section 7.4 on page 291.

KEY FACT

Suppose a random sample of size $n \geq 30$ is to be taken from a population with mean μ and standard deviation σ. Then the standardized random variable

$$z = \frac{\bar{x} - \mu}{\sigma/\sqrt{n}}$$

has approximately the *standard normal distribution*. That is, probabilities for that random variable are approximately equal to areas under the *standard normal curve*.

EXAMPLE   8.8   **Illustrates how to obtain confidence intervals with a specified confidence level**

In 1980, a telephone company in the Southwest undertook a study of telephone usage. According to the media relations manager, the company randomly selected $n = 15,000$ local telephone calls of residential customers in Phoenix. The mean duration of the calls sampled was $\bar{x} = 3.8$ minutes. Use this information to determine a 95% confidence interval for the true mean duration, μ, of all telephone calls made by residential customers in Phoenix. [Assume $\sigma = 4.0$ minutes.[†]]

SOLUTION
We first note that the sample size is $n = 15,000$, which exceeds 30. Thus, by the previous Key Fact, the random variable

$$z = \frac{\bar{x} - \mu}{\sigma/\sqrt{n}} = \frac{\bar{x} - \mu}{4.0/\sqrt{15,000}} = \frac{\bar{x} - \mu}{0.03}$$

has (approximately) the *standard normal distribution*.

We want a 95% confidence interval for μ. In Example 8.7 we found that the area under the standard normal curve between -1.96 and 1.96 equals 0.95, or 95%. See Figure 8.8 on page 309. Since probabilities for the random variable $(\bar{x} - \mu)/0.03$ are equal to areas under the standard normal curve, we deduce that

$$P(-1.96 \leq \frac{\bar{x} - \mu}{0.03} \leq 1.96) = 0.95$$

[†] We will consider problems in which σ is unknown in the next section.

We want to rewrite the expression on the left so that μ stands alone in the middle. The algebraic steps are as follows:

$$-1.96 \leq \frac{\bar{x} - \mu}{0.03} \leq 1.96$$

$$-1.96 \leq \frac{\mu - \bar{x}}{0.03} \leq 1.96^\dagger$$

$$-1.96 \cdot 0.03 \leq \mu - \bar{x} \leq 1.96 \cdot 0.03$$

$$-0.06 \leq \mu - \bar{x} \leq 0.06$$

or

$$\bar{x} - 0.06 \leq \mu \leq \bar{x} + 0.06$$

Consequently, we can rewrite

$$P(-1.96 \leq \frac{\bar{x} - \mu}{0.03} \leq 1.96) = 0.95$$

as

$$P(\bar{x} - 0.06 \leq \mu \leq \bar{x} + 0.06) = 0.95$$

In other words, if $\bar{x}$ denotes the mean duration of 15,000 randomly selected phone calls, then the probability is 0.95 that the interval from $\bar{x} - 0.06$ to $\bar{x} + 0.06$ will contain μ.

Since the mean duration of the 15,000 calls actually obtained turned out to be $\bar{x} = 3.8$ minutes, we have

$$\bar{x} - 0.06 = 3.8 - 0.06 = 3.74$$
$$\bar{x} + 0.06 = 3.8 + 0.06 = 3.86$$

Consequently, the interval from 3.74 to 3.86 minutes is a 95% confidence interval for μ. The telephone company can be 95% confident that the mean duration, μ, of all telephone calls made by Phoenix residential customers is somewhere between 3.74 and 3.86 minutes. ∎

Exercises 8.2

8.9 Determine $z_{0.33}$; that is, find the z-value with area 0.33 to its right. Illustrate your work with a picture similar to Figure 8.6 on page 309.

8.10 Determine $z_{0.015}$; that is, find the z-value with area 0.015 to its right. Illustrate your work with a picture similar to Figure 8.6 on page 309.

8.11 Find the following z-values and draw graphs to illustrate your work.
a) $z_{0.03}$ b) $z_{0.005}$

8.12 Obtain the z-values below and draw graphs illustrating your work.
a) $z_{0.20}$ b) $z_{0.06}$

8.13 Find the two z-values that divide the area under the standard normal curve into a middle 0.90 area and two outside 0.05 areas. Illustrate your work with pictures similar to Figures 8.7 and 8.8.

8.14 Determine the two z-values that divide the area under the standard normal curve into a middle 0.99

$\dagger$ This first step uses the fact that if a number, b, is between -1.96 and 1.96, then so is $-b$.

area and two outside 0.005 areas. Illustrate your work with graphs similar to the ones in Figures 8.7 and 8.8.

In each of Exercises 8.15 through 8.18, find the designated confidence interval by applying the same procedure, step-by-step, as used in Example 8.8.

8.15 The Gallup Organization conducts annual national surveys on home gardening. Results are published by the National Association for Gardening. [In 1983, about 42 percent of the households in the U.S. had vegetable gardens!] Suppose $n = 250$ households with vegetable gardens are selected at random and that the average size of their gardens is $\bar{x} = 643$ sq. ft.
a) Determine a 90% confidence interval for the true mean size, μ, of household vegetable gardens in the U.S. [Assume $\sigma = 247$ sq. ft.]
b) Interpret your results in words.

8.16 A quality control engineer in a bakery goods plant needs to estimate the mean weight, μ, of bags of potato chips that are packed by machine. He knows from experience that $\sigma = 0.1$ oz for this machine. He takes a random sample of $n = 36$ bags and finds the sample mean weight to be 16.01 oz.
a) Use this data to find a 99% confidence interval for μ.
b) Interpret your results in words.

8.17 The U.S. Bureau of Labor Statistics collects and publishes data on employment and hourly earnings in the aircraft industry. Suppose $n = 30$ people working in the aircraft industry are selected at random and that their hourly earnings are as follows:

$15.20	5.94	12.28	14.99	5.99
16.42	15.71	4.01	8.90	9.25
13.77	18.22	9.18	11.60	13.20
14.05	15.17	13.57	12.52	14.49
9.43	8.85	18.82	5.83	14.82
17.02	11.84	12.16	7.73	9.30

a) Use this data to find a 95% confidence interval for the true mean hourly earnings, μ, of all persons employed in the aircraft industry. Assume $\sigma = \$3.25$. [*Note:* The sum of the data is $\Sigma x = \$360.26$.]
b) Interpret your results in words.

8.18 A sociologist wants information on the number of children per farm family in her native state of Nebraska. Forty randomly selected farm families have the following number of children:

3	5	2	1	1	0	2	3
1	1	2	1	2	0	1	5
4	1	0	1	3	1	0	1
0	1	8	0	1	2	2	2
2	1	5	3	1	4	1	0

a) Find a 90% confidence interval for the true mean number of children, μ, per farm family in Nebraska. Assume $\sigma = 1.95$. [*Note:* The sum of the data is $\Sigma x = 74$.]
b) Interpret your results in words.

8.3   Confidence intervals for a population mean (large samples)

In Example 8.8, we found a 95% confidence interval. As we have noted, the number 0.95 is called the *confidence level.* It is frequently useful to express confidence levels in a different form. For instance, the confidence level 0.95 is often written as $1 - 0.05$. The Greek letter α (alpha) is used to denote the number that must be subtracted from 1, in this case 0.05, to get the confidence level. Thus, for a 95% confidence interval, the confidence level is 0.95 and $\alpha = 0.05$. For a 90% confidence interval, the confidence level is 0.90 and $\alpha = 0.10$.

Confidence intervals for μ—large samples

If we carefully study the method used to solve Example 8.8 on pages 310–311, then we can write down a quick and simple procedure for determining confidence intervals for a population mean μ. The procedure applies when the sample size is large ($n \geq 30$).

PROCEDURE 8.1 To find a confidence interval for a population mean, μ.

Assumption
Sample size is at least 30 ($n \geq 30$).

STEP 1 *For a confidence level of $1 - \alpha$, use Table II to find $z_{\alpha/2}$.*
STEP 2 *The confidence interval for μ is*

$$\bar{x} - z_{\alpha/2} \cdot \frac{s}{\sqrt{n}} \quad to \quad \bar{x} + z_{\alpha/2} \cdot \frac{s}{\sqrt{n}}$$

where $z_{\alpha/2}$ is found in Step 1, n is the sample size, and $\bar{x}$ and s are computed from the actual sample data obtained.

Note: In Step 2 of Procedure 8.1 we have used the sample standard deviation, s, in the formula for the confidence interval. Theoretically, we should use the population standard deviation, σ, as we did in Example 8.8. However, σ is rarely known and so we have used s in its place. This is acceptable because, for large samples, the sample standard deviation is likely to be a good approximation of the population standard deviation. In the rare case where the population standard deviation is known, it should always be used in place of the sample standard deviation in Step 2 of Procedure 8.1.

EXAMPLE 8.9 **Illustrates Procedure 8.1**

The U.S. Bureau of Labor Statistics collects and publishes data on the ages of people in the civilian labor force. Suppose $n = 50$ such people are randomly selected and that their ages are as displayed in Table 8.6.

TABLE 8.6

Ages of 50 randomly selected people in the civilian labor force.

22	58	40	42	43	32	34	45	38	19
33	16	49	29	30	43	37	19	21	62
60	41	28	35	37	51	37	65	57	26
27	31	33	24	34	28	39	43	26	38
42	40	31	34	38	35	29	33	32	33

Find a 90% confidence interval for the mean age, μ, of all people in the civilian labor force.

SOLUTION
Since $n \geq 30$, we use Procedure 8.1 to find the confidence interval.

STEP 1 *For a confidence level of $1 - \alpha$, use Table II to find $z_{\alpha/2}$.*

We need a 90% confidence interval, so the confidence level is $0.90 = 1 - 0.10$. This means that $\alpha = 0.10$. Consulting Table II we see that

$$z_{\alpha/2} = z_{0.05} = 1.645$$

STEP 2 *The confidence interval for μ is*

$$\bar{x} - z_{\alpha/2} \cdot \frac{s}{\sqrt{n}} \quad to \quad \bar{x} + z_{\alpha/2} \cdot \frac{s}{\sqrt{n}}$$

We have $n = 50$ and, from Step 1, $z_{a/2} = 1.645$. To compute $\bar{x}$ and s for the data in Table 8.6, we apply the usual formulas:

$$\bar{x} = \frac{\Sigma x}{n} = \frac{1819}{50} = 36.38$$

$$s = \sqrt{\frac{n(\Sigma x^2) - (\Sigma x)^2}{n(n-1)}} = \sqrt{\frac{50(72179) - (1819)^2}{50 \cdot 49}} = 11.07$$

Consequently, a 90% confidence interval for μ is from

$$36.38 - 1.645 \cdot \frac{11.07}{\sqrt{50}} \quad \text{to} \quad 36.38 + 1.645 \cdot \frac{11.07}{\sqrt{50}}$$

or

$$33.8 \quad \text{to} \quad 39.0$$

We can be 90% confident that the mean age, μ, of all people in the civilian labor force is somewhere between 33.8 and 39.0 years. ∎

Relation of confidence level to length of confidence interval

The confidence level of a confidence interval for a population mean, μ, signifies the confidence of the estimate. That is, it expresses the confidence we have that the value of μ actually lies in the confidence interval. The length of the confidence interval, on the other hand, indicates the precision of the estimate. Long confidence intervals indicate poor precision, while short confidence intervals indicate good precision.

How does the confidence level affect the length of the confidence interval? Look back to Example 8.9 on page 313, where we found a 90% confidence interval for the mean age, μ, of people in the civilian labor force. The confidence level there is 0.90, and the confidence interval we computed is from 33.8 to 39.0 years. We can be 90% confident that μ is somewhere between 33.8 and 39.0 years.

What happens to the length of the confidence interval if we change the confidence level from 0.90 to, say, 0.95? Then $z_{a/2}$ changes from $z_{0.05} = 1.645$ to $z_{0.025} = 1.96$. The resulting confidence interval (using the same sample data of Table 8.6) is then from

$$36.38 - 1.96 \cdot \frac{11.07}{\sqrt{50}} \quad \text{to} \quad 36.38 + 1.96 \cdot \frac{11.07}{\sqrt{50}}$$

or from 33.3 to 39.4 years. We picture both the 90% and 95% confidence intervals in Figure 8.9.

FIGURE 8.9
90% and 95% confidence intervals for μ using Table 8.6 data.

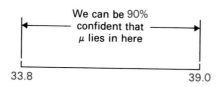

We can be 90% confident that μ lies in here	We can be 95% confident that μ lies in here
33.8 39.0	33.3 39.4
(90% Confidence Interval)	(95% Confidence Interval)

Thus, increasing the confidence level increases the length of the confidence interval. This makes sense. If we want to be more confident that the true value of μ lies in our confidence interval, then we must naturally have a more extensive interval. In summary, we have the following important fact.

KEY FACT

For a fixed sample size, the greater the required level of confidence, the greater the length of the confidence interval.

Exercises 8.3

8.19 Find the confidence level and α for
a) a 90% confidence interval.
b) a 99% confidence interval.

8.20 Find the confidence level and α for
a) an 85% confidence interval.
b) a 95% confidence interval.

In Exercises 8.15 through 8.18 you were asked to find a confidence interval for a population mean, μ, by applying the "long method" illustrated in Example 8.8 on pages 310–311. For purposes of comparison and practice, use the "short method" (Procedure 8.1, page 313) in Exercises 8.21 through 8.24 to solve the same problems.

8.21 The Gallup Organization conducts annual national surveys on home gardening. Results are published by the National Association for Gardening. Suppose $n = 250$ households with vegetable gardens are selected at random and that the average size of their gardens is $\bar{x} = 643$ sq. ft.
a) Use Procedure 8.1 to determine a 90% confidence interval for the true mean size, μ, of household vegetable gardens in the U.S. [Use $\sigma = 247$ sq. ft.]
b) Interpret your results in words.

8.22 A quality control engineer in a bakery goods plant needs to estimate the mean weight, μ, of bags of potato chips that are packed by machine. He knows from experience that $\sigma = 0.1$ oz for this machine. He takes a random sample of $n = 36$ bags and finds the sample mean weight to be 16.01 oz.
a) Use Procedure 8.1 to find a 99% confidence interval for μ.
b) Interpret your results in words.

8.23 The U.S. Bureau of Labor Statistics collects and publishes data on employment and hourly earnings

in the aircraft industry. Suppose $n = 30$ people working in the aircraft industry are selected at random and that their hourly earnings are as follows:

$15.20	5.94	12.28	14.99	5.99
16.42	15.71	4.01	8.90	9.25
13.77	18.22	9.18	11.60	13.20
14.05	15.17	13.57	12.52	14.49
9.43	8.85	18.82	5.83	14.82
17.02	11.84	12.16	7.73	9.30

a) Apply Procedure 8.1 to find a 95% confidence interval for the true mean hourly earnings, μ, of all people employed in the aircraft industry. Assume $\sigma = \$3.25$. [*Note:* The sum of the data is $\Sigma x = \$360.26$.]
b) Interpret your results in words.

8.24 A sociologist wants information on the number of children per farm family in her native state of Nebraska. Forty randomly selected farm families have the following number of children:

3	5	2	1	1	0	2	3
1	1	2	1	2	0	1	5
4	1	0	1	3	1	0	1
0	1	8	0	1	2	2	2
2	1	5	3	1	4	1	0

a) Apply Procedure 8.1 to find a 90% confidence interval for the true mean number of children, μ, per farm family in Nebraska. Assume $\sigma = 1.95$. [*Note:* The sum of the data is $\Sigma x = 74$.]
b) Interpret your results in words.

8.25 The U.S. National Center for Health Statistics estimates mean weights and heights of U.S. adults by age and sex. Suppose that 40 women, 5 ft 4 inches tall and aged 18–24, are randomly selected and that their weights, in pounds, are as shown here.

140	136	147	138	143	122	115	125
136	152	130	134	150	153	148	132
116	159	128	136	134	126	120	146
131	167	145	132	138	137	115	145
154	139	139	147	123	154	127	116

a) Find a 90% confidence interval for the mean weight, μ, of all U.S. women 5 ft 4 inches tall and in the age group 18–24. [*Note:* $\Sigma x = 5475$, $\Sigma x^2 = 755,749$.]
b) Interpret your results in words.

8.26 A research physician wants to estimate the average age of persons with diabetes. She takes a random sample of $n = 35$ such people and obtains the following ages:

48	41	57	83	41	55	59
61	38	48	79	75	77	7
54	23	47	56	79	68	61
64	45	53	82	68	38	70
10	60	83	76	21	65	47

a) Determine a 95% confidence interval for the mean age, μ, of people with diabetes. [*Note:* $\Sigma x = 1939$, $\Sigma x^2 = 120,861$.]
b) Interpret your results in words.

8.27 Refer to Exercise 8.25.
a) Find a 99% confidence interval for μ.
b) Why is the confidence interval you just found in part (a) longer than the one in Exercise 8.25?
c) Draw a graph similar to Figure 8.9 on page 314 that displays both confidence intervals.

8.28 Refer to Exercise 8.26.
a) Determine an 80% confidence interval for μ.
b) Why is the confidence interval you just found in part (a) shorter than the one in Exercise 8.26?
c) Draw a graph similar to Figure 8.9 on page 314 that displays both confidence intervals.

8.29 The U.S. Bureau of the Census collects and publishes data on family size. Suppose $n = 500$ U.S. families are randomly selected in order to estimate the mean size, μ, of all U.S. families and that the results are as shown in the following frequency distribution:

Size of family x	Frequency f
2	198
3	118
4	101
5	59
6	12
7	3
8	8
9	1

a) Find a 95% confidence interval for the mean size, μ, of all U.S. families. *Hint:* Use Formula 3.2 on page 83 to compute $\bar{x}$ and Formula 3.3 on page 84 to compute s.
b) Interpret your results in words.

8.30 In this exercise we will mathematically verify the Key Fact on page 315—namely, that for a fixed sample size, the greater the required level of confidence, the greater the length of the confidence interval. We will assume σ is known, so that the form of the confidence interval is

$$\bar{x} - z_{\alpha/2} \cdot \frac{\sigma}{\sqrt{n}} \quad \text{to} \quad \bar{x} + z_{\alpha/2} \cdot \frac{\sigma}{\sqrt{n}}$$

a) Show that the length of the confidence interval is

$$L = 2z_{\alpha/2} \cdot \frac{\sigma}{\sqrt{n}}$$

b) Show that increasing the confidence level decreases the value of α.
c) Verify that decreasing the value of α increases the value of $z_{\alpha/2}$.
d) Deduce from parts (a) through (c) that increasing the confidence level increases the length of the confidence interval.

8.4 Sample size considerations

In this section we will examine in detail how the sample size affects the precision of estimating a population mean by a sample mean. We have already alluded to the fact that "the larger the sample size, the greater the likelihood that the sampling error will be small." (See page 281.) Now that we have studied confidence intervals, we can determine exactly how the sample size affects the accuracy of the estimate. We can also learn how to find the sample size required in order to attain a specified precision. To begin, let us consider the following example.

EXAMPLE 8.10   **Introduces the maximum error of the estimate**

The U.S. Energy Information Administration surveys households in order to obtain information on household transportation. Among other things, data is collected to estimate the average monthly fuel expenditure, μ, per vehicle. Suppose 30 monthly fuel expenditures are selected at random and that their mean is $\bar{x} = \$58.56$.

a) Find a 95% confidence interval for the true mean monthly fuel expenditure, μ. [Assume $\sigma = \$20.65$.]

b) Discuss the precision with which $\bar{x}$ estimates μ.

SOLUTION

a) For a 95% confidence interval, $z_{a/2} = z_{0.025} = 1.96$. Also, we have $n = 30$, $\sigma = \$20.65$, and $\bar{x} = \$58.56$. Applying Procedure 8.1 (page 313), we find that a 95% confidence interval for μ is from

$$\bar{x} - z_{a/2} \cdot \frac{\sigma}{\sqrt{n}} \quad \text{to} \quad \bar{x} + z_{a/2} \cdot \frac{\sigma}{\sqrt{n}}$$

or

$$58.56 - 1.96 \cdot \frac{20.65}{\sqrt{30}} \quad \text{to} \quad 58.56 + 1.96 \cdot \frac{20.65}{\sqrt{30}}$$

or

$$58.56 - 7.39 \quad \text{to} \quad 58.56 + 7.39$$

or

$$51.17 \quad \text{to} \quad 65.95$$

We can be 95% confident that the true mean monthly fuel expenditure, μ, is somewhere between $51.17 and $65.95.

b) It may have struck you that the confidence interval we just found gives a rather wide range for the possible values of μ, and that it would be desirable to narrow the range, that is, to decrease the length of the confidence interval. It is desirable to decrease the length of the confidence interval because that length determines the precision with which $\bar{x}$ estimates μ.

As we learned in Section 8.3, one way to decrease the length of the confidence interval is to lower the confidence level from 95% to some lower level. But suppose we want to retain the same level of confidence and still narrow the confidence interval. What can be done to accomplish this?

To answer this question, let us first look more closely at the confidence interval in part (a). That confidence interval is displayed graphically in Figure 8.10.

FIGURE 8.10
95% confidence
interval for mean
monthly fuel
expenditure, μ, per
vehicle.

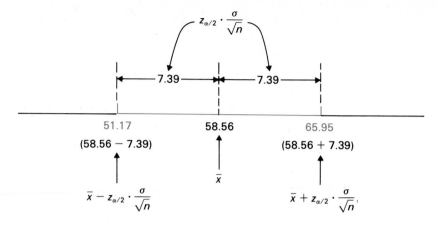

If you examine the picture carefully or look back at the computations in part (a), you will find that the length of the confidence interval is determined by the quantity

$$E = z_{\alpha/2} \cdot \frac{\sigma}{\sqrt{n}}$$

which is just half the length of the confidence interval. In this case, $E = 7.39$. That quantity is called the **maximum error of the estimate.** We use this terminology because we can be 95% confident that μ lies in the confidence interval, or equivalently, that the maximum error made in using $\bar{x}$ to estimate μ is $E = \$7.39$. (See Figure 8.10.)

In any event, we see that to narrow the confidence interval and thereby increase the precision of our estimate, we need only decrease the value of E. Since n is in the denominator in the formula for E, *we can decrease E by increasing the sample size n.* This makes sense, of course, because we expect to get more accurate information from larger samples. ∎

Below we summarize the discussion in part (b) of Example 8.10.

DEFINITION 8.3 Maximum error of the estimate for μ

The **maximum error of the estimate, E,** is defined by

$$E = z_{\alpha/2} \cdot \frac{\sigma}{\sqrt{n}}$$

and is equal to half the length of the confidence interval.

FIGURE 8.11
$E = z_{\alpha/2} \cdot \frac{\sigma}{\sqrt{n}}$

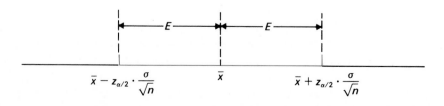

KEY FACT

The length of a confidence interval, and hence the precision of estimating μ by $\bar{x}$, is determined by the *maximum error of the estimate, E.* For a given confidence level, we can increase the precision of our estimate by increasing the sample size, n.

Determining the required sample size

As we have seen, the maximum error of the estimate, E, indicates the accuracy with which a sample mean, $\bar{x}$, estimates a population mean, μ. Furthermore, the value of E can be controlled by the sample size n.

In many cases, the precision required and confidence desired are stipulated in advance. From this information we must then determine the sample size necessary to meet the given specifications. The formula for the required sample size can be obtained by solving for n in the expression $E = z_{\alpha/2} \cdot \sigma/\sqrt{n}$. The result is Formula 8.1.

FORMULA 8.1 Sample size

The sample size required for a $(1 - \alpha)$-level confidence interval for μ with a specified maximum error of the estimate, E, is given by

$$n = \left[\frac{z_{\alpha/2} \cdot \sigma}{E} \right]^2$$

EXAMPLE 8.11 **Illustrates Formula 8.1**

Consider again the problem of estimating the mean monthly fuel expenditure, μ, per household vehicle.

a) Determine the sample size required to insure that we can be 95% confident that our estimate, $\bar{x}$, will be within \$0.50 of μ.

b) Find a 95% confidence interval for μ if a sample of the size indicated in part (a) yields an $\bar{x} = \$59.02$.

SOLUTION

a) To find the required sample size, we apply Formula 8.1. Since the desired confidence level is 0.95, we have $\alpha = 0.05$. Consulting Table II, we see that $z_{\alpha/2} = z_{0.025} = 1.96$. The maximum error of the estimate is specified at $E = \$0.50$. Recalling that $\sigma = \$20.65$, we get

$$n = \left[\frac{z_{\alpha/2} \cdot \sigma}{E} \right]^2 = \left[\frac{1.96 \cdot 20.65}{0.5} \right]^2 = 6552.58$$

Obviously, we cannot take a fractional sample size, so to be conservative, we *round up* to 6553. Consequently, if 6553 monthly fuel expenditures are randomly selected, then we can be 95% confident that their mean, $\bar{x}$, is within \$0.50 of the true mean monthly fuel expenditure, μ. [By the way, the U.S. Energy Information Administration takes a sample of size 6841.]

b) For a sample of size $n = 6553$ with a sample mean of $\bar{x} = \$59.02$, a 95% confidence interval for μ is from

$$\bar{x} - z_{a/2} \cdot \frac{\sigma}{\sqrt{n}} \quad \text{to} \quad \bar{x} + z_{a/2} \cdot \frac{\sigma}{\sqrt{n}}$$

or

$$59.02 - 1.96 \cdot \frac{20.65}{\sqrt{6553}} \quad \text{to} \quad 59.02 + 1.96 \cdot \frac{20.65}{\sqrt{6553}}$$

or

$$59.02 - 0.50 \quad \text{to} \quad 59.02 + 0.50$$

or

$$\$58.52 \quad \text{to} \quad \$59.52$$

We can be 95% confident that the mean monthly fuel expenditure, μ, is somewhere between $58.52 and $59.52. ∎

Note: In order to use Formula 8.1,

$$n = \left[\frac{z_{a/2} \cdot \sigma}{E} \right]^2$$

we need to know the population standard deviation σ. If σ is unknown and we want to apply the formula, then we must first estimate σ. We can do this by taking a preliminary sample of size 30 or more, and then computing the sample standard deviation, s, of the data. The value of s so obtained is an estimate of σ and can be used in place of σ in Formula 8.1.

Exercises 8.4

8.31   A 90% confidence interval for the mean size, μ, of household vegetable gardens was found in Exercise 8.21 of Section 8.3 (page 315). The confidence interval is from 617.3 to 668.7 square feet.
a) Determine the maximum error of the estimate.
b) Explain the meaning of E in regard to the estimation of μ.
c) Determine the sample size required to have the same maximum error of the estimate as in part (a), but with a 95% confidence level. [Use $\sigma = 247$ sq. ft.]

8.32   In Exercise 8.24 of Section 8.3, you were asked to find a 90% confidence interval for the mean number of children, μ, per farm family in Nebraska based on a sample of size $n = 40$ (given on page 315). The 90% confidence interval for μ is from 1.3 to 2.4 children per farm family.

a) Determine the maximum error of the estimate.
b) Explain the meaning of E, in this context, as far as the accuracy of the estimate is concerned.
c) Determine the sample size required to insure that we can be 90% confident that our estimate, $\bar{x}$, will be within 0.1 children of μ. [Recall that $\sigma = 1.95$.]
d) Find a 90% confidence interval for μ, if a sample of the size indicated in part (c) yields an $\bar{x} = 1.9$ children.

8.33   In Exercise 8.23 of Section 8.3, you were asked to find a 95% confidence interval for the mean hourly earnings of persons employed in the aircraft industry based on a sample of size $n = 30$ (given on page 315). The 95% confidence interval is from $10.85 to $13.17.
a) Determine the maximum error of the estimate.

b) Explain the meaning of E, in this context, as far as the accuracy of the estimate is concerned.

c) Determine the sample size required to insure that we can be 95% confident that our estimate, $\bar{x}$, will be within $0.50 of μ. [Recall that $\sigma = 3.25.]

d) Find a 95% confidence interval for μ, if a sample of the size indicated in part (c) yields an $\bar{x} = 12.87.

8.34 You were asked, in Exercise 8.26 of Section 8.3, to find a 95% confidence interval for the average age of people with diabetes based on a sample of size $n = 35$ (given on page 316.) The confidence interval is from 48.8 to 62.0 years.

a) Determine the maximum error of the estimate.

b) Determine the sample size required to have a maximum error of the estimate of 0.5 years and a 95% confidence level. [*Note:* The sample standard deviation of the sample of 35 ages from Exercise 8.26 is $s = 19.88$ years.]

c) Why did you use the sample standard deviation, $s = 19.88$, in place of σ in your solution to part (b)?

d) Find a 95% confidence interval for the mean age, μ, of people with diabetes if a sample of the size indicated in part (b) gives the statistics $\bar{x} = 58.3$ years and $s = 19.0$ years.

8.35 In Exercise 8.25 of Section 8.3 (page 315), you were given a sample of weights obtained from the random selection of $n = 40$ women 5 ft 4 inches tall and aged 18–24. Based on that data, a 90% confidence interval for the mean weight, μ, of all such women is from 133.6 to 140.2 lb.

a) Determine the maximum error of the estimate.

b) Determine the sample size required to have a maximum error of the estimate of 2 lb and a 99% confidence level. [*Note:* $s = 12.77$ lb for the sample data in Exercise 8.25.]

c) Why did you use the sample standard deviation, $s = 12.77$, in place of σ in your solution to part (b)?

d) Find a 99% confidence interval for the mean weight, μ, of all women aged 18–24 who are 5 ft 4 inches tall, if a sample of the size indicated in part (b) gives $\bar{x} = 134.2$ lb with $s = 13.0$ lb.

8.36 Why can a sample standard deviation, s, be used in place of σ in the formula $n = [z_{\alpha/2} \cdot \sigma / E]^2$ if σ is unknown and n is large?

8.37 The U.S. Bureau of the Census gives estimates for the mean value, μ, of the land and buildings per corporate farm. Suppose that an estimate, $\bar{x}$, is obtained and that the maximum error of the estimate is $E = 1000. Does this mean that the estimate is within $1000 of the true mean μ? Explain.

8.38 Use Formula 8.1 (page 319) to establish the following fact: For a fixed confidence level, it is necessary to (approximately) *quadruple* the sample size, in order to *double* the accuracy of the estimate.

8.5   Confidence intervals for a population mean (small samples)

In Section 8.3, we learned how to find confidence intervals for a population mean, μ, when dealing with large samples ($n \geq 30$). However, there are many cases in which large samples are either unavailable, extremely expensive, or undesirable. For example, tests of automobiles to analyze the impact of collisions often involve wrecking the cars. Clearly, it is desirable to employ small samples in this and similar cases.

In this section, we will learn how to find confidence intervals for a population mean, μ, when dealing with small samples ($n < 30$). We shall assume that the population standard deviation, σ, is unknown since that is usually the case.

Student's *t*-distribution

As we said, our next objective is to obtain a small-sample confidence interval procedure for μ when σ is unknown. To develop such a procedure, we first need to identify the probability distribution of the random variable

$$\frac{\overline{x} - \mu}{s/\sqrt{n}}$$

The probability distribution of the above random variable depends on the distribution of the population being sampled and so, in general, not much can be said. Since so many populations are, at least approximately, normally distributed, we shall confine our attention to that case.

For normally distributed populations, the random variable

$$\frac{\overline{x} - \mu}{s/\sqrt{n}}$$

has what is called **Student's *t*-distribution.** This was discovered by W. S. Gosset in 1908. Gosset did much of his research while working for the Guiness Brewery of Ireland. The company prohibited its employees from publishing research, so Gosset presented his finding using the pen name "Student." Thus the name "Student" in "Student's *t*-distribution." For brevity, we shall just use the term ***t*-distribution.**

As you know, probabilities for a random variable having a normal distribution are equal to areas under an appropriate normal curve. Probabilities for a random variable having a *t*-distribution are also equal to areas under a curve, suitably called a ***t*-curve.** Actually, there are infinitely many *t*-curves, and the one that is used depends on the sample size. If the sample size is *n,* then we identify the *t*-curve in question by saying that it is the *t*-curve with $n - 1$ **degrees of freedom.** The mathematical concepts involved in defining degrees of freedom are somewhat complex. Therefore, we define degrees of freedom simply as a number that identifies the appropriate *t*-curve or *t*-distribution. For convenience we will usually write **df $= n - 1$** to indicate $n - 1$ degrees of freedom. Before examining the *t*-distribution and *t*-curves further, we summarize the preceding discussion.

KEY FACT

Suppose a random sample of size *n* is to be taken from a *normally distributed* population with mean μ. Then the random variable

$$t = \frac{\overline{x} - \mu}{s/\sqrt{n}}$$

has the *t-distribution with n $-$ 1 degrees of freedom.* That is, probabilities for this random variable are equal to areas under the *t*-curve with df $= n - 1$.

Let us look at some specific examples of *t*-curves and then note some general properties of such curves. As we mentioned, there is a different *t*-curve for each number of degrees of freedom. Nonetheless, all *t*-curves look quite alike and are very similar to the standard normal curve. For illustration we have drawn in Figure 8.12 the standard normal curve along with two *t*-curves. One *t*-curve has 1 degree of freedom and the other has 6 degrees of freedom.

FIGURE 8.12

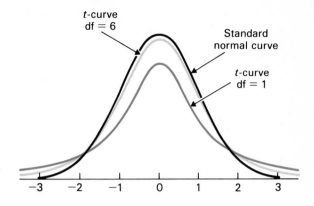

As illustrated by Figure 8.12, *t*-curves have the following properties:

KEY FACT Basic properties of *t*-curves

PROPERTY 1 *The total area under any t-curve is 1.*
PROPERTY 2 *A t-curve extends indefinitely in both directions, approaching the horizontal axis as it does so.*
PROPERTY 3 *A t-curve is symmetric about 0.*
PROPERTY 4 *As the number of degrees of freedom gets larger, t-curves look increasingly like the standard normal curve.*

In Section 6.1 we learned how to find areas under the standard normal curve by using Table II. We will now see how to find areas under *t*-curves by using Table III. For our purposes, one of which is obtaining confidence intervals for a population mean, we do not need a complete *t*-table for each *t*-curve. There are only certain areas that will be important for us to know. In Table 8.7 we have reproduced part of the Student's *t*-distribution table, Table III, which can be found on the right front inside cover of the book.

TABLE 8.7

Student's *t*-distribution
(values of t_a).

df	$t_{0.10}$	$t_{0.05}$	$t_{0.025}$	$t_{0.01}$	$t_{0.005}$	df
.	.	.	.	.	.	.
.	.	.	.	.	.	.
.	.	.	.	.	.	.
11	1.363	1.796	2.201	2.718	3.106	11
12	1.356	1.782	2.179	2.681	3.055	12
13	1.350	1.771	2.160	2.650	3.012	13
14	1.345	1.761	2.145	2.624	2.977	14
15	1.341	1.753	2.131	2.602	2.947	15
16	1.337	1.746	2.120	2.583	2.921	16
17	1.333	1.740	2.110	2.567	2.898	17
.	.	.	.	.	.	.
.	.	.	.	.	.	.
.	.	.	.	.	.	.

The table permits us to find *t*-values that have certain specified areas to their *right.* The two outside columns, labelled df, give the number of degrees of freedom. We will consider a *t*-curve with df = 16. Look at the row of Table 8.7 labelled "16." The first entry in that row is 1.337. That entry is also in the column labelled $t_{0.10}$. This means that for a *t*-curve with df = 16, the *t*-value with area 0.10 to its right is $t = 1.337$. See Figure 8.13.

FIGURE 8.13

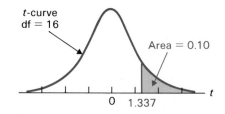

We further illustrate the use of the *t*-table in Example 8.12.

EXAMPLE 8.12   **Illustrates how to find the *t*-value for a specified area**

For a *t*-curve with 13 degrees of freedom, find the *t*-value with area 0.05 to its right. See Figure 8.14.

FIGURE 8.14

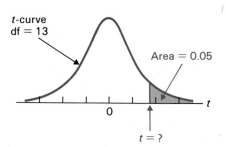

SOLUTION

To determine the *t*-value in question, we use Table III. In this example, the number of degrees of freedom is 13, so we concentrate on the row of the table with df = 13. If you go across that row until you are under the column headed $t_{0.05}$, you will see the number 1.771. This is the desired *t*-value. That is, for a *t*-curve with df = 13, the *t*-value with area 0.05 to its right is $t = 1.771$ See Figure 8.15.

FIGURE 8.15

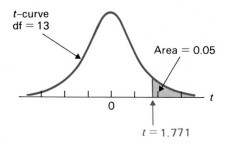

We will use the notation $t_{0.05}$ to denote the t-value with area 0.05 to its *right*. So, as we have just seen, for a t-curve with df = 13, $t_{0.05} = 1.771$. ∎

Confidence intervals for μ—small samples

Having discussed the t-distribution, we can now develop a procedure to find confidence intervals for a population mean, μ, when dealing with small samples. As we learned on page 322, when sampling from a normally distributed population, the random variable

$$t = \frac{\bar{x} - \mu}{s/\sqrt{n}}$$

has the t-distribution with df = $n - 1$. In other words, probabilities for that random variable are equal to areas under the t-curve with df = $n - 1$. From this fact it follows that the large-sample procedure, Procedure 8.1, given on page 313 can be applied to small samples, provided that we use $t_{\alpha/2}$ instead of $z_{\alpha/2}$.

PROCEDURE 8.2 To find a confidence interval for a population mean, μ.

Assumption
Normal population.

STEP 1 *For a confidence level of $1 - \alpha$, use Table III to find $t_{\alpha/2}$ for df = $n - 1$.*
STEP 2 *The confidence interval for μ is*

$$\bar{x} - t_{\alpha/2} \cdot \frac{s}{\sqrt{n}} \quad to \quad \bar{x} + t_{\alpha/2} \cdot \frac{s}{\sqrt{n}}$$

where $t_{\alpha/2}$ is found in Step 1, n is the sample size, and $\bar{x}$ and s are computed from the actual sample data obtained.

EXAMPLE   8.13   **Illustrates Procedure 8.2**

The gestation period of domestic dogs is *normally distributed*. To estimate the mean gestation period, 15 randomly selected dogs are observed during pregnancy. The gestation periods of the 15 dogs are displayed in Table 8.8.

TABLE 8.8		
Gestation periods, in days, of 15 randomly selected domestic dogs.		
62.0	61.4	59.8
62.2	60.3	60.4
59.4	60.2	60.4
60.8	61.8	59.2
61.1	60.4	60.9

Find a 95% confidence interval for the true mean gestation period, μ, of the domestic dog.

SOLUTION

Since the population (gestation periods) is normally distributed, we apply Procedure 8.2.

STEP 1 *For a confidence level of $1 - \alpha$, use Table III to find $t_{\alpha/2}$ for $df = n - 1$.*

The specified confidence level is 0.95, so that $\alpha = 0.05$. Since $n = 15$, we have $df = n - 1 = 15 - 1 = 14$. Table III shows that for $df = 14$,

$$t_{\alpha/2} = t_{0.025} = 2.145$$

STEP 2 *The confidence interval for μ is*

$$\bar{x} - t_{\alpha/2} \cdot \frac{s}{\sqrt{n}} \quad \text{to} \quad \bar{x} + t_{\alpha/2} \cdot \frac{s}{\sqrt{n}}$$

From Step 1, $t_{\alpha/2} = 2.145$. Applying the usual formulas for $\bar{x}$ and s to the data in Table 8.8, we find that

$$\bar{x} = \frac{\Sigma x}{n} = \frac{910.3}{15} = 60.69$$

$$s = \sqrt{\frac{n(\Sigma x^2) - (\Sigma x)^2}{n(n-1)}} = \sqrt{0.81} = 0.90$$

Consequently, a 95% confidence interval for μ is from

$$60.69 - 2.145 \cdot \frac{0.90}{\sqrt{15}} \quad \text{to} \quad 60.69 + 2.145 \cdot \frac{0.90}{\sqrt{15}}$$

or

$$60.19 \quad \text{to} \quad 61.18$$

We can be 95% confident that the mean gestation period of the domestic dog is somewhere between 60.19 and 61.18 days. ∎

MTB
SPSS

Note: Procedure 8.2 permits us to find confidence intervals for a population mean, μ, when dealing with small samples. An assumption for its use is that the population being sampled is normally distributed. Actually, the procedure works reasonably well even for a population that is only *approximately* normally distributed.

Exercises 8.5

8.39 Suppose a random sample of size n is to be taken from a *normally distributed* population with mean μ and standard deviation σ. What is the probability distribution of each of the following random variables?

a) $\dfrac{\bar{x} - \mu}{\sigma/\sqrt{n}}$

b) $\dfrac{\bar{x} - \mu}{s/\sqrt{n}}$

c) How do we determine probabilities for the random variable in (a)?

d) How do we determine probabilities for the random variable in (b)?

8.40 IQs are normally distributed with a mean of $\mu = 100$ and a standard deviation of $\sigma = 16$. Suppose a random sample of $n = 250$ IQs is to be selected. What is the probability distribution of each of the following random variables?

a) $\dfrac{\bar{x} - 100}{16/\sqrt{250}}$

b) $\dfrac{\bar{x} - 100}{s/\sqrt{250}}$

c) Referring to the random variable in part (b), we note that it has a t-distribution with df $= 249$. Thus, probabilities for that random variable are equal to areas under the t-curve with df $= 249$. But our t-table, Table III, does not have df $= 249$. So, for instance, how would you find the t-value with area 0.025 to its right for a t-curve with df $= 249$?

8.41 For a t-curve with df $= 21$, find the following t-values and illustrate your results with pictures.
   a) The t-value with area 0.10 to its right.
   b) $t_{0.01}$.
   c) The t-value with area 0.025 to its *left*.
   d) The two t-values that divide the area under the curve into a middle 0.90 area and two outside 0.05 areas.

8.42 For a t-curve with df $= 8$, determine the following t-values and illustrate your results with pictures.
   a) The t-value with area 0.05 to its right.
   b) $t_{0.10}$.
   c) The t-value with area 0.01 to its *left*.
   d) The two t-values that divide the area under the curve into a middle 0.95 area and two outside 0.025 areas.

8.43 According to the Salt River Project (SRP), a supplier of electricity to the greater Phoenix area, the mean annual electric bill in 1984 was \$852.31. An economist wants to estimate the mean for last year. He takes a random sample of $n = 18$ SRP customers and obtains the following amounts for their last year's electric bills:

\$1175	778	1506	1040	1130	816
1038	786	1241	908	1094	1128
564	1299	1098	1094	898	868

   a) Assume that, for a given year, annual electric bills are normally distributed. Find a 95% confidence interval for last year's mean annual electric bill, μ, for SRP customers. [The sample mean and sample standard deviation of the data are $\bar{x} = \$1025.61$ and $s = \$222.45$.]
   b) Does it appear that the mean annual electric bill has increased from the 1984 figure of \$852.31? Explain.

8.44 A city planner working on bikeways needs information about local bicycle commuters. He designs a questionnaire. One of the questions asks how many minutes it takes the rider to pedal from home to his or her destination. A random sample of 22 local bicycle commuters yields the following times:

22	19	24	31
29	29	21	15
27	23	37	31
30	26	16	26
12	23	48	
22	29	28	

   a) Find a 90% confidence interval for the true mean commuting time, μ, for local bicycle commuters in the city. Assume that the times are normally distributed. [The sample mean and sample standard deviation of the data are $\bar{x} = 25.82$ and $s = 7.71$.]
   b) Interpret your results in part (a).

8.45 According to the U.S. Department of Agriculture, the mean yield of oats for U.S. farms is about 58.4 bushels per acre. A farmer wants to estimate his mean yield using a newly developed fertilizer. He uses the fertilizer on a random sample of $n = 15$ one-acre plots and obtains the yields, in bushels, below.

67	65	55	57	58
61	61	61	64	62
62	60	62	60	67

   a) Find a 99% confidence interval for the true mean yield per acre, μ, that this farmer will get on his land with the new fertilizer.
   b) What assumption did you make in solving part (a)?
   c) Does it appear that, by using the new fertilizer, the farmer can get a better mean yield than the national average? Explain your answer.

8.46 According to the R. R. Bowker Company of New York, the mean annual subscription rate to law periodicals was \$29.66 in 1983. A random sample of $n = 12$ law periodicals yields the following annual subscription rates, to the nearest dollar, for this year:

\$20	36	34	37
32	28	52	45
42	38	33	44

   a) Find a 95% confidence interval for this year's true mean annual subscription rate, μ, for law periodicals.

b) What assumption did you make in solving part (a)?

c) Does your result from part (a) indicate an increase in the mean annual subscription rate over that in 1983? Justify your answer.

8.47 The random variable

$$t = \frac{\bar{x} - \mu}{s/\sqrt{n}}$$

is used for analysis of small samples from normally distributed populations when σ is unknown. If the population standard deviation σ is known, then we can use the random variable

$$z = \frac{\bar{x} - \mu}{\sigma/\sqrt{n}}$$

which has the standard normal distribution, even if the sample size is small.

a) Rework Exercise 8.45(a) under the assumption that σ is known, and $\sigma = 3.38$.

b) Compare your confidence interval from part (a) with the one found in Exercise 8.45.

c) What is the general effect on a $(1 - \alpha)$-level confidence interval if σ is known and a z-statistic can be used instead of a t-statistic?

8.48 Suppose a random sample of size n is to be taken from a normally distributed population with mean μ.

a) Verify that

$$P\left(-t_{\alpha/2} \le \frac{\bar{x} - \mu}{s/\sqrt{n}} \le t_{\alpha/2}\right) = 1 - \alpha$$

b) Use your result from part (a) to show that

$$P\left(\bar{x} - t_{\alpha/2} \cdot \frac{s}{\sqrt{n}} \le \mu \le \bar{x} + t_{\alpha/2} \cdot \frac{s}{\sqrt{n}}\right) = 1 - \alpha$$

c) Use your result from part (b) to explain why the interval from

$$\bar{x} - t_{\alpha/2} \cdot \frac{s}{\sqrt{n}} \quad \text{to} \quad \bar{x} + t_{\alpha/2} \cdot \frac{s}{\sqrt{n}}$$

is a $(1 - \alpha)$-level confidence interval for μ.

8.6 Confidence intervals for a population proportion (large samples)

Many statistical studies are concerned with the proportion (percentage) of a population that has a specified attribute. For example, we might be interested in the percentage of voters who will vote Republican in the next presidential election, or in the percentage of adults who drink alcoholic beverages, or in the percentage of unemployed workers.

In most cases it is either impractical or impossible to determine the proportion in question exactly. For instance, try to imagine interviewing *every* adult in order to ascertain the percentage of adults who drink alcoholic beverages. Thus, we generally resort to *sampling* and use the sample results to make inferences about the entire population. Let us now look at an example for the purpose of introducing some notation and terminology.

EXAMPLE 8.14 Introduces proportion terminology

The U.S. Energy Information Administration conducts annual surveys to obtain information on appliances possessed and generally used by U.S. households. In this year's survey, 6841 U.S. households are to be randomly selected and asked, among other things, whether or not they have a dishwasher. The proportion of the 6841 households sampled that have a dishwasher will then be used to estimate the proportion of *all* U.S. households that have a dishwasher.

We employ the letter p to denote the proportion of *all* U.S. households having a dishwasher. This is the **population proportion** and is the parameter whose value is to be estimated. The proportion of the 6841 U.S. households *surveyed* that have a dishwasher is called the **sample proportion** and is designated by the symbol $\bar{p}$ (read *p*-bar). This is the statistic that will be used to estimate the unknown population proportion, p. In summary, we will employ the following notation:

$$p = \text{population proportion}$$
$$\bar{p} = \text{sample proportion}$$

We should emphasize that the population proportion, p, although unknown, is a fixed number. On the other hand, the sample proportion, $\bar{p}$, is a random variable. Its value depends on chance—namely, on which households are selected for the survey. For example, if out of the 6841 households sampled, 2564 have a dishwasher, then $\bar{p} = 2564/6841 = 0.375$, or 37.5%. But, if out of the 6841 households sampled, 2448 have a dishwasher, then $\bar{p} = 2448/6841 = 0.358$, or 35.8%.

These two calculations also indicate how the sample proportion is computed: Divide the number of households sampled that have a dishwasher, x, by the total number of households sampled, n. In symbols,

$$\bar{p} = \frac{x}{n}$$

∎

We have used a specific example (Example 8.14) to introduce some notation and terminology required for statistical inferences concerning proportions. In general, we make the following definitions.

DEFINITION 8.4 Population proportion and sample proportion

Consider a population in which each member is classified as either having or not having a specified attribute—a so-called *two-category* population.

Population proportion, p: The proportion (percentage) of the entire population that has the specified attribute.

Sample proportion, $\bar{p}$: The proportion (percentage) of a sample from the population that has the specified attribute.

In Example 8.14, the specified attribute is "having a dishwasher." The population proportion, p, is the proportion of *all* U.S. households that have a dishwasher, and the sample proportion, $\bar{p}$, is the proportion of the households *sampled* that have a dishwasher. As we saw in Example 8.14, the sample proportion, $\bar{p}$, is computed using the following formula.

FORMULA 8.2 Sample proportion

The sample proportion, $\bar{p}$, is given by

$$\bar{p} = \frac{x}{n}$$

where x denotes the number of members sampled having the specified attribute and n denotes the sample size.

Note: For convenience, we will refer to x—the number of members sampled having the specified attribute—as the "number of successes."

Before proceeding, it might be helpful to draw some parallels between our discussion of proportions and our previous work with means. Table 8.9 shows the correspondence between the notation for means and the notation for proportions.

TABLE 8.9
Correspondence between notation for means and proportions.

	Population parameter	Sample statistic
Means	μ	$\bar{x}$
Proportions	p	$\bar{p}$

Just as we use a sample mean, $\bar{x}$, to make statistical inferences about a population mean, μ, we also use a sample proportion, $\bar{p}$, to make statistical inferences about a population proportion, p.

We have seen that in order to make statistical inferences for a population mean, μ, it is necessary to know the sampling distribution of the mean, that is, the probability distribution of the random variable $\bar{x}$. The same is true for proportions. To make statistical inferences for a population proportion, p, we need to know the **sampling distribution of a proportion,** that is, the probability distribution of the random variable $\bar{p}$. The following Key Fact can be derived from our knowledge of the sampling distribution of the mean, since we can always consider a proportion to be a mean. (See Exercise 8.59.)

KEY FACT The sampling distribution of a proportion—large samples

Suppose a large random sample of size n is to be taken from a two-category population with population proportion p. Then the random variable $\bar{p}$ is approximately *normally distributed* and has mean $\mu_{\bar{p}} = p$ and standard deviation $\sigma_{\bar{p}} = \sqrt{p(1-p)/n}$. In other words, probabilities for $\bar{p}$ are approximately equal to areas under the normal curve with parameters p and $\sqrt{p(1-p)/n}$.

Confidence intervals for a proportion

We will now learn how to find a confidence interval for a population proportion, p, based on the proportion, $\bar{p}$, of a sample from the population. Basically, the procedure is just a special case of Procedure 8.1 on page 313, which provides large-sample confidence intervals for μ when σ is unknown. Specifically, our confidence-interval procedure for proportions relies on the following result, which can be derived from the previous Key Fact. (See Exercise 8.58.)

KEY FACT

Suppose a random sample of size n is to be taken from a two-category population with population proportion p. Then the random variable

$$\frac{\overline{p} - p}{\sqrt{\overline{p}(1 - \overline{p})/n}}$$

has approximately the *standard normal distribution*. The approximation is adequate provided the number of successes, x, and the number of failures, $n - x$, are both at least 5.

Using this Key Fact and reasoning similar to that in Example 8.8 on page 310, we obtain the following confidence-interval procedure for proportions:

> **PROCEDURE 8.3** **To find a confidence interval for a population proportion, p.**
>
> *Assumption*
> The number of successes, x, and the number of failures, $n - x$, are both at least 5.
>
> **STEP 1** *For a confidence level of $1 - \alpha$, use Table II to find $z_{\alpha/2}$.*
> **STEP 2** *The confidence interval for p is*
>
> $$\overline{p} - z_{\alpha/2} \cdot \sqrt{\overline{p}(1 - \overline{p})/n} \quad to \quad \overline{p} + z_{\alpha/2} \cdot \sqrt{\overline{p}(1 - \overline{p})/n}$$
>
> *where $z_{\alpha/2}$ is found in Step 1, n is the sample size, and $\overline{p} = x/n$ is the sample proportion.*

EXAMPLE 8.15 **Illustrates Procedure 8.3**

Let us return to the situation of Example 8.14. A random sample of $n = 6841$ U.S. households is selected in order to estimate the proportion, p, of all U.S. households that have a dishwasher. If 2470 of the 6841 households selected have a dishwasher, find a 95% confidence interval for p.

S O L U T I O N
We will apply Procedure 8.3, but first we need to check that the conditions for its use are satisfied. The attribute in question is "having a dishwasher." The sample size is $n = 6841$. Since 2470 of the households sampled have a dishwasher,

$$x = 2470$$

and

$$n - x = 6841 - 2470 = 4371$$

Thus, both x and $n - x$ are at least 5, and so the conditions for using Procedure 8.3 are met.

STEP 1 *For a confidence level of $1 - \alpha$, use Table II to find $z_{\alpha/2}$.*

We want a 95% confidence interval. This means $\alpha = 0.05$ and, hence,

$$z_{\alpha/2} = z_{0.025} = 1.96$$

STEP 2 *The confidence interval for p is*

$$\bar{p} - z_{\alpha/2} \cdot \sqrt{\bar{p}(1 - \bar{p})/n} \quad to \quad \bar{p} + z_{\alpha/2} \cdot \sqrt{\bar{p}(1 - \bar{p})/n}$$

We have $n = 6841$ and, from Step 1, $z_{\alpha/2} = 1.96$. Also, since 2470 of the 6841 households sampled have a dishwasher, the sample proportion is

$$\bar{p} = \frac{x}{n} = \frac{2470}{6841} = 0.361$$

Consequently, a 95% confidence interval for p is

$$0.361 \pm 1.96 \cdot \sqrt{(0.361)(1 - 0.361)/6841}$$

or

$$0.350 \quad to \quad 0.372$$

We can be 95% confident that the percentage of U.S. households that have an automatic dishwasher is somewhere between 35.0% and 37.2%. ∎

Exercises 8.6

8.49 Is a population proportion, p, a parameter or a statistic? What about a sample proportion, $\bar{p}$? Justify your answers.

In Exercises 8.50 through 8.55 apply Procedure 8.3 on page 331 to determine the required confidence interval. Be sure to check that the conditions for applying the procedure are met.

8.50 Studies are performed to estimate the percentage of the nation's 10 million asthmatics who are allergic to sulfites. Suppose $n = 500$ asthmatics are randomly selected and that 38 of the 500 are found to be allergic to sulfites.
a) Find a 95% confidence interval for the true proportion, p, of U.S. asthmatics who are allergic to sulfites.
b) Interpret your results from part (a) in words.

8.51 A Reader's Digest/Gallup Survey on the drinking habits of Americans estimated the percentage of adults across the country who drink beer, wine, or hard liquor, at least occasionally. Of the 1516 adults interviewed, 985 said they drank.
a) Find a 95% confidence interval for the true pro-

portion, p, of Americans who drink beer, wine, or hard liquor, at least occasionally.
b) Interpret your results in words.

8.52 A Gallup Poll conducted in the spring of 1984 asked public school teachers nationwide to grade their own performance. The poll found that 634 of the 813 teachers surveyed gave their fellow educators a grade of A or B for performance.
a) Determine a 90% confidence interval for the actual percentage of all teachers who would give their fellow educators a grade of A or B for performance.
b) Interpret your results in words.

8.53 A 1985 Gallup Poll estimated the support among Americans for "right to die" laws. For the survey, 1528 adults, 18 and older, were asked if they were in favor of *voluntary* withholding of life support systems from the terminally ill. 1238 said they were.
a) Find a 99% confidence interval for the true percentage, p, of adult Americans who are in favor of "right to die" laws.
b) Interpret your results in words.

8.54 Dr. Charles Kuntzleman directed a study in which he estimated the percentage of schoolchildren who have at least one of the significant factors contributing to heart disease. The three-year study involved 24,000 children. Dr. Kuntzleman found that 98% of the children examined had at least one of the significant factors.

a) Use the information above to determine a 99% confidence interval for the proportion, p, of all schoolchildren who have at least one of the significant factors contributing to heart disease.
b) Interpret your results in part (a) in terms of percentages.

8.55 A Los Angeles research group publishes an annual report on trends in real estate.

a) Suppose 200 randomly selected real estate experts are asked whether they believe next year will be a good time to buy real estate. If 78% reply in the affirmative, determine a 90% confidence interval for the proportion of *all* real estate experts who believe next year will be a good time to buy real estate.
b) Interpret your results in words.

8.56 Suppose you have been hired to estimate the percentage of adults in your state who are literate. You take a random sample of 100 adults and find that 96 of them are literate. You then determine a 95% confidence interval as follows:

$$0.96 \pm 1.96 \cdot \sqrt{(0.96)(0.04)/100}$$

or

$$0.922 \quad \text{to} \quad 0.998$$

From this you conclude that we can be 95% confident that the percentage of all adults in your state who are literate is somewhere between 92.2% and 99.8%. Is there anything wrong with your reasoning?

8.57 Suppose that I have been commissioned to estimate the infant mortality rate in Norway. From a random sample of 500 live births, I find that 0.8% of them resulted in infant deaths. I then construct a 90% confidence interval:

$$0.008 \pm 1.645 \cdot \sqrt{(0.008)(0.992)/500}$$

or

$$0.001 \quad \text{to} \quad 0.015$$

Then I conclude: "We can be 90% confident that the proportion of infant deaths in Norway is somewhere between 0.001 and 0.015." How did I do?

In Exercises 8.58 and 8.59, we will derive the Key Fact on page 330, which gives the *sampling distribution of a proportion* (the probability distribution of $\bar{p}$). In Exercise 8.58, we will also verify the Key Fact on page 331 and justify the use of Procedure 8.3.

8.58 Formula 8.2 indicates that

$$(1) \qquad \bar{p} = \frac{x}{n}$$

where $x =$ number of successes and $n =$ sample size. The random variable x has, at least approximately, the *binomial distribution* with parameters n and p. Formulas 5.3 and 5.4 (pages 210 and 211) give the mean and standard deviation of x in terms of n and p:

$$\mu_x = np \qquad \sigma_x = \sqrt{np(1-p)}$$

a) Use these formulas, along with Formula (1) and Exercise 5.27(f) on page 190, to show that
$$\mu_{\bar{p}} = p \quad \text{and} \quad \sigma_{\bar{p}} = \sqrt{p(1-p)/n}$$

b) Recall from Section 6.5 that, for large n, a binomially distributed random variable, such as x, is approximately normally distributed. From this it follows that $\bar{p}$ is also approximately normally distributed. Use this last fact along with part (a) to deduce that, for large n, the random variable
$$z = \frac{\bar{p} - p}{\sqrt{p(1-p)/n}}$$
has approximately the *standard normal distribution*.

c) Use part (b) to conclude that, for large n, the random variable
$$\frac{\bar{p} - p}{\sqrt{\bar{p}(1-\bar{p})/n}}$$
has approximately the standard normal distribution. This statement is the Key Fact on page 331.

d) Use part (c) to show that, for large n,
$$P\left(-z_{\alpha/2} \leq \frac{\bar{p} - p}{\sqrt{\bar{p}(1-\bar{p})/n}} \leq z_{\alpha/2}\right) \approx 1 - \alpha$$

e) Deduce from part (d) that, for large n, the interval with endpoints
$$\bar{p} \pm z_{\alpha/2} \cdot \sqrt{\bar{p}(1-\bar{p})/n}$$
is a $(1 - \alpha)$-level confidence interval for p. [This last fact justifies Procedure 8.3.]

8.59 In Exercise 8.58 we examined the sampling distribution of a proportion and some of its important consequences. This exercise provides an alternate approach to deriving the sampling distribution of a proportion. Consider a finite, two-category population in which the proportion of members with the specified attribute is p. We can think of such a population as consisting of 1's and 0's. A member of the population is a "1" if it has the specified attribute and is a "0" otherwise.

a) If the size of the population is N, how many 1's are in the population?
b) Use part (a) and Definition 3.6 on page 95 to show that the mean of this population of 1's and

0's is p—that is, $\mu = p$.

c) Use part (b) and the shortcut formula on page 97 to show that the standard deviation of this population of 1's and 0's is $\sqrt{p(1-p)}$—that is, $\sigma = \sqrt{p(1-p)}$.
d) Suppose a sample of size n is to be taken from the population. Verify that $\bar{x} = \bar{p}$.
e) Use parts (b) through (d) along with the Key Fact on page 287 to obtain the sampling distribution of a proportion, that is, the Key Fact on page 330.

8.7   Computer packages*

In this chapter we have presented methods for constructing confidence intervals for a population mean, μ. The procedure to use in a given situation depends on sample size and the distribution of the population. Computer packages, such as Minitab, have programs that will determine confidence intervals for us. In this section, we will illustrate two such programs by applying them to examples that we have done previously by hand.

ZINTERVAL

Procedure 8.1 on page 313 gives a method for constructing confidence intervals for μ when the sample size is large ($n \geq 30$). Minitab has a program, called **ZINTERVAL,** that will do the work in constructing such confidence intervals. Let us see how it applies to Example 8.9 (page 313).

EXAMPLE 8.16   Illustrates the use of ZINTERVAL

The U.S. Bureau of Labor Statistics collects and publishes data on the ages of people in the civilian labor force. Suppose 50 such people are randomly selected and that their ages are as displayed in Table 8.10.

TABLE 8.10
Ages of 50 randomly selected people in the civilian labor force.

22	58	40	42	43	32	34	45	38	19
33	16	49	29	30	43	37	19	21	62
60	41	28	35	37	51	37	65	57	26
27	31	33	24	34	28	39	43	26	38
42	40	31	34	38	35	29	33	32	33

Find a 90% confidence interval for the mean age, μ, of all people in the civilian labor force.

SOLUTION
To obtain the desired confidence interval using Minitab, we first enter the sample data in Table 8.10 into the computer. The program ZINTERVAL requires the

user to input the population standard deviation, σ. However, σ is not given here. But since the sample size is large ($n = 50$), we can use the sample standard deviation, s, in place of σ. The sample standard deviation of the data in Table 8.10 is $s = 11.07$ (this can be obtained by using a calculator or by using Minitab).

Now we use the command ZINTERVAL followed by the desired confidence level (90%), the estimated value of σ (11.07), and the storage location of the sample data (C4). The command and its results are displayed in Printout 8.1.

PRINTOUT 8.1
Minitab output for
ZINTERVAL.

```
MTB > ZINTERVAL 90, 11.07, C4

THE ASSUMED SIGMA =11.07


           N      MEAN    STDEV   SE MEAN     90.0 PERCENT C.I.
C4         50     36.38   11.07     1.57    (   33.80,    38.96)
```

The first line of the output displays the value used for σ—THE ASSUMED SIGMA =11.07. Then come the sample size, sample mean, sample standard deviation, and estimated standard error of the mean ($s/\sqrt{n}$). The final item is the confidence interval. Hence, a 90% confidence interval is from 33.80 to 38.96. We can be 90% confident that the mean age, μ, of all people in the civilian labor force is somewhere between 33.80 and 38.96 years. ∎

TINTERVAL

Procedure 8.2 on page 325 gives a procedure for finding confidence intervals for μ when the sample size is small. Actually, the procedure applies equally well to small samples and large samples. An assumption for using Procedure 8.2 is that the population being sampled is normally distributed.

A Minitab program called **TINTERVAL** (the first "T" is for t-distribution) does the computations required in applying Procedure 8.2. Below we will employ TINTERVAL to obtain the confidence interval in Example 8.13 on page 325.

EXAMPLE 8.17 **Illustrates the use of TINTERVAL**

The gestation period of domestic dogs is *normally distributed*. To estimate the mean gestation period, 15 randomly selected dogs are observed during pregnancy. Their gestation periods are displayed in Table 8.11.

TABLE 8.11
Gestation periods, in days,
of 15 randomly selected
domestic dogs.

62.0	61.4	59.8
62.2	60.3	60.4
59.4	60.2	60.4
60.8	61.8	59.2
61.1	60.4	60.9

Find a 95% confidence interval for the true mean gestation period, μ, of the domestic dog.

S O L U T I O N

To use TINTERVAL, we first enter the sample data into the computer. Then we type the command TINTERVAL, followed by the confidence level (95%), and the storage location of the sample data (C5). The result is shown in Printout 8.2.

PRINTOUT 8.2
Minitab output for
TINTERVAL.

```
MTB > TINTERVAL, 95, C5

          N      MEAN    STDEV   SE MEAN      95.0 PERCENT C.I.
C5       15    60.687    0.898    0.232   ( 60.190,   61.184)
```

The printout first displays the sample size, sample mean, sample standard deviation, and estimated standard error of the mean ($s/\sqrt{n}$). Then comes the confidence interval. Thus, we can be 95% confident that the mean gestation period, μ, of the domestic dog is somewhere between 60.190 and 61.184 days. ∎

◆ Chapter Review

KEY TERMS

confidence interval, 302
confidence level, 302
degrees of freedom (df), 322
maximum error of the estimate
  for μ, 318
point estimate, 299
population proportion (p), 329
sample proportion ($\bar{p}$), 329

sampling distribution of the mean, 300
sampling distribution of a proportion, 330
Student's t-distribution, 322
t-curve, 322
t_a, 325
TINTERVAL*, 335
z_a, 308
ZINTERVAL*, 334

FORMULAS

In the formulas below,

μ = population mean
σ = population standard deviation
$\bar{x}$ = sample mean
s = sample standard deviation

n = sample size
$a = 1 -$ confidence level
p = population proportion
$\bar{p}$ = sample proportion

Confidence interval for μ—large sample, 313

$$\bar{x} - z_{a/2} \cdot \frac{s}{\sqrt{n}} \quad \text{to} \quad \bar{x} + z_{a/2} \cdot \frac{s}{\sqrt{n}}$$

Maximum error of the estimate for μ, 318

$$E = z_{a/2} \cdot \frac{\sigma}{\sqrt{n}}$$

Sample size for estimating μ, 319

$$n = \left[\frac{z_{a/2} \cdot \sigma}{E} \right]^2$$

(E = maximum error of the estimate)

Confidence interval for μ—normal population, 325

$$\bar{x} - t_{a/2} \cdot \frac{s}{\sqrt{n}} \quad \text{to} \quad \bar{x} + t_{a/2} \cdot \frac{s}{\sqrt{n}}$$

($df = n - 1$)

Sample proportion, 329

$$\bar{p} = \frac{x}{n}$$

(x = number of "successes")

Confidence interval for p—large sample, 331

$$\bar{p} - z_{a/2} \cdot \sqrt{\bar{p}(1 - \bar{p})/n} \quad \text{to} \quad \bar{p} + z_{a/2} \cdot \sqrt{\bar{p}(1 - \bar{p})/n}$$

YOU SHOULD BE ABLE TO . . .

1 use and understand the preceding formulas.

2 use Table II to find $z_{a/2}$ for any given value of α.

3 find a large-sample confidence interval for μ.

4 compute the maximum error of the estimate for μ.

5 understand the relationship between the sample size, standard deviation, confidence level, and maximum error of the estimate for a confidence interval for a population mean.

6 determine the sample size required for a specified confidence level and maximum error of the esti-

mate for μ.

7 use Table III to find $t_{a/2}$ for df = $n - 1$.

8 find a confidence interval for μ when sampling from a normal population.

9 decide on the appropriate confidence interval procedure for μ for any given problem.

10 find a large-sample confidence interval for p.

11 interpret a confidence interval for a mean or proportion.

REVIEW TEST

1 The U.S. Bureau of the Census surveys households in order to estimate the mean household income of various groups. Suppose $n = 774$ households of Spanish origin are to be randomly selected. The mean income, $\bar{x}$, of the households obtained is to be used as an estimate of the mean household income, μ, of all households of Spanish origin.

   a) What is the probability that $\bar{x}$ will be within $1000 of the true mean household income, μ, of all households of Spanish origin? [Assume $\sigma = \$15,900$.]

   b) Suppose that the mean household income of the 774 households obtained turns out to be $\bar{x} = \$17,642$ (to the nearest dollar). Fill in the blanks. A _____ % confidence interval for μ is the interval from $ _____ to $ _____ .

   c) Interpret your results in part (b) in words.

   d) Does the actual mean household income, μ, lie in your confidence interval from part (b)? Explain.

   e) Suppose that the mean household income, $\bar{x}$, is computed for each possible sample of 774 households of Spanish origin. What percentage of the intervals $\bar{x} - 1000$ to $\bar{x} + 1000$ will contain the true mean household income, μ?

2 Use Table II to find the following z-values and draw graphs to illustrate your work.

   a) $z_{0.063}$ b) $z_{0.40}$ c) $z_{0.007}$

3 Dr. Thomas Stanley, a marketing professor at Georgia State University, has surveyed millionaires since 1973. Among other things, Dr. Stanley ob-

tains estimates for the mean age, μ, of all U.S. millionaires. Suppose 36 millionaires are randomly selected and that their ages are as follows:

31	45	79	64	48	38	39	68	52
59	68	79	42	79	53	74	66	66
71	61	52	47	39	54	67	55	71
77	64	60	75	42	69	48	57	48

Determine a 95% confidence interval for the mean age, μ, of all U.S. millionaires. [*Note:* The sample mean and sample standard deviation of the data are $\bar{x} = 58.5$ years and $s = 13.4$ years.]

4 From Problem 3, we know that "a 95% confidence interval for the mean age of all U.S. millionaires is from 54.1 to 62.9 years." Decide which of the following provide a correct interpretation of the statement in quotes, and give a reason for each of your answers.
a) 95% of all U.S. millionaires are between the ages of 54.1 and 62.9 years old.
b) There is a 95% chance that the mean age of all U.S. millionaires is between 54.1 and 62.9 years old.
c) We can be 95% confident that the mean age of all U.S. millionaires is between 54.1 and 62.9 years old.
d) The probability is 0.95 that the mean age of all U.S. millionaires is between 54.1 and 62.9 years old.

5 A hand-held computer runs on four 1.5V, size AAA batteries. To estimate the average battery life, 50 computers are tested. The mean battery life for the 50 computers is $\bar{x} = 60.1$ hours. Assume $\sigma = 4.3$ hours.
a) Find a 99% confidence interval for the true mean battery life, μ.
b) Interpret your results in words.

6 Refer to Problem 5.
a) Find the maximum error of the estimate, E, for your confidence interval in Problem 5.
b) Explain the meaning of E, in this context, as far as the accuracy of the estimate is concerned.
c) Determine the sample size required to have a maximum error of the estimate of 0.5 hours and a 99% confidence level.
d) Find a 99% confidence interval for μ, if a sample of the size indicated in part (c) gives $\bar{x} = 59.8$ hours.

7 The figure below shows the standard normal curve and two t-curves. Which of the two t-curves has the larger degrees of freedom? Explain.

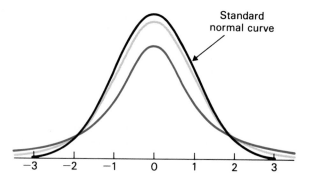

8 Suppose a random sample of size n is to be taken from a *normally distributed* population with mean μ and standard deviation σ.
a) What is the probability distribution of the random variable $\bar{x}$?
b) Determine

$$P\left(-2 \le \frac{\bar{x} - \mu}{\sigma/\sqrt{n}} \le 2\right)$$

c) How do we find probabilities for the random variable

$$\frac{\bar{x} - \mu}{s/\sqrt{n}}?$$

9 For a t-curve with $\mathrm{df} = 18$, find the following t-values and illustrate your results with pictures.
a) The t-value with area 0.025 to its right.
b) $t_{0.05}$
c) The t-value with area 0.10 to its *left*.
d) The two t-values that divide the area under the curve into a middle 0.99 area and two outside 0.005 areas.

10 The A. C. Nielsen Co. estimates the mean TV viewing time per day in American households. Suppose a random sample of 20 American households yields the following daily viewing times, in hours:

7.5	2.5	6.8	5.0	7.9
5.3	8.9	8.8	10.3	8.8
9.5	9.5	6.1	9.4	8.4
8.2	6.5	9.0	6.4	8.4

a) Assuming that daily viewing times for households are approximately normally distributed, find a 95% confidence interval for the true mean daily viewing time, μ, of all American households.

b) Interpret your results in words.

c) In 1983 the average American household watched seven hours and two minutes of TV per day. Does your result in part (a) provide evidence of an increase in average daily viewing time? Explain.

11 On Sunday, October 7, 1984, presidential candidates Ronald Reagan and Walter Mondale debated on national television. A Newsweek poll, conducted by the Gallup Organization, was taken to determine which candidate the public thought did a better job in the debate. The Gallup Organization telephoned 379 registered voters who watched the debate and found that 205 thought that Mondale did a better job.

a) Find a 95% confidence interval for the proportion, p, of all registered voters watching the debate who thought that Mondale did a better job.

b) Interpret your results in words.

c) Does the poll convince you that a majority of the registered voters watching the debate thought Mondale did a better job? Explain.

In Chapter 8 we studied problems involving the *estimation* of certain parameters. We examined methods of finding confidence intervals for population means and population proportions. Such interval estimates are based on the appropriate statistics—$\bar{x}$ for means and $\bar{p}$ for proportions. Now we will learn how those statistics can be employed to make decisions regarding *hypothesized* values for the corresponding parameters. For example, according to the Motor Vehicle Manufacturers Association, the mean age of all trucks in use in 1983 was 8.1 years. We will see how the mean age, $\bar{x}$, of a sample of trucks in use this year can be used to decide whether the mean age, μ, of all trucks in use this year has changed from the 1983 mean. Statistical inferences of this kind are called *hypothesis tests*. In this chapter, we will study hypothesis tests for means and proportions.

Hypothesis tests for means and proportions

9.1

The nature of hypothesis testing

Quite frequently it is necessary to use inferential statistics to make *decisions* about the value of a parameter, such as a population mean or a population proportion. For example, we might need to decide whether the mean age, μ, of juveniles being held in public custody has decreased from the 1982 mean of 15.4 years; or whether the proportion, p, of adult Americans who approve of the way the President is handling his job exceeds 50%.

One of the most commonly used procedures for making such decisions is to perform a **hypothesis test.** A **hypothesis** is simply a statement that something is true. For instance, the statement "more than 50% of adult Americans approve of the way the President is handling his job" is a hypothesis.

Typically, there are two hypotheses in a hypothesis test. One hypothesis is called the **null hypothesis** and the other is called the **alternative hypothesis.** These can be defined as follows.

DEFINITION 9.1 Null and alternative hypotheses

Null hypothesis: A hypothesis to be tested. We use the symbol H_0 to stand for null hypothesis.

Alternative hypothesis: A hypothesis to be considered as an alternate to the null hypothesis. We use the symbol H_a to stand for alternative hypothesis.

Originally, the term "null" in null hypothesis stood for "no difference" or "the difference is null." However, nowadays null hypothesis has come to mean simply a hypothesis to be tested. The problem in a hypothesis test is to decide whether or not the null hypothesis should be rejected in favor of the alternative hypothesis.

In hypothesis tests concerning one parameter, the null hypothesis generally specifies a *single value* for the parameter in question. Consider the following two examples.

EXAMPLE   9.1   Illustrates Definition 9.1

A snack food company produces a 454 gram (one pound) bag of pretzels. Although the actual net weights deviate somewhat from 454 grams and vary from one bag to another, it is nevertheless important to the company that the *mean* net weight, μ, of the bags be kept at 454 grams. Consequently, the quality assurance (QA) department periodically performs a statistical analysis to test whether the packaging machine is working properly. That is, the QA department tests whether the true mean weight, μ, being packaged is 454 grams. Determine the null and alternative hypotheses for such a test.

SOLUTION
The null hypothesis for the test is that the packaging machine is working properly—that is, the mean weight, μ, being packaged equals 454 grams. We write this concisely as

$$H_0: \mu = 454 \text{ g}$$

Note that the null hypothesis specifies a single value for the parameter μ, namely, 454.

The alternative hypothesis for the test is that the packaging machine is not working properly—that is, the mean weight, μ, being packaged is not 454 grams. We write this as

$$H_a: \mu \neq 454 \text{ g}$$

∎

EXAMPLE   9.2   Illustrates Definition 9.1

The manufacturer of a new model car, called the Orion, claims that a "typical" car gets 26 mpg (miles per gallon). An independent consumer group, skeptical of this claim, thinks that the mean gas mileage of all Orions may very well be *less than* 26 mpg. In other words, the consumer group wants to test the hypothesis that the mean gas mileage, μ, equals 26 mpg against the hypothesis that the mean gas mileage, μ, is less than 26 mpg. What are the null and alternative hypotheses for this hypothesis test?

SOLUTION
The null hypothesis is the manufacturer's claim that the mean gas mileage of all Orions is 26 mpg:

$$H_0: \mu = 26 \text{ mpg}$$

Note again that the null hypothesis specifies a single value for the parameter μ; in this case, 26.

The alternative hypothesis is the consumer group's conjecture that the mean gas mileage of all Orions is less than 26 mpg:

$$H_a: \mu < 26 \text{ mpg}$$ ∎

The decision rule

Once the null and alternative hypotheses are formulated, the next question is: How do we decide whether or not the null hypothesis should be rejected? We need to have a **decision rule.**

DEFINITION 9.2 Decision rule

A **decision rule** is a criterion that specifies whether or not the null hypothesis should be rejected in favor of the alternative hypothesis.

To understand the reasoning behind the selection of a decision rule, we return to the gas-mileage illustration of Example 9.2. The null and alternative hypotheses are

$$H_0: \mu = 26 \text{ mpg (manufacturer's claim)}$$
$$H_a: \mu < 26 \text{ mpg (consumer group's conjecture)}$$

To try to justify its contention, the consumer group plans to perform mileage tests on a random sample of $n = 30$ Orions. The sample mean gas mileage, $\bar{x}$, of the 30 cars tested will then be used as a basis for deciding whether or not the null hypothesis should be rejected in favor of the alternative hypothesis.

The question now is, just how will the sample mean gas mileage, $\bar{x}$, be used to make the decision? The idea is simply this: We know that the mean gas mileage, $\bar{x}$, of the 30 Orions tested should be roughly equal to the mean gas mileage, μ, of *all* Orions. Consequently, if the null hypothesis is true ($\mu = 26$ mpg), then we would expect the mean gas mileage, $\bar{x}$, of the 30 Orions tested to be about 26 mpg. To put it another way, if $\bar{x}$ is "too much smaller" than 26 mpg, then we would tend to doubt that the null hypothesis ($\mu = 26$ mpg) is true. We would be more willing to believe that the alternative hypothesis ($\mu < 26$ mpg) is true. But, how much smaller than 26 mpg is "too much smaller"?

For instance, suppose the mean gas mileage of the 30 Orions tested turns out to be $\bar{x} = 15$ mpg. We would undoubtedly consider this to be "too much smaller" than 26 mpg and reject the null hypothesis that $\mu = 26$ mpg in favor of the alternative hypothesis that $\mu < 26$ mpg. On the other hand, suppose the mean gas mileage of the 30 Orions tested turns out to be $\bar{x} = 25.9$ mpg. Then we would probably *not* consider this to be "too much smaller" than 26 mpg. We would probably be willing to accept as reasonable the null hypothesis that $\mu = 26$ mpg, and attribute to sampling error the discrepancy between the sample mean of $\bar{x} = 25.9$ mpg and the null hypothesis value of $\mu = 26$ mpg. [Remember that although $\bar{x}$ is generally "close" to μ, it will rarely equal μ.]

To summarize, if the sample mean gas mileage, $\bar{x}$, of the 30 Orions tested turns out to be much smaller than 26 mpg, then we would reject the null hypothesis in favor of the alternative hypothesis. However, if the sample mean gas mileage turns out to be only slightly smaller (or larger) than 26 mpg, then we would most likely not reject the null hypothesis. The problem in making a decision arises when the sample mean gas mileage is neither much smaller nor slightly smaller than 26 mpg.

So, our question remains: How much smaller than 26 mpg is "too much smaller?" Is a sample mean of $\bar{x} = 25.8$ mpg too much smaller? How about a sample mean of $\bar{x} = 25$ mpg? We need to establish a cutoff point. In other words, we must decide which values of $\bar{x}$ we will consider "too much smaller" than 26 mpg and which values we will not consider "too much smaller" than 26 mpg. Mathematically speaking, we need to select a number c, so that if the mean gas mileage, $\bar{x}$, of the 30 Orions tested is *less than c*, then we will reject the null hypothesis in favor of the alternative hypothesis, whereas, if the mean gas mileage, $\bar{x}$, of the 30 Orions tested is *at least c*, then we will not reject the null hypothesis. See Figure 9.1.

FIGURE 9.1

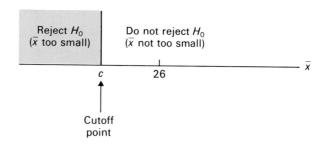

Cutoff
point

Before we discuss how to determine the cutoff point, c, let us consider some possible outcomes of the consumer group's testing. For simplicity, we will assume that the population standard deviation of the gas mileages for all Orions is known to be $\sigma = 1.5$ mpg.

Now suppose, for example, that the sample mean gas mileage, $\bar{x}$, of the 30 Orions tested turns out to be 25.8 mpg. Let us compute the probability of obtaining a value for $\bar{x}$ of 25.8 mpg or less, *if the null hypothesis ($\mu = 26$ mpg) is true.*

To do this, recall that the random variable $\bar{x}$ has mean $\mu_{\bar{x}} = \mu = 26$ and standard deviation, $\sigma_{\bar{x}} = \sigma/\sqrt{n} = 1.5/\sqrt{30} = 0.27$. Moreover, since $n \geq 30$, the central limit theorem implies that $\bar{x}$ is approximately normally distributed. Thus, $P(\bar{x} \leq 25.8)$ is equal to the area under the normal curve with parameters 26 and 0.27 that lies to the left of 25.8. We compute that area in the usual way in Figure 9.2.

Consequently, if the true mean gas mileage is $\mu = 26$ mpg, then the probability is about 0.23 that the sample mean gas mileage of the 30 Orions tested will be 25.8 mpg or less. In other words, if $\mu = 26$ mpg, then the chances of obtaining a value for $\bar{x}$ of 25.8 mpg or less are roughly one in four. This is not very strong evidence against the null hypothesis.

FIGURE 9.2
Determination of
the area under the
normal curve with
parameters $\mu_{\bar{x}} = 26$
and $\sigma_{\bar{x}} = 0.27$ that
lies to the left of
25.8.

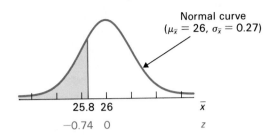

z-score computation:

Area between 0 and z:

$$\bar{x} = 25.8 \longrightarrow z = \frac{25.8 - 26}{0.27} = -0.74 \qquad 0.2704$$

Shaded area = 0.5000 − 0.2704 = 0.2296

On the other hand, suppose that the sample mean gas mileage, $\bar{x}$, of the 30 Orions tested turns out to be 25 mpg. As before, we can compute the probability of obtaining a value for $\bar{x}$ of 25 mpg or less, if the null hypothesis ($\mu = 26$ mpg) is true. See Figure 9.3.

FIGURE 9.3
Determination of
the area under the
normal curve with
parameters $\mu_{\bar{x}} = 26$
and $\sigma_{\bar{x}} = 0.27$ that
lies to the left of 25.

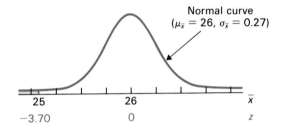

z-score computation:

Area between 0 and z:

$$\bar{x} = 25 \longrightarrow z = \frac{25 - 26}{0.27} = -3.70 \qquad 0.4999$$

Shaded area = 0.5000 − 0.4999 = 0.0001

Consequently, if the true mean gas mileage is $\mu = 26$ mpg, then the probability is about 0.0001 that the sample mean gas mileage of the 30 Orions tested will be 25 mpg or less. That is, if $\mu = 26$ mpg, then the chances of obtaining a value for $\bar{x}$ of 25 mpg or less are about one in 10,000.

We can therefore make the following statement: If the sample mean gas mileage, $\bar{x}$, of the 30 Orions tested turns out to be 25 mpg, then either

(1) the null hypothesis is true, and an extremely unlikely event occurred, *or*
(2) the null hypothesis is false.

In such a case, we would surely select (2) as the more reasonable choice. Thus, if the mean gas mileage, $\bar{x}$, of the 30 Orions tested turns out to be 25 mpg, then this constitutes strong evidence that the null hypothesis $\mu = 26$ mpg is false.

The previous two computations indicate how we will select the cutoff point c (pictured in Figure 9.1) for the gas-mileage test. Basically, we must decide on how unlikely a value of $\bar{x}$ we will tolerate before we feel compelled to reject the null hypothesis. Statisticians commonly use 1%, 5%, or 10% as limits for how unlikely a value of $\bar{x}$ they will tolerate before rejecting the null hypothesis. The percentage limit selected, or its decimal equivalent, is called the **significance level** of the hypothesis test and is denoted by the Greek letter α (alpha).

Suppose, for example, we decide on a significance level of 5%, that is, $\alpha = 0.05$. Then we choose the cutoff point, c, so that if the null hypothesis is true, only 5% of the possible $\bar{x}$-values are less than c. See Figure 9.4.

FIGURE 9.4

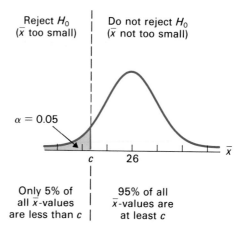

To actually determine the value of c, we first recall that if the null hypothesis ($\mu = 26$ mpg) is true, then $\bar{x}$ is approximately normally distributed, with mean $\mu_{\bar{x}} = \mu = 26$ and standard deviation $\sigma_{\bar{x}} = \sigma/\sqrt{n} = 1.5/\sqrt{30} = 0.27$. Hence the curve in Figure 9.4 is the normal curve with parameters 26 and 0.27. Next we note from Figure 9.4 that the area under the normal curve to the left of c is 0.05 (5%). This means that the z-score corresponding to c is the z-value with area 0.05 to its left. Using Table II, we find that z-value to be $z = -1.645$. We can picture our work as in Figure 9.5.

FIGURE 9.5

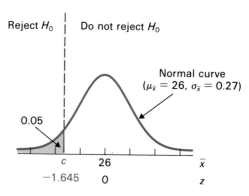

Finally, since the $\bar{x}$-values and z-values are related by the equation

$$z = \frac{\bar{x} - 26}{0.27}$$

we see that

$$-1.645 = \frac{c - 26}{0.27}$$

or

$$c = 26 - 1.645 \cdot 0.27 = 25.6$$

(to one decimal place).

In summary, if we select a significance level of 5% ($\alpha = 0.05$), then the cutoff point is $c = 25.6$ mpg. This gives us the following *decision rule* for deciding whether or not the null hypothesis should be rejected:

Decision rule for gas-mileage test using a 5% significance level

$\begin{cases} \text{If } \bar{x} < 25.6 \text{ mpg, reject the null hypothesis} \\ \text{in favor of the alternative hypothesis. If } \bar{x} \geq \\ 25.6 \text{ mpg, do not reject the null hypothesis.} \end{cases}$

See Figure 9.6.

FIGURE 9.6
Graphical display of decision rule for gas-mileage illustration. Significance level: $\alpha = 0.05$. Sample size: $n = 30$.

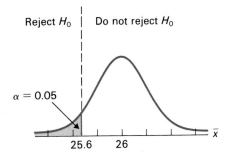

The hypothesis test

Let us illustrate the use of the above decision rule, portrayed graphically in Figure 9.6, as it applies to a specific sample.

E X A M P L E   9.3   Illustrates the use of the decision rule

Recall that the manufacturer of the Orion claims that a "typical" car gets 26 mpg, but that an independent consumer group thinks the mean gas mileage may be less than 26 mpg. The consumer group performs mileage tests on 30 randomly selected Orions. The results are displayed in Table 9.1.

TABLE 9.1
Gas mileages (mpg) of the 30 Orions tested by the consumer group.

25.3	25.1	29.6	24.6	26.0	26.0
26.3	23.6	26.0	25.4	26.1	23.8
25.1	24.1	25.8	26.4	23.4	24.8
22.6	26.6	25.1	26.6	28.0	23.3
23.8	25.4	26.2	25.1	25.3	21.5

Employ the data in Table 9.1 to perform the hypothesis test

$$H_0: \mu = 26 \text{ mpg (manufacturer's claim)}$$
$$H_a: \mu < 26 \text{ mpg (consumer group's conjecture)}$$

where μ is the true mean gas mileage of *all* Orions. Use a significance level of 5% ($\alpha = 0.05$).

SOLUTION

We have already done most of the work required to perform the hypothesis test by determining the decision rule for the hypothesis test at the 5% significance level. (See page 347 and Figure 9.6.) All that remains is to compute the mean gas mileage, $\bar{x}$, of the 30 cars tested and apply the decision rule. From Table 9.1 we find that

$$\bar{x} = \frac{\Sigma x}{n} = \frac{756.9}{30} = 25.23 \text{ mpg}$$

In Figure 9.7 we have reproduced the graph from Figure 9.6 and have located the value $\bar{x} = 25.23$ with a dot.

FIGURE 9.7

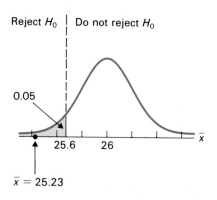

The decision rule is: "If $\bar{x} < 25.6$ mpg, reject the null hypothesis in favor of the alternative hypothesis. If $\bar{x} \geq 25.6$ mpg, do not reject the null hypothesis." Since the mean gas mileage of the 30 Orions tested is $\bar{x} = 25.23$ mpg, which is less than 25.6 mpg, we reject the null hypothesis in favor of the alternative hypothesis. In other words, the results of the mileage tests on the 30 Orions indicate that the mean gas mileage, μ, of *all* Orions is less than the manufacturer's claim of 26 mpg. This does not prove that the manufacturer's claim is false, but it provides strong evidence to that effect. ∎

Exercises 9.1

9.1 Explain the meaning of the term *hypothesis* as used in inferential statistics.

9.2 What role does the decision rule play in a hypothesis test?

9.3 A soft-drink bottler sells "one-liter" bottles of soda. The Food and Drug Administration (FDA) is concerned that the bottler may be short changing the customers. Specifically, the bottler claims that the mean content, μ, of the bottles of soda is 1000 ml and the FDA is concerned that μ is less than 1000 ml. The FDA plans to take a random sample of $n = 30$ bottles of soda in order to perform the hypothesis test

$$H_0: \mu = 1000 \text{ ml (bottler's claim)}$$
$$H_a: \mu < 1000 \text{ ml (FDA's concern)}$$

a) Suppose the sample mean content, $\bar{x}$, of the 30 bottles obtained turns out to be 996.8 ml. Assuming the null hypothesis, $\mu = 1000$ ml, is true, determine the probability of getting a value for $\bar{x}$ of 996.8 ml or less. [Use $\sigma = 33.3$ ml.]

b) Does a sample mean of $\bar{x} = 996.8$ ml constitute evidence against the null hypothesis $\mu = 1000$ ml? Explain.

c) Repeat parts (a) and (b) if the sample mean content turns out to be $\bar{x} = 981.2$ ml.

9.4   The U.S. National Center for Health Statistics reports that, in 1982, the average hospital stay was 7.1 days. A researcher thinks that this year's average will be less. The researcher wants to test the hypotheses

$$H_0: \mu = 7.1 \text{ days}$$
$$H_a: \mu < 7.1 \text{ days}$$

where μ is the true mean hospital stay for this year.

a) Suppose that the mean stay, $\bar{x}$, for 40 randomly selected patients turns out to be 6.6 days. Assuming that the null hypothesis, $\mu = 7.1$ days, is true, determine the probability of obtaining a value for $\bar{x}$ of 6.6 days or less. [Use $\sigma = 7.7$ days.]

b) Does a sample mean of $\bar{x} = 6.6$ days constitute evidence against the null hypothesis $\mu = 7.1$ days? Explain.

c) Repeat parts (a) and (b) if the mean stay, $\bar{x}$, of the 40 randomly selected patients turns out to be 2.84 days.

9.5   According to the Motor Vehicle Manufacturers Association, the average age of the trucks in use in 1983 was 8.1 years. The trend over the past 10 years has been an increase in the average age. Say that you want to decide whether this year's average age for trucks in use exceeds the 1983 mean of 8.1 years. That is, you wish to test the hypotheses

$$H_0: \mu = 8.1 \text{ years}$$
$$H_a: \mu > 8.1 \text{ years}$$

where μ is the actual mean age of trucks in use this year.

a) Suppose you randomly select 37 trucks presently in use and find their mean age to be

$\bar{x} = 10.6$ years. Assuming the null hypothesis, $\mu = 8.1$ years, is true, find the probability of getting a value for $\bar{x}$ of 10.6 years or greater. [Use $\sigma = 5.2$ years.]

b) Does a sample mean of $\bar{x} = 10.6$ years provide evidence that the null hypothesis $\mu = 8.1$ years is false? Explain.

c) Repeat parts (a) and (b) if the mean age, $\bar{x}$, of the 37 trucks you select turns out to be 8.4 years.

9.6   A Louisiana cotton farmer has used a certain brand of fertilizer for the past five years. His experience with the fertilizer indicates that the mean yield is about 623 lb/acre. Recently, a new brand of fertilizer came out on the market that will supposedly increase cotton yield. The farmer intends to try the new fertilizer on 80 one-acre plots in order to decide for himself whether it increases the mean yield of cotton on his land. If we let μ denote the mean yield of cotton that the farmer will get using the new fertilizer, then the hypotheses to be tested are

$$H_0: \mu = 623 \text{ lb/acre (no improvement)}$$
$$H_a: \mu > 623 \text{ lb/acre (improvement)}$$

a) Suppose the mean yield of cotton, $\bar{x}$, for the 80 one-acre plots turns out to be 631.6 lb/acre. If the null hypothesis, $\mu = 623$ lb/acre, is true, find the probability of getting a value for $\bar{x}$ of 631.6 lb/acre or greater. [Assume $\sigma = 20.8$ lb/acre.]

b) Does a sample mean of $\bar{x} = 631.6$ lb/acre provide evidence that the null hypothesis $\mu = 623$ lb/acre is false? Explain.

c) Repeat parts (a) and (b) if the mean yield of cotton, $\bar{x}$, for the 80 one-acre plots turns out to be 624.4 lb/acre.

9.7   The U.S. Department of Agriculture reports that the mean annual consumption of beef per person in 1983 was 106.5 lb. To test whether last year's mean differs from the 1983 figure, 50 people are to be randomly selected. Their mean beef consumption, $\bar{x}$, for last year will be used to test the hypotheses

$$H_0: \mu = 106.5 \text{ lb}$$
$$H_a: \mu \neq 106.5 \text{ lb}$$

where μ is last year's mean beef consumption per person.

a) Suppose that the mean beef consumption per person for the 50 randomly selected people turns out to be $\bar{x} = 107.3$ lb. If the null hypothesis, $\mu = 106.5$ lb, is true, find the probability of getting a value for $\bar{x}$ of 107.3 lb or greater. [Assume $\sigma = 16.7$ lb.]

b) Does a sample mean of $\bar{x} = 107.3$ lb suggest that the null hypothesis $\mu = 106.5$ lb is false? Explain.

c) Does a sample mean of $\bar{x} = 101.0$ lb suggest that the null hypothesis $\mu = 106.5$ lb is false? *Hint:* Compute $P(\bar{x} \leq 101.0)$, given that the null hypothesis is true.

9.8 The general partner of a limited partnership firm has told a potential investor that the mean monthly rent for three-bedroom homes in the area is $525. The investor wants to check out this claim on her own. She obtains the monthly rental charges for a random sample of 35 three-bedroom homes in order to test the hypotheses

$$H_0: \mu = \$525$$
$$H_a: \mu \neq \$525$$

where μ is the actual mean monthly rent for three-bedroom homes in the area.

a) Suppose that the mean monthly rent for the 35 randomly selected rentals turns out to be $581.65. Assuming that the null hypothesis, $\mu = \$525$, is true, determine the probability of obtaining a value for $\bar{x}$ of $581.65 or more. [Use $\sigma = \$118.73$.]

b) Does a sample mean of $\bar{x} = \$581.65$ suggest that the null hypothesis is false? Explain.

c) Does a sample mean of $\bar{x} = \$512.23$ suggest that the null hypothesis is false?

9.9 Refer to Exercise 9.3.

a) Find the cutoff point, c, for a test of the hypotheses

$$H_0: \mu = 1000 \text{ ml (bottler's claim)}$$
$$H_a: \mu < 1000 \text{ ml (FDA's concern)}$$

at the 5% significance level. [Recall that $n = 30$ and $\sigma = 33.3$ ml.] Construct graphs similar to Figures 9.4 and 9.5 on page 346.

b) State the decision rule for the hypothesis test with $\alpha = 0.05$. Draw a graph like Figure 9.6 on page 347 that gives a visual portrayal of the decision rule.

c) The contents of the 30 bottles of soda selected by the FDA are as follows:

1025	977	1018	975	977
990	959	957	1031	964
986	914	1010	988	1028
989	1001	984	974	1017
1060	1030	991	999	997
996	1014	946	995	987

Use the data, along with the decision rule from part (b), to perform the hypothesis test of part (a) at the 5% significance level. [The sum of the data is $\Sigma x = 29{,}779$ ml.]

d) Interpret your results in part (c) in words.

9.10 Refer to Exercise 9.4.

a) Find the cutoff point, c, for a test of the hypotheses

$$H_0: \mu = 7.1 \text{ days}$$
$$H_a: \mu < 7.1 \text{ days}$$

at the 10% significance level. [Recall that $n = 40$ and $\sigma = 7.7$ days.] Construct graphs similar to Figures 9.4 and 9.5 on page 346.

b) State the decision rule for the hypothesis test with $\alpha = 0.10$. Draw a graph like Figure 9.6 on page 347 that provides a visual display of the decision rule.

c) The stays, in days, for the 40 randomly selected patients are given below.

13	1	2	3	7	21	3	11
7	2	1	4	1	1	7	3
17	5	3	4	9	1	2	9
1	6	12	14	37	5	2	6
10	1	13	12	1	8	8	1

Use the data, along with the decision rule obtained in part (b), to perform the hypothesis test of part (a).

d) State your conclusion from part (c) in words.

9.11 Refer to Exercise 9.5.

a) Determine the cutoff point, c, for a test of the hypotheses

$$H_0: \mu = 8.1 \text{ years}$$
$$H_a: \mu > 8.1 \text{ years}$$

at the 10% significance level. Note that the cutoff point will be to the *right* of 8.1, since the alternative hypothesis is $\mu > 8.1$ years. Also, recall that $n = 37$ and $\sigma = 5.2$ years. Draw graphs similar to Figures 9.4 and 9.5 on page 346, but remember that c will be to the right of 8.1.

b) State the decision rule for the hypothesis test with $\alpha = 0.10$. Draw a graph similar to Figure 9.6 on page 347.

c) Suppose that the ages of the 37 trucks you select are

8	12	14	16	15	5	11	13
4	12	12	15	12	3	10	9
11	3	18	4	9	11	17	
7	4	12	12	8	9	10	
9	9	1	7	6	9	7	

Apply the decision rule in part (b) to these data in order to perform the hypothesis test indicated in part (a). [*Note:* $\Sigma x = 354$.]

d) State your conclusion from part (c) in words.

9.12 Refer to Exercise 9.6.
a) Determine the cutoff point, c, for the hypothesis test

$$H_0: \mu = 623 \text{ lb/acre (no improvement)}$$
$$H_a: \mu > 623 \text{ lb/acre (improvement)}$$

at the 5% significance level. Note that the cutoff point will be to the right of 623, since the alternative hypothesis is $\mu > 623$ lb/acre. Also, recall that $n = 80$ and $\sigma = 20.8$ lb/acre. Draw graphs similar to Figures 9.4 and 9.5 on page 346, but remember that c will be to the right of 623.

b) State the decision rule for the hypothesis test with $\alpha = 0.05$, and draw a graph like Figure 9.6 on page 347.

c) The cotton yields resulting from the use of the new fertilizer on the 80 one-acre plots are as follows:

639	653	631	590	628	638	618	602
622	667	614	644	611	652	598	627
637	613	591	627	637	604	620	618
621	642	654	643	634	630	621	560
637	634	645	630	605	617	598	641
636	599	616	578	620	639	608	615
587	601	629	627	626	613	568	638
599	627	630	615	620	658	629	629
626	670	642	647	637	593	617	654
627	645	646	643	648	654	651	613

Apply the decision rule in part (b) to the data in order to perform the hypothesis test in part (a). [The sum of the data is 50,013 lb.]

d) If the new fertilizer costs more than the one the farmer presently uses, would you buy the new fertilizer if you were the farmer? Why?

9.13 The hypothesis test considered in Exercise 9.7 is

$$H_0: \mu = 106.5 \text{ lb}$$
$$H_a: \mu \neq 106.5 \text{ lb}$$

where μ is last year's mean beef consumption per person. To carry out the hypothesis test, 50 people are to be randomly selected and their last year's mean beef consumption, $\bar{x}$, determined. Since the alternative hypothesis is $\mu \neq 106.5$ lb, the null hypothesis will be rejected if $\bar{x}$ is either "too much smaller" or "too much larger" than 106.5 lb. In other words, there will be *two* cutoff points—one to the left of 106.5, which we call ℓ, and one to the right of 106.5, which we call r.

a) Find the cutoff points, ℓ and r, for a test at the 5% significance level. Recall that $\sigma = 16.7$ lb. [*Hint:* Choose ℓ and r so that, if the null hypothesis is true, the area to the left of ℓ is 0.025 and the area to the right of r is 0.025.] Draw graphs similar to Figures 9.4 and 9.5 on page 346, but remember that there are *two* cutoff points here.

b) State the decision rule for the hypothesis test with $\alpha = 0.05$. Draw a picture similar to Figure 9.6 on page 347.

c) Suppose that last year's beef consumptions for the 50 randomly selected people are as shown below.

119	91	81	106	119	129	142	119	99	103
138	115	102	79	48	148	107	122	127	132
96	114	102	126	110	130	118	109	69	126
113	102	99	83	101	100	132	96	106	128
116	134	120	109	135	93	118	85	116	119

Apply the decision rule obtained in part (b) to the data in order to perform the hypothesis test. [The sum of the data is 5531.]

d) State your decision regarding the hypothesis test in words.

9.14 In Exercise 9.8, a potential investor in a limited partnership wants to test the hypotheses

$$H_0: \mu = \$525$$
$$H_a: \mu \neq \$525$$

where μ is the mean monthly rent for three-bedroom homes in her area.

a) Find the cutoff points, ℓ and r, for a test at the 10% significance level. Recall that $n = 35$ and $\sigma = \$118.73$. [If necessary, refer to the discussion regarding ℓ and r in Exercise 9.13.] Draw graphs similar to Figures 9.4 and 9.5 on page 346.

b) State the decision rule for the hypothesis test with $\alpha = 0.10$, and draw a graph similar to Figure 9.6 on page 347.

c) The monthly rents for the 35 homes randomly selected by the investor are

$532	547	514	740	622	652	621
511	676	497	493	682	550	502
721	621	567	341	757	677	530
381	611	736	598	453	756	455
616	505	584	594	624	584	610

Use this sample data and the decision rule obtained in part (b) to perform the hypothesis test. [*Note:* $\Sigma x = \$20,460$.]

d) State your decision in part (c) in words.

9.2   A hypothesis testing procedure

Now that we have gone through the reasoning and computations for an entire hypothesis test, we can outline the basic steps in the procedure.

> **PROCEDURE 9.1 To perform a hypothesis test.**
>
> **STEP 1** *State the null and alternative hypotheses.*
> **STEP 2** *Decide on the significance level, α.*
> **STEP 3** *Determine the decision rule.*
> **STEP 4** *Apply the decision rule to the sample data and make the decision.*
> **STEP 5** *State the conclusion in words.*

EXAMPLE   9.4   Illustrates Procedure 9.1

According to the R. R. Bowker Company, the average retail price of hardcover history books in 1983 was $24.96. A random sample of 40 history books yielded the following data on this year's prices:

TABLE 9.2
This year's retail prices for 40 randomly selected history books. Prices are given to the nearest dollar.

$28	25	26	16	21
41	36	32	17	27
31	22	25	30	29
28	23	33	25	19
24	25	26	17	16
48	33	30	23	18
26	34	31	15	39
18	23	17	24	23

At the 10% significance level, do the data indicate that this year's mean price for history books has *increased* over the 1983 mean of $24.96? [Assume $\sigma = \$8.33$.]

SOLUTION
We use Procedure 9.1.

STEP 1 *State the null and alternative hypotheses.*

Let μ denote this year's mean retail price for history books. Then the null and alternative hypotheses are

$$H_0: \mu = \$24.96 \text{ (mean price has not increased)}$$
$$H_a: \mu > \$24.96 \text{ (mean price has increased)}$$

STEP 2 *Decide on the significance level, α.*

We are to perform the hypothesis test at the 10% significance level. Thus, $\alpha = 0.10$.

STEP 3 *Determine the decision rule.*

If the sample mean price, $\bar{x}$, of the 40 history books obtained is "close" to $24.96, then we will not reject the null hypothesis ($\mu = \$24.96$). On the other hand, if $\bar{x}$ is "too much larger" than $24.96, then we will reject the null hypothesis ($\mu = \$24.96$) in favor of the alternative hypothesis ($\mu > \$24.96$). Thus, the cutoff point, c, is to the right of $24.96. Moreover, since the significance level is $\alpha = 0.10$, we choose c so that, if the null hypothesis is true, only 10% of the possible $\bar{x}$-values are greater than c. See Figure 9.8.

FIGURE 9.8

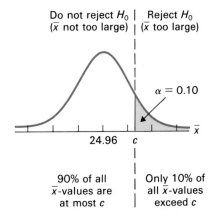

Do not reject H_0 | Reject H_0
($\bar{x}$ not too large) | ($\bar{x}$ too large)

$\alpha = 0.10$

24.96 c

90% of all $\bar{x}$-values are at most c | Only 10% of all $\bar{x}$-values exceed c

To determine the cutoff point, c, we proceed as we did in the gas-mileage illustration. First, we see from Figure 9.8 that the z-score corresponding to c is the z-value with area 0.10 to its right (that is, $z_{0.10}$). Using Table II we find that z-value to be $z = 1.28$. Next, we note that the normal curve in Figure 9.8 has parameters $\mu_{\bar{x}} = \mu = 24.96$ and $\sigma_{\bar{x}} = \sigma/\sqrt{n} = 8.33/\sqrt{40} = 1.32$. Consequently, the $\bar{x}$-values and z-values are related by the equation

$$z = \frac{\bar{x} - 24.96}{1.32}$$

Thus,

$$1.28 = \frac{c - 24.96}{1.32}$$

or

$$c = 24.96 + 1.28 \cdot 1.32 = 26.65$$

Therefore, the decision rule is:

> If $\bar{x} > \$26.65$, reject the null hypothesis in favor of the alternative hypothesis. If $\bar{x} \leq \$26.65$, do not reject the null hypothesis.

We depict the decision rule graphically in Figure 9.9 at the top of page 354.

STEP 4 *Apply the decision rule to the sample data and make the decision.*

To apply the decision rule, we must first compute the mean price, $\bar{x}$, of the 40 randomly selected history books. From Table 9.2 we get

$$\bar{x} = \frac{\Sigma x}{n} = \frac{\$1044}{40} = \$26.10$$

FIGURE 9.9

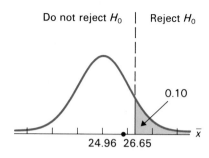

We have marked this value of $\bar{x}$ with a dot in Figure 9.9. Since $\bar{x} \leq \$26.65$, we do not reject the null hypothesis.

STEP 5 *State the conclusion in words.*

Based on the sample of 40 history books, we do *not* have sufficient evidence to conclude that the average retail price has increased from the 1983 average of $24.96. ∎

EXAMPLE   9.5   Illustrates Procedure 9.1

A company that produces snack foods uses a packaging machine to package 454-gram bags of pretzels. The QA department periodically samples packages of pretzels in order to check whether the machine is working properly—that is, to check whether the mean net weight of pretzels being packaged by the machine is 454 g. In Table 9.3, we have displayed the net weights of a random sample of 50 bags of pretzels obtained by the QA department.

TABLE 9.3
Net weights, in grams, of a random sample of $n = 50$ bags of pretzels.

463.8	450.2	449.9	456.1	451.8
432.8	445.6	446.1	449.6	447.4
442.5	438.4	451.8	447.4	459.7
450.4	453.4	455.6	445.9	433.3
447.6	450.1	438.6	452.3	459.2
453.8	456.2	454.4	451.7	449.2
462.8	448.7	447.2	466.1	445.8
447.3	450.1	449.3	457.2	463.8
467.5	446.8	433.2	464.1	469.0
456.8	453.6	450.8	452.7	443.0

Test, at the 5% significance level, whether the data provide evidence that the packaging machine is not working properly. [Assume $\sigma = 7.9$ g.]

SOLUTION
We employ the step-by-step method given in Procedure 9.1.

STEP 1 *State the null and alternative hypotheses.*

Let μ be the actual mean net weight of pretzels being packaged by the machine. Then the null and alternative hypotheses are

$$H_0: \mu = 454 \text{ g (machine is working properly)}$$
$$H_a: \mu \neq 454 \text{ g (machine is not working properly)}$$

STEP 2 *Decide on the significance level, α.*

The hypothesis test is to be performed at the 5% significance level, so $\alpha = 0.05$.

STEP 3 *Determine the decision rule.*

If the sample mean weight, $\bar{x}$, of the 50 bags of pretzels selected is "close" to 454 g, then we will not reject the null hypothesis ($\mu = 454$ g). However, if $\bar{x}$ is either "too much smaller" or "too much larger" than 454 g, then we will reject the null hypothesis ($\mu = 454$ g) in favor of the alternative hypothesis ($\mu \neq 454$ g). This is because values of $\bar{x}$ that are too far away from 454 *in either direction* indicate that the machine is not packaging a mean net weight of 454 g per bag.

Consequently, in this case, we will have *two* cutoff points, one to the left of 454 and one to the right of 454. Let us designate these two cutoff points by the letters ℓ (for left) and r (for right). Since the significance level is $\alpha = 0.05$, or 5%, we choose ℓ and r so that, if the null hypothesis is true, then 2.5% of the possible $\bar{x}$-values are less than ℓ and 2.5% are greater than r. [We split the 5% in two—half goes to the left of ℓ and half to the right of r.] See Figure 9.10.

FIGURE 9.10

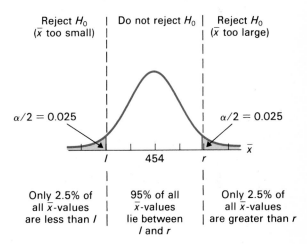

To determine the cutoff point ℓ, we first note from Figure 9.10 that the z-score corresponding to ℓ is the z-value with area 0.025 to its left. Table II shows that z-value to be $z = -1.96$. Next, we note that the normal curve in Figure 9.10 has parameters $\mu_{\bar{x}} = \mu = 454$ and $\sigma_{\bar{x}} = \sigma/\sqrt{n} = 7.9/\sqrt{50} = 1.12$. Consequently, the $\bar{x}$-values and z-values are related by the equation

$$z = \frac{\bar{x} - 454}{1.12}$$

Thus,

$$-1.96 = \frac{\ell - 454}{1.12}$$

or

$$\ell = 454 - 1.96 \cdot 1.12 = 451.80$$

Similarly, we find that

$$r = 454 + 1.96 \cdot 1.12 = 456.20$$

Therefore, the decision rule is:

> If $\bar{x} < 451.80$ g or $\bar{x} > 456.20$ g, reject the null hypothesis in favor of the alternative hypothesis. If 451.80 g $\leq \bar{x} \leq 456.20$ g, do not reject the null hypothesis.

Figure 9.11 provides a visual display of the decision rule.

FIGURE 9.11

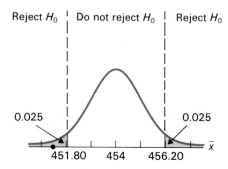

STEP 4 *Apply the decision rule to the sample data and make the decision.*

We must compute the sample mean of the data in Table 9.3 and then apply the decision rule. The sum of the data in Table 9.3 is 22,560.5 and consequently,

$$\bar{x} = \frac{\Sigma x}{n} = \frac{22,560.5}{50} = 451.21 \text{ g}$$

This value of $\bar{x}$ is marked with a dot in Figure 9.11. Since $\bar{x} < 451.80$ g, the decision rule requires us to reject H_0 and conclude that H_a is true.

STEP 5 *State the conclusion in words.*

On the basis of the sample data, we conclude that the packaging machine is *not* working properly—that is, the mean net weight, μ, being packaged is *not* 454 grams.   ■

Exercises 9.2

In Exercises 9.15 through 9.20, use Procedure 9.1 on page 352 to perform the hypothesis test.

9.15 According to the Philadelphia CPA firm Laventhol and Horwath, the average room rate for full-service lodging establishments was $46.67 in 1983. A random sample of 35 full-service lodging establishments yields the following data on this year's room rates:

$64	71	83	61	62	48	48
51	65	43	80	51	59	43
28	50	66	82	58	59	51
75	47	86	41	54	36	39
71	87	58	67	67	69	21

At the 10% significance level, do the data indicate that this year's mean room rate, μ, has *increased* over the 1983 mean of $46.67? To answer that question, perform the hypothesis test

H_0: $\mu = \$46.67$ (mean has not increased)
H_a: $\mu > \$46.67$ (mean has increased)

with $\alpha = 0.10$. Assume $\sigma = \$16.52$. [The sum of the data is $2,041.]

9.16 A consumer advocacy group thinks that Wheat Flakes brand cereal contains, on the average, *less* than the advertised weight of 15 ounces per box. A random sample of 40 boxes of Wheat Flakes gives the following weights:

15.8	15.1	15.2	15.4	14.8	15.6	15.7	14.5
14.8	15.4	15.3	15.5	15.2	14.6	15.4	15.4
15.5	14.7	14.7	15.1	14.7	15.3	15.3	15.5
14.0	14.2	14.6	15.0	15.1	14.9	14.9	15.8
15.0	14.4	15.4	14.3	15.4	15.9	15.2	15.6

Do the weights of these 40 randomly selected boxes of cereal indicate that the consumer group's contention is correct? To answer that question, perform the hypothesis test

H_0: $\mu = 15$ oz (advertised weight)
H_a: $\mu < 15$ oz (consumer group's contention)

at the 5% significance level. Assume $\sigma = 0.5$ oz. [*Note:* $\Sigma x = 604.2$.]

9.17 The Internal Revenue Service reports that, in 1982, the average income tax per taxable return was $3603. A random sample of 30 returns from last year gives the following data:

$ 431	6316	974	2790	57	1704
280	615	305	639	1763	1186
2730	1355	2424	7532	1498	2729
1712	6950	3160	1627	1582	8392
2307	2648	665	710	2001	113

Do these tax amounts suggest that the average income tax per taxable return, μ, for last year has *decreased* from the 1982 mean of $3603? Test the hypotheses

H_0: $\mu = \$3603$ (mean has not decreased)
H_a: $\mu < \$3603$ (mean has decreased)

at the 5% significance level. Assume $\sigma = \$3400$. [The sum of the data is $67,195.]

9.18 The owner of a menswear store knows that, in the past, average weekly sales have been $1700, with a standard deviation of $250. A couple of years ago, he decided to advertise in local papers in an attempt to improve weekly sales. Of course, he does not wish to continue spending money on such ad-

vertising if sales do not increase. Let μ denote the true mean weekly sales *with* the advertising. By testing the hypotheses

H_0: $\mu = \$1700$ (no improvement)
H_a: $\mu > \$1700$ (improvement)

the owner can decide whether to continue advertising. The weekly sales for 32 randomly selected weeks with advertising are

$1531	1850	1778	1630	1992	1091	1982	1879
1910	1803	1999	1857	1995	1075	1572	1749
1685	2903	2151	1486	1383	2122	2133	2101
1654	1434	1547	2352	1359	1955	1204	2485

a) Perform the hypothesis test at the 5% significance level. Use $\sigma = \$250$. [The sum of the data is $\Sigma x = \$57,647$.]
b) If you were the store owner, would you continue advertising? Why?

9.19 According to the U.S. Bureau of Economic Analysis, in 1980 the average length of stay in Europe and the Mediterranean by U.S. travelers was 21 days. A random sample of 36 U.S. residents who have traveled to Europe and the Mediterranean this year yielded the following data, in days:

41	16	6	21	1	21
5	31	20	27	17	10
3	32	2	48	8	12
21	44	1	56	5	12
3	13	15	10	18	3
1	11	14	12	64	10

At the 10% significance level, do the data indicate that the mean stay by this year's travelers *differs* from that in 1980? To answer that question test the hypotheses

H_0: $\mu = 21$ days
H_a: $\mu \neq 21$ days

where μ is this year's mean stay in Europe and the Mediterranean by U.S. travelers. Use $\sigma = 16$ days. [The sum of the data is 634.]

9.20 The U.S. Energy Information Administration publishes data on energy production and consumption. In 1982, the average energy consumed per household in the U.S. was 114 million BTU. For that same year, data on annual energy consumed by a random sample of 50 households in the South are as follows:

130	55	45	64	155	66	60	80	102	62
58	101	75	111	151	139	81	55	66	90
97	77	51	67	125	50	136	55	83	91
54	86	100	78	93	113	111	104	96	113
96	87	129	109	69	94	99	97	83	97

Perform the hypothesis test

H_0: $\mu = 114$ million BTU
H_a: $\mu \neq 114$ million BTU

to decide, at the 5% significance level, whether the 1982 average energy consumed by southern households *differed* from that of all U.S. households. Assume $\sigma = 25$ million BTU. [*Note:* $\Sigma x = 4486$.]

9.3 Terms, errors, and hypotheses

To fully understand the nature of hypothesis testing, we need to learn some additional terms and concepts. In this section we will define several more terms, discuss the two types of errors that can be made in hypothesis tests, and explain how to choose the null and alternative hypotheses in hypothesis tests.

Some additional terminology

To introduce some further terminology that is used in hypothesis testing, we return to the gas-mileage illustration. Recall that the manufacturer of the Orion claims a mean gas mileage, μ, of 26 mpg, while a consumer group contends that the mean gas mileage is less. The null and alternative hypotheses are

H_0: $\mu = 26$ mpg (manufacturer's claim)
H_a: $\mu < 26$ mpg (consumer group's conjecture)

To decide whether the null hypothesis should be rejected in favor of the alternative hypothesis, the consumer group performs mileage tests on 30 randomly selected Orions. If the mean gas mileage, $\bar{x}$, of the 30 cars sampled is "close" to 26 mpg, then we do not reject the null hypothesis. On the other hand, if the mean gas mileage, $\bar{x}$, of the 30 cars sampled is "too much smaller" than 26 mpg, then we reject the null hypothesis in favor of the alternative hypothesis. The mean gas mileage, $\bar{x}$, of the 30 cars sampled is called the **test statistic,** since it is the statistic upon which we will base our decision about the hypothesis test.

In Section 9.1 we found that, for a test at the 5% significance level, the cutoff point is $c = 25.6$ mpg. That is, if $\bar{x} < 25.6$ mpg, we reject the null hypothesis in favor of the alternative hypothesis; whereas, if $\bar{x} \geq 25.6$ mpg, we do not reject the null hypothesis. Figure 9.6, which graphically displays this decision rule, is repeated as Figure 9.12.

FIGURE 9.12
Graphical display of decision rule for gas-mileage illustration. Significance level: $\alpha = 0.05$. Sample size: $n = 30$.

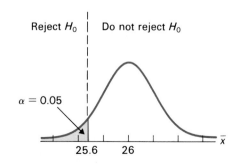

The cutoff point, *c,* which here is 25.6, is given the technical name **critical value.**

The set of values for the test statistic, $\bar{x}$, that lead us to reject the null hypothesis is called the **rejection region.**[†] In this case, the rejection region consists of all $\bar{x}$-values to the left of the critical value 25.6 or, equivalently, that part of the horizontal axis under the shaded area in Figure 9.12.

On the other hand, the set of values for the test statistic, $\bar{x}$, that lead us not to reject the null hypothesis is called the **nonrejection region.**[††] Here the nonrejection region consists of all $\bar{x}$-values to the right of, and including, the critical value 25.6. Figure 9.13 summarizes the above discussion pictorially.

FIGURE 9.13

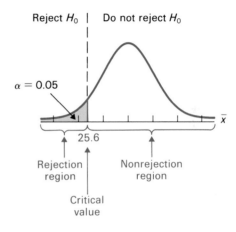

In general, we have the following definitions.

DEFINITION 9.3 Test statistic, rejection region, nonrejection region, critical value

Test statistic: The statistic upon which we base our decision to either reject or not reject the null hypothesis.
Rejection region: The set of values for the test statistic that lead us to reject the null hypothesis.
Nonrejection region: The set of values for the test statistic that lead us not to reject the null hypothesis.
Critical value(s): The value(s) of the test statistic that separates the rejection and nonrejection regions.

EXAMPLE 9.6 Illustrates Definition 9.3

In Example 9.4 we performed a hypothesis test to decide whether the mean retail price, μ, of history books has increased from the 1983 mean of $24.96. The

---

[†] The term **critical region** is also used.
[††] The terms **noncritical region** and **acceptance region** are also used.

null and alternative hypotheses are

$$H_0: \mu = \$24.96 \text{ (mean price has not increased)}$$
$$H_a: \mu > \$24.96 \text{ (mean price has increased)}$$

The decision rule for a test at the 10% significance level using a sample size of 40 is portrayed in Figure 9.9. We repeat that graph here as Figure 9.14.

FIGURE 9.14
Graphical display of decision rule for history-book price illustration. Significance level: $\alpha = 0.10$. Sample size: $n = 40$.

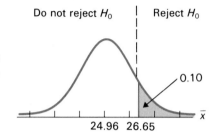

For this hypothesis test, determine the
a) test statistic.
b) rejection region.
c) nonrejection region.
d) critical value.
e) Construct a graph similar to Figure 9.13 to display the results for parts (a) through (d).

S O L U T I O N
a) The test statistic is the mean retail price, $\bar{x}$, of the sample of 40 history books.
b) The rejection region consists of all values for the test statistic that lead us to reject the null hypothesis. Referring to Figure 9.14, we see that the rejection region consists of all $\bar{x}$-values to the right of 26.65.
c) The nonrejection region consists of all values for the test statistic that lead us not to reject the null hypothesis. From Figure 9.14, we find that the nonrejection region consists of all $\bar{x}$-values to the left of, and including, 26.65.
d) The critical value is 26.65.
e) A graph displaying the results of parts (a) through (d) is shown in Figure 9.15 at the top of the next page.   ■

Type I and Type II errors

Whenever statistical inference methods are employed, it is always possible that the decision reached will be incorrect. This is because partial information (obtained from the sample) is used to make conclusions about the entire population. In hypothesis testing, two types of errors are possible.

Type I error: Rejecting a true null hypothesis.
Type II error: Not rejecting a false null hypothesis.

See Table 9.4.

FIGURE 9.15

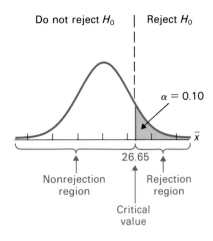

TABLE 9.4

		H_0 is:	
		True	*False*
Decision	*Do not reject H_0*	Correct decision	Type II error
	Reject H_0	Type I error	Correct decision

EXAMPLE 9.7 Illustrates Type I and Type II errors

Consider again the pretzel-packaging illustration of Example 9.5 on page 354. The null and alternative hypotheses are

$$H_0: \mu = 454 \text{ g (machine is working properly)}$$
$$H_a: \mu \neq 454 \text{ g (machine is not working properly)}$$

where μ is the actual mean net weight of pretzels per bag.

As we have seen, the decision rule for this hypothesis test at the 5% significance level using a sample size of $n = 50$ is the following: "If $\bar{x} < 451.80$ g or $\bar{x} > 456.20$ g, reject H_0 in favor of H_a. If $451.80 \text{ g} \leq \bar{x} \leq 456.20$ g, do not reject H_0." See Figure 9.16.

FIGURE 9.16
Graphical display of decision rule for pretzel-packaging illustration. Significance level: $\alpha = 0.05$. Sample size: $n = 50$.

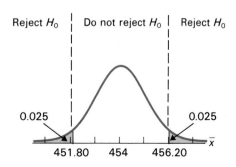

Explain what each of the following would mean.

    a) A Type I error.

    b) A Type II error.

    c) A correct decision.

S O L U T I O N

    a) A Type I error occurs when a true null hypothesis is rejected. In this case a Type I error would occur if the machine *is* working properly (that is, it is packaging an average of 454 g per bag) but the results of the sampling lead to the conclusion that the machine is *not* working properly (that is, it is not packaging an average of 454 g per bag). Specifically, a Type I error occurs if and only if the mean weight, μ, being packaged is in fact 454 g, but the mean weight, $\bar{x}$, of the 50 bags sampled turns out to be either less than 451.80 g or greater than 456.20 g (see Figure 9.16).

    b) A Type II error occurs when a false null hypothesis is not rejected. Here that means the machine is *not* working properly ($\mu \neq 454$ g) but the results of the sampling fail to lead to that conclusion. Thus, a Type II error occurs if and only if the mean weight, μ, being packaged is in fact not 454 g, but the mean weight, $\bar{x}$, of the 50 bags sampled turns out to be between 451.80 g and 456.20 g (see Figure 9.16).

    c) A correct decision can occur in either of two ways. First, a correct decision occurs when a true null hypothesis is not rejected (upper left-hand corner of Table 9.4). In the present situation this means that the machine is working properly *and* the results of the sampling do not lead to the rejection of that fact. That is, the mean weight, μ, being packaged is in fact 454 g, and the mean weight, $\bar{x}$, of the 50 bags sampled turns out to be between 451.80 g and 456.20 g.

    A correct decision also occurs when a false null hypothesis is rejected (lower right-hand corner of Table 9.4). Here this means that the machine is not working properly *and* the results of the sampling lead to rejection of the null hypothesis. In other words, the mean weight, μ, being packaged is in fact not 454 g, and the mean weight, $\bar{x}$, of the 50 bags sampled turns out to be either less than 451.80 g or greater than 456.20 g.   ∎

    As indicated, we use the following terminology to describe the two possible kinds of errors that can occur in hypothesis testing.

DEFINITION 9.4 Type I and Type II errors

Type I error: The mistake of rejecting the null hypothesis when it is in fact true.
Type II error: The mistake of not rejecting the null hypothesis when it is in fact false.

E X A M P L E   **9.8**   **Illustrates error types in hypothesis testing**

Recall that the results of sampling 50 bags of pretzels turned out to be as displayed in Table 9.3 on page 354. As we have seen, the mean of that data is $\bar{x} = 451.21$ g, which leads to rejection of the null hypothesis, since 451.21 g is less than the left-hand critical value of 451.80 g. Thus, the QA department rejects the null hypothesis ($\mu = 454$ g) that the machine is working properly and concludes that the machine is not working properly ($\mu \neq 454$ g). Classify the conclusion of the QA department by error type or as a correct decision if, in fact, the mean weight, μ, being packaged

    a) *is* 454 g.

    b) is *not* 454 g.

S O L U T I O N

    a) If the mean weight, μ, being packaged is 454 g, then the machine is working properly and so the null hypothesis is, in fact, true. Therefore, by concluding that the machine is not working properly, the QA department has made a Type I error—it has rejected a true null hypothesis.

    b) If the mean weight, μ, being packaged is not 454 g, then the machine is not working properly and so the null hypothesis is false. Consequently, by concluding that the machine is not working properly, the QA department has made a correct decision—it has rejected a false null hypothesis. ∎

Probabilities of Type I and Type II errors

The probability of making a Type I error is the probability of rejecting a true null hypothesis. In other words, it is the probability that the value of the test statistic falls in the rejection region when, in fact, the null hypothesis is true. However, a look at Figure 9.12 on page 358, Figure 9.14 on page 360, or Figure 9.16 on page 361 shows that, if the null hypothesis is true, then the probability that the test statistic, $\bar{x}$, takes a value in the rejection region is just the *significance level* of the test.

KEY FACT

The probability of a Type I error (that is, of rejecting a true null hypothesis) is equal to the significance level, α, of the hypothesis test. In symbols,

$$P(\text{Type I error}) = \text{Significance level} = \alpha$$

Thus, the probability of a Type I error for the gas-mileage illustration is $\alpha = 0.05$, since the test is at the 5% significance level. For the history-book price illustration, the probability of a Type I error is $\alpha = 0.10$, because that test is at the 10% significance level.

The probability of a Type I error indicates the likelihood that a true null hypothesis will be rejected. Consequently, in the gas-mileage illustration, there is only a 5% chance that the consumer group will decide that the mean gas mileage of all Orions is less than 26 mpg, when in fact the mean *is* 26 mpg. Similarly, in the history-book price illustration, there is only a 10% chance that we will conclude that the mean price of history books has increased over the 1983 mean of $24.96, when in fact it has not.

The probability of making a Type II error is the probability of not rejecting a false null hypothesis. The Greek letter β (beta) is used to denote the probability of a Type II error: $\beta = P(\text{Type II error})$. The probability, β, of a Type II error depends on what the true value of μ is.

Ideally, we would like the probabilities of both Type I and Type II errors to be as small as possible. For then the chances of making an incorrect decision would be small, regardless of whether the null hypothesis is true. Since the probability of a Type I error is just the significance level, α, of the hypothesis test, we can control the size of the Type I error probability simply by selecting the appropriate significance level. If it is important that we do not reject a true null hypothesis, then we should specify a small value for α. However, in choosing the significance level, α, we must keep in mind the following fact:

KEY FACT

For a fixed sample size, the smaller the Type I error probability, α, of rejecting a true null hypothesis, the larger the Type II error probability, β, of not rejecting a false null hypothesis.

Consequently, it is always necessary to assess the risks involved in committing both types of errors and to use that assessment as a method for balancing the Type I and Type II error probabilities.

Two-tailed and one-tailed tests

Hypothesis tests can be broadly classified as either two-tailed or one-tailed. A **two-tailed test** is a hypothesis test for which the rejection region consists of two parts—one on the left and one on the right. The pretzel-packaging illustration provides an example of a two-tailed test. The null and alternative hypotheses are

$$H_0: \mu = 454 \text{ g (machine is working properly)}$$
$$H_a: \mu \neq 454 \text{ g (machine is not working properly)}$$

The rejection and nonrejection regions are given in Figure 9.11 and are repeated here in Figure 9.17.

FIGURE 9.17
Rejection and nonrejection regions for pretzel-packaging illustration. Significance level: $\alpha = 0.05$. Sample size: $n = 50$.

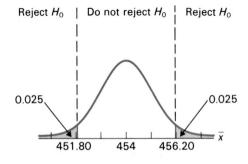

As you can see from the figure, the rejection region consists of two parts. The part on the left consists of all $\bar{x}$-values less than 451.80. The part on the right consists of all $\bar{x}$-values greater than 456.20.

There is a simple way to determine whether a hypothesis test is two-tailed without even having to look at the rejection region. A two-tailed test has a *not-equal sign* ($\neq$) in the alternative hypothesis. This indicates that the null hypothesis will be rejected if the test statistic is either too small or too large and, hence, that the test is two-tailed.

A **one-tailed test** is a hypothesis test for which the rejection region consists of only one part. It is useful to categorize one-tailed tests as left-tailed and right-tailed. A **left-tailed test** is a one-tailed test for which the rejection region is on the left. Our gas-mileage illustration is an example of a left-tailed test. Its null and alternative hypotheses are

$$H_0: \mu = 26 \text{ mpg (manufacturer's claim)}$$
$$H_a: \mu < 26 \text{ mpg (consumer group's conjecture)}$$

The rejection and nonrejection regions are reproduced in Figure 9.18.

FIGURE 9.18
Rejection and nonrejection regions for gas-mileage illustration. Significance level: $\alpha = 0.05$. Sample size: $n = 30$.

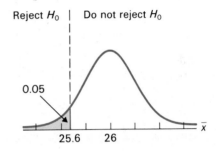

From Figure 9.18, we see that the rejection region consists of only one part and that the part is on the left. It consists of all $\bar{x}$-values less than 25.6.

We can determine whether a hypothesis test is left-tailed simply by observing the form of the alternative hypothesis—a left-tailed test has a *less-than sign* ($<$) in the alternative hypothesis. This signifies that the null hypothesis will be rejected if the test statistic is too small and, therefore, that the rejection region is on the left.

Finally, a **right-tailed test** is a one-tailed test for which the rejection region is on the right. The history-book price illustration is an example of a right-tailed test. The null and alternative hypotheses are

$$H_0: \mu = \$24.96 \text{ (mean price has not increased)}$$
$$H_a: \mu > \$24.96 \text{ (mean price has increased)}$$

In Figure 9.19, we have duplicated the graph of the rejection and nonrejection regions from Figure 9.9.

FIGURE 9.19
Rejection and nonrejection regions for history-book price illustration. Significance level: $\alpha = 0.10$. Sample size: $n = 40$.

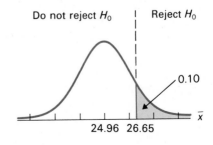

Figure 9.19 shows that the rejection region is made up of only one part and that the part is on the right. It consists of all $\bar{x}$-values greater than 26.65.

We can find whether a hypothesis test is right-tailed by just looking at the form of the alternative hypothesis. A right-tailed test has a *greater-than sign* ($>$) in the alternative hypothesis. This indicates that the null hypothesis will be rejected if the test statistic is too large and, hence, that the rejection region is on the right.

Table 9.5 summarizes the discussion about two-tailed, left-tailed, and right-tailed tests. See also Figure 9.20.

TABLE 9.5

	Two-tailed test	Left-tailed test	Right-tailed test
Sign in H_a	$\neq$	$<$	$>$
Rejection region	Both sides	Left side	Right side

FIGURE 9.20

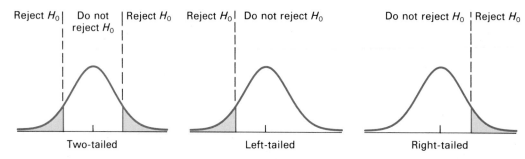

Two-tailed   Left-tailed   Right-tailed

Choosing the hypotheses

An important question in setting up a hypothesis test is how to choose the hypotheses. That is, how does one decide what the null hypothesis should be and what the alternative hypothesis should be? Unfortunately, there is no simple answer to this question. The choices of null and alternative hypotheses depend on such subjective things as who the tester is and what the tester believes, suspects, or wants to show.

In most problems in this text, it should be fairly clear how to make such choices, but there is often room for disagreement. The more experience you have with hypothesis testing, the more you will understand the rationale for selecting the null and alternative hypotheses. However, in order to help you now, we offer the following *guidelines*. Although the guidelines refer specifically to hypothesis tests for a single population mean μ, the same general principles apply to any hypothesis test concerning one or two parameters.

1 Null hypothesis

In this book, the null hypothesis for a test concerning a population mean, μ, should always specify a single value for that parameter. This means that there should always be an *equal sign* ($=$) in the null hypothesis and that the null hypothesis should be of the form

$$H_0: \mu = \mu_0$$

where μ_0 is some specified number. For instance, in the gas-mileage illustration the null hypothesis is H_0: $\mu = 26$ mpg; in the history-book price illustration the null hypothesis is H_0: $\mu = \$24.96$; and in the pretzel-packaging illustration the null hypothesis is H_0: $\mu = 454$ g. Thus, the choice of the null hypothesis is almost always clear.

2 Alternative hypothesis

The choice of the alternative hypothesis depends on the tester's purpose in performing the hypothesis test.

a) *Two-tailed test:* If the tester is primarily concerned with deciding whether the true value of a population mean, μ, is *different from* a value μ_0, then the test should be two-tailed. This means that there should be a *not-equal sign* ($\neq$) in the alternative hypothesis and that the alternative hypothesis should be of the form H_a: $\mu \neq \mu_0$. For example, in the pretzel-packaging illustration, the QA department wants to decide whether the mean net weight, μ, being packaged is *different from* 454 g. Consequently, the appropriate alternative hypothesis is H_a: $\mu \neq 454$ g.

b) *Left-tailed test:* If the tester is primarily concerned with deciding whether the true value of a population mean, μ, is *less than* a value μ_0, then the test should be left-tailed. This means that there should be a *less-than sign* ($<$) in the alternative hypothesis and that the alternative hypothesis should be of the form H_a: $\mu < \mu_0$. For instance, in the gas-mileage illustration, the consumer group contends that the true mean gas mileage, μ, of all Orions is *less than* the manufacturer's claim of 26 mpg. Hence the appropriate alternative hypothesis is H_a: $\mu < 26$ mpg.

c) *Right-tailed test:* If the tester is primarily concerned with deciding whether the true value of a population mean, μ, is *greater than* a value μ_0, then the test should be right-tailed. This means that there should be a *greater-than sign* ($>$) in the alternative hypothesis and that the alternative hypothesis should be of the form H_a: $\mu > \mu_0$. For example, in the history-book price illustration, we felt that the mean price, μ, of history books may have *risen* since 1983. Thus, we took as our alternative hypothesis H_a: $\mu > \$24.96$.

Possible conclusions for a hypothesis test

It is important to be aware of the following points about the possible conclusions for a hypothesis test. First recall that the significance level, α, is the probability of making a Type I error, that is, of rejecting a true null hypothesis. Thus, if the hypothesis test is performed at a small significance level (for example, $\alpha = 0.05$), then it is unlikely that the null hypothesis will be rejected when it is in fact true. Since, in this text, we will generally use a small significance level, we can make the following statement: If we *do* reject the null hypothesis in a hypothesis test, then we can be reasonably confident that the alternative hypothesis is true.

On the other hand, we will usually not know the probability, β, of making a Type II error, that is, of not rejecting a false null hypothesis. Thus, if we *do not* reject the null hypothesis in a hypothesis test, then we simply reserve judgment about which hypothesis is true. In other words, if we do not reject the null hypothesis, we conclude only that the data did not provide sufficient evidence to support the alternative hypothesis. We do *not* conclude that the data provided sufficient evidence to support the null hypothesis. We summarize the above discussion in the following Key Fact.

KEY FACT

The possible conclusions for a hypothesis test are:

1 If the null hypothesis is rejected, conclude that the alternative hypothesis is probably true.
2 If the null hypothesis is not rejected, conclude that the data do not provide sufficient evidence to support the alternative hypothesis.

MTB
SPSS

Exercises 9.3

In Exercises 9.21 through 9.26 we have repeated the null and alternative hypotheses from Exercises 9.15 through 9.20, respectively. We have also drawn a graph displaying the decision rule for each test. Determine, for each exercise, the

a) test statistic.
b) rejection region.
c) nonrejection region.
d) critical value(s).
e) Construct a graph similar to Figure 9.13 on page 359 to depict your results from parts (a) through (d).

9.21 The null and alternative hypotheses for Exercise 9.15 are

H_0: $\mu = \$46.67$ (mean has not increased)
H_a: $\mu > \$46.67$ (mean has increased)

where μ is this year's mean room rate for full-service lodging establishments, and $46.67 is the mean room rate for 1983. A graphical display of the decision rule for a test at the 10% significance level using a sample size of $n = 35$ is

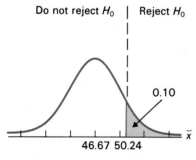

9.22 The null and alternative hypotheses for Exercise 9.16 are

H_0: $\mu = 15$ oz (advertised weight)
H_a: $\mu < 15$ oz (consumer group's contention)

where μ is the true mean weight of boxes of Wheat Flakes brand cereal. A graphical display of the decision rule for a test at the 5% significance level using a sample size of $n = 40$ is

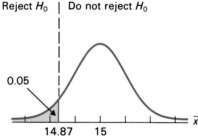

9.23 The null and alternative hypotheses for Exercise 9.17 are

H_0: $\mu = \$3603$ (mean has not decreased)
H_a: $\mu < \$3603$ (mean has decreased)

where μ denotes the mean income tax per taxable return for last year, and $3603 is the mean for 1982. The decision rule for a test at the 5% significance level with a sample size of 30 is pictured below.

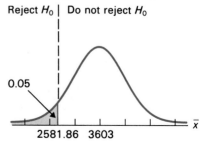

9.24 In Exercise 9.18, the null and alternative hypotheses are

$$H_0: \mu = \$1700 \text{ (no improvement)}$$
$$H_a: \mu > \$1700 \text{ (improvement)}$$

Here μ is the true mean weekly sales for the menswear store with advertising, and $1700 is the mean without advertising. For a test at the 5% significance level with a sample of size $n = 32$, the decision rule is as shown below.

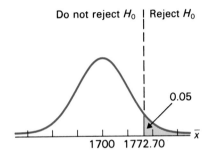

9.25 The null and alternative hypotheses for Exercise 9.19 are

$$H_0: \mu = 21 \text{ days}$$
$$H_a: \mu \neq 21 \text{ days}$$

where μ is this year's mean stay in Europe and the Mediterranean by U.S. travelers, and 21 days is the mean stay in 1980. The decision rule for a test at the 10% significance level, using a sample of size $n = 36$, is displayed graphically below.

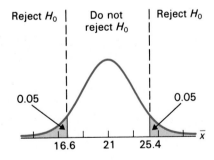

9.26 For Exercise 9.20, the null and alternative hypotheses are

$$H_0: \mu = 114 \text{ million BTU}$$
$$H_a: \mu \neq 114 \text{ million BTU}$$

where μ is the mean energy consumed by southern households in 1982, and 114 million BTU is the national mean for 1982. The following is a graphical display of the decision rule for a test at the 5% significance level using a sample of size $n = 50$.

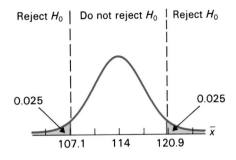

9.27 Refer to Exercise 9.21. Explain what each of the following would mean:
a) A Type I error.
b) A Type II error.
c) A correct decision.

Now, recall that the sample of 35 full-service lodging establishments led to rejection of the null hypothesis. (See Exercise 9.15.) Classify that decision by error type or as a correct decision if, in fact, this year's mean room rate for full-service lodging establishments
d) has not increased over the 1983 mean of $46.67.
e) has increased over the 1983 mean of $46.67.

9.28 Refer to Exercise 9.22. Explain what each of the following would mean:
a) A Type I error.
b) A Type II error.
c) A correct decision.

Recall that the sample of 40 boxes of Wheat Flakes led us not to reject the null hypothesis. (See Exercise 9.16.) Classify that decision by error type or as a correct decision if, in fact, the true mean weight
d) is 15 oz.
e) is less than 15 oz.

9.29 Refer to Exercise 9.25. Explain what each of the following would mean:
a) A Type I error.
b) A Type II error.
c) A correct decision.

Recall that the sample of 36 U.S. travelers led us not to reject the null hypothesis. (See Exercise 9.19.) Classify that decision by error type or as a correct decision if, in fact, the true mean stay
d) is 21 days.
e) is not 21 days.

9.30 Refer to Exercise 9.26. Explain what each of the following would mean:
a) A Type I error.
b) A Type II error.
c) A correct decision.

Now, recall that the sample of 50 southern households led to rejection of the null hypothesis. (See Exercise 9.20.) Classify that decision by error type or as a correct decision if, in fact, the mean energy consumed by southern households in 1982
d) was the same as the national mean of 114 million BTU.
e) was not the same as the national mean of 114 million BTU.

9.31 Classify the hypothesis tests in the exercises listed here as two-tailed, left-tailed, or right-tailed.
a) Exercise 9.21.
b) Exercise 9.23.
c) Exercise 9.25.

9.32 Classify the hypothesis tests in the exercises listed here as two-tailed, left-tailed, or right-tailed.
a) Exercise 9.22.
b) Exercise 9.24.
c) Exercise 9.26.

9.33 True or false: If it is important *not* to make a Type I error, then the hypothesis test should be performed at a small significance level.

9.34 True or false: For a fixed sample size, decreasing the significance level for a hypothesis test results in an increase in the probability of making a Type II error.

9.35 Suppose that we choose the significance level of a hypothesis test to be $\alpha = 0$.
a) What is the probability of a Type I error, that is, of rejecting a true null hypothesis?
b) What is the probability of a Type II error, that is, of not rejecting a false null hypothesis?

9.36 Find an example or exercise in this section in which it is important to have
a) a small α probability.
b) a small β probability.
c) both α and β probabilities be small.

9.37 A brewer produces a 355 ml (12 fluid oz) can of beer. To ensure that the machine used to fill the cans is working properly, periodic checks are made. Specifically, the following hypothesis test is performed:

$$H_0: \mu = 355 \text{ ml (working properly)}$$
$$H_a: \mu \neq 355 \text{ ml (not working properly)}$$

where μ is the actual mean amount of beer being dispensed into the cans. For the test, $n = 30$ cans of beer are randomly selected and their contents carefully measured. Explain what each of the following would mean.
a) A Type I error.
b) A Type II error.

Assume that the test is performed at the 5% significance level.
c) Find the critical values and draw a graph portraying the decision rule. [Assume $\sigma = 5.9$ ml.]
d) Determine the probability, α, of a Type I error.
e) Determine the probability, β, of a Type II error, if the true mean amount of beer being dispensed into the cans is

(i) 350 ml	(ii) 351 ml	(iii) 352 ml
(iv) 353 ml	(v) 354 ml	(vi) 356 ml
(vii) 357 ml	(viii) 358 ml	(ix) 359 ml
(x) 360 ml		

Hint: For each part you must compute the probability that $\bar{x}$ falls in the nonrejection region, given that the true value of μ is as specified.
f) Use your results from part (e) to draw a graph of the true value of μ versus the probability of a Type II error. What does the graph tell you?

9.4

Hypothesis tests for a population mean (large samples)

Procedure 9.1 on page 352 gives a step-by-step method for performing a hypothesis test. The procedure applies to any hypothesis testing situation, not just to hypothesis tests for means. In this section we will develop a simple and efficient hypothesis-testing procedure designed specifically for large-sample tests about a population mean.

Thus far we have based the decision rule for such hypothesis tests on the statistic $\bar{x}$. For a specified significance level and sample size, we compute the critical value(s) by "destandardizing" the appropriate z-score(s). For instance, consider the history-book price illustration of Example 9.4. In Step 3 of the hypothesis test on page 353, we computed the critical value, c, for a right-tailed test at the 10% significance level. The z-score corresponding to c is $z_{0.10} = 1.28$, as indicated in Figure 9.8 on page 353. Using the fact that the $\bar{x}$-values and z-values are related by the equation

$$z = \frac{\bar{x} - \mu_0}{\sigma/\sqrt{n}} = \frac{\bar{x} - 24.96}{1.32}$$

we concluded that

$$1.28 = \frac{c - 24.96}{1.32}$$

or

$$c = 24.96 + 1.28 \cdot 1.32 = 26.65$$

In this last step the critical value, c, is obtained by "destandardizing" the z-score, $z = 1.28$.

Generally, it is simpler to use the standardized version of $\bar{x}$,

$$z = \frac{\bar{x} - \mu_0}{\sigma/\sqrt{n}}$$

as the test statistic. Then the entire hypothesis test can be carried out in terms of z-values. With that in mind, we can now state the following procedure for performing a large-sample hypothesis test for a population mean.

PROCEDURE 9.2 To perform a hypothesis test for a population mean, with null hypothesis H_0: $\mu = \mu_0$.

Assumption
Sample size is at least 30 ($n \geq 30$).

STEP 1 *State the null and alternative hypotheses.*
STEP 2 *Decide on the significance level, α.*
STEP 3 *The critical value(s)*
  a) *for a two-tailed test are $\pm z_{\alpha/2}$.*
  b) *for a left-tailed test is $-z_\alpha$.*
  c) *for a right-tailed test is z_α.*
Use Table II to find the critical value(s).

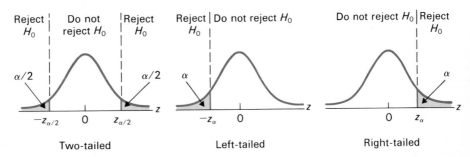

| Two-tailed | Left-tailed | Right-tailed |

STEP 4 *Compute the value of the test statistic*

$$z = \frac{\bar{x} - \mu_0}{s/\sqrt{n}}$$

STEP 5 *If the value of the test statistic falls in the rejection region, reject H_0; otherwise, do not reject H_0.*

STEP 6 *State the conclusion in words.*

Note: In Step 4 of Procedure 9.2 we have used the sample standard deviation, s, in the denominator of the test statistic. Theoretically, we should use the population standard deviation, σ. However, σ is rarely known and so we have used s in its place. This is acceptable because, for large samples, the sample standard deviation is likely to be a good approximation of the population standard deviation. In the rare case where σ is known, it should always be used in place of s in Step 4 of Procedure 9.2.

EXAMPLE   9.9   Illustrates Procedure 9.2

Calcium is the most abundant and one of the most important minerals in the body. It works with phosphorus in building and maintaining bones and teeth. According to the Food and Nutrition Board of the National Academy of Sciences, the recommended daily allowance (RDA) of calcium for adults is 800 milligrams.

A nutritionist thinks that people with incomes below the poverty level average *less* than the RDA of 800 mg. To test her claim, she determines the daily intake of calcium for a random sample of $n = 50$ such people. The results are displayed in Table 9.6. Data are in milligrams.

TABLE 9.6

879	1096	701	986	828	1077	703	633	1119	951
555	422	997	473	702	508	530	688	691	943
513	720	944	673	574	707	864	748	498	881
1199	743	1325	655	1043	599	1008	792	915	456
705	180	287	542	893	1052	473	739	642	915

Use the data to decide whether the mean daily intake of calcium, μ, for people with incomes below the poverty level is *less than* the RDA of 800 mg. Perform the test at the 5% significance level.

SOLUTION

Since the sample size is at least 30 ($n = 50$), we apply Procedure 9.2.

STEP 1 *State the null and alternative hypotheses.*

The null and alternative hypotheses are

$$H_0: \mu = 800 \text{ mg}$$
$$H_a: \mu < 800 \text{ mg}$$

where μ is the mean daily intake of calcium for people with incomes below the poverty level. Note that the test is left-tailed since there is a less-than sign ($<$) in the alternative hypothesis.

STEP 2 *Decide on the significance level, α.*

We are to perform the test at the 5% significance level. Thus, $\alpha = 0.05$.

STEP 3 *The critical value for a left-tailed test is $-z_\alpha$.*

Since $\alpha = 0.05$, the critical value is $-z_{0.05}$. From Table II we find that $z_{0.05} = 1.645$. So, the critical value is

$$-z_{0.05} = -1.645$$

See Figure 9.21.

FIGURE 9.21

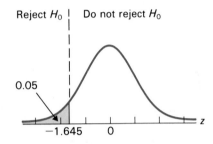

STEP 4 *Compute the value of the test statistic*

$$z = \frac{\bar{x} - \mu_0}{s/\sqrt{n}}$$

We have $\mu_0 = 800$ and $n = 50$. From the data in Table 9.6 we get

$$\bar{x} = \frac{\Sigma x}{n} = \frac{37{,}767}{50} = 755.3 \text{ mg}$$

and

$$s = \sqrt{\frac{n(\Sigma x^2) - (\Sigma x)^2}{n(n-1)}} = \sqrt{\frac{50(31{,}333{,}629) - (37{,}767)^2}{50 \cdot 49}} = 239.3 \text{ mg}$$

Thus, the value of the test statistic is

$$z = \frac{\bar{x} - \mu_0}{s/\sqrt{n}} = \frac{755.3 - 800}{239.3/\sqrt{50}} = -1.32$$

STEP 5 *If the value of the test statistic falls in the rejection region, reject H_0; otherwise, do not reject H_0.*

The value of the test statistic, found in Step 4, is $z = -1.32$. As we see from Figure 9.21, this does not fall in the rejection region and so we do not reject H_0.

STEP 6 *State the conclusion in words.*

The sample of 50 daily calcium intakes does *not* provide sufficient evidence for us to conclude that the average daily calcium intake of people with incomes below the poverty level is less than the RDA of 800 mg. ■

MTB
SPSS

Exercises 9.4

9.38 Why is it permissible to use the sample standard deviation, *s*, in place of an unknown population standard deviation, σ, when performing a large-sample hypothesis test for a population mean μ?

9.39 According to the U.S. Bureau of the Census, the 1983 mean annual expenditure per residential property owner for maintenance and repairs was $280. A random sample of 40 residential property owners revealed the following expenditures for 1986:

$158	110	185	136	167	84	437	420
230	57	942	287	259	123	712	170
531	347	505	148	111	442	145	38
49	188	107	134	223	36	360	253
321	821	89	202	759	151	1176	31

Do the data indicate an *increase* in the mean amount spent in 1986 for maintenance and repairs over the 1983 mean of $280? Assume $\sigma = \$265$ and perform the hypothesis test at the 5% significance level. [The sum of the data is $11,664.]

9.40 *The World Almanac, 1985* reports that the average travel time to work in 1980 for residents of South Dakota was 13 minutes. For this year, a random sample of 35 travel times for South Dakota residents yielded the data below (in minutes).

29	40	0	12	10	6	41
25	21	5	4	19	2	7
10	8	3	6	52	4	12
0	33	6	2	17	21	8
38	2	13	8	14	11	2

Test, at the 5% significance level, whether this year's mean travel time to work for South Dakota residents appears to have *changed* from the 1980 mean of 13 minutes. Assume $\sigma = 11.6$ minutes. [The sum of the data is 491.]

9.41 The Food and Nutrition Board of the National Academy of Sciences states that the recommended daily allowance (RDA) of iron for adult females under the age of 51 is 18 mg. The amounts of iron intake during a 24-hour period for a sample of 45 such females are as follows (data are in mg):

15.0	18.1	14.4	14.6	10.9	18.1	18.2	18.3	15.0
16.0	12.6	16.6	20.7	19.8	11.6	12.8	15.6	11.0
15.3	9.4	19.5	18.3	14.5	16.6	11.5	16.4	12.5
14.6	11.9	12.5	18.6	13.1	12.1	10.7	17.3	12.4
17.0	6.3	16.8	12.5	16.3	14.7	12.7	16.3	11.5

At the 1% significance level, do the data suggest that adult females under the age of 51 are, on the average, getting *less* than the RDA of 18 mg of iron? [$\Sigma x = 660.6$, $\Sigma x^2 = 10,115.88$.]

9.42 According to the U.S. Office of Juvenile Justice and Delinquency Prevention, the mean age of juveniles held in public custody in 1982 was 15.4 years. The mean age of 250 randomly selected juveniles being held in public custody this year is $\bar{x} = 15.26$ years, with a standard deviation of $s = 1.01$ years. Does it appear that the mean age, μ, of all juveniles being held in public custody has *decreased* since 1982? Perform the appropriate hypothesis test at the 10% significance level.

9.43 A dog food manufacturer sells "50-lb" bags of dog food. Suppose that you randomly select $n = 75$ of these bags and weigh the contents of each bag.
   a) Assume that you get a mean weight for the 75 bags of $\bar{x} = 50.11$ lb with a standard deviation of $s = 0.84$ lb. Would you be inclined to believe that the actual mean weight, μ, of all bags of this dog food *differs from* the advertised weight of 50 lb? Perform your hypothesis test at the 5% significance level.
   b) Repeat part (a) if the mean weight of the 75 bags you weigh is 50.21 lb instead of 50.11 lb.

9.44 The Health Insurance Association of America reports that, in 1983, the average daily room charge for a semi-private room in U.S. hospitals was $195. In the same year, a random sample of 30 Massachusetts hospitals yielded a mean semi-private room charge of $\bar{x} = \$202.68$ with a standard deviation of $s = \$12.77$. Does this suggest that the average daily room charge in Massachusetts for 1983 *exceeded* the national average of $195? Perform the hypothesis test at the 5% significance level.

9.45 The U.S. Federal Highway Administration reports that the average passenger vehicle was driven 8.9 thousand miles in 1982. A random sample of $n = 500$ passenger vehicles gave an average of $\bar{x} = 8.7$ thousand miles driven for last year, with a standard deviation of $s = 6.0$ thousand miles. Let μ denote the true mean distance driven per passenger vehicle for last year.
   a) Perform the hypothesis test

H_0: $\mu = 8.9$ thousand miles

H_a: $\mu \neq 8.9$ thousand miles

at the 5% significance level.

b) Use Procedure 8.1 on page 313 to find a 95% confidence interval for μ.

c) Does the mean of 8.9 used in the null hypothesis in part (a) lie in your confidence interval from part (b)?

d) Repeat parts (a) through (c) if the 500 passenger vehicles sampled were driven an average of 9.5 thousand miles last year.

e) Based on your observations in parts (a) through (d), complete the following statements concerning a two-tailed hypothesis test

H_0: $\mu = \mu_0$

H_a: $\mu \neq \mu_0$

at the significance level α, and a $(1 - \alpha)$-level confidence interval for μ.

  (i) If μ_0 lies in the $(1 - \alpha)$-level confidence interval for μ, (*reject, do not reject*) the null hypothesis.

  (ii) If μ_0 lies outside the $(1 - \alpha)$-level confidence interval for μ, (*reject, do not reject*) the null hypothesis.

9.46 **Hypothesis testing based on confidence intervals.**

a) Show that the inequality

$$\bar{x} - z_{\alpha/2} \cdot \frac{s}{\sqrt{n}} \leq \mu_0 \leq \bar{x} + z_{\alpha/2} \cdot \frac{s}{\sqrt{n}}$$

is equivalent to

$$-z_{\alpha/2} \leq \frac{\bar{x} - \mu_0}{s/\sqrt{n}} \leq z_{\alpha/2}$$

b) Deduce the following fact from part (a). For a two-tailed hypothesis test

H_0: $\mu = \mu_0$

H_a: $\mu \neq \mu_0$

at the significance level α, we will not reject the null hypothesis if μ_0 lies in the $(1 - \alpha)$-level confidence interval for μ, and we will reject the null hypothesis if μ_0 does not lie in the $(1 - \alpha)$-level confidence interval for μ.

9.5   Hypothesis tests for a population mean (small samples)

Up until now we have discussed procedures for performing hypothesis tests for a population mean when dealing with large samples ($n \geq 30$). However, as we mentioned in Section 8.5, there are many instances in which large samples are either unavailable, extremely expensive, or undesirable.

For instance, suppose a new type of car bumper is being tested to determine its effectiveness in protecting a car in a front-end collision. Testing the bumper might actually involve driving cars into a brick wall. Obviously, we would like to use a small sample size here because of the costs involved in such testing. We would also employ small samples if we were testing a new drug that might have harmful side effects.

In this section we will consider hypothesis tests for a population mean, μ, when the sample size is small ($n < 30$). We will assume that the population being sampled is, at least approximately, *normally distributed*. We will also assume that the population standard deviation, σ, is unknown since, in practice, that is usually the case.

As we learned in Section 8.5, when sampling from a normally distributed population with mean μ, the random variable

$$t = \frac{\bar{x} - \mu}{s/\sqrt{n}}$$

has the *t*-distribution with df $= n - 1$. This implies that we can perform a hypothesis test for μ in such situations by employing that random variable as our test statistic and using the *t*-table to obtain the critical value(s) for the test.

PROCEDURE 9.3 To perform a hypothesis test for a population mean, with null hypothesis $H_0: \mu = \mu_0$.

Assumption
Normal population.

STEP 1 *State the null and alternative hypotheses.*
STEP 2 *Decide on the significance level, α.*
STEP 3 *The critical value(s)*
    a) *for a two-tailed test are* $\pm t_{\alpha/2}$,
    b) *for a left-tailed test is* $-t_\alpha$,
    c) *for a right-tailed test is* t_α,
with df = n − 1. Use Table III to find the critical value(s).

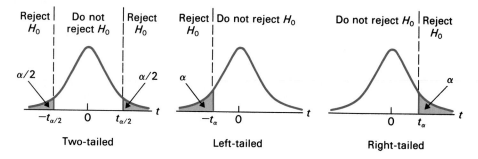

STEP 4 *Compute the value of the test statistic*

$$t = \frac{\overline{x} - \mu_0}{s/\sqrt{n}}$$

STEP 5 *If the value of the test statistic falls in the rejection region, reject H_0; otherwise, do not reject H_0.*
STEP 6 *State the conclusion in words.*

EXAMPLE 9.10 Illustrates Procedure 9.3

According to the U.S. Energy Information Administration, the average residential energy expenditure for 1982 was $1022 per household. That same year, 15 randomly selected upper-income families reported the energy expenditures shown in Table 9.7.

TABLE 9.7
1982 energy expenditures
for 15 randomly selected
upper-income families.

$1153	1249	1126	1053	1689
1514	1420	807	1130	1268
1610	1192	1104	1250	1084

At the 5% significance level, do the data indicate that, in 1982, upper-income families spent *more,* on the average, for energy than the average of $1022 spent by families in general? [Assume that the 1982 energy expenditures for upper-income families are approximately normally distributed.]

SOLUTION

Since the population under consideration is normally distributed, we employ Procedure 9.3.

STEP 1 *State the null and alternative hypotheses.*

If we let μ denote the mean energy expenditure for upper-income families in 1982, then the null and alternative hypotheses are

$$H_0: \mu = \$1022$$
$$H_a: \mu > \$1022$$

Note that the test is right-tailed since there is a greater-than sign ($>$) in the alternative hypothesis.

STEP 2 *Decide on the significance level, α.*

We are to perform the hypothesis test at the 5% significance level. Thus, $\alpha = 0.05$.

STEP 3 *The critical value for a right-tailed test is t_α, with df $= n - 1$.*

Here $n = 15$ and $\alpha = 0.05$. Table III shows that for df $= 15 - 1 = 14$,

$$t_{0.05} = 1.761$$

See Figure 9.22.

FIGURE 9.22

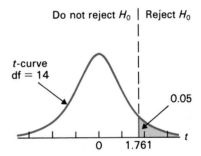

Do not reject H_0 | Reject H_0

t-curve
df = 14

0.05

0   1.761   t

STEP 4 *Compute the value of the test statistic*

$$t = \frac{\bar{x} - \mu_0}{s/\sqrt{n}}$$

We have $\mu_0 = \$1022$ and $n = 15$. Furthermore, the mean and standard deviation of the sample data in Table 9.7 are

$$\bar{x} = \frac{\Sigma x}{n} = \frac{\$18,649}{15} = \$1243.27$$

and

$$s = \sqrt{\frac{n(\Sigma x^2) - (\Sigma x)^2}{n(n-1)}} = \sqrt{\frac{15(23,932,721) - (18,649)^2}{15 \cdot 14}} = \$231.00$$

Consequently, the value of the test statistic is

$$t = \frac{\bar{x} - \mu_0}{s/\sqrt{n}} = \frac{1243.27 - 1022}{231.00/\sqrt{15}} = 3.710$$

STEP 5 *If the value of the test statistic falls in the rejection region, reject H_0; otherwise, do not reject H_0.*

The value of the test statistic, found in Step 4, is $t = 3.710$. As we see from Figure 9.22, this falls in the rejection region. Hence, we **reject** H_0.

STEP 6 *State the conclusion in words.*

It appears that, in 1982, upper income families spent more, on the average, for energy than the average of $1022 spent by families in general.   ∎

MTB
SPSS

Exercises 9.5

9.47 A paint manufacturer claims that the average drying time for its new latex paint is two hours. To test this claim, the drying times are obtained for $n = 20$ randomly selected cans of paint. The results are displayed below in minutes.

123	109	115	121	130
127	106	120	116	136
131	128	139	110	133
122	133	119	135	109

Assuming that the drying times are normally distributed, do the sample data suggest that the mean drying time is actually *greater* than the manufacturer's claim of 120 minutes? Use $\alpha = 0.05$. [The sample mean and sample standard deviation of the data are $\bar{x} = 123.1$ and $s = 10.0$.]

9.48 The U.S. Energy Information Administration reports that the average annual motor fuel expenditure per U.S. household was $1317 in 1983. That same year, a random sample of 16 households within metropolitan areas gave the following annual motor fuel expenditures:

$1390	1459	2043	1551
415	1359	1778	1537
1167	1716	1463	904
560	1592	1710	638

Assume that the annual motor fuel expenditures for households within metropolitan areas are normally distributed. At the 5% significance level, do the data provide evidence that the 1983 mean annual fuel expenditure for households within metropolitan areas *differed* from the national average of $1317? [The sample mean and sample standard deviation of the data are $\bar{x} = \$1330.13$ and $s = \$470.36$.]

9.49 A retailer has received a large shipment of automobile batteries from a new supplier. The supplier claims that the batteries have a mean life of 36 months. A test on $n = 10$ batteries randomly sampled from the shipment gave the following results, in months.

27.6	28.7	34.7	29.0	22.9
29.6	29.4	30.2	36.5	34.7

a) Do the data indicate that the mean life of the supplier's batteries is *less* than the claimed 36 months? Perform the test at the 1% significance level. [*Note:* $\Sigma x = 303.3$ and $\Sigma x^2 = 9343.85$.]

b) What assumptions are you making about battery life?

9.50 According to the College Entrance Examination Board, the mean verbal score on the Scholastic Aptitude Test (SAT) in 1983 was 425 points out of a possible 800. The scores on the SAT are approximately normally distributed. A random sample of 25 verbal scores for last year yielded the following results:

346	496	352	378	315
491	360	385	500	558
381	303	434	562	496
420	485	446	479	422
494	289	436	516	615

At the 10% significance level, does it appear that last year's mean for verbal SAT scores has *increased* over the 1983 mean of 425 points? [*Note:* $\Sigma x = 10,959$ and $\Sigma x^2 = 4,979,401$.]

9.51 The average retail price for oranges in 1983 was 38.5 cents per pound, as reported by the U.S. Department of Agriculture. Recently, a random sample of 15 markets reported the following prices for oranges in cents per pound:

43.0	40.0	42.6	40.2	37.5
44.1	45.2	41.8	35.6	34.6
37.9	44.2	44.5	38.2	42.4

Assuming that the retail prices for oranges are normally distributed, can we conclude that the mean retail price for oranges now is *different* from the 1983 mean of 38.5 cents per pound? Use $\alpha = 0.05$.

9.52 Atlas Fishing Line, Inc. manufactures a 10-lb test line. Twelve randomly selected spools are subjected to tensile-strength tests. The results are

9.8	10.2	9.8	9.4
9.7	9.7	10.1	10.1
9.8	9.6	9.1	9.7

Use the fact that tensile strengths are normally distributed to decide if Atlas Fishing Line's 10-lb test line is not up to specifications. Perform the test at the 5% significance level.

9.53 A manufacturer of light bulbs produces a 60-watt bulb with an average life of 1000 hours. Let us sup-

pose that the research and development (R&D) department claims to have developed a new bulb, which costs the same to produce but outlasts the present type. To justify its claim, R&D tests 10 of the new type of bulbs. Its results show an average life of $\bar{x} = 1050.2$ hours for the 10 bulbs, with a standard deviation of $s = 65.8$ hours.

a) Do the tests performed by R&D support their claim at the 1% significance level? [Assume that bulb life is normally distributed.]

b) Suppose that in part (a) you mistakenly conclude that the test statistic

$$\frac{\bar{x} - 1000}{s/\sqrt{10}}$$

has the standard normal distribution.

(i) What critical value (z-value) would you then have used?

(ii) What critical value (t-value) did you actually use?

(iii) In general, does the mistaken use of a z critical value, when a t critical value should be used, make it more or less likely that the null hypothesis will be rejected?

Hypothesis tests for a population proportion (large samples)

Large-sample hypothesis tests concerning a population proportion follow essentially the same pattern as those for a population mean. Before proceeding, however, let us review the terminology and notation used in studying proportions. Recall that a *two-category population* is one in which each member of the population is classified as either having or not having a specified attribute. For a human population, that attribute might be "female," it might be "owning a home," or it might be "over 40." The proportion (percentage) of the entire population that has the specified attribute is called the *population proportion* and is denoted by the letter p. The proportion of a sample from the population that has the specified attribute is called a *sample proportion* and is denoted by the symbol $\bar{p}$. Thus,

$$p = \text{population proportion}$$
$$\bar{p} = \text{sample proportion}$$

For a sample of size n, the sample proportion is computed from the formula

$$\bar{p} = \frac{x}{n}$$

where x is the number of members sampled that have the specified attribute—the number of "successes."

Now that we have reviewed the terminology and notation used in statistical inferences for a population proportion, we return to the development of a hypothesis testing procedure. The essential fact used in designing a large-sample hypothesis testing procedure for proportions is that the probability distribution of $\bar{p}$ (the sampling distribution of a proportion) is approximately normally distributed. More precisely, as we learned on page 330 of Section 8.6, the following is true.

KEY FACT The sampling distribution of a proportion—large samples

Suppose a large random sample of size n is to be taken from a two-category population with population proportion p. Then the random variable $\bar{p}$ is approximately *normally distributed* and has mean $\mu_{\bar{p}} = p$ and standard deviation $\sigma_{\bar{p}} = \sqrt{p(1-p)/n}$.

As with hypothesis tests for means, it is generally more convenient to use standardized test statistics and critical values. Thus, we will take as the test statistic the standardized random variable

$$z = \frac{\bar{p} - p}{\sqrt{p(1-p)/n}}$$

which, by the Key Fact, has approximately the standard normal distribution. The critical value(s) for the hypothesis test can then be obtained from the standard normal table.

PROCEDURE 9.4 To perform a hypothesis test for a population proportion, with null hypothesis H_0: $p = p_0$.

Assumption
Both np_0 and $n(1 - p_0)$ are at least 5.

STEP 1 *State the null and alternative hypotheses.*
STEP 2 *Decide on the significance level, α.*
STEP 3 *The critical value(s)*
   a) *for a two-tailed test are $\pm z_{\alpha/2}$.*
   b) *for a left-tailed test is $-z_\alpha$.*
   c) *for a right-tailed test is z_α.*
Use Table II to find the critical value(s).

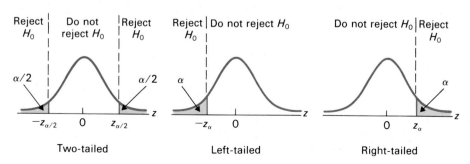

| Two-tailed | Left-tailed | Right-tailed |

STEP 4 *Compute the value of the test statistic*

$$z = \frac{\bar{p} - p_0}{\sqrt{p_0(1 - p_0)/n}}$$

STEP 5 *If the value of the test statistic falls in the rejection region, reject H_0; otherwise, do not reject H_0.*

STEP 6 *State the conclusion in words.*

EXAMPLE 9.11   Illustrates Procedure 9.4

On Sunday, October 7, 1984, presidential candidates Ronald Reagan and Walter Mondale debated on national television. A Newsweek poll, conducted by the Gallup Organization, was taken to determine which candidate the public thought did a better job in the debate. The Gallup Organization telephoned 379 registered voters who watched the debate and found that 205 thought Mondale did a better job. Can we infer from the poll, at the 5% significance level, that more than 50% of *all* registered voters who watched the debate thought Mondale did a better job?

SOLUTION

We will apply Procedure 9.4, but first it is necessary to check that the conditions for its use are satisfied. We have $n = 379$ and $p_0 = 0.50$ (50%). Therefore,

$$np_0 = 379 \cdot 0.50 = 189.5$$

and

$$n(1 - p_0) = 379 \cdot (1 - 0.50) = 189.5$$

Since both np_0 and $n(1 - p_0)$ are at least 5, we can use Procedure 9.4 to perform the hypothesis test.

STEP 1 *State the null and alternative hypotheses.*

H_0: $p = 0.50$ (not more than 50% felt that Mondale did a better job)
H_a: $p > 0.50$ (more than 50% felt that Mondale did a better job)

Note that the test is right-tailed since there is a greater-than sign ($>$) in the alternative hypothesis.

STEP 2 *Decide on the significance level, α.*

We are to perform the test at the 5% significance level. So, $\alpha = 0.05$.

STEP 3 *The critical value for a right-tailed test is z_α.*

Since $\alpha = 0.05$, the critical value is

$$z_{0.05} = 1.645$$

See Figure 9.23 at the top of page 382.

FIGURE 9.23

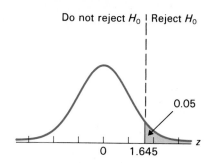

STEP 4 *Compute the value of the test statistic*

$$z = \frac{\bar{p} - p_0}{\sqrt{p_0(1 - p_0)/n}}$$

We have $n = 379$ and $p_0 = 0.50$. The number of registered voters sampled who thought Mondale did a better job is $x = 205$. Thus, the proportion of those sampled who thought Mondale did a better job is

$$\bar{p} = \frac{x}{n} = \frac{205}{379} = 0.541 \ (54.1\%)$$

Consequently, the value of the test statistic is

$$z = \frac{\bar{p} - p_0}{\sqrt{p_0(1 - p_0)/n}} = \frac{0.541 - 0.50}{\sqrt{(0.50) \ (1 - 0.50)/379}} = 1.59$$

STEP 5 *If the value of the test statistic falls in the rejection region, reject H_0; otherwise, do not reject H_0.*

From Step 4, the value of the test statistic is $z = 1.59$. A look at Figure 9.23 shows that this does not fall in the rejection region. Thus, we do not reject H_0.

STEP 6 *State the conclusion in words.*

Based on the survey done by the Gallup Organization, we *cannot* infer that more than 50% of all registered voters who watched the debate thought Mondale did a better job than Reagan. ∎

Exercises 9.6

In Exercises 9.54 through 9.61, use Procedure 9.4 to perform each of the hypothesis tests. Be sure to check whether the conditions for applying that procedure are met.

9.54 According to a report by the Motor Vehicle Man-ufacturers Association, in 1962 foreign cars made up 4.8% of all U.S. car sales. In 1983, the percentage was 27.8%. U.S. automakers have implemented several policies to cut costs in order to challenge the imports, and want to know if the 1983 foreign car percentage of 27.8% has *decreased* during the present year. From a random sample of $n = 500$ car sales, it is determined that 128 are imports. Do the data suggest, at the 5% significance level, that the percentage of foreign car sales has declined from the 1983 figure of 27.8%?

9.55 In 1984, when a political incumbent ran for city council, 63.8% of the voters thought that repair of city streets was an important priority. Since then, repairs have been effected in several areas of the city. The councillor wonders if attitudes on the issue have *changed.* Out of a sample of 350 voters, 204 still think that repair of city streets is an important issue. Perform the appropriate hypothesis test at the 5% significance level.

9.56 According to the Arizona Real Estate Commission, in 1982 only 10% of the people holding real estate licenses were active in the industry. An independent agency has been asked by the Commission to see whether this year's percentage is *higher*. A random sample of 150 licensed people reveals that 24 are currently active. Perform the appropriate hypothesis test at the 1% significance level.

9.57 According to the American Medical Association, 9.0% of all physicians in 1975 were women; in 1980, 11.6% were women; and in 1981, 12.2% were women. For this year, out of a sample of 125 randomly selected physicians, 19 are women. Do the data indicate, at the 10% significance level, that the proportion of female physicians is *higher* now than in 1981?

9.58 A 1974 study by the U.S. National Institute on Drug Abuse showed that about 48.8% of young adults (18–25 years old) smoked cigarettes. In 1982, 39.5% of 1283 young adults sampled smoked cigarettes. Do the data provide evidence that the percentage of young adult smokers *declined* between 1974 and 1982? Use $\alpha = 0.05$.

9.59 A direct-mail firm wants to conduct a test-market offering for one of its new products. A random sample of 800 people is chosen to receive advertising material describing the new product. It is decided that additional advertising and promotion will occur only if the sample results provide strong evidence that the actual (population) response rate, *p*, will exceed 6.5%. What decision will be made if 70 out of the 800 people make a purchase? Use $\alpha = 0.01$.

9.60 Of the 38 numbers on a roulette wheel, 18 are red, 18 are black, and two are green. If the wheel is true, the probability of the ball landing on red is $18/38 = 0.474$. A gambler has been studying a roulette wheel, and wonders if he can find an imperfection in the wheel so that he can improve his odds of winning. The gambler observes 200 spins of the wheel and finds that, out of these 200 trials, the ball lands on red 93 times. At the 10% significance level, does it appear that red is not coming up the appropriate percentage of the time?

9.61 In 1983, according to the U.S. Bureau of the Census, 12.3% of all families earned incomes below the poverty level. During that same year, a sociologist randomly selected 250 families in the Northeast. Twenty-eight of these 250 families earned incomes below the poverty level. Does the study indicate that in 1983 the proportion of northeastern families below the poverty level was *less* than the national proportion? Use $\alpha = 0.05$.

9.62 The test statistic used in Procedure 9.4 is

(1)
$$z = \frac{\bar{p} - p_0}{\sqrt{p_0(1 - p_0)/n}}$$

In this exercise we will show that the test statistic given in (1) is the familiar

(2)
$$z = \frac{\bar{x} - \mu_0}{\sigma/\sqrt{n}}$$

when the latter is specialized to the situation of proportions. In other words, we will prove that Procedure 9.4 is really just a special case of Procedure 9.2.

Consider a finite, two-category population in which the proportion of members with the specified attribute is *p*. We can think of such a population as consisting of 1's and 0's. A member of the population is a "1" if it has the specified attribute and is a "0" otherwise.

a) If the size of the population is *N*, how many 1's are in the population?

b) Use part (a) and Definition 3.6 on page 95 to show that the mean of this population of 1's and 0's is *p*—that is, $\mu = p$.

c) Use part (b) and the shortcut formula on page 97 to show that the standard deviation of this population of 1's and 0's is $\sqrt{p(1-p)}$—that is, $\sigma = \sqrt{p(1-p)}$.

d) Suppose a sample of size *n* is to be taken from the population. Verify that $\bar{x} = \bar{p}$.

e) Deduce from parts (b) through (d) that the test statistics in (2) becomes the one in (1), when specialized to the situation of proportions.

9.7   Computer packages*

In this section we will examine how computer packages can be used to perform hypothesis tests for a population mean, μ. Specifically, we will illustrate two Minitab programs by applying them to examples we have already done by hand.

ZTEST

Procedure 9.2 on page 371 gives a hypothesis testing procedure that applies when the sample size is large ($n \geq 30$). A Minitab program called **ZTEST** will perform Procedure 9.2 for us. We will apply ZTEST to the illustration in Example 9.9 on page 372.

EXAMPLE 9.12   **Illustrates the use of ZTEST**

Calcium is the most abundant and one of the most important minerals in the body. It works with phosphorus in building and maintaining bones and teeth. According to the Food and Nutrition Board of the National Academy of Sciences, the recommended daily allowance (RDA) of calcium for adults is 800 mg.

A nutritionist thinks that people with incomes below the poverty level average *less* than the RDA of 800 mg. To test her claim, she determines the daily intake of calcium for a random sample of $n = 50$ such people. The results are displayed in Table 9.8. Data are in milligrams.

TABLE 9.8

879	1096	701	986	828	1077	703	633	1119	951
555	422	997	473	702	508	530	688	691	943
513	720	944	673	574	707	864	748	498	881
1199	743	1325	655	1043	599	1008	792	915	456
705	180	287	542	893	1052	473	739	642	915

Use the data to decide whether the mean daily intake of calcium, μ, for people with incomes below the poverty level is *less than* the RDA of 800 mg. Perform the appropriate hypothesis test at the 5% significance level.

SOLUTION
The problem is to perform the left-tailed hypothesis test

$$H_0: \mu = 800 \text{ mg}$$
$$H_a: \mu < 800 \text{ mg}$$

with $\alpha = 0.05$. In Example 9.9 on page 372, we used Procedure 9.2 to perform the hypothesis test. Here we will employ ZTEST to do the hypothesis test.

First we enter the sample data in Table 9.8 into the computer. The program ZTEST requires the user to input the population standard deviation, σ. However, σ is not given here. But since the sample size is large ($n = 50$), we can use the sample standard deviation, s, in place of σ. The sample standard deviation of the data in Table 9.8 is $s = 239.3$ (this can be obtained by using a calculator or by using Minitab).

Now we use the command ZTEST followed by the null hypothesis (MU=800), the estimated value of σ (SIGMA =239.3), and the storage location of the sample data (C6). Directly beneath this we type ALTERNATIVE = −1. This tells the computer that we want to perform a left-tailed test. [For a right-tailed test we would type ALTERNATIVE = 1, and for a two-tailed test we do not need to type anything because the program is already set for a two-tailed test.] The command and its results are shown in Printout 9.1.

PRINTOUT 9.1
Minitab output for ZTEST.

```
MTB > ZTEST, MU = 800, SIGMA = 239.3, C6;
SUBC> ALTERNATIVE = −1.


TEST OF MU = 800.000 VS MU L.T. 800.000
THE ASSUMED SIGMA = 239.3


        N      MEAN      STDEV     SE MEAN         Z    P VALUE
C6     50   755.340   239.332      33.847     −1.32      0.094
```

The output first displays a statement of the null and alternative hypotheses—TEST OF MU = 800.000 VS MU L.T. 800.000 (L.T. stands for "less than"). Next it shows the value given for σ—THE ASSUMED SIGMA =239.3. Then it displays the sample size, sample mean, sample standard deviation, and estimated standard error of the mean. The next to last entry, Z, gives the value of the test statistic

$$z = \frac{\bar{x} - \mu_0}{s/\sqrt{n}} = \frac{755.340 - 800}{239.3/\sqrt{50}}$$

So, $z = -1.32$.

The final entry, P VALUE, is really the only quantity that we need to make our decision concerning the hypothesis test. The **P-value** is the smallest significance level at which we can reject H_0. As the output indicates, the P-value here is 0.094. Since this exceeds the designated significance level of $\alpha = 0.05$, we do not reject H_0. In other words, the data do *not* provide sufficient evidence to conclude that the mean daily intake of calcium for people with incomes below the poverty level is less than the RDA of 800 mg. ∎

We have gone through the computer solution in great detail so that you will understand all of the computer output in Printout 9.1. In practice, however, we just want to know the result of the test and how to interpret it. This is extremely easy to accomplish using the computer. The pertinent facts are summarized below.

DEFINITION 9.5 *P*-value

The **P-value** for a hypothesis test is the smallest significance level at which the null hypothesis can be rejected with the observed sample data.

KEY FACT *Decision rule using P-values*

The decision rule for a hypothesis test using *P*-values is as follows: If the *P*-value is less than the significance level, α, reject H_0; otherwise, do not reject H_0.

In Example 9.12, the computer output shows that the *P*-value is 0.094. Since this is not less than the significance level $\alpha = 0.05$, we do not reject H_0. That's all there is to it.

TTEST

Procedure 9.3 on page 376 provides a hypothesis testing procedure that can be employed when the population being sampled is normally distributed. Minitab's analogue of Procedure 9.3 is a program called **TTEST**. The implementation of TTEST is almost identical to that of ZTEST. We will apply TTEST to perform the hypothesis test in Example 9.10 on page 376, which we did previously by using Procedure 9.3.

EXAMPLE 9.13 **Illustrates the use of TTEST**

According to the U.S. Energy Information Administration, the average residential energy expenditure for 1982 was $1022 per household. That same year, 15 randomly selected upper-income families reported the energy expenditures shown in Table 9.9.

TABLE 9.9
1982 energy expenditures
for 15 randomly selected
upper-income families.

$1153	1249	1126	1053	1689
1514	1420	807	1130	1268
1610	1192	1104	1250	1084

At the 5% significance level, do the data suggest that, in 1982, upper-income families spent *more* on the average for energy than the average of $1022 spent by families in general? [Assume that 1982 energy expenditures for upper-income families are approximately normally distributed.]

SOLUTION
We need to perform the right-tailed hypothesis test

$$H_0: \mu = \$1022$$
$$H_a: \mu > \$1022$$

with $\alpha = 0.05$, where μ is the true 1982 mean energy expenditure for upper-income families.

To apply TTEST we first enter the sample data from Table 9.9 into the computer. Then we use the command TTEST followed by the null hypothesis (MU=1022) and the storage location of the sample data (C7). Directly beneath this we type ALTERNATIVE=1. This tells the computer that we want to perform a right-tailed test. The command and its results are displayed in Printout 9.2.

PRINTOUT 9.2
Minitab output for
TTEST.

```
MTB > TTEST, MU=1022, C7;
SUBC> ALTERNATIVE=1.

TEST OF MU = 1022.0 VS MU G.T. 1022.0

          N      MEAN     STDEV    SE MEAN       T    P VALUE
C7       15    1243.3     231.0       59.6    3.71     0.0012
```

The computer first prints the null and alternative hypotheses—TEST OF MU = 1022.0 VS MU G.T. 1022.0 (G.T. stands for "greater than"). Then it prints the sample size, sample mean, sample standard deviation, and estimated standard error of the mean. The next to last entry, T, gives the value of the test statistic

$$t = \frac{\bar{x} - \mu_0}{s/\sqrt{n}} = \frac{1243.3 - 1022}{231.0/\sqrt{15}}$$

Thus, we see that $t = 3.71$.

The final entry is the P-value, which, as we know, is the only thing we really need in order to make our decision concerning the hypothesis test. Since the P-value, 0.0012, is less than the significance level $\alpha = 0.05$, we reject H_0. In other words, the data indicate that upper-income families spent more on the average for energy in 1982 than the average of \$1022 spent by families in general. ■

◆ Chapter Review

KEY TERMS

alternative hypothesis, 341
critical value, 359
decision rule, 343
hypothesis, 341
hypothesis test, 341
left-tailed test, 365
nonrejection region, 359
null hypothesis, 341
one-tailed test, 365
P-value*, 385
population proportion (p), 379

rejection region, 359
right-tailed test, 365
sample proportion ($\bar{p}$), 379
significance level (α), 346
test statistic, 359
TTEST*, 386
two-category population, 379
two-tailed test, 364
Type I error, 362
Type II error, 362
ZTEST*, 384

FORMULAS

In the formulas below,

μ = population mean
$\bar{x}$ = sample mean
s = sample standard deviation
n = sample size
p = population proportion
$\bar{p}$ = sample proportion

Test statistic for H_0: $\mu = \mu_0$—large sample, 372

$$z = \frac{\bar{x} - \mu_0}{s/\sqrt{n}}$$

Test statistic for H_0: $\mu = \mu_0$—normal population, 376

$$t = \frac{\bar{x} - \mu_0}{s/\sqrt{n}}$$

with df $= n - 1$.

Test statistic for H_0: $p = p_0$—large sample, 381

$$z = \frac{\bar{p} - p_0}{\sqrt{p_0(1 - p_0)/n}}$$

YOU SHOULD BE ABLE TO . . .

1 use and understand the preceding formulas.

2 explain the logic behind hypothesis testing.

3 define and apply the concepts of Type I and Type II errors.

4 perform a large-sample hypothesis test for μ.

5 perform a hypothesis test for μ when sampling from a normal population.

6 perform a large-sample hypothesis test for p.

REVIEW TEST

1 The U.S. Department of Agriculture reports that the average American consumed 20.6 lb of cheese in 1983. There has been a steady increase in cheese consumption since 1960, when the average American ate only 8.3 lb of cheese. A researcher thinks that the trend of increasing cheese consumption is still continuing. He wants to determine whether last year's mean, μ, is greater than the 1983 mean of 20.6 lb. The researcher plans to randomly select 35 people and use their last year's mean cheese consumption, $\bar{x}$, to test the hypotheses

$$H_0: \mu = 20.6 \text{ lb (mean has not increased)}$$
$$H_a: \mu > 20.6 \text{ lb (mean has increased)}$$

a) Suppose that last year's mean cheese consumption, $\bar{x}$, for the 35 people selected turns out to be 23.8 lb. If the null hypothesis $\mu = 20.6$ lb is true,

find the probability of getting a value for $\bar{x}$ of 23.8 lb or greater. [Use $\sigma = 6.9$ lb.]

b) Does a sample mean of $\bar{x} = 23.8$ lb provide evidence that the null hypothesis is false? Explain.

c) Repeat parts (a) and (b) if last year's mean cheese consumption for the 35 people selected turns out to be 21.3 lb.

2 Refer to Problem 1. Suppose the researcher chooses $\bar{x}$ for his test statistic.

a) Find the critical value, c, for a test at the 5% significance level. [Recall that $n = 35$ and $\sigma = 6.9$ lb.]

b) State the decision rule for the hypothesis test with $\alpha = 0.05$. Draw a graph that gives a visual portrayal of the decision rule.

c) Use the decision rule in part (b) to perform the hypothesis test, if last year's cheese consumption for the 35 randomly selected people is as follows:

33	16	20	25	29	23	27
21	20	19	23	9	15	34
13	29	23	31	19	32	11
26	15	23	20	31	20	13
24	16	14	18	23	24	24

[The sum of the data is 763 lb.]

d) State your decision regarding the hypothesis test in words.

3 Refer to Problems 1 and 2. Identify the
 a) test statistic.
 b) rejection region.
 c) nonrejection region.
 d) critical value.

4 Refer to Problems 1 and 2. Explain what each of the following would mean.
 a) A Type I error.
 b) A Type II error.
 c) A correct decision.

Now, recall that the sample of 35 people led us not to reject the null hypothesis (part (c) of Problem 2). Classify that decision by error type or as a correct decision if, in fact, last year's mean consumption of cheese
 d) has not increased over the 1983 mean of 20.6 lb.
 e) has increased over the 1983 mean of 20.6 lb.

Finally, is the hypothesis test that we have been considering
 f) two-tailed, left-tailed, or right-tailed?

5 Between May 17 and 20, 1985, the Gallup Organization conducted a survey to estimate President Ronald Reagan's overall standing with the public. Out of a sample of 1528 adults, 840 approved of the way Ronald Reagan was handling his job as President. Do these results indicate, at the 1% significance level, that a majority of all Americans were satisfied with President Reagan's performance?

6 Each year, manufacturers perform mileage tests on new car models and submit the results to the Environmental Protection Agency (EPA). The EPA then tests the vehicles to determine whether the manufacturers are correct. In 1985, one company reported that a particular model equipped with a four-speed manual transmission averaged 29 miles per gallon (mpg) on the highway. Let us suppose that the EPA tested 15 of the cars and obtained the gas mileages given below.

27.3	31.2	29.4	31.6	28.6
30.9	29.7	28.5	27.8	27.3
25.9	28.8	28.9	27.8	27.6

At the 5% significance level, what decision would you make regarding the company's report on the gas mileage of the car? [Assume that the gas mileages are approximately normally distributed.]

7 The FBI reports that in 1983 the mean value lost due to purse snatching was $178. For this year, 40 randomly selected purse-snatching offenses have a mean value lost of $\bar{x} = \$159$, with a standard deviation of $s = \$84$. Do the data suggest that the mean value lost due to purse snatching has *decreased* from the 1983 mean of $178? [Use $\alpha = 0.05$.]

In Chapters 8 and 9 we examined methods of obtaining confidence intervals and performing hypothesis tests for a single population mean, μ, or a single population proportion, p. Frequently, however, inferential statistics is used to compare the means or proportions of *two or more* populations. For instance, we might perform a hypothesis test to decide whether the mean starting salary of business majors is greater than the mean starting salary of liberal arts majors. Or we might need to find a confidence interval for the difference between the percentages of unemployed white-collar workers and unemployed blue-collar workers. In this chapter we will study methods of making such statistical inferences.

Inferences concerning two or more populations

CHAPTER OUTLINE

10.1 Large-sample inferences for two population means (independent samples)

In this section we will examine methods of making statistical inferences concerning the means of two populations when dealing with large samples. The methods require that the samples selected from the two populations are not only random, but are also *independent*. Two samples are **independent** if the sample selected from one of the populations has no effect on the sample selected from the other population. The following example introduces the basic ideas involved in performing a hypothesis test to compare two population means.

EXAMPLE 10.1 Introduces hypothesis tests for comparing two population means

Suppose we want to perform a hypothesis test in order to determine whether there is a *difference* between the mean salary of faculty teaching in public institutions and the mean salary of faculty teaching in private institutions. We can formulate the problem statistically as follows: First, let the two populations in question be designated as Population 1 and Population 2:

Population 1: All salaries of faculty teaching in public institutions.
Population 2: All salaries of faculty teaching in private institutions.

Next, denote the mean salary of faculty teaching in public institutions by μ_1 and the mean salary of faculty teaching in private institutions by μ_2. Then the hypothesis test that we want to perform can be stated as

$$H_0: \mu_1 = \mu_2 \text{ (mean salaries are the same)}$$
$$H_a: \mu_1 \neq \mu_2 \text{ (mean salaries are different)}$$

Roughly speaking, the hypothesis test can be carried out as follows:

1 Take a random sample from each of the two populations.
2 Compute the mean, $\bar{x}_1$, of the sample from Population 1 and the mean, $\bar{x}_2$, of the sample from Population 2.
3 Reject the null hypothesis if $\bar{x}_1$ and $\bar{x}_2$ differ by too much; otherwise, do not reject the null hypothesis.

This process is pictured in Figure 10.1.

FIGURE 10.1

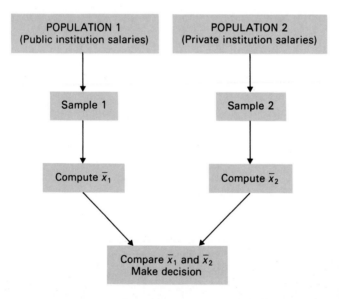

Suppose we obtain a sample of $n_1 = 30$ salaries from Population 1 (public institution salaries) and a sample of $n_2 = 35$ salaries from Population 2 (private institution salaries); and that the results are as depicted in Table 10.1.

TABLE 10.1

Sample 1 (public institutions)					
$23,565	26,497	43,309	41,419	28,958	43,321
21,789	65,433	30,238	20,240	13,227	41,816
19,706	27,131	39,757	21,958	40,748	59,877
31,650	51,693	20,390	27,268	52,503	12,920
26,833	30,755	26,565	45,337	36,400	28,753

Sample 2 (private institutions)						
$24,666	46,108	27,258	75,204	37,502	35,401	23,968
48,015	35,741	47,872	48,341	26,843	33,300	65,725
20,706	27,303	32,691	24,611	32,813	23,486	15,490
39,217	39,494	23,607	31,853	13,211	11,775	17,845
31,687	64,767	36,566	36,633	39,740	44,927	60,198

The sample means for the two data sets in Table 10.1 are

$$\bar{x}_1 = \frac{\Sigma x}{n_1} = \frac{\$1,000,056}{30} = \$33,335.20$$

$$\bar{x}_2 = \frac{\Sigma x}{n_2} = \frac{\$1,244,564}{35} = \$35,558.97$$

The question now is whether the difference of $2223.77 between these two sample means can reasonably be attributed to sampling error, or whether it indicates that the two populations have different means. To answer this question, we need to know how likely it would be to get such a difference in sample means if, in fact, the two population means are equal. In other words, we need to have the probability distribution of the difference, $\bar{x}_1 - \bar{x}_2$, between two sample means— the **sampling distribution of the difference between two means.** Let us therefore examine that sampling distribution. We will complete the solution to the salary problem shortly. ∎

The sampling distribution of the difference between two means—large, independent samples

First we need to become familiar with the notation for parameters and statistics when considering two populations. Let us call the two populations Population 1 and Population 2. Then, as indicated in Example 10.1, we use a subscript "1" when referring to parameters or statistics for Population 1, and a subscript "2" when referring to parameters or statistics for Population 2. This notation is displayed in Table 10.2.

TABLE 10.2
Notation for parameters and statistics when considering two populations.

	Population 1	Population 2
Population mean	μ_1	μ_2
Population std. dev.	σ_1	σ_2
Sample mean	$\bar{x}_1$	$\bar{x}_2$
Sample std. dev.	s_1	s_2
Sample size	n_1	n_2

Next we will obtain formulas that relate the mean and standard deviation of the random variable $\bar{x}_1 - \bar{x}_2$ to the means, μ_1 and μ_2, and standard deviations, σ_1 and σ_2, of the two populations. It can be shown that the mean of the difference of the random variables $\bar{x}_1$ and $\bar{x}_2$ is simply the difference of the means: $\mu_{\bar{x}_1 - \bar{x}_2} = \mu_{\bar{x}_1} - \mu_{\bar{x}_2}$. Additionally, it can be shown that for *independent samples,* the standard deviation of the random variable $\bar{x}_1 - \bar{x}_2$ is related to the standard deviations of the random variables $\bar{x}_1$ and $\bar{x}_2$ by the equation $\sigma_{\bar{x}_1 - \bar{x}_2} = \sqrt{\sigma_{\bar{x}_1}^2 + \sigma_{\bar{x}_2}^2}$. Combining these results with the formulas $\mu_{\bar{x}} = \mu$ and $\sigma_{\bar{x}} = \sigma/\sqrt{n}$ from Section 7.3, we obtain the formulas

$$\mu_{\bar{x}_1 - \bar{x}_2} = \mu_1 - \mu_2$$

and

$$\sigma_{\bar{x}_1 - \bar{x}_2} = \sqrt{(\sigma_1^2/n_1) + (\sigma_2^2/n_2)}$$

These two formulas express the mean and standard deviation of the random variable $\bar{x}_1 - \bar{x}_2$ in terms of the population parameters and the sample sizes. See Exercise 10.15 for details.

Finally, as we learned in Section 7.4, the random variable $\bar{x}$ is approximately normally distributed for large samples. Thus, if both n_1 and n_2 are large, the random variables $\bar{x}_1$ and $\bar{x}_2$ are both approximately normally distributed. From this it can be proved mathematically that, for independent samples, the random variable $\bar{x}_1 - \bar{x}_2$ is also approximately normally distributed. Consequently, we have the following fundamental fact:

KEY FACT The sampling distribution of the difference between two means —large, independent samples

Suppose that a random sample of size $n_1 \geq 30$ is to be taken from a population with mean μ_1 and standard deviation σ_1; and that a random sample of size $n_2 \geq 30$ is to be taken from a population with mean μ_2 and standard deviation σ_2. Further suppose that the two samples are to be selected independently of one another. Then the random variable $\bar{x}_1 - \bar{x}_2$ is approximately *normally distributed* and has mean $\mu_{\bar{x}_1 - \bar{x}_2} = \mu_1 - \mu_2$ and standard deviation $\sigma_{\bar{x}_1 - \bar{x}_2} = \sqrt{(\sigma_1^2/n_1) + (\sigma_2^2/n_2)}$. Thus, the standardized random variable

$$z = \frac{(\bar{x}_1 - \bar{x}_2) - (\mu_1 - \mu_2)}{\sqrt{(\sigma_1^2/n_1) + (\sigma_2^2/n_2)}}$$

has approximately the *standard normal distribution.*

Hypothesis tests for two means —large, independent samples

Now that we have obtained the sampling distribution of the difference between two means when dealing with large and independent samples, we can state a procedure for performing a hypothesis test. The null hypothesis will be

$$H_0: \mu_1 = \mu_2 \text{ (population means are equal)}$$

Note that if the null hypothesis is true, then $\mu_1 - \mu_2 = 0$, and so the standardized random variable

$$z = \frac{(\bar{x}_1 - \bar{x}_2) - (\mu_1 - \mu_2)}{\sqrt{(\sigma_1^2/n_1) + (\sigma_2^2/n_2)}}$$

becomes simply

(1) $$z = \frac{\bar{x}_1 - \bar{x}_2}{\sqrt{(\sigma_1^2/n_1) + (\sigma_2^2/n_2)}}$$

This last random variable can serve as the test statistic for comparing two means in the same way as the random variable

$$z = \frac{\bar{x} - \mu_0}{\sigma/\sqrt{n}}$$

does for tests concerning a single mean. The *numerator* of the test statistic for two means, given in Equation (1), measures the difference between the sample means. The *denominator* standardizes the test statistic so that we can use the standard normal table to obtain the critical value(s).

If the null hypothesis H_0: $\mu_1 = \mu_2$ is true, then we would expect the sample means $\bar{x}_1$ and $\bar{x}_2$ to be roughly equal (why?). Thus, for a two-tailed test, we will reject the null hypothesis if $\bar{x}_1$ and $\bar{x}_2$ differ by too much, that is, if $\bar{x}_1 - \bar{x}_2$ is too far away from zero. Since $\bar{x}_1 - \bar{x}_2$ is just the numerator of our test statistic, we see that, for a two-tailed test, values of z too far away from zero lead to rejection of the null hypothesis. In other words, the rejection region will consist of two "tails" of the standard normal distribution, just as in two-tailed tests involving a single mean. Similar reasoning applies to one-tailed tests. Consequently, we have the following procedure for performing hypothesis tests to compare two population means.

PROCEDURE 10.1 **To perform a hypothesis test for two population means, with null hypothesis H_0: $\mu_1 = \mu_2$.**

Assumptions
1 Independent samples.
2 Sample sizes are at least 30 ($n_1 \geq 30$, $n_2 \geq 30$).

STEP 1 *State the null and alternative hypotheses.*
STEP 2 *Decide on the significance level, α.*
STEP 3 *The critical value(s)*
   a) *for a two-tailed test are $\pm z_{\alpha/2}$.*
   b) *for a left-tailed test is $-z_\alpha$.*
   c) *for a right-tailed test is z_α.*
Use Table II to find the critical value(s).

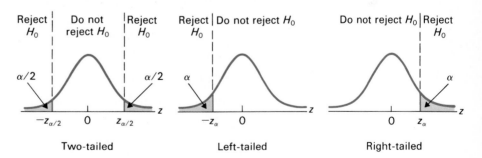

| Two-tailed | Left-tailed | Right-tailed |

STEP 4 *Compute the value of the test statistic*

$$z = \frac{\bar{x}_1 - \bar{x}_2}{\sqrt{(s_1^2/n_1) + (s_2^2/n_2)}}$$

STEP 5 *If the value of the test statistic falls in the rejection region, reject H_0; otherwise, do not reject H_0.*
STEP 6 *State the conclusion in words.*

Note: In Step 4 of Procedure 10.1 we have used the sample standard deviations, s_1 and s_2, in the denominator of the test statistic. Theoretically, we should use the population standard deviations, σ_1 and σ_2. However, σ_1 and σ_2 are rarely known and so we have used s_1 and s_2 in their place. This is acceptable because for large samples the sample standard deviations are likely to be a good approximation of the population standard deviations. In the rare case where the population standard deviations are known, they should always be used in place of the sample standard deviations in Step 4 of Procedure 10.1.

EXAMPLE 10.2   **Illustrates Procedure 10.1**

We now return to the salary problem posed in Example 10.1. Recall that we want to perform a hypothesis test to determine whether there is a difference between the mean salaries of faculty teaching in public and private institutions. Random samples of $n_1 = 30$ faculty teaching in public institutions and $n_2 = 35$ faculty teaching in private institutions yield the data in Table 10.1 on page 392. Do the data suggest, at the 5% significance level, that there is a *difference* in mean salaries for faculty teaching in public and private institutions?

S O L U T I O N
First note that both sample sizes are at least 30—$n_1 = 30$ and $n_2 = 35$. Next note that the description of the problem implies that the samples are independent. Thus, we can employ Procedure 10.1 to perform the hypothesis test.

STEP 1 *State the null and alternative hypotheses.*

The null and alternative hypotheses are

$$H_0: \mu_1 = \mu_2 \text{ (mean salaries are the same)}$$
$$H_a: \mu_1 \neq \mu_2 \text{ (mean salaries are different)}$$

where μ_1 and μ_2 are the mean salaries for faculty in public and private institutions, respectively. Note that the test is two-tailed since there is a not-equal sign ($\neq$) in the alternative hypothesis.

STEP 2 *Decide on the significance level, α.*

We are to perform the hypothesis test at the 5% significance level. So, $\alpha = 0.05$.

STEP 3 *The critical values for a two-tailed test are $\pm z_{\alpha/2}$.*

Since $\alpha = 0.05$, the critical values are

$$\pm z_{0.05/2} = \pm z_{0.025} = \pm 1.96$$

See Figure 10.2 at the top of the following page.

STEP 4 *Compute the value of the test statistic*

$$z = \frac{\overline{x}_1 - \overline{x}_2}{\sqrt{(s_1^2/n_1) + (s_2^2/n_2)}}$$

FIGURE 10.2

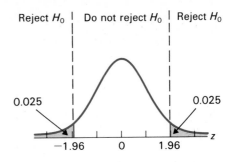

We have $n_1 = 30$ and $n_2 = 35$. To compute the sample mean and sample standard deviation for each of the two data sets in Table 10.1, we proceed in the usual way:

$$\bar{x}_1 = \frac{\Sigma x}{n_1} = \$33,335.20, \quad s_1 = \sqrt{\frac{n_1(\Sigma x^2) - (\Sigma x)^2}{n_1(n_1 - 1)}} = \$13,129.09$$

and

$$\bar{x}_2 = \frac{\Sigma x}{n_2} = \$35,558.97, \quad s_2 = \sqrt{\frac{n_2(\Sigma x^2) - (\Sigma x)^2}{n_2(n_2 - 1)}} = \$14,940.88$$

Thus, the value of the test statistic is

$$z = \frac{\bar{x}_1 - \bar{x}_2}{\sqrt{(s_1^2/n_1) + (s_2^2/n_2)}} = \frac{33,335.20 - 35,558.97}{\sqrt{(13,129.09^2/30) + (14,940.88^2/35)}} = -0.64$$

STEP 5 *If the value of the test statistic falls in the rejection region, reject H_0; otherwise, do not reject H_0.*

From Step 4 we see that the value of the test statistic is $z = -0.64$, which does not fall in the rejection region. Hence we do not reject H_0.

STEP 6 *State the conclusion in words.*

Based on the sample data, we have insufficient evidence to conclude that there is a difference in mean salaries for faculty in public and private institutions.   ■

MTB
SPSS

Confidence intervals for the difference between two means—large, independent samples

We can also use the Key Fact on page 394 to derive the following confidence-interval procedure for the difference between two means when dealing with large, independent samples. See Exercise 10.16 for details of the derivation.

PROCEDURE 10.2 To find a confidence interval for the difference between two population means.

Assumptions
1 Independent samples.
2 Sample sizes are at least 30 ($n_1 \geq 30$, $n_2 \geq 30$).

STEP 1 *For a confidence level of $1 - \alpha$, use Table II to find $z_{\alpha/2}$.*
STEP 2 *The endpoints of the confidence interval for $\mu_1 - \mu_2$ are*

$$(\bar{x}_1 - \bar{x}_2) \pm z_{\alpha/2} \cdot \sqrt{(s_1^2/n_1) + (s_2^2/n_2)}$$

EXAMPLE 10.3 **Illustrates Procedure 10.2**

Consider once more the situation of Example 10.1. Find a 95% confidence interval for the difference, $\mu_1 - \mu_2$, between the mean salaries of faculty teaching in public and private institutions.

SOLUTION
We apply Procedure 10.2.

STEP 1 *For a confidence level of $1 - \alpha$, use Table II to find $z_{\alpha/2}$.*

For a 95% confidence interval, the confidence level is $0.95 = 1 - 0.05$. So, $\alpha = 0.05$. From Table II we find that

$$z_{\alpha/2} = z_{0.05/2} = z_{0.025} = 1.96$$

STEP 2 *The endpoints of the confidence interval for $\mu_1 - \mu_2$ are*

$$(\bar{x}_1 - \bar{x}_2) \pm z_{\alpha/2} \cdot \sqrt{(s_1^2/n_1) + (s_2^2/n_2)}$$

From Step 1, $z_{\alpha/2} = 1.96$. Also, $n_1 = 30$, $n_2 = 35$, and from Example 10.2 (page 397) we know that $\bar{x}_1 = \$33,335.20$, $s_1 = \$13,129.09$ and $\bar{x}_2 = \$35,558.97$, $s_2 = \$14,940.88$. Consequently, the endpoints of the confidence interval for $\mu_1 - \mu_2$ are

$$(33,335.20 - 35,558.97) \pm 1.96 \cdot \sqrt{(13,129.09^2/30) + (14,940.88^2/35)}$$

or

$$-2223.77 \pm 6824.56$$

Thus, a 95% confidence interval for $\mu_1 - \mu_2$ is from

$$-9048.33 \quad \text{to} \quad 4600.79$$

MTB
SPSS

We can be 95% confident that the difference, $\mu_1 - \mu_2$, between the mean salaries of faculty teaching in public institutions and faculty teaching in private institutions is somewhere between $-\$9,048.33$ and $\$4,600.79$. ∎

Exercises 10.1

10.1 Consider the quantities $\mu_1, \sigma_1, \bar{x}_1, s_1, \mu_2, \sigma_2, \bar{x}_2, s_2$.
   a) Which quantities represent parameters and which represent statistics?
   b) Which quantities are fixed numbers and which are random variables?

In Exercises 10.2 through 10.7 use Procedure 10.1 to perform each of the hypothesis tests.

10.2 The National Education Association does surveys on starting salaries of college graduates. Independent random samples of $n_1 = 32$ accounting graduates and $n_2 = 35$ liberal arts graduates yield the following starting salaries. Data are in thousands of dollars, rounded to the nearest hundred dollars.

Accounting				Liberal arts				
24.9	20.4	23.3	20.9	20.0	16.8	21.3	20.6	21.0
23.4	25.3	26.8	22.6	18.7	23.8	18.3	17.7	18.0
24.0	22.0	21.7	23.0	19.1	21.1	19.6	19.0	18.7
26.2	25.9	24.8	25.3	20.8	20.4	17.1	19.5	
25.5	20.4	22.4	24.5	19.9	19.2	19.1	17.2	
24.9	25.4	25.8	24.2	18.3	22.7	20.0	19.2	
22.2	22.1	21.0	24.7	20.9	19.1	20.8	20.5	
21.5	22.5	25.4	26.6	21.4	16.3	16.3	20.5	
(*Note:* $\Sigma x = 759.6$)				(*Note:* $\Sigma x = 682.9$)				

At the 5% significance level, does it appear that accounting graduates have a *higher* mean starting salary than liberal arts graduates? [Assume $\sigma_1 = 1.73$ and $\sigma_2 = 1.82$.]

10.3 The U.S. National Center for Health Statistics collects data on the length of stay by patients in short-term hospitals. Independent random samples of $n_1 = 40$ male patients and $n_2 = 35$ female patients gave the following data on length of stay. The data are in days.

Male								Female						
4	4	12	18	9	6	12	10	14	7	15	1	12	1	3
3	6	15	7	3	55	1	2	7	21	4	1	5	4	4
10	13	5	7	1	23	9	2	3	5	18	12	5	1	7
1	17	2	24	11	14	6	2	7	2	15	4	9	10	7
1	8	1	3	19	3	1	13	3	6	5	9	6	2	14
(*Note:* $\Sigma x = 363$)								(*Note:* $\Sigma x = 249$)						

At the 10% significance level, do the data suggest that, on the average, males stay in the hospital *longer* than females? [Assume $\sigma_1 = 7.5$ days and $\sigma_2 = 6.8$ days.]

10.4 An agronomist wants to know whether a larger corn crop can be obtained if sterilized males of an insect pest are introduced to control the pest population instead of using insecticides. The insecticide is used on 40 randomly selected one-acre plots of corn, and the sterilized male insects on another 40 randomly selected one-acre plots of corn. The yields, in bushels, are as follows:

Insecticide					Sterilized males				
109	101	97	89	100	105	109	110	118	109
98	98	94	99	104	113	111	111	99	112
103	88	108	102	106	106	117	99	107	119
97	105	102	104	101	110	111	103	110	108
101	100	105	110	96	104	102	111	114	114
102	95	100	95	109	122	117	101	109	109
91	98	113	91	95	102	109	103	109	106
106	98	101	99	96	107	107	111	128	109
$\begin{pmatrix} \Sigma x = 4006 \\ \Sigma x^2 = 402,480 \end{pmatrix}$					$\begin{pmatrix} \Sigma x = 4381 \\ \Sigma x^2 = 481,261 \end{pmatrix}$				

Do the data provide evidence that the use of sterilized male insects is *more effective* than insecticides in controlling the insect pest? Use $\alpha = 0.01$.

10.5 The U.S. Energy Information Administration publishes figures on residential energy consumption and expenditures. Suppose you want to decide whether last year's mean annual fuel expenditure for households using natural gas is *different* from that for households using only electricity. What conclusion would you draw given the data below? Use $\alpha = 0.05$.

Natural gas			Electricity			
2002	1456	1394	1376	1452	1235	1480
1541	1321	1338	1185	1327	1059	1400
1495	1526	1358	1227	1102	1168	1070
1801	1478	1376	1180	1221	1351	1014
1579	1375	1664	1461	1102	976	1394
1305	1458	1369	1379	987	1002	1532
1495	1507	1636	1450	1177	1150	
1698	1249	1377	1352	1266	1109	
1648	1557	1491	949	1351	1259	
1505	1355	1574	1179	1393	1456	
$\begin{pmatrix} \Sigma x = 44,928 \\ \Sigma x^2 = 68,029,844 \end{pmatrix}$			$\begin{pmatrix} \Sigma x = 44,771 \\ \Sigma x^2 = 56,633,389 \end{pmatrix}$			

10.6   Researchers in obesity wanted to test the effectiveness of dieting with exercise against dieting without exercise. Seventy-three patients were randomly divided into two groups. Group 1, numbering 37 patients, was put on a program of dieting with exercise. Group 2, numbering 36 patients, dieted only. The results for weight loss after two months are summarized here.

Diet with exercise group	Diet only group
$\bar{x}_1 = 16.8$ lb	$\bar{x}_2 = 17.1$ lb
$s_1 = 3.5$ lb	$s_2 = 5.2$ lb

Determine, at the 0.05 significance level, whether there is a *difference* between the two treatments.

10.7   A regional sales manager chooses two similar offices to study the effectiveness of a new training program aimed at increasing sales. One office institutes the training program and the other does not. The office that does not, Office 1, has $n_1 = 47$ sales people. The mean sales per person over the next month turns out to be $\bar{x}_1 = \$3,197$, with a standard deviation of $s_1 = \$102$. Office 2, the office that does institute the training program, has $n_2 = 51$ sales people; and the mean sales per person over the next month turns out to be $\bar{x}_2 = \$3,229$, with a standard deviation of $s_2 = \$107$.
a) At the 5% significance level, does the training program appear to increase sales?
b) Repeat part (a) at the 10% significance level.

In Exercises 10.8 through 10.13 use Procedure 10.2 to determine the required confidence interval.

10.8   Refer to Exercise 10.2.
a) Determine a 90% confidence interval for the difference, $\mu_1 - \mu_2$, between the mean starting salaries of accounting and liberal arts graduates.
b) Interpret your results in words.

10.9   Refer to Exercise 10.3.
a) Determine an 80% confidence interval for the difference, $\mu_1 - \mu_2$, between the mean lengths of stay in short-term hospitals by males and females.
b) Interpret your results in words.

10.10   Refer to Exercise 10.4.
a) Determine a 98% confidence interval for the difference, $\mu_1 - \mu_2$, between the mean yields of corn when the insecticide is used to control the insect pest and when sterilized males are used.
b) Interpret your results in words.

10.11   Refer to Exercise 10.5.
a) Find a 95% confidence interval for the difference, $\mu_1 - \mu_2$, between last year's mean fuel expenditures for households using natural gas and those using only electricity.
b) Interpret your results in words.

10.12   Refer to Exercise 10.6.
a) Find a 95% confidence interval for the difference, $\mu_1 - \mu_2$, between the mean weight losses after two months for those using the diet-with-exercise method and those using the diet-only method.
b) Interpret your results in words.

10.13   Refer to Exercise 10.7.
a) Determine a 90% confidence interval for the difference, $\mu_1 - \mu_2$, between the mean sales per person per month for those that do not take the training program and those that do.
b) Repeat part (a) using an 80% confidence level.

10.14   Use your results from Exercises 10.5, 10.6, 10.11, and 10.12 to complete the following statement: A hypothesis test of the form $H_0: \mu_1 = \mu_2$ versus $H_a: \mu_1 \neq \mu_2$, at the significance level α, will lead to rejection of the null hypothesis if and only if the number _____ does not lie in the $(1 - \alpha)$-level confidence interval for $\mu_1 - \mu_2$.

10.15   In this exercise we will derive the formulas for the mean and standard deviation of the random variable $\bar{x}_1 - \bar{x}_2$.
a) Use the results of Exercises 5.26(e) and 5.27(f) on page 190 to show that

$$\mu_{\bar{x}_1 - \bar{x}_2} = \mu_{\bar{x}_1} - \mu_{\bar{x}_2}$$

and

$$\sigma_{\bar{x}_1 - \bar{x}_2} = \sqrt{\sigma_{\bar{x}_1}^2 + \sigma_{\bar{x}_2}^2}$$

b) Apply the formulas $\mu_{\bar{x}} = \mu$ and $\sigma_{\bar{x}} = \sigma/\sqrt{n}$ to the results in part (a) to derive the formulas

$$\mu_{\bar{x}_1 - \bar{x}_2} = \mu_1 - \mu_2$$

and

$$\sigma_{\bar{x}_1 - \bar{x}_2} = \sqrt{(\sigma_1^2/n_1) + (\sigma_2^2/n_2)}$$

10.16   This exercise justifies Procedures 10.1 and 10.2. Suppose that independent random samples of sizes $n_1 \geq 30$ and $n_2 \geq 30$ are to be taken from populations with means μ_1 and μ_2 and standard deviations σ_1 and σ_2. Assume as known that the random variable $\bar{x}_1 - \bar{x}_2$ is approximately normally distributed.

a) Use the results of Exercise 10.15(b) to show that the random variable

$$z = \frac{(\bar{x}_1 - \bar{x}_2) - (\mu_1 - \mu_2)}{\sqrt{(\sigma_1^2/n_1) + (\sigma_2^2/n_2)}}$$

has approximately the standard normal distribution.

b) Deduce from part (a) that the random variable

$$\frac{(\bar{x}_1 - \bar{x}_2) - (\mu_1 - \mu_2)}{\sqrt{(s_1^2/n_1) + (s_2^2/n_2)}}$$

also has approximately the standard normal distribution. [This justifies Procedure 10.1.]

c) Use part (b) to show that

$$P\left(-z_{a/2} \leq \frac{(\bar{x}_1 - \bar{x}_2) - (\mu_1 - \mu_2)}{\sqrt{(s_1^2/n_1) + (s_2^2/n_2)}} \leq z_{a/2}\right) \approx 1 - \alpha$$

d) Use part (c) to verify Procedure 10.2.

10.2   Small-sample inferences for two population means (independent samples)

In the previous section we examined inferential methods for comparing the means of two populations when dealing with large and independent samples. This section is concerned with those same statistical inferences, but in the case where the sample sizes may be small. As in our previous small-sample considerations, we will assume that the populations being sampled are approximately *normally distributed*. We will also assume that the population standard deviations are unknown since, in practice, that is usually the case.

We know that when sampling from a normally distributed population, the random variable $\bar{x}$ is normally distributed, regardless of the sample size n. From this it can be proved that, when independent sampling is done from two normally distributed populations, the random variable $\bar{x}_1 - \bar{x}_2$ is also normally distributed, regardless of the sample sizes n_1 and n_2. Combining this fact with the formulas given on page 393 for the mean and standard deviation of $\bar{x}_1 - \bar{x}_2$, we obtain the following result:

KEY FACT   The sampling distribution of the difference between two means —normal populations and independent samples

Suppose that a random sample of size n_1 is to be taken from a normally distributed population with mean μ_1 and standard deviation σ_1; and that a random sample of size n_2 is to be taken from a normally distributed population with mean μ_2 and standard deviation σ_2. Further suppose that the two samples are to be selected independently. Then the random variable $\bar{x}_1 - \bar{x}_2$ is also *normally distributed* and has mean $\mu_{\bar{x}_1 - \bar{x}_2} = \mu_1 - \mu_2$ and standard deviation $\sigma_{\bar{x}_1 - \bar{x}_2} = \sqrt{(\sigma_1^2/n_1) + (\sigma_2^2/n_2)}$. Thus the standardized random variable

$$z = \frac{(\bar{x}_1 - \bar{x}_2) - (\mu_1 - \mu_2)}{\sqrt{(\sigma_1^2/n_1) + (\sigma_2^2/n_2)}}$$

has the *standard normal distribution* (irrespective of sample sizes).

When dealing with small samples and the population standard deviations are unknown, it is necessary to consider two cases. One case is where the population standard deviations are required to be equal—$\sigma_1 = \sigma_2$; and the other case is where they are not required to be equal, although they may in fact be equal.

Hypothesis tests for two means—normal populations, independent samples, and σs unknown but assumed equal

We will now develop a hypothesis testing procedure for comparing two population means when the populations are normally distributed and have equal, but unknown, standard deviations. Our immediate goal is to find a test statistic that can be used to perform such a hypothesis test.

Let us use σ to denote the common standard deviation of the two populations. As we have seen, when dealing with independent samples from normal populations, the random variable

$$z = \frac{(\bar{x}_1 - \bar{x}_2) - (\mu_1 - \mu_2)}{\sqrt{(\sigma_1^2 / n_1) + (\sigma_2^2 / n_2)}}$$

has the standard normal distribution. If we replace σ_1 and σ_2 in the above expression by their common value σ, we can use algebra to obtain the relationship

$$(2) \qquad z = \frac{(\bar{x}_1 - \bar{x}_2) - (\mu_1 - \mu_2)}{\sigma \sqrt{(1/n_1) + (1/n_2)}}$$

Of course, this random variable cannot be used as our test statistic since σ is unknown. Consequently, we need to use sample information to estimate the unknown population standard deviation σ, or equivalently, the population variance σ^2. The best way to do this is to use the sample variances s_1^2 and s_2^2 to get two estimates of σ^2 and then **pool** these estimates by weighting them according to sample size (actually by degrees of freedom). Thus, our estimate for the common population variance, σ^2, is

$$s_p^2 = \frac{(n_1 - 1)s_1^2 + (n_2 - 1)s_2^2}{n_1 + n_2 - 2}$$

and, hence, for the common population standard deviation, σ, is

$$s_p = \sqrt{\frac{(n_1 - 1)s_1^2 + (n_2 - 1)s_2^2}{n_1 + n_2 - 2}}$$

The subscript "p" stands for "pooled," and the quantity s_p is called the **pooled sample standard deviation.** In summary, we use the pooled sample standard deviation, s_p, as our estimate for the common, but unknown standard deviation, σ, of the two populations.

Replacing the σ in Equation (2) by its estimate s_p, we get the random variable

$$\frac{(\bar{x}_1 - \bar{x}_2) - (\mu_1 - \mu_2)}{s_p \sqrt{(1/n_1) + (1/n_2)}}$$

which can be used as our test statistic. However, unlike the random variable in Equation (2), this random variable does not have the standard normal distribution. But, its distribution is a familiar one—the t-distribution.

KEY FACT

Suppose that independent random samples of sizes n_1 and n_2 are to be taken from two *normally distributed* populations with means μ_1 and μ_2, respectively. Further suppose that the standard deviations of the two populations are equal. Then the random variable

$$t = \frac{(\bar{x}_1 - \bar{x}_2) - (\mu_1 - \mu_2)}{s_p\sqrt{(1/n_1) + (1/n_2)}}$$

has the *t-distribution with df* $= n_1 + n_2 - 2$.

As before, the null hypothesis for our tests of two means will be $H_0: \mu_1 = \mu_2$. Thus, if the null hypothesis is true, then $\mu_1 - \mu_2 = 0$, and our test statistic becomes simply

$$t = \frac{\bar{x}_1 - \bar{x}_2}{s_p\sqrt{(1/n_1) + (1/n_2)}}$$

Consequently, we have the following procedure for performing hypothesis tests to compare two population means, when the populations being sampled are normally distributed and have equal, but unknown, standard deviations.

PROCEDURE 10.3 **To perform a hypothesis test for two population means, with null hypothesis $H_0: \mu_1 = \mu_2$.**

Assumptions
1 Independent samples.
2 Normal populations.
3 Equal, but unknown, population standard deviations.

STEP 1 *State the null and alternative hypotheses.*
STEP 2 *Decide on the significance level, α.*
STEP 3 *The critical value(s)*
   a) *for a two-tailed test are $\pm t_{\alpha/2}$,*
   b) *for a left-tailed test is $-t_{\alpha}$,*
   c) *for a right-tailed test is t_{α},*
with df $= n_1 + n_2 - 2$. Use Table III to find the critical value(s).

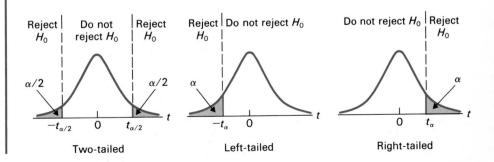

Two-tailed   Left-tailed   Right-tailed

STEP 4 *Compute the value of the test statistic*

$$t = \frac{\bar{x}_1 - \bar{x}_2}{s_p\sqrt{(1/n_1) + (1/n_2)}}$$

where

$$s_p = \sqrt{\frac{(n_1 - 1)s_1^2 + (n_2 - 1)s_2^2}{n_1 + n_2 - 2}}$$

STEP 5 *If the value of the test statistic falls in the rejection region, reject H_0; otherwise, do not reject H_0.*

STEP 6 *State the conclusion in words.*

Note: In Step 4 we need to calculate the pooled sample standard deviation, s_p. The pooled sample standard deviation always lies between the two sample standard deviations, s_1 and s_2. If you calculate s_p and it does not lie between s_1 and s_2, then you made an error.

EXAMPLE 10.4   Illustrates Procedure 10.3

The U.S. National Center for Health Statistics gathers and publishes data on the daily intake of selected nutrients by race and income level. Suppose we are considering protein intake and want to compare the mean daily intake of people with incomes above the poverty level to that of people with incomes below the poverty level. The data in Table 10.3 give the protein intakes, in grams, over a 24-hour period of independent random samples of people with incomes above and below the poverty level.

TABLE 10.3

Above poverty level		Below poverty level		
86.0	69.0	51.4	49.7	72.0
59.7	80.2	76.7	65.8	55.0
68.6	78.1	73.7	62.1	79.7
98.6	69.8	66.2	75.8	65.4
87.7	77.2	65.5	62.0	73.3

At the 5% significance level, do the data suggest that people with incomes above the poverty level have a *greater* mean daily intake of protein than those with incomes below the poverty level? [Assume that daily intakes of protein for both people with incomes above and below the poverty level are approximately normally distributed and that the standard deviations for both are equal.]

SOLUTION
We apply Procedure 10.3.

STEP 1 *State the null and alternative hypotheses.*

If we let μ_1 denote the true mean daily intake of protein for people with incomes above the poverty level and μ_2 denote the true mean daily intake of protein for people with incomes below the poverty level, then the null and alternative hypotheses are

$$H_0: \mu_1 = \mu_2 \text{ (above-poverty mean is not greater)}$$
$$H_a: \mu_1 > \mu_2 \text{ (above-poverty mean is greater)}$$

Note that the test is right-tailed since there is a greater-than sign ($>$) in the alternative hypothesis.

STEP 2 *Decide on the significance level, α.*

The test is to be done at the 5% significance level, so $\alpha = 0.05$.

STEP 3 *The critical value for a right-tailed test is t_α with df $= n_1 + n_2 - 2$.*

From Step 2, $\alpha = 0.05$. Also $n_1 = 10$ and $n_2 = 15$, so that df $= 10 + 15 - 2 = 23$. Consulting Table III, we find that the critical value is

$$t_{0.05} = 1.714$$

See Figure 10.3.

FIGURE 10.3

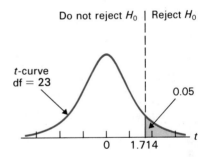

STEP 4 *Compute the value of the test statistic*

$$t = \frac{\bar{x}_1 - \bar{x}_2}{s_p\sqrt{(1/n_1) + (1/n_2)}}$$

where

$$s_p = \sqrt{\frac{(n_1 - 1)s_1^2 + (n_2 - 1)s_2^2}{n_1 + n_2 - 2}}$$

We first determine the pooled sample standard deviation, s_p. The standard deviations s_1 and s_2 of the two data sets in Table 10.3 are calculated in the usual way. We find that $s_1 = 11.34$ g and $s_2 = 9.17$ g. Consequently, since $n_1 = 10$ and $n_2 = 15$,

$$s_p = \sqrt{\frac{(10 - 1) \cdot 11.34^2 + (15 - 1) \cdot 9.17^2)}{10 + 15 - 2}} = 10.07 \text{ g}$$

The sample means of the two data sets in Table 10.3 are $\bar{x}_1 = 77.49$ g and $\bar{x}_2 = 66.29$ g. Therefore, the value of the test statistic is

$$t = \frac{\bar{x}_1 - \bar{x}_2}{s_p\sqrt{(1/n_1) + (1/n_2)}} = \frac{77.49 - 66.29}{10.07\sqrt{(1/10) + (1/15)}} = 2.724$$

STEP 5 *If the value of the test statistic falls in the rejection region, reject H_0; otherwise, do not reject H_0.*

From Step 4, the value of the test statistic is $t = 2.724$, which falls in the rejection region (see Figure 10.3). Thus, we reject H_0.

STEP 6 *State the conclusion in words.*

Evidently, people with incomes above the poverty level have a greater mean daily intake of protein than people with incomes below the poverty level. ■

MTB
SPSS

Confidence intervals for the difference between two means —normal populations, independent samples, and σs unknown but assumed equal

We can also use the Key Fact on page 403 to derive the following confidence-interval procedure for the difference between two means. Note carefully the conditions required for using this procedure.

PROCEDURE 10.4 To find a confidence interval for the difference between two population means.

Assumptions
1 Independent samples.
2 Normal populations.
3 Equal, but unknown, population standard deviations.

STEP 1 *For a confidence level of $1 - \alpha$, use Table III to find $t_{\alpha/2}$ for df $= n_1 + n_2 - 2$.*
STEP 2 *The endpoints of the confidence interval for $\mu_1 - \mu_2$ are*

$$(\bar{x}_1 - \bar{x}_2) \pm t_{\alpha/2} \cdot s_p \sqrt{(1/n_1) + (1/n_2)}$$

EXAMPLE   10.5   Illustrates Procedure 10.4

Consider again the situation of Example 10.4 on page 404. Use the sample data in Table 10.3 to determine a 95% confidence interval for the difference, $\mu_1 - \mu_2$, between the mean daily intakes of protein for people with incomes above and below the poverty level.

SOLUTION
We apply Procedure 10.4.

STEP 1 *For a confidence level of $1 - \alpha$, use Table III to find $t_{\alpha/2}$ for df $= n_1 + n_2 - 2$.*

For a 95% confidence interval, the confidence level is $0.95 = 1 - 0.05$. So, $\alpha = 0.05$. From Table III we find that for $df = n_1 + n_2 - 2 = 10 + 15 - 2 = 23$,

$$t_{\alpha/2} = t_{0.05/2} = t_{0.025} = 2.069$$

STEP 2 *The endpoints of the confidence interval for* $\mu_1 - \mu_2$ *are*

$$(\bar{x}_1 - \bar{x}_2) \pm t_{\alpha/2} \cdot s_p \sqrt{(1/n_1) + (1/n_2)}$$

From Step 1, $t_{\alpha/2} = 2.069$. Also, $n_1 = 10$, $n_2 = 15$, and from Example 10.4 (page 404) we know that $\bar{x}_1 = 77.49$ g, $\bar{x}_2 = 66.29$ g, and $s_p = 10.07$ g. Consequently, the endpoints of the confidence interval for $\mu_1 - \mu_2$ are

$$(77.49 - 66.29) \pm 2.069 \cdot 10.07\sqrt{(1/10) + (1/15)}$$

or

$$11.20 \pm 8.51$$

Thus, a 95% confidence interval for $\mu_1 - \mu_2$ is from

$$2.69 \quad \text{to} \quad 19.71$$

We can be 95% confident that the difference, $\mu_1 - \mu_2$, between the mean daily intakes of protein for people with incomes above and below the poverty level is somewhere between 2.69 and 19.71 grams. ∎

MTB
SPSS

Hypothesis tests for two means—normal populations, independent samples, and σs unknown but not assumed equal

Procedure 10.3 enables us to perform a hypothesis test to compare two population means using independent samples when the populations are normal and have equal, but unknown, standard deviations. What if the two populations do not have equal standard deviations, or if we simply do not know whether they are equal? We need to develop a statistical inference procedure for comparing two population means that works for small samples but does not require the populations to have equal standard deviations.

As before, we begin with the fact that for independent samples from normal populations, the random variable

$$z = \frac{(\bar{x}_1 - \bar{x}_2) - (\mu_1 - \mu_2)}{\sqrt{(\sigma_1^2/n_1) + (\sigma_2^2/n_2)}}$$

has the standard normal distribution. Since we are assuming that the population standard deviations σ_1 and σ_2 are unknown, we cannot use the above random variable as our test statistic. We therefore replace σ_1 and σ_2 by their corresponding sample estimates, s_1 and s_2, and obtain the random variable

$$\frac{(\bar{x}_1 - \bar{x}_2) - (\mu_1 - \mu_2)}{\sqrt{(s_1^2/n_1) + (s_2^2/n_2)}}$$

which *can* be used as our test statistic. This random variable does not have the standard normal distribution, but has a *t*-distribution.

KEY FACT

Suppose that independent random samples of sizes n_1 and n_2 are to be taken from two *normally distributed* populations with means μ_1 and μ_2, respectively. Then the random variable

$$t = \frac{(\bar{x}_1 - \bar{x}_2) - (\mu_1 - \mu_2)}{\sqrt{(s_1^2/n_1) + (s_2^2/n_2)}}$$

has the *t-distribution* with degrees of freedom given by

$$df = \frac{[(s_1^2/n_1) + (s_2^2/n_2)]^2}{\frac{(s_1^2/n_1)^2}{n_1 - 1} + \frac{(s_2^2/n_2)^2}{n_2 - 1}}$$

rounded down to the nearest integer.

For a hypothesis test with null hypothesis H_0: $\mu_1 = \mu_2$, our test statistic will therefore be

$$t = \frac{\bar{x}_1 - \bar{x}_2}{\sqrt{(s_1^2/n_1) + (s_2^2/n_2)}}$$

Thus, we have the following procedure for performing hypothesis tests to compare two population means when the populations being sampled are normally distributed.

PROCEDURE 10.5 **To perform a hypothesis test for two population means, with null hypothesis H_0: $\mu_1 = \mu_2$.**

Assumptions
1 Independent samples.
2 Normal populations.

STEP 1 *State the null and alternative hypotheses.*
STEP 2 *Decide on the significance level, α.*
STEP 3 *The critical value(s)*
   a) *for a two-tailed test are* $\pm t_{\alpha/2}$,
   b) *for a left-tailed test is* $-t_\alpha$,
   c) *for a right-tailed test is* t_α,
with degrees of freedom given by

$$df = \frac{[(s_1^2/n_1) + (s_2^2/n_2)]^2}{\frac{(s_1^2/n_1)^2}{n_1 - 1} + \frac{(s_2^2/n_2)^2}{n_2 - 1}}$$

rounded down to the nearest integer. Use Table III to find the critical value(s).
STEP 4 *Compute the value of the test statistic*

$$t = \frac{\bar{x}_1 - \bar{x}_2}{\sqrt{(s_1^2/n_1) + (s_2^2/n_2)}}$$

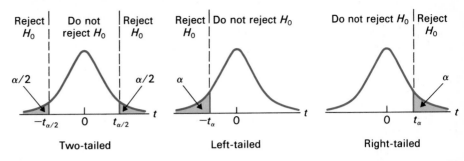

STEP 5 *If the value of the test statistic falls in the rejection region, reject H_0; otherwise, do not reject H_0.*
STEP 6 *State the conclusion in words.*

EXAMPLE 10.6 **Illustrates Procedure 10.5**

A general contractor wants to compare the lifetimes of two major brands of electric water heaters—Eagle and National. The contractor obtained the sample data displayed in Table 10.4.

TABLE 10.4
Lifetimes, in years, of water heaters sampled.

Eagle	National
6.9	8.7
7.3	7.0
7.8	8.7
7.4	6.7
7.2	7.8
6.6	8.6
6.2	6.1
8.2	7.5
7.6	7.7
5.7	7.5
5.5	11.2
6.9	6.1
	6.3
	7.0
	10.7

Determine, at the 5% significance level, whether the two brands of water heaters have different mean lifetimes. [Assume that the lifetimes of each brand are approximately normally distributed.]

SOLUTION
We will apply Procedure 10.5. Before doing so, however, we give the values of the statistics that will be required to perform the hypothesis test. These are obtained from the sample data in Table 10.4 in the usual way. See Table 10.5.

TABLE 10.5

Eagle	National
$\bar{x}_1 = 6.94$ yr	$\bar{x}_2 = 7.84$ yr
$s_1 = 0.82$ yr	$s_2 = 1.53$ yr
$n_1 = 12$	$n_2 = 15$

STEP 1 *State the null and alternative hypotheses.*

Let μ_1 and μ_2 denote the true mean lifetimes for Eagle and National water heaters. Then the null and alternative hypotheses are

$$H_0: \mu_1 = \mu_2 \text{ (mean lifetimes are the same)}$$
$$H_a: \mu_1 \neq \mu_2 \text{ (mean lifetimes are different)}$$

Note that this test is two–tailed since there is a not–equal sign ($\neq$) in the alternative hypothesis.

STEP 2 *Decide on the significance level, α.*

The test is to be performed at the 5% significance level; thus, $\alpha = 0.05$.

STEP 3 *The critical values for a two-tailed test are $\pm t_{\alpha/2}$ with degrees of freedom given by*

$$df = \frac{[(s_1^2/n_1) + (s_2^2/n_2)]^2}{\dfrac{(s_1^2/n_1)^2}{n_1 - 1} + \dfrac{(s_2^2/n_2)^2}{n_2 - 1}}$$

rounded down to the nearest integer.

We have

$$df = \frac{[(0.82^2/12) + (1.53^2/15)]^2}{\dfrac{(0.82^2/12)^2}{12 - 1} + \dfrac{(1.53^2/15)^2}{15 - 1}} = 22 \text{ (rounded down)}$$

Therefore, the critical values are

$$\pm t_{0.05/2} = \pm t_{0.025} = \pm 2.074$$

See Figure 10.4.

FIGURE 10.4

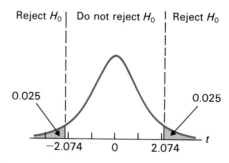

STEP 4 *Compute the value of the test statistic*

$$t = \frac{\overline{x}_1 - \overline{x}_2}{\sqrt{(s_1^2/n_1) + (s_2^2/n_2)}}$$

The required statistics for computing the test statistic have already been calculated and are displayed in Table 10.5. We have

$$t = \frac{\overline{x}_1 - \overline{x}_2}{\sqrt{(s_1^2/n_1) + (s_2^2/n_2)}} = \frac{6.94 - 7.84}{\sqrt{(0.82^2/12) + (1.53^2/15)}} = -1.954$$

STEP 5 *If the value of the test statistic falls in the rejection region, reject H_0; otherwise, do not reject H_0.*

From Step 4, the value of the test statistic is $t = -1.954$ which, as we see from Figure 10.4, does not fall in the rejection region. Thus, we do not reject H_0.

STEP 6 *State the conclusion in words.*

The data do not provide sufficient evidence to conclude that the two brands of water heaters have different mean lifetimes. ∎

MTB
SPSS

Confidence intervals for the difference between two means —normal populations, independent samples, and σs unknown but not assumed equal

Suppose we want to obtain a confidence interval for the difference between two means when dealing with independent samples from two normally distributed populations. As we have seen, if the standard deviations of the two populations are equal, then we can apply Procedure 10.4 to obtain a confidence interval. However, if the standard deviations of the two populations are not equal or if we simply do not know whether they are equal, then Procedure 10.4 should not be used. Instead, the following procedure should be employed. This procedure can be derived using the Key Fact on page 408.

PROCEDURE 10.6 **To find a confidence interval for the difference between two population means.**

Assumptions
1 Independent samples.
2 Normal populations.

STEP 1 *For a confidence level of $1 - \alpha$, use Table III to find $t_{\alpha/2}$ with*

$$df = \frac{[(s_1^2/n_1) + (s_2^2/n_2)]^2}{\dfrac{(s_1^2/n_1)^2}{n_1 - 1} + \dfrac{(s_2^2/n_2)^2}{n_2 - 1}}$$

rounded down to the nearest integer.
STEP 2 *The endpoints of the confidence interval for $\mu_1 - \mu_2$ are*

$$(\bar{x}_1 - \bar{x}_2) \pm t_{\alpha/2} \cdot \sqrt{(s_1^2/n_1) + (s_2^2/n_2)}$$

EXAMPLE 10.7 **Illustrates Procedure 10.6**

Refer to Example 10.6 on page 409. Use the sample data in Table 10.4 to determine a 95% confidence interval for the difference, $\mu_1 - \mu_2$, between the mean lifetimes of Eagle and National water heaters.

S O L U T I O N
We apply Procedure 10.6.

STEP 1 *For a confidence level of $1 - \alpha$, use Table III to find $t_{\alpha/2}$ with*

$$df = \frac{[(s_1^2/n_1) + (s_2^2/n_2)]^2}{\dfrac{(s_1^2/n_1)^2}{n_1 - 1} + \dfrac{(s_2^2/n_2)^2}{n_2 - 1}}$$

rounded down to the nearest integer.

For a 95% confidence interval, $\alpha = 0.05$. As we saw in Example 10.6, df = 22. Consulting Table III we find that for df = 22,

$$t_{\alpha/2} = t_{0.05/2} = t_{0.025} = 2.074$$

STEP 2 *The endpoints of the confidence interval for $\mu_1 - \mu_2$ are*

$$(\bar{x}_1 - \bar{x}_2) \pm t_{\alpha/2} \cdot \sqrt{(s_1^2/n_1) + (s_2^2/n_2)}$$

From Step 1, $t_{\alpha/2} = 2.074$. Referring to Table 10.5 on page 409, we conclude that the endpoints of the confidence interval for $\mu_1 - \mu_2$ are

$$(6.94 - 7.84) \pm 2.074 \cdot \sqrt{(0.82^2/12) + (1.53^2/15)}$$

or

$$-0.90 \pm 0.96$$

Thus, a 95% confidence interval for $\mu_1 - \mu_2$ is from

$$-1.86 \quad \text{to} \quad 0.06$$

MTB
SPSS

We can be 95% confident that the difference, $\mu_1 - \mu_2$, between the mean lifetimes of Eagle and National water heaters is somewhere between -1.86 and 0.06 years. ■

Pooled versus nonpooled

Suppose that we want to perform a small-sample inference to compare the means, μ_1 and μ_2, of two normally distributed populations with unknown standard deviations. If we take independent samples, then we need to use either the pooled procedure or the nonpooled procedure.

The pooled procedure is used when the population standard deviations, σ_1 and σ_2, are equal (but unknown). What if the pooled procedure is employed when, in fact, the population standard deviations are *not* equal? The answer to this question depends on several factors. If the population standard deviations are unequal, but not too unequal, and the sample sizes, n_1 and n_2, are about the same, then using the pooled procedure will not cause any serious difficulties. However, if the population standard deviations are actually quite different, applying the pooled procedure can result in a significantly larger Type I error probability than the one specified.

On the other hand, the nonpooled procedure does not require the population standard deviations to be equal—it applies whether or not they are equal. Then why use the pooled procedure at all? The answer is that, if the population standard deviations *are* equal, then, on the average, the pooled procedure is slightly more powerful. That is, there is a somewhat smaller probability of making a Type II error.

We summarize this discussion by suggesting a method for deciding between the pooled and nonpooled procedures.

KEY FACT

Suppose you want to perform a small-sample inference using independent samples to compare the means of two normally distributed populations whose standard deviations are unknown. If you are reasonably sure that the populations have equal standard deviations, then use the pooled procedure. Otherwise, use the nonpooled procedure.

Exercises 10.2

In Exercises 10.17 through 10.20 assume that the populations being sampled are approximately normally distributed with equal population standard deviations.

10.17 A highway official wants to compare two brands of paint used for striping roads. Ten stripes of each paint are run across the highway. The number of months that each stripe lasts is given below.

Brand A		Brand B	
35.6	36.1	37.2	36.4
37.0	35.8	39.7	37.5
34.9	34.9	37.2	40.5
36.0	38.8	38.8	38.2
36.6	36.5	37.7	36.6
$\left(\begin{array}{l} \Sigma x = 362.2 \\ \Sigma x^2 = 13{,}130.48 \end{array} \right)$		$\left(\begin{array}{l} \Sigma x = 379.8 \\ \Sigma x^2 = 14{,}440.76 \end{array} \right)$	

Based on the sample data, does there appear to be a *difference* in mean lasting times between the two paints? Use $\alpha = 0.05$.

10.18 In a packing plant, a machine packs cartons with jars. A salesperson claims that the machine she is selling will pack faster. To test that claim, the time it takes each machine to pack 10 cartons is re-

corded. The results, in seconds, are as follows:

New machine		Present machine	
42.0	41.0	42.7	43.6
41.3	41.8	43.8	43.3
42.4	42.8	42.5	43.5
43.2	42.3	43.1	41.7
41.8	42.7	44.0	44.1
$\left(\begin{array}{l} \Sigma x = 421.3 \\ \Sigma x^2 = 17{,}753.59 \end{array} \right)$		$\left(\begin{array}{l} \Sigma x = 432.3 \\ \Sigma x^2 = 18{,}693.39 \end{array} \right)$	

Do the data suggest that the new machine packs *faster*, on the average? Use $\alpha = 0.05$.

10.19 The U.S. Bureau of Labor Statistics does monthly surveys to estimate hourly earnings of nonsupervisory employees in various industry groups. Suppose that random samples of 14 mine workers and 17 construction workers yield the following statistics:

Mining	Construction
$\bar{x}_1 = \$12.93$	$\bar{x}_2 = \$13.42$
$s_1 = \$ \ 2.25$	$s_2 = \$ \ 2.36$

At the 5% significance level, do the data provide evidence that mine workers earn *less* on the average than construction workers?

10.20 The U.S. Energy Information Administration does surveys to estimate the number of miles driven annually by U.S. households. Independent random samples of 15 midwestern households and 14 southern households produced the following statistics on the number of miles driven last year:

Midwest	South
$\bar{x}_1 = 16{,}229$ mi	$\bar{x}_2 = 17{,}689$ mi
$s_1 = \ \ 4{,}057$ mi	$s_2 = \ \ 4{,}420$ mi

At the 5% significance level, does there appear to be a *difference* in the average number of miles driven by midwestern and southern households?

10.21 Refer to Exercise 10.17.
a) Determine a 95% confidence interval for the difference, $\mu_1 - \mu_2$, between the mean lasting times of Brand A and Brand B.
b) Interpret your results in words.

10.22 Refer to Exercise 10.18.
a) Determine a 90% confidence interval for the difference, $\mu_1 - \mu_2$, between the mean times to pack 10 cartons for the new machine and the present machine.
b) Interpret your results in words.

10.23 Refer to Exercise 10.19.
a) Determine a 90% confidence interval for the difference, $\mu_1 - \mu_2$, between the mean hourly earnings of nonsupervisory mine and construction workers.
b) Interpret your results in words.

10.24 Refer to Exercise 10.20.
a) Find a 95% confidence interval for the difference, $\mu_1 - \mu_2$, between last year's mean number of miles driven by midwestern and southern households.
b) Interpret your results in words.

In Exercises 10.25 through 10.28 assume that the populations being sampled are approximately normally distributed. Do not assume that their standard deviations are equal, although they may be.

10.25 Independent random samples of 17 sophomores and 13 juniors attending a large state university gave the statistics below for cumulative grade point averages (GPA).

Sophomores	Juniors
$\bar{x}_1 = 2.54$	$\bar{x}_2 = 2.68$
$s_1 = 0.42$	$s_2 = 0.38$

Can you conclude from these data that there is a *difference* in mean GPA for sophomores and juniors at the university? Use $\alpha = 0.05$.

10.26 In past years, college-bound males have outperformed college-bound females on the mathematics portion of tests given by the American College Testing (ACT) Program. Random samples of this year's scores yield the data below.

Males	Females
$\bar{x}_1 = 18.3$	$\bar{x}_2 = 16.2$
$s_1 = \ \ 3.8$	$s_2 = \ \ 4.0$
$n_1 = 15$	$n_2 = 15$

Does it appear that college-bound males are, on the average, still outperforming college-bound females on the mathematics portion of ACT tests?
a) Use $\alpha = 0.05$.
b) Use $\alpha = 0.10$.

10.27 The owner of a chain of car washes needs to decide between two brands of hot waxes. One of the brands, Sureglow, costs less than the other brand, Mirror-Sheen. Therefore, unless there is strong evidence that the second brand outlasts the first, the owner will purchase the first brand. With the cooperation of several local automobile dealers, 30 cars are randomly selected to take part in the test. Fifteen of the 30 cars are waxed with Sureglow and 15 with Mirror-Sheen. The cars are then exposed to the same environmental conditions. The test results shown below give the effectiveness times, in days.

Sureglow			Mirror-Sheen		
87	90	88	92	92	91
93	90	92	91	92	91
91	89	93	93	93	92
88	87	89	92	93	94
91	91	90	93	94	91
($\Sigma x = 1349$, $\Sigma x^2 = 121{,}373$)			($\Sigma x = 1384$, $\Sigma x^2 = 127{,}712$)		

a) At the 1% level of significance, does Mirror-Sheen seem to have a longer effectiveness time, on the average, than Sureglow?
b) Do the data provide *strong* evidence that Mirror-Sheen outlasts Sureglow? Explain.

10.28 The marketing manager of a firm that produces laundry products decides to test market a new laundry product in each of the firm's two sales regions. He wants to determine whether there will be a difference in mean sales per market per month between the two regions. Supermarkets from each region are randomly selected to take part in the test, 10 from Region 1 and 15 from Region 2. The data below give the number of cases sold in each store during the testing month.

Region 1		Region 2		
74	87	84	86	95
96	94	87	89	94
78	77	92	93	88
83	83	85	81	93
86	80	92	92	85
($\Sigma x = 838$, $\Sigma x^2 = 70,684$)		($\Sigma x = 1336$, $\Sigma x^2 = 119,248$)		

At the 10% significance level, does the test marketing reveal a *difference* in potential mean sales per market in the two regions?

10.29 Refer to Exercise 10.25.
a) Determine a 95% confidence interval for the difference, $\mu_1 - \mu_2$, between the mean GPAs of sophomores and juniors at the university.
b) Interpret your results in words.

10.30 Refer to Exercise 10.26.
a) Determine a 90% confidence interval for the difference, $\mu_1 - \mu_2$, between this year's mean mathematics ACT scores of males and females.
b) Repeat part (a) using a confidence level of 80%.

10.31 Refer to Exercise 10.27.
a) Determine a 98% confidence interval for the difference, $\mu_1 - \mu_2$, between the mean effectiveness times of Sureglow and Mirror-Sheen.
b) Interpret your results in words.

10.32 Refer to Exercise 10.28.
a) Find a 90% confidence interval for the difference, $\mu_1 - \mu_2$, between potential mean monthly sales per market in Region 1 and Region 2.

b) Interpret your results in words. [Be extra careful here!]

10.33 Let
$$z = \frac{(\bar{x}_1 - \bar{x}_2) - (\mu_1 - \mu_2)}{\sqrt{(\sigma_1^2/n_1) + (\sigma_2^2/n_2)}}$$
Show that if $\sigma_1 = \sigma_2 = \sigma$, we can rewrite the expression for z as
$$z = \frac{(\bar{x}_1 - \bar{x}_2) - (\mu_1 - \mu_2)}{\sigma\sqrt{(1/n_1) + (1/n_2)}}$$
[This verifies Equation (2) on page 402.]

10.34 The formula given on page 402 for the pooled variance is
$$s_p^2 = \frac{(n_1 - 1)s_1^2 + (n_2 - 1)s_2^2}{n_1 + n_2 - 2}$$
Show that, if the sample sizes n_1 and n_2 are equal, then s_p^2 is just the mean of s_1^2 and s_2^2.

10.35 We have presented two t-statistics for performing small-sample hypothesis tests to compare the means of two normally distributed populations. The first one,
$$t = \frac{\bar{x}_1 - \bar{x}_2}{s_p\sqrt{(1/n_1) + (1/n_2)}}$$
is used when the standard deviations of the two populations are unknown but assumed equal. The second one
$$t = \frac{\bar{x}_1 - \bar{x}_2}{\sqrt{(s_1^2/n_1) + (s_2^2/n_2)}}$$
is used when the standard deviations of the two populations are unknown and not assumed equal.
a) Show that, if the sample sizes n_1 and n_2 are equal, then both test statistics will give exactly the same value.
b) Does part (a) imply that the two t-tests are equivalent when the sample sizes are equal?

10.3 Inferences for two population means (paired samples)

Suppose we want to decide whether a newly developed gasoline additive increases mileage. Let μ_1 denote the mean gas mileage of all cars when the additive is used and μ_2 the mean gas mileage of all cars when the additive is not used.

Then we want to test the hypotheses

H_0: $\mu_1 = \mu_2$ (mean mileage with additive is not greater)
H_a: $\mu_1 > \mu_2$ (mean mileage with additive is greater)

One method that can be used to perform the hypothesis test is as follows: Randomly and independently select two groups of, say, 10 cars each; have one group drive with the additive and the other drive without the additive; and then, as described in Section 10.2, use the two samples of 10 mileages obtained to perform the hypothesis test. This procedure employs *independent samples.*

Instead of independent samples, it is often more appropriate to use *paired samples.* For instance, to perform the hypothesis test for gas mileage, we can randomly select a single group of 10 cars; have each of the 10 cars drive both with and without the additive; and then use the 10 *pairs* of mileages obtained, one pair for each car, to perform the hypothesis test. This procedure employs **paired samples.** Each item in the sample consists of a pair of numbers, in this case, the gas mileage of a given car both with and without the additive.

By pairing the samples when it is appropriate, we can remove extraneous sources of variation, such as the variation due to cars and drivers. As a consequence, the sampling error made in estimating the difference between the population means is generally smaller. This fact, in turn, makes it more likely that we will detect differences between the means, when such differences exist.

EXAMPLE 10.8   **Introduces paired-difference tests**

A major oil company has developed a new gasoline additive that is supposed to increase mileage. To test that hypothesis, 10 cars are randomly selected. The cars are driven both with and without the additive. The results are displayed in the second and third columns of Table 10.6.

TABLE 10.6
Results of mileage tests, with and without additive, for 10 randomly selected cars. Data are in miles per gallon.

Car	With additive x_1	Without additive x_2	Paired difference $d = x_1 - x_2$
1	25.7	24.9	0.8
2	20.0	18.8	1.2
3	28.4	27.7	0.7
4	13.7	13.0	0.7
5	18.8	17.8	1.0
6	12.5	11.3	1.2
7	28.4	27.8	0.6
8	8.1	8.2	−0.1
9	23.1	23.1	0.0
10	10.4	9.9	0.5

In the final column of Table 10.6 we have recorded the difference in gas mileage with and without the additive, for each of the 10 cars tested. Each difference is referred to as a **paired difference,** since it is the difference of a pair of numbers. For example, the first car on the list got 25.7 mpg with the additive and 24.9 mpg without the additive, giving a paired difference of $25.7 - 24.9 = 0.8$ mpg—an improvement in gas mileage of 0.8 mpg with the additive.

We want to use the sample data to test the hypotheses

$H_0: \mu_1 = \mu_2$ (mean mileage with additive is not greater)
$H_a: \mu_1 > \mu_2$ (mean mileage with additive is greater)

where μ_1 denotes the mean gas mileage of all cars when the additive is used and μ_2 denotes the mean gas mileage of all cars when the additive is not used.

The idea of the paired-difference test is this: If the null hypothesis is true, we would expect the paired differences in gas mileages (with and without the additive) for the cars sampled to average out to about zero. In other words, we would expect the sample mean, $\bar{d}$, of the paired differences in the final column of Table 10.6 to be near zero. To put it another way, if $\bar{d}$ is too much greater than zero, we would take this as evidence that the null hypothesis is false and conclude that the mean gas mileage, μ_1, with the additive is greater than the mean gas mileage, μ_2, without the additive.

We computed the sample mean of the paired differences and found that

$$\bar{d} = \frac{\Sigma d}{n} = \frac{6.6}{10} = 0.66 \text{ mpg,}$$

an average improvement in gas mileage of 0.66 mpg with the additive for the cars sampled.

The question now is whether this value of $\bar{d}$ can reasonably be attributed to sampling error or whether it indicates that the mean gas mileage with the additive is greater than the mean gas mileage without the additive. To answer this question we need to know the sampling distribution of $\bar{d}$. We will discuss that and then finish solving the gas mileage problem. ∎

The sampling distribution of the difference between two means—paired samples and normal differences

Suppose a random sample of n pairs is to be selected from populations with means μ_1 and μ_2. Let $\bar{d}$ denote the sample mean paired difference of the pairs obtained, where each difference is taken by subtracting the second number in the pair from the first ($d = x_1 - x_2$). Also, let s_d denote the sample standard deviation of the paired differences.

We can think of the paired differences of the pairs obtained as a random sample from the population of *all* possible paired differences. Let us denote the mean of that population by μ_d. Then it can be shown that

$$\mu_d = \mu_1 - \mu_2$$

That is, the mean of the population of paired differences is equal to the difference of the means of the two populations. (See Exercise 10.51.)

In what follows, we will assume that the population of all paired differences is *normally distributed*—so-called **normal differences.**[†] Applying the Key Fact on page 322 to the population of paired differences, we obtain the following result.

[†] For large samples, that assumption is unnecessary.

KEY FACT

Suppose a random sample of n pairs is to be taken from populations with means μ_1 and μ_2. Further suppose that the population of all paired differences is *normally distributed.* Then the random variable

$$t = \frac{\overline{d} - (\mu_1 - \mu_2)}{s_d/\sqrt{n}}$$

has the *t-distribution with df = n − 1.*

Hypothesis tests for two means—paired samples and normal differences

From the previous Key Fact, we know that if the population of all paired differences is normally distributed, then the random variable

$$t = \frac{\overline{d} - (\mu_1 - \mu_2)}{s_d/\sqrt{n}}$$

has the *t-distribution with df = n − 1.* This means that for a hypothesis test with null hypothesis H_0: $\mu_1 = \mu_2$, we can use the random variable

$$t = \frac{\overline{d}}{s_d/\sqrt{n}}$$

as our test statistic and find the critical value(s) from the *t*-table. Thus, we have

---

PROCEDURE 10.7 To perform a hypothesis test for two population means, with null hypothesis H_0: $\mu_1 = \mu_2$.

Assumptions
1 Paired samples.
2 Normal differences.

STEP 1 *State the null and alternative hypotheses.*
STEP 2 *Decide on the significance level, α.*
STEP 3 *The critical value(s)*
   a) *for a two-tailed test are* $\pm t_{\alpha/2}$,
   b) *for a left-tailed test is* $-t_\alpha$,
   c) *for a right-tailed test is* t_α,
   with df = n − 1. Use Table III to find the critical value(s).

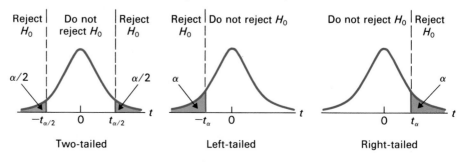

Two-tailed   Left-tailed   Right-tailed

STEP 4 *Calculate the paired differences, $d = x_1 - x_2$, of the sample pairs.*

STEP 5 *Compute the value of the test statistic*

$$t = \frac{\bar{d}}{s_d/\sqrt{n}}$$

STEP 6 *If the value of the test statistic falls in the rejection region, reject H_0; otherwise, do not reject H_0.*

STEP 7 *State the conclusion in words.*

EXAMPLE 10.9 **Illustrates Procedure 10.7**

Recall from Example 10.8 that a major oil company has developed a new gasoline additive that is supposed to increase mileage. The gas mileages are determined for each of 10 randomly selected cars, both with and without the additive. The results are displayed in the second and third columns of Table 10.6 on page 416. Do the data suggest that, on the average, the gasoline additive improves gas mileage? Perform the test at the 5% significance level. [Assume that changes in gas mileage due to the gasoline additive are normally distributed.]

SOLUTION

Note that we are dealing here with paired samples. Each pair consists of the gas mileage of a car both with and without the additive. Since, by assumption, the population of all possible paired differences is normally distributed, we can apply Procedure 10.7 to perform the hypothesis test.

STEP 1 *State the null and alternative hypotheses.*

The null and alternative hypotheses are

$H_0: \mu_1 = \mu_2$ (mean mileage with additive is not greater)
$H_a: \mu_1 > \mu_2$ (mean mileage with additive is greater)

where μ_1 is the mean gas mileage of all cars when the additive is used and μ_2 is the mean gas mileage of all cars when the additive is not used. Note that the test is right-tailed since there is a greater-than sign ($>$) in the alternative hypothesis.

STEP 2 *Decide on the significance level, α.*

The test is to be performed at the 5% significance level. Thus, $\alpha = 0.05$.

STEP 3 *The critical value for a right-tailed test is t_α with $df = n - 1$.*

From Step 2, $\alpha = 0.05$. Also, since there are 10 pairs in the sample, we have $df = 10 - 1 = 9$. So, the critical value is

$$t_{0.05} = 1.833$$

See Figure 10.5 at the top of the next page.

FIGURE 10.5

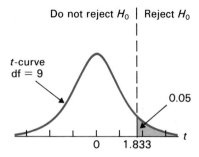

STEP 4 *Calculate the paired differences, $d = x_1 - x_2$, of the sample pairs.*

We have already done this in the final column of Table 10.6 on page 416.

STEP 5 *Compute the value of the test statistic*

$$t = \frac{\overline{d}}{s_d/\sqrt{n}}$$

We first need to compute the sample mean and sample standard deviation of the d-values in the final column of Table 10.6. This is accomplished in the usual manner:

$$\overline{d} = \frac{\Sigma d}{n} = \frac{6.6}{10} = 0.66$$

and

$$s_d = \sqrt{\frac{n(\Sigma d^2) - (\Sigma d)^2}{n(n-1)}} = \sqrt{\frac{10(6.12) - (6.6)^2}{10 \cdot 9}} = 0.44$$

Consequently the value of the test statistic is

$$t = \frac{\overline{d}}{s_d/\sqrt{n}} = \frac{0.66}{0.44/\sqrt{10}} = 4.74$$

STEP 6 *If the value of the test statistic falls in the rejection region, reject H_0; otherwise, do not reject H_0.*

From Step 5, the value of the test statistic is $t = 4.74$, which falls in the rejection region. Hence, we reject H_0.

STEP 7 *State the conclusion in words.*

The data indicate that the mean gas mileage of all cars when the additive is used is greater than the mean gas mileage of all cars when the additive is not used. In other words, it appears that the additive is effective in increasing gas mileage.   ■

MTB
SPSS

Confidence intervals for the difference between two means —paired samples and normal differences

We can also use the Key Fact on page 418 to derive the following confidence-interval procedure for the difference between two means. The procedure applies in the case where the samples are paired and the population of all paired differences is normally distributed.

PROCEDURE 10.8 **To find a confidence interval for the difference between two population means.**

Assumptions
1 Paired samples.
2 Normal differences.

STEP 1 *For a confidence level of $1 - \alpha$, use Table III to find $t_{\alpha/2}$ for df $= n - 1$.*
STEP 2 *The endpoints of the confidence interval for $\mu_1 - \mu_2$ are*

$$\overline{d} \pm t_{\alpha/2} \cdot \frac{s_d}{\sqrt{n}}$$

EXAMPLE 10.10   **Illustrates Procedure 10.8**

Consider again the situation of Example 10.8 on page 416. Use the sample data in Table 10.6 to determine a 90% confidence interval for the difference, $\mu_1 - \mu_2$, between the mean gas mileage of all cars when the additive is used and the mean gas mileage of all cars when the additive is not used.

S O L U T I O N
We apply Procedure 10.8.

STEP 1 *For a confidence level of $1 - \alpha$, use Table III to find $t_{\alpha/2}$ for df $= n - 1$.*

For a 90% confidence interval, we have $\alpha = 0.10$. From Table III we find that for df $= n - 1 = 10 - 1 = 9$,

$$t_{\alpha/2} = t_{0.10/2} = t_{0.05} = 1.833$$

STEP 2 *The endpoints of the confidence interval for $\mu_1 - \mu_2$ are*

$$\overline{d} \pm t_{\alpha/2} \cdot \frac{s_d}{\sqrt{n}}$$

From Step 1, $t_{\alpha/2} = 1.833$. Also, $n = 10$ and, from Example 10.9, we know that $\overline{d} = 0.66$ and $s_d = 0.44$. Consequently, the endpoints of the confidence interval for $\mu_1 - \mu_2$ are

$$0.66 \pm 1.833 \cdot \frac{0.44}{\sqrt{10}}$$

or

$$0.66 \pm 0.26$$

Thus, a 90% confidence interval for $\mu_1 - \mu_2$ is from

$$0.40 \quad \text{to} \quad 0.92$$

MTB
SPSS

We can be 90% confident that the difference, $\mu_1 - \mu_2$, between the mean gas mileage of all cars when the additive is used and the mean gas mileage of all cars when the additive is not used is somewhere between 0.40 and 0.92 mpg. In particular, therefore, we can be 90% confident that, on the average, the additive increases gas mileage by at least 0.40 mpg. ∎

Exercises 10.3

For Exercises 10.36 through 10.41, assume that the population of all possible paired differences is approximately normally distributed.

10.36 A pediatrician began measuring the blood cholesterol levels of her young patients. She was surprised to find that many of them had levels over 200, indicating increased risk of artery disease. Ten such patients were randomly selected to take part in a nutritional program designed to lower blood cholesterol. Two months following the commencement of the program, the pediatrician measured the blood cholesterol levels of the 10 patients again. The results are as follows:

Patient	Before program	After program
1	210	212
2	217	210
3	208	210
4	215	213
5	202	200
6	209	208
7	207	203
8	210	199
9	221	218
10	218	214

Do the data suggest that the nutritional program is, on the average, effective in reducing cholesterol levels? Perform the appropriate test at the 1% level of significance.

10.37 An exercise physiologist measured the heart rates of 15 randomly selected people. The people were then placed on a running program. One year later their heart rates were measured again. The results are as follows:

Person	Before program	After program
1	68	67
2	76	77
3	74	74
4	71	74
5	71	69
6	72	70
7	75	71
8	83	77
9	75	71
10	74	74
11	76	73
12	77	68
13	78	71
14	75	72
15	75	77

Do the data provide evidence that the running program will, on the average, reduce heart rates? Use $\alpha = 0.01$.

10.38 The A.C. Nielsen Company collects data in order to estimate the TV viewing habits of Americans. Suppose that 20 married couples are randomly selected and that their weekly viewing times, in hours, are as follows:

Husband	Wife	Husband	Wife	Husband	Wife
21	24	38	45	36	35
56	55	27	29	20	34
34	55	30	41	43	32
30	34	31	37	4	13
41	32	30	35	16	9
35	38	32	48	21	23
26	38	15	17		

At the 5% level of significance, does it appear that married men watch less TV, on the average, than married women? [Note: $\bar{d} = -4.4$ and $s_d = 8.15$.]

10.39 In Exercise 10.19 we considered a hypothesis test using *independent samples* to decide whether nonsupervisory mine workers earn a smaller average hourly wage than nonsupervisory construction workers. Now we will look at that same hypothesis test using *paired samples*. Suppose mine and construction workers are paired by matching workers with similar experience and job classification. Further suppose that a random sample of 15 pairs yields the following paired differences for hourly wages (mining *minus* construction).

0.80	1.03	0.57
−2.38	0.89	−2.16
−1.36	−0.05	−1.89
−0.63	1.20	−1.40
−0.67	−1.23	0.47

Use this data to test whether nonsupervisory mine workers have a smaller average hourly wage than nonsupervisory construction workers. Perform the test at the 5% significance level. [$\Sigma d = -6.81$ and $\Sigma d^2 = 24.5517$.]

10.40 A college algebra teacher wants to compare two methods of instruction. One is the lecture method and the other is the personalized system of instruction (PSI) method. Students are paired by matching those with similar mathematics background and performance. A random sample of 11 pairs is selected. From each pair, one student is randomly chosen to take the lecture course; the other student takes the PSI course. Both courses are taught by the college algebra teacher. The final grades for the 11 pairs of students turn out to be the following:

Lecture	PSI	Lecture	PSI
66	67	80	79
93	93	73	70
36	35	74	67
84	85	83	79
60	64	52	50
66	57		

Do the data provide evidence that there is a *difference* in mean student performance between the two instructional methods? Use $\alpha = 0.05$.

10.41 The U.S. Bureau of the Census collects data on the ages of married people. Suppose that 10 married couples are randomly selected and have the ages given here.

Husband	Wife	Husband	Wife
54	53	33	35
21	22	68	67
32	33	32	28
78	74	54	41
70	64	52	44

Do the data suggest that the mean age of married males is *greater* than the mean age of married females? Perform the appropriate hypothesis test at the 5% significance level.

10.42 Refer to Exercise 10.36.
a) Determine a 98% confidence interval for the difference, $\mu_1 - \mu_2$, between the mean blood cholesterol levels of high-level patients before and after the nutritional program.
b) Interpret your results in words.

10.43 Refer to Exercise 10.37.
a) Determine a 98% confidence interval for the difference, $\mu_1 - \mu_2$, between the mean heart rates of people before and after the running program.
b) Interpret your results in words.

10.44 Refer to Exercise 10.38.
a) Determine a 90% confidence interval for the difference, $\mu_1 - \mu_2$, between the mean weekly TV viewing times of married men and married women.
b) Interpret your results in words.

10.45 Refer to Exercise 10.39.
a) Determine a 90% confidence interval for the difference, $\mu_1 - \mu_2$, between the mean hourly earnings of nonsupervisory mine and construction workers.
b) Interpret your results in words.

10.46 Refer to Exercise 10.40.
a) Determine a 95% confidence interval for the difference, $\mu_1 - \mu_2$, between the mean final grades of all students who take college algebra from the teacher using the lecture method and those who take college algebra from the teacher using the PSI method.
b) Interpret your results in words.

10.47 Refer to Exercise 10.41.
a) Determine a 90% confidence interval for the difference, $\mu_1 - \mu_2$, between the mean ages of married males and females.
b) Interpret your results in words.

10.48   This exercise shows what can happen when a test designed for use with independent samples is applied to perform a test in which the samples are in fact paired. In Example 10.9 on page 419, we performed a paired-difference test to determine whether a gasoline additive is effective in increasing gas mileage. Specifically, if we let μ_1 and μ_2 denote, respectively, the mean gas mileage of all cars when the additive is and is not used, then the hypothesis test is

$$H_0: \mu_1 = \mu_2$$
$$H_a: \mu_1 > \mu_2$$

a) Apply Procedure 10.5 on page 408 to the sample data in Table 10.6 on page 416 to perform this hypothesis test. Use $\alpha = 0.05$.
b) Why is it inappropriate to perform the test you did in part (a)?
c) Compare your result in part (a) to that of Example 10.9.

In Exercises 10.49 and 10.50 we will consider **large-sample inferences for two population means using paired samples.** The procedures used for such inferences are similar to the corresponding small-sample procedures. The only differences are the followng: (1) We do not need to assume the population of all paired differences is normally distributed; (2) We use the standard normal table instead of the t-table.

10.49   A tire company has developed two new processes, Process A and Process B, for making longer-wearing steel-belted radials. To compare the two processes, 50 tires made using Process A and 50 tires made using Process B are randomly selected. Each of the 50 Process A tires is randomly assigned to either the front left or front right of one of 50 cars. If a Process A tire gets assigned to be the left front tire for a given car, then a Process B tire gets assigned to be the right front tire, and vice-versa. [The rear of each of the 50 cars is equipped with two tires currently manufactured by the tire company.] The differences in tire life (Process A tire life *minus* Process B tire life) for the 50 pairs of tires are given below in thousands of miles, to the nearest hundred miles.

−0.8	−2.4	−0.2	1.6	−3.6
−0.6	−1.7	−1.4	−2.5	−1.8
−0.9	0.9	2.8	−2.7	−0.6
0.0	−2.0	1.8	−1.5	1.9
−4.2	−2.1	3.0	0.0	0.4

0.9	0.9	1.6	−2.7	0.9
1.0	−2.3	−0.4	0.5	2.4
−0.3	0.9	−0.6	0.8	3.6
−0.4	0.5	1.3	−4.0	2.4
−4.3	2.4	−0.5	1.5	3.1

a) At the 10% significance level, does there appear to be a *difference* in the mean lifetimes of Process A and Process B tires? [The sum of the data is −7.4 and the sum of the squares of the data is 197.90.]
b) Find a 90% confidence interval for the difference, $\mu_1 - \mu_2$, between the mean lifetimes of Process A and Process B tires.

10.50   In Example 10.2 on page 396 we used independent samples to test whether there is a *difference* in mean salaries for faculty teaching in public and private institutions. Suppose that we want to use paired samples to perform the same hypothesis test. Pairs are formed by matching faculty in public and private institutions by rank and speciality. A random sample of $n = 30$ pairs yields the following data:

Public	Private	Public	Private	Public	Private
$38,080	35,752	66,380	65,296	21,139	22,434
32,104	42,431	35,496	34,656	23,152	23,867
29,875	33,496	20,426	25,824	22,750	24,559
26,564	30,184	41,638	43,208	21,749	25,032
53,326	64,784	21,724	33,345	28,893	27,908
35,568	34,875	29,877	32,636	20,889	21,349
39,190	42,421	32,738	28,158	24,906	32,198
27,689	28,008	33,294	33,744	32,546	34,247
25,606	25,770	16,406	16,720	48,669	47,045
44,025	50,914	51,607	52,350	57,468	56,695

a) Perform the appropriate hypothesis test at the 5% significance level.
b) Compare your result in part (a) with that in Example 10.2.
c) Use the above data to find a 95% confidence interval for the difference, $\mu_1 - \mu_2$, between the mean salaries of faculty teaching in public and private institutions.
d) Compare your result in part (c) with that in Example 10.3 on page 398.

10.51   On page 417 we gave the formula

$$(3) \qquad\qquad \mu_d = \mu_1 - \mu_2$$

which states that the mean of the population of paired differences is equal to the difference of the two population means.

a) Suppose that the two populations of interest are *finite* and each have *N* members. Prove Formula (3).

b) In general, let (x_1, x_2) denote a randomly selected pair from the population of all pairs, and let $d = x_1 - x_2$ denote the paired difference. Note that x_1, x_2, and *d* are random variables. Use the results of Exercises 5.26(e) and 5.27(f) on page 190 to prove Formula (3). *Hint:* Refer to the Key Fact on page 183.

10.52 This exercise considers paired-difference tests with a null hypothesis of the form

$$H_0: \mu_1 = \mu_2 + \mu_0$$

where μ_0 is not necessarily equal to zero. We will assume that the population of all possible paired differences is approximately normally distributed.

a) Show that the test statistic for such a hypothesis test is

$$t = \frac{\bar{d} - \mu_0}{s_d/\sqrt{n}}$$

where $df = n - 1$. *Hint:* See the Key Fact on page 418.

b) The distributor of a nutritional meal replacement (NMR) claims that the average overweight person will lose more than 12 lb during the first month of its use. To test that claim, 15 overweight people are randomly selected. Their weights are recorded now and will be recorded again in one month. In the interim, each person will use the NMR. Suppose the results are as depicted here.

Before	After	Before	After	Before	After
152	132	178	162	170	164
168	155	160	147	140	128
146	130	172	157	137	116
157	148	168	150	132	113
151	126	158	141	149	128

Do the data support the distributor's claim? Use $\alpha = 0.05$.

10.4 Large-sample inferences for two population proportions (independent samples)

In Section 9.6 we studied inferences for one population proportion, *p*. We now examine inferences concerning two population proportions, p_1 and p_2.

Recall that a *two-category population* is one in which each member is classified as either having or not having a specified attribute. The proportion (percentage) of the entire population that has the specified attribute is called the *population proportion.* The proportion of a sample from the population that has the specified attribute is called a *sample proportion.*

Statistical inferences for two population proportions are concerned with comparing the proportions of *two* two-category populations that have a specified attribute. Consider the following example.

EXAMPLE 10.11 **Introduces hypothesis tests for two proportions**

Suppose that 132 out of 400 randomly selected adult males are cigarette smokers, and 93 out of 350 randomly selected adult females are cigarette smokers. Do the data indicate that the percentage of males who smoke exceeds the percentage of females who smoke?

SOLUTION
We begin by noting that the specified attribute here is "smokes cigarettes." Next we let p_1 denote the population proportion of males who smoke cigarettes and p_2

denote the population proportion of females who smoke cigarettes. Then we want to perform the hypothesis test

$$H_0: p_1 = p_2 \text{ (percentage of male smokers is not higher)}$$
$$H_a: p_1 > p_2 \text{ (percentage of male smokers is higher)}$$

Roughly speaking, the hypothesis test is carried out as follows:

1 Compute the sample proportion, $\bar{p}_1$, of males sampled who smoke cigarettes and the sample proportion, $\bar{p}_2$, of females sampled who smoke cigarettes.
2 If $\bar{p}_1$ is too much bigger than $\bar{p}_2$, then reject H_0; otherwise, do not reject H_0.

The first step is easy. Since 132 out of the 400 males sampled are smokers, the sample proportion, $\bar{p}_1$, is

$$\bar{p}_1 = \frac{x_1}{n_1} = \frac{132}{400} = 0.330 \ (33.0\%)$$

and since 93 out of the 350 females sampled are smokers, the sample proportion, $\bar{p}_2$, is

$$\bar{p}_2 = \frac{x_2}{n_2} = \frac{93}{350} = 0.266 \ (26.6\%)$$

For the second step, we need to decide whether the sample proportion $\bar{p}_1 = 0.330$ exceeds the sample proportion $\bar{p}_2 = 0.266$ by a sufficient amount to warrant rejection of the null hypothesis in favor of the alternative hypothesis. In other words, we need to decide whether the difference in sample proportions can be reasonably attributed to sampling error, or whether it indicates that the proportion of all males who smoke exceeds the proportion of all females who smoke. To make this decision, we must have the probability distribution of the difference, $\bar{p}_1 - \bar{p}_2$, between two sample proportions—the **sampling distribution of the difference between two proportions.** We will discuss that sampling distribution and then complete the hypothesis test. ∎

The sampling distribution of the difference between two proportions—large, independent samples

To begin our discussion of the sampling distribution of the difference between two proportions, we present a summary of the required notation in Table 10.7.

TABLE 10.7   Notation for parameters and statistics when considering two two-category populations.

	Population 1	Population 2
Population proportion	p_1	p_2
Sample proportion	$\bar{p}_1$	$\bar{p}_2$
Number of successes	x_1	x_2
Sample size	n_1	n_2

Recall that the "number of successes" refers to the number of members sampled that have the specified attribute. Consequently, the sample proportions are computed from the formulas

$$\bar{p}_1 = \frac{x_1}{n_1} \quad \text{and} \quad \bar{p}_2 = \frac{x_2}{n_2}$$

To continue, we will present formulas that relate the mean and standard deviation of the random variable $\bar{p}_1 - \bar{p}_2$ to the population proportions p_1 and p_2. We have

$$\mu_{\bar{p}_1 - \bar{p}_2} = p_1 - p_2$$

and

$$\sigma_{\bar{p}_1 - \bar{p}_2} = \sqrt{p_1(1 - p_1)/n_1 + p_2(1 - p_2)/n_2}$$

These formulas are derived in almost exactly the same way as the formulas for the mean and standard deviation of the random variable $\bar{x}_1 - \bar{x}_2$. (See Exercise 10.66.)

Finally, as we learned in Section 8.6 (page 330), the random variables $\bar{p}_1$ and $\bar{p}_2$ are both approximately normally distributed for large sample sizes. From this it can be shown that the random variable $\bar{p}_1 - \bar{p}_2$ is also approximately normally distributed. Hence, we can state the following fact:

KEY FACT The sampling distribution of the difference between two proportions—large and independent samples

Suppose that a random sample of size n_1 is to be taken from a two-category population with population proportion p_1; and that a random sample of size n_2 is to be taken from a two-category population with population proportion p_2. Further suppose that the two samples are to be selected independently of one another. Then, for large samples, the random variable $\bar{p}_1 - \bar{p}_2$ is approximately *normally distributed* and has mean $\mu_{\bar{p}_1 - \bar{p}_2} = p_1 - p_2$ and standard deviation $\sigma_{\bar{p}_1 - \bar{p}_2} = \sqrt{p_1(1 - p_1)/n_1 + p_2(1 - p_2)/n_2}$. Thus, the standardized random variable

$$z = \frac{(\bar{p}_1 - \bar{p}_2) - (p_1 - p_2)}{\sqrt{p_1(1 - p_1)/n_1 + p_2(1 - p_2)/n_2}}$$

has approximately the *standard normal distribution.*

Hypothesis tests for two proportions—large and independent samples

We can now develop a hypothesis-testing procedure for comparing two population proportions. The null hypothesis will be

$$H_0: p_1 = p_2 \text{ (population proportions are equal)}$$

If the null hypothesis is true, then $p_1 - p_2 = 0$ and the standardized random variable

$$z = \frac{(\bar{p}_1 - \bar{p}_2) - (p_1 - p_2)}{\sqrt{p_1(1-p_1)/n_1 + p_2(1-p_2)/n_2}}$$

becomes

$$z = \frac{\bar{p}_1 - \bar{p}_2}{\sqrt{p(1-p)/n_1 + p(1-p)/n_2}}$$

where p denotes the common value of p_1 and p_2. Factoring $p(1-p)$ out of the denominator of this last expression gives

(4)
$$z = \frac{\bar{p}_1 - \bar{p}_2}{\sqrt{p(1-p)} \sqrt{(1/n_1) + (1/n_2)}}$$

This random variable cannot be used as a test statistic since p is unknown. Consequently, we must estimate p using the sample information. The best estimate for p is obtained by *pooling* the data to get the proportion of successes in both samples combined. That is, we estimate p by

$$\bar{p} = \frac{x_1 + x_2}{n_1 + n_2}$$

We call $\bar{p}$ the **pooled sample proportion.**

Replacing the p in Equation (4) by its estimate $\bar{p}$ yields the random variable

$$\frac{\bar{p}_1 - \bar{p}_2}{\sqrt{\bar{p}(1-\bar{p})} \sqrt{(1/n_1) + (1/n_2)}}$$

This random variable *can* be used as our test statistic and, like the random variable in Equation (4), has approximately the standard normal distribution for large samples. The numerator of the test statistic measures the difference between the two sample proportions, and the denominator standardizes the test statistic so that the standard normal table can be used to obtain the critical value(s) for a hypothesis test. Consequently, we have the following hypothesis-testing procedure for comparing two population proportions.

PROCEDURE 10.9 To perform a hypothesis test for two population proportions, with null hypothesis H_0: $p_1 = p_2$.

Assumptions
1 Independent samples.
2 Large samples.

STEP 1 *State the null and alternative hypotheses.*
STEP 2 *Decide on the significance level, α.*
STEP 3 *The critical value(s)*
   a) *for a two-tailed test are $\pm z_{\alpha/2}$.*
   b) *for a left-tailed test is $-z_\alpha$.*
   c) *for a right-tailed test is z_α.*
Use Table II to find the critical value(s).

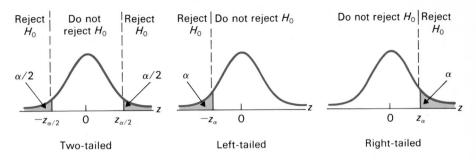

STEP 4 *Compute the value of the test statistic*

$$z = \frac{\bar{p}_1 - \bar{p}_2}{\sqrt{\bar{p}(1 - \bar{p})}\ \sqrt{(1/n_1) + (1/n_2)}}$$

where

$$\bar{p} = \frac{x_1 + x_2}{n_1 + n_2}$$

STEP 5 *If the value of the test statistic falls in the rejection region, reject H_0; otherwise, do not reject H_0.*

STEP 6 *State the conclusion in words.*

EXAMPLE 10.12 **Illustrates Procedure 10.9**

We now return to the problem stated in Example 10.11. Out of 400 randomly selected adult males, it is found that 132 smoke; and out of 350 randomly selected adult females, it is found that 93 smoke. At the 5% significance level, do the data indicate that the percentage of males who smoke is greater than the percentage of females who smoke?

SOLUTION
We apply Procedure 10.9.

STEP 1 *State the null and alternative hypotheses.*

Let p_1 denote the proportion of all adult males who smoke and p_2 denote the proportion of all adult females who smoke. Then the null and alternative hypotheses are

$$H_0: p_1 = p_2 \text{ (percentage of male smokers is not higher)}$$
$$H_a: p_1 > p_2 \text{ (percentage of male smokers is higher)}$$

Note that the test is right-tailed since there is a greater-than sign ($>$) in the alternative hypothesis.

STEP 2 *Decide on the significance level, α.*

The test is to be performed at the 5% significance level. Thus, $\alpha = 0.05$.

STEP 3 *The critical value for a right-tailed test is z_α.*

Since $\alpha = 0.05$, the critical value is

$$z_{0.05} = 1.645$$

See Figure 10.6.

FIGURE 10.6

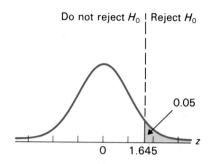

STEP 4 *Compute the value of the test statistic*

$$z = \frac{\bar{p}_1 - \bar{p}_2}{\sqrt{\bar{p}(1 - \bar{p})}\,\sqrt{(1/n_1) + (1/n_2)}}$$

where

$$\bar{p} = \frac{x_1 + x_2}{n_1 + n_2}$$

We first compute $\bar{p}_1$, $\bar{p}_2$, and $\bar{p}$. Since 132 of the 400 males sampled are smokers and 93 of the 350 females sampled are smokers, we have $x_1 = 132$, $n_1 = 400$ and $x_2 = 93$, $n_2 = 350$. Therefore,

$$\bar{p}_1 = \frac{x_1}{n_1} = \frac{132}{400} = 0.330$$

$$\bar{p}_2 = \frac{x_2}{n_2} = \frac{93}{350} = 0.266$$

and

$$\bar{p} = \frac{x_1 + x_2}{n_1 + n_2} = \frac{132 + 93}{400 + 350} = \frac{225}{750} = 0.300$$

Consequently, the value of the test statistic is

$$z = \frac{\bar{p}_1 - \bar{p}_2}{\sqrt{\bar{p}(1 - \bar{p})}\,\sqrt{(1/n_1) + (1/n_2)}}$$

$$= \frac{0.330 - 0.266}{\sqrt{(0.3)(1 - 0.3)}\,\sqrt{(1/400) + (1/350)}} = 1.92$$

STEP 5 *If the value of the test statistic falls in the rejection region, reject H_0; otherwise, do not reject H_0.*

From Step 4, the value of the test statistic is $z = 1.92$, which falls in the rejection region. Thus, we reject H_0.

STEP 6 *State the conclusion in words.*

The data provide sufficient evidence to conclude that the percentage of males who smoke is greater than the percentage of females who smoke. ∎

Confidence intervals for the difference between two proportions—large and independent samples

The Key Fact on page 427 can be used to derive a confidence-interval procedure for the difference between two population proportions. That procedure is presented below. For details of the derivation of the procedure, see Exercise 10.67.

> **PROCEDURE 10.10** To find a confidence interval for the difference between two population proportions.
>
> *Assumptions*
> 1 Independent samples.
> 2 Large samples.
>
> **STEP 1** *For a confidence level of $1 - \alpha$, use Table II to find $z_{\alpha/2}$.*
> **STEP 2** *The endpoints of the confidence interval for $p_1 - p_2$ are*
> $$(\bar{p}_1 - \bar{p}_2) \pm z_{\alpha/2} \cdot \sqrt{\bar{p}_1(1 - \bar{p}_1)/n_1 + \bar{p}_2(1 - \bar{p}_2)/n_2}$$

EXAMPLE 10.13 **Illustrates Procedure 10.10**

Consider again the smoking illustration of Example 10.11. Use the sample data to determine a 90% confidence interval for the difference, $p_1 - p_2$, between the proportion of all males who smoke and the proportion of all females who smoke.

SOLUTION
We apply Procedure 10.10.

STEP 1 *For a confidence level of $1 - \alpha$, use Table II to find $z_{\alpha/2}$.*

For a 90% confidence interval, $\alpha = 0.10$. Consulting Table II we find that
$$z_{\alpha/2} = z_{0.10/2} = z_{0.05} = 1.645$$

STEP 2 *The endpoints of the confidence interval for $p_1 - p_2$ are*
$$(\bar{p}_1 - \bar{p}_2) \pm z_{\alpha/2} \cdot \sqrt{\bar{p}_1(1 - \bar{p}_1)/n_1 + \bar{p}_2(1 - \bar{p}_2)/n_2}$$

From Step 1, $z_{\alpha/2} = 1.645$. Referring to Example 10.12 on page 429, we see that $\bar{p}_1 = 0.330$, $n_1 = 400$ and $\bar{p}_2 = 0.266$, $n_2 = 350$. Therefore, the endpoints of the confidence interval for $p_1 - p_2$ are

$$(0.330 - 0.266) \pm 1.645 \cdot \sqrt{0.330(1 - 0.330)/400 + 0.266(1 - 0.266)/350}$$

or

$$0.064 \pm 0.055$$

Thus, a 90% confidence interval for $p_1 - p_2$ is from

$$0.009 \quad \text{to} \quad 0.119$$

We can be 90% confident that the difference, $p_1 - p_2$, between the proportion of males who smoke and the proportion of females who smoke is somewhere between 0.009 and 0.119. In other words, we can be 90% confident that the percentage of males who smoke exceeds the percentage of females who smoke by at least 0.9% but by no more than 11.9%.   ∎

Exercises 10.4

10.53   Consider the quantities $p_1, p_2, x_1, x_2, \overline{p}_1, \overline{p}_2, \overline{p}$.
a) Which quantities represent parameters and which represent statistics?
b) Which quantities are fixed numbers and which are random variables?

10.54   The Gallup Organization conducts frequent surveys in order to estimate the percentage of Americans who approve of the way the President is doing his job. In April of 1985, 795 adults out of a random sample of 1528 adults said that they approved of the way Ronald Reagan was handling his job as President. In May of 1985, 840 adults out of another random sample of 1528 adults said that they approved. At the 5% significance level, do the data suggest that the percentage of Americans approving of President Reagan increased from April to May?

10.55   The U.S. Energy Information Administration surveys American households to estimate the percentage that own various appliances. Euromonitor Publications Limited conducts similar surveys for other countries. Suppose that out of a random sample of 500 American households, 370 own washing machines; and that out of a random sample of 450 French households, 365 own washing machines. Do the data provide evidence that there is a difference in the percentages of American and French households that own washing machines? Use $\alpha = 0.01$.

10.56   The Organization for Economic Cooperation and Development (Paris, France) publishes statistics on labor force participation rates. Suppose that 300 American women and 250 Canadian women are randomly selected, and that 184 of the American women and 148 of the Canadian women are in their respective labor forces. At the 5% significance level, do the data suggest that there is a difference in labor force participation rates between American and Canadian women?

10.57   The U.S. Bureau of the Census conducts annual surveys to obtain information on the percentage of the voting-age population that has registered to vote. Suppose that 400 employed persons and 450 unemployed persons are independently and randomly selected, and that 262 of the employed persons and 224 of the unemployed persons have registered to vote. Can we conclude that the percentage of employed workers who have registered to vote exceeds the percentage of unemployed workers who have registered to vote? Use $\alpha = 0.05$.

10.58   Suppose we tell you that the percentage of U.S. adult males who are married exceeds the percentage of U.S. adult females who are married. Further suppose that to check this claim you randomly select 550 adult males and 575 adult females. You find that 367 of the males selected are married and 353 of the females selected are married. Do your data provide sufficient evidence, at the 5% significance level, to support our claim? Explain.

10.59   A random sample of 300 American dentists contains 77 who practice in the Northeast, and a random sample of 300 American physicians contains 82 who practice in the Northeast. Do these data suggest that the percentage of American dentists practicing in the Northeast is smaller than the percentage of American physicians practicing in the Northeast? Use $\alpha = 0.10$.

10.60 Refer to Exercise 10.54.
a) Determine a 90% confidence interval for the difference, $p_1 - p_2$, between the proportion of Americans approving of President Reagan in April of 1985 and the proportion of Americans approving of President Reagan in May of 1985.
b) Interpret your results in words.

10.61 Refer to Exercise 10.55.
a) Determine a 99% confidence interval for the difference, $p_1 - p_2$, between the proportions of American and French households that own washing machines.
b) Interpret your results in words.

10.62 Refer to Exercise 10.56.
a) Find a 95% confidence interval for the difference, $p_1 - p_2$, between the labor force participation rates of American and Canadian women.
b) Interpret your results in words.

10.63 Refer to Exercise 10.57.
a) Find a 90% confidence interval for the difference, $p_1 - p_2$, between the proportions of employed and unemployed workers who have registered to vote.
b) Interpret your results in words.

10.64 Refer to Exercise 10.58.
a) Find a 90% confidence interval for the difference, $p_1 - p_2$, between the proportion of U.S. adult males who are married and the proportion of U.S. adult females who are married.
b) Interpret your results in words.

10.65 Refer to Exercise 10.59.
a) Find an 80% confidence interval for the difference, $p_1 - p_2$, between the proportions of American dentists and physicians practicing in the Northeast.
b) Interpret your results in words.

10.66 In this exercise we will establish the formulas presented on page 427 for the mean and standard deviation of the random variable $\bar{p}_1 - \bar{p}_2$.
a) Use the results of Exercises 5.26(e) and 5.27(f) on page 190 to show that

$$\mu_{\bar{p}_1 - \bar{p}_2} = \mu_{\bar{p}_1} - \mu_{\bar{p}_2}$$

and

$$\sigma_{\bar{p}_1 - \bar{p}_2} = \sqrt{\sigma_{\bar{p}_1}^2 + \sigma_{\bar{p}_2}^2}$$

b) Apply the formulas $\mu_{\bar{p}} = p$ and $\sigma_{\bar{p}} = \sqrt{p(1-p)/n}$ to the results in part (a) to derive the formulas

$$\mu_{\bar{p}_1 - \bar{p}_2} = p_1 - p_2$$

and

$$\sigma_{\bar{p}_1 - \bar{p}_2} = \sqrt{p_1(1-p_1)/n_1 + p_2(1-p_2)/n_2}$$

10.67 This exercise justifies Procedure 10.10.
a) Use the Key Fact on page 427 to explain why the random variable

$$\frac{(\bar{p}_1 - \bar{p}_2) - (p_1 - p_2)}{\sqrt{\bar{p}_1(1-\bar{p}_1)/n_1 + \bar{p}_2(1-\bar{p}_2)/n_2}}$$

has approximately the standard normal distribution for large samples.
b) Use part (a) to derive the confidence-interval formula given in Step 2 of Procedure 10.10.

10.5 Inferences for more than two population means (ANOVA)*

In previous sections of this chapter we studied inferences for comparing the means of two populations. **Analysis of variance (ANOVA)** provides methods for comparing the means of more than two populations. The reason for the word "variance" in "analysis of variance" is that the procedure for comparing the means involves analyzing the *variation* in the sample data.

In this section we will examine **one-way analysis of variance** which is the generalization to more than two populations of the pooled-*t* procedure discussed in Section 10.2. As in the pooled-*t* procedure, we make the following assumptions.

Assumptions for one-way ANOVA

1 *Independent samples:*   The samples taken from the various populations are independent of one another.
2 *Normal populations:*   The populations being sampled are, at least approximately, normally distributed.
3 *Equal standard deviations:*   The standard deviations of the populations being sampled are equal.

ANOVA procedures utilize a class of continuous probability distributions called **F-distributions** (named in honor of Ronald Fisher). We will begin our study of ANOVA by discussing the F-distribution.

The F-distribution

Probabilities for a random variable having an F-distribution are equal to areas under a curve that, not surprisingly, is called an **F-curve.** Recall that a t-distribution or a chi-square distribution depends on the number of degrees of freedom, df. An F-distribution also depends on the number of degrees of freedom, but there are *two* numbers of degrees of freedom instead of one. Figure 10.7 depicts two different F-curves.

FIGURE 10.7
F-curves for
df = (10, 2)
and df = (9, 50).

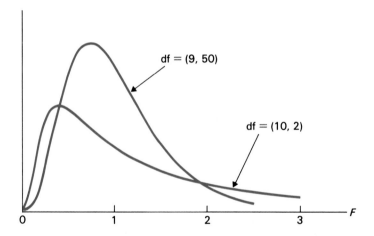

The first number of degrees of freedom for an F-curve is called the **degrees of freedom for the numerator,** and the second number of degrees of freedom is called the **degrees of freedom for the denominator.** [You will see shortly why that terminology is used.] Thus, for the F-curve in Figure 10.7 with df = (10, 2), we have

$$df = (10, 2)$$

degrees of freedom   degrees of freedom
for numerator   for denominator

Some of the fundamental properties of F-curves are as follows:

KEY FACT Basic properties of *F*-curves

PROPERTY 1 *The total area under an F-curve is 1.*
PROPERTY 2 *An F-curve starts at 0 on the horizontal axis and extends indefinitely to the right, approaching the horizontal axis as it does so.*
PROPERTY 3 *An F-curve is not symmetrical, but is* skewed *to the right. That is, it climbs to its high point rapidly and comes back to the axis more slowly.*

Areas under *F*-curves have been compiled and put into tables. These tables give areas that are likely to be used as significance levels, such as 0.01 and 0.05. For *F*-curves, there are entirely separate tables for each area value because critical values are needed for each different combination of degrees of freedom for the numerator and degrees of freedom for the denominator.

As you might expect, the notation F_α will be used to denote the *F*-value with area α to its right. In this book, tables for $F_{0.05}$ and $F_{0.01}$ are given. Let us now consider Table V in the appendix, which gives values for $F_{0.05}$. As you can see, the values of $F_{0.05}$ are displayed inside the table, and the degrees of freedom for the numerator and denominator on the top and side, respectively. The following example illustrates the use of *F*-tables.

EXAMPLE 10.14   **Illustrates how to find the *F*-value for a specified area**

For an *F*-curve with df = (4, 12), find $F_{0.05}$. That is, find the *F*-value with area 0.05 to its right. See Figure 10.8.

FIGURE 10.8

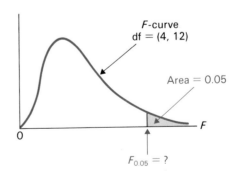

SOLUTION
To determine the *F*-value in question, we use Table V. As we said, the top of the table gives the degrees of freedom for the numerator, which here is 4. The left side of the table gives the degrees of freedom for the denominator, which, in this case, is 12. If you go down the column labelled df = 4 until you are in line with the row labelled df = 12, you will see the number 3.26. This is the desired *F*-value. That is, for an *F*-curve with df = (4, 12), the *F*-value with area 0.05 to its right is 3.26—$F_{0.05}$ = 3.26. ∎

MTB
SPSS

The ideas behind one-way ANOVA

Let us now consider an example for the purpose of presenting the essential ideas behind the one-way analysis of variance procedure and introducing some of the pertinent terminology and notation.

EXAMPLE 10.15   **Introduces one-way ANOVA**

The U.S. Energy Information Administration gathers data on residential energy consumption and expenditures. A researcher wants to know if there is a difference in mean annual energy consumption among the four regions of the United States. Let μ_1, μ_2, μ_3, and μ_4 denote, respectively, last year's mean energy consumption for households in the Northeast, Midwest, South, and West. Then the hypotheses to be tested are

H_0: $\mu_1 = \mu_2 = \mu_3 = \mu_4$ (means of energy consumption are all the same)
H_a: Not all the means are the same.

The basic strategy for carrying out the hypothesis test is as follows:
1 Take a random sample of last year's energy consumption for households in each of the four populations (regions).
2 Compute the sample mean energy consumption $\bar{x}_1$, $\bar{x}_2$, $\bar{x}_3$, and $\bar{x}_4$ for the four samples.
3 Reject the null hypothesis if the sample means differ by too much; otherwise, do not reject the null hypothesis.

This process is depicted in Figure 10.9.

FIGURE 10.9

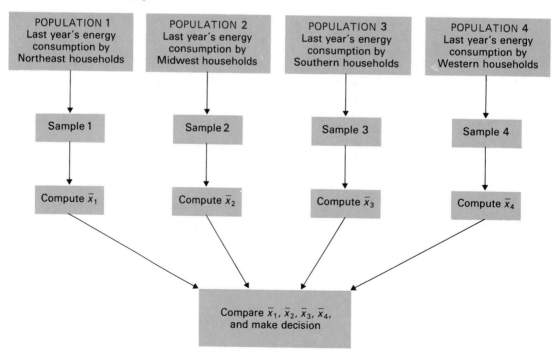

Steps (1) and (2) entail collecting the sample data and computing the sample means. Suppose the results of those steps are as given in Table 10.8.

TABLE 10.8 Samples and their means of last year's energy consumption for households in the four U.S. regions. Data are given to the nearest 10 million BTU.

	Northeast	Midwest	South	West
	15	17	11	10
	10	12	7	12
	13	18	9	8
	14	13	13	7
	13	15		9
		12		
$\bar{x}$	13.0	14.5	10.0	9.2

Step (3) involves comparing the four sample means given at the bottom of Table 10.8. Specifically, we must decide whether the variation among the four sample means can be reasonably attributed to sampling error or whether that variation is large enough to indicate that the population means are not all the same.

In hypothesis tests for *two* population means, we measure the variation between the two sample means by computing their difference, $\bar{x}_1 - \bar{x}_2$. When more than two populations are involved, as in the problem at hand, we cannot measure the variation among the sample means by simply taking a difference. However, we *can* measure the variation among the sample means by computing something like the standard deviation or variance of the sample means or, for that matter, any descriptive statistic that measures the variation among the sample means.

In analysis of variance, we measure the variation among the sample means by a weighted average of their squared deviations about the mean, $\bar{\bar{x}}$, of all the sample data. That measure of variation is called the **treatment mean square, MSTR,** and is defined by

$$MSTR = \frac{SSTR}{k - 1}$$

where k denotes the number of populations being sampled (in this case, $k = 4$) and

$$SSTR = n_1(\bar{x}_1 - \bar{\bar{x}})^2 + n_2(\bar{x}_2 - \bar{\bar{x}})^2 + \cdots + n_k(\bar{x}_k - \bar{\bar{x}})^2$$

The quantity $SSTR$ is called the **treatment sum of squares.**

$MSTR$ is similar to the sample variance of the sample means. In fact, if the sample sizes are all equal, then $MSTR$ is equal to the common sample size times the sample variance of the sample means (see Exercise 10.99).

If the null hypothesis of equal population means is true, then we would expect the sample means to be roughly equal, resulting in a small value for $MSTR$. In other words, if $MSTR$ is too large, then this gives us evidence that the null hypothesis of equal population means is false.

Let us determine $MSTR$ for the sample data in Table 10.8. We have $k = 4$ and, referring to Table 10.8, we see that $n_1 = 5$, $n_2 = 6$, $n_3 = 4$, $n_4 = 5$, and $\bar{x}_1 = 13.0$,

$\bar{x}_2 = 14.5$, $\bar{x}_3 = 10.0$, $\bar{x}_4 = 9.2$. To obtain the overall mean, $\bar{\bar{x}}$, we need to divide the sum, Σx, of all the data in Table 10.8 by the total number, n, of pieces of data:

$$\bar{\bar{x}} = \frac{\Sigma x}{n} = \frac{15 + 10 + 13 + \cdots + 7 + 9}{20} = \frac{238}{20} = 11.9$$

Therefore,

$$
\begin{aligned}
SSTR &= n_1(\bar{x}_1 - \bar{\bar{x}})^2 + n_2(\bar{x}_2 - \bar{\bar{x}})^2 + n_3(\bar{x}_3 - \bar{\bar{x}})^2 + n_4(\bar{x}_4 - \bar{\bar{x}})^2 \\
&= 5(13.0 - 11.9)^2 + 6(14.5 - 11.9)^2 + 4(10.0 - 11.9)^2 + 5(9.2 - 11.9)^2 \\
&= 97.5
\end{aligned}
$$

and so

$$MSTR = \frac{SSTR}{k-1} = \frac{97.5}{4-1} = 32.5$$

This is our measure of variation among the four sample means shown at the bottom of Table 10.8.

The question now, of course, is whether this value of $MSTR$ is large enough to indicate that the null hypothesis of equal population means is false. To decide, we compare $MSTR$ to a measure of variation within the samples. This latter measure is simply the pooled estimate of the common population variance, σ^2. It is called the **error mean square, MSE,** and is defined by

$$MSE = \frac{SSE}{n-k}$$

where k denotes the number of populations under consideration, n denotes the total number of pieces of sample data, and

$$SSE = (n_1 - 1)s_1^2 + (n_2 - 1)s_2^2 + \cdots + (n_k - 1)s_k^2$$

The quantity SSE is called the **error sum of squares.**[†]

For the sample data in Table 10.8, we have $k = 4$, $n_1 = 5$, $n_2 = 6$, $n_3 = 4$, $n_4 = 5$, and $n = 20$. Computing the sample variance for each of the four data sets in Table 10.8, we find that $s_1^2 = 3.5$, $s_2^2 = 6.7$, $s_3^2 = 6.\overline{6}$, and $s_4^2 = 3.7$. Consequently,

$$
\begin{aligned}
SSE &= (n_1 - 1)s_1^2 + (n_2 - 1)s_2^2 + (n_3 - 1)s_3^2 + (n_4 - 1)s_4^2 \\
&= (5 - 1) \cdot 3.5 + (6 - 1) \cdot 6.7 + (4 - 1) \cdot 6.\overline{6} + (5 - 1) \cdot 3.7 \\
&= 82.3
\end{aligned}
$$

and so

$$MSE = \frac{SSE}{n-k} = \frac{82.3}{20-4} = 5.144$$

This is our measure of variation within the samples.

[†] The terms *treatment* and *error* arose from the fact that many ANOVA techniques were first developed to analyze agricultural experiments. In any case, the treatments refer to the different populations, whereas the errors pertain to the variability inherent within the populations.

As we said, we will decide whether the variation among the sample means, *MSTR*, is large enough to indicate that the null hypothesis of equal population means is false by comparing it to the variation within samples, *MSE*. Specifically, we use the test statistic

$$F = \frac{MSTR}{MSE}$$

to make the comparison. Large values of F indicate that *MSTR* is large relative to *MSE* and therefore that the null hypothesis should be rejected. For the energy-consumption data in Table 10.8, we have seen that $MSTR = 32.5$ and $MSE = 5.144$. Thus, the value of the *F*-statistic is

$$F = \frac{MSTR}{MSE} = \frac{32.5}{5.144} = 6.32$$

Is this value of F large enough to suggest that the null hypothesis of equal population means is false? To answer that question we need to know the probability distribution of F. We will discuss that and a few other concepts and then return to complete the hypothesis test considered in this example. ∎

KEY FACT

Suppose that independent random samples of sizes $n_1, n_2, \ldots, n_k$ are to be taken from k normally distributed populations with means $\mu_1, \mu_2, \ldots, \mu_k$, respectively. Further suppose that the standard deviations of the k populations are equal. If $\mu_1 = \mu_2 = \cdots = \mu_k$, then the random variable

$$F = \frac{MSTR}{MSE}$$

has the *F-distribution with df = (k − 1, n − k)*, where n is the total number of pieces of data.

We have now studied all of the elements necessary to construct a procedure for performing a one-way analysis of variance. However, it will be helpful to consider two additional concepts before presenting such a procedure.

One-way ANOVA identity

To begin, we will define another sum of squares. This sum of squares gives a measure of total variation among all the sample data. It is called the **total sum of squares, SST,** and is defined by

$$SST = \Sigma(x - \bar{\bar{x}})^2$$

where the sum extends over all n pieces of sample data. If we divide *SST* by $n - 1$, then we get the sample variance of all the data. So, *SST* really does give us a measure of total variation.

For the sample data in Table 10.8, we have $\bar{\bar{x}} = 11.9$ and so

$$SST = \Sigma(x - \bar{\bar{x}})^2 = (15 - 11.9)^2 + (10 - 11.9)^2 + \cdots + (9 - 11.9)^2$$
$$= 9.61 + 3.61 + \cdots + 8.41$$
$$= 179.8$$

A fundamental identity in one-way ANOVA is that the total sum of squares equals the treatment sum of squares plus the error sum of squares:

$$SST = SSTR + SSE$$

This identity is called the **one-way ANOVA identity.** The one-way ANOVA identity shows that we can partition the total variation in the data into a component representing variation among the sample means and a component representing variation within the samples.

Let us verify the one-way ANOVA identity for the sample data in Table 10.8. We have just seen that $SST = 179.8$ and previously we found that $SSTR = 97.5$ and $SSE = 82.3$. Since

$$179.8 = 97.5 + 82.3$$

we see that $SST = SSTR + SSE$.

ANOVA tables

Next, we will discuss the **one-way ANOVA table.** This table is useful for organizing and summarizing the calculations required to perform a one-way analysis of variance. The general format of a one-way ANOVA table is shown in Table 10.9.

TABLE 10.9
ANOVA table format
for a one-way analysis
of variance.

Source	df	SS	MS = SS/df	F-statistic
Treatment	$k-1$	$SSTR$	$MSTR = \dfrac{SSTR}{k-1}$	$F = \dfrac{MSTR}{MSE}$
Error	$n-k$	SSE	$MSE = \dfrac{SSE}{n-k}$	
Total	$n-1$	SST		

For the energy-consumption data in Table 10.8, we have already computed all of the quantities that appear in the one-way ANOVA table. Table 10.10 displays the one-way ANOVA table for that data.

TABLE 10.10
ANOVA table for
energy-consumption
data.

Source	df	SS	MS = SS/df	F-statistic
Treatment	3	97.5	32.500	6.32
Error	16	82.3	5.144	
Total	19	179.8		

The one-way ANOVA procedure

We now present a step-by-step procedure that can be used to perform a one-way analysis of variance. The procedure utilizes shortcut formulas to compute the sums of squares since those formulas are easier to work with and reduce the possibility of roundoff error. Note that the hypothesis test is always *right-tailed* since the null hypothesis of equal means is rejected only when F is too large.

PROCEDURE 10.11 To perform a one-way ANOVA test for k population means.

Assumptions
1 Independent samples.
2 Normal populations.
3 Equal population standard deviations.

STEP 1 *State the null and alternative hypotheses.*

STEP 2 *Decide on the significance level, α.*

STEP 3 *The critical value is F_α, with $df = (k-1, n-k)$, where n is the total number of pieces of data.*

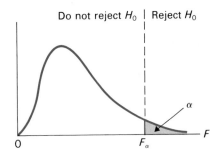

STEP 4 *Calculate the sums of squares using the shortcut formulas*

$$SST = \Sigma x^2 - \frac{(\Sigma x)^2}{n}$$

$$SSTR = \left(\frac{T_1^2}{n_1} + \frac{T_2^2}{n_2} + \cdots + \frac{T_k^2}{n_k}\right) - \frac{(\Sigma x)^2}{n}$$

$$SSE = SST - SSTR$$

where
  n_j = *sample size from Population j,*
  T_j = *sum of sample data from Population j.*

STEP 5 *Construct an ANOVA table.*

Source	df	SS	MS = SS/df	F-statistic
Treatment	$k-1$	SSTR	$MSTR = \dfrac{SSTR}{k-1}$	$F = \dfrac{MSTR}{MSE}$
Error	$n-k$	SSE	$MSE = \dfrac{SSE}{n-k}$	
Total	$n-1$	SST		

STEP 6 *If the value of the F-statistic falls in the rejection region, reject H_0; otherwise, do not reject H_0.*

STEP 7 *State the conclusion in words.*

Note: In Step 4 of Procedure 10.11, we need to compute Σx, the sum of all the data. Generally, it is easiest to do this by summing the T_js; that is, by using the formula $\Sigma x = T_1 + T_2 + \cdots + T_k$.

EXAMPLE 10.16   **Illustrates Procedure 10.11**

We will apply Procedure 10.11 to perform the hypothesis test proposed in Example 10.15. Recall that independent random samples of households in the four U.S. regions gave the following data on last year's energy consumption.

TABLE 10.11

	Northeast	Midwest	South	West
	15	17	11	10
	10	12	7	12
	13	18	9	8
	14	13	13	7
	13	15		9
		12		
Σ	65	87	40	46

At the 5% significance level, do the data provide evidence of a difference in last year's mean energy consumption among households in the four U.S. regions?

SOLUTION

STEP 1 *State the null and alternative hypotheses.*

Let μ_1, μ_2, μ_3, and μ_4 denote last year's true mean energy consumption for households in the Northeast, Midwest, South, and West, respectively. Then the null and alternative hypotheses are

H_0: $\mu_1 = \mu_2 = \mu_3 = \mu_4$ (means of energy consumption are equal)
H_a: Not all the means are the same.

STEP 2 *Decide on the significance level, α.*

We will perform the test at the 5% significance level—$\alpha = 0.05$.

STEP 3 *The critical value is F_a, with df $= (k - 1, n - k)$.*

From Step 2, $\alpha = 0.05$. The number of populations under consideration is four, so $k = 4$. The total number of pieces of data is 20, as we can see from Table 10.11; thus, $n = 20$. Consequently, df $= (k - 1, n - k) = (4 - 1, 20 - 4) = (3, 16)$. We find, therefore, from Table V that the critical value is

$$F_a = F_{0.05} = 3.24$$

See Figure 10.10.

FIGURE 10.10

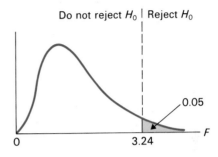

STEP 4 *Calculate the sums of squares using the shortcut formulas.*

We computed the sum of the data for each sample and recorded the results at the bottom of Table 10.11. From the table we see that

$$k = 4$$

$n_1 = 5$	$n_2 = 6$	$n_3 = 4$	$n_4 = 5$
$T_1 = 65$	$T_2 = 87$	$T_3 = 40$	$T_4 = 46$

and

$$n = 5 + 6 + 4 + 5 = 20$$
$$\Sigma x = 65 + 87 + 40 + 46 = 238$$

Summing the squares of all the data in Table 10.11 gives

$$\Sigma x^2 = 15^2 + 10^2 + 13^2 + \cdots + 7^2 + 9^2 = 3012$$

Consequently,

$$SST = \Sigma x^2 - \frac{(\Sigma x)^2}{n} = 3012 - \frac{(238)^2}{20}$$

$$= 3012 - 2832.2 = 179.8$$

and

$$SSTR = \left(\frac{T_1^2}{n_1} + \frac{T_2^2}{n_2} + \frac{T_3^2}{n_3} + \frac{T_4^2}{n_4}\right) - \frac{(\Sigma x)^2}{n}$$

$$= \left(\frac{65^2}{5} + \frac{87^2}{6} + \frac{40^2}{4} + \frac{46^2}{5}\right) - \frac{(238)^2}{20}$$

$$= 2929.7 - 2832.2 = 97.5$$

and

$$SSE = SST - SSTR = 179.8 - 97.5 = 82.3$$

STEP 5 *Construct an ANOVA table.*

The ANOVA table for the energy-consumption data was constructed earlier in Table 10.10 on page 440. We repeat it below.

Source	df	SS	MS = SS/df	F-statistic
Treatment	3	97.5	32.500	6.32
Error	16	82.3	5.144	
Total	19	179.8		

STEP 6 *If the value of the F-statistic falls in the rejection region, reject H_0; otherwise, do not reject H_0.*

From Step 5, the value of the F-statistic is $F = 6.32$. A glance at Figure 10.10 shows that this value falls in the rejection region. Thus we reject H_0.

STEP 7 *State the conclusion in words.*

The data indicate that last year's means of energy consumption for households in the four U.S. regions are not all the same.   ■

MTB
SPSS

Concluding remarks

There are two important issues regarding the one-way analysis-of-variance procedure that we need to discuss. The first issue concerns the assumptions for one-way ANOVA, which are given on page 434—(1) independent samples, (2) normal populations, and (3) equal standard deviations. Assumption (1) on independent samples is absolutely essential to the one-way ANOVA procedure. Assumption (2) on normality is not too critical as long as the populations are not too far from being normally distributed. Assumption (3) on equal standard deviations is also not so important provided that the sample sizes are roughly the same.

The second issue involving the one-way analysis-of-variance procedure may already have occurred to you; namely, suppose we perform an ANOVA test and *reject* the null hypothesis. Then we can conclude that the means of the populations under consideration are *not* all the same. But how can we decide on such things as which means are different, which mean is largest, and so forth? There are methods to answer these and other related questions. The methods are called **multiple comparisons.** We will not cover those procedures here, but instead refer the reader to any of the more advanced books on inferential statistics.

Finally, we should remark that we have barely scratched the surface of ANOVA. Entire books and courses are devoted to the study of analysis of variance. Here we have presented only the simplest type of analysis of variance—one-way ANOVA. However, it is useful to note that the essential idea in all ANOVA procedures is the partitioning of the total sum of squares, *SST,* into several components representing different sources of variation.

Exercises 10.5

10.68 State the three assumptions required for one-way ANOVA.

10.69 One-way ANOVA provides a procedure for comparing the means of several populations. It is the generalization of what procedure for comparing the means of two populations?

10.70 Explain the reason for the word "variance" in "analysis of variance."

10.71 An F-curve has df $= (12, 7)$. What is the number of degrees of freedom for the
a) numerator?
b) denominator?

10.72 An F-curve has df $= (8, 19)$. What is the number of degrees of freedom for the
a) denominator?
b) numerator?

In Exercises 10.73 through 10.78, use Tables IV and V to determine the indicated F-values. Illustrate your work with pictures similar to Figure 10.8.

10.73 For an F-curve with df $= (24, 40)$, find the F-value with
a) area 0.05 to its right.
b) area 0.01 to its right.

10.74 For an F-curve with df $= (12, 5)$, find the F-value with
a) area 0.01 to its right.
b) area 0.05 to its right.

10.75 For an F-curve with df $= (20, 21)$, find
a) $F_{0.01}$
b) $F_{0.05}$

10.76 For an F-curve with df $= (6, 10)$, find
a) $F_{0.05}$
b) $F_{0.01}$

10.77 An F-curve has df $= (30, 15)$. Determine the F-value with
a) area 0.95 to its left.
b) area 0.99 to its left.

10.78 An F-curve has df $= (9, 8)$. Determine the F-value with
a) area 0.99 to its left.
b) area 0.95 to its left.

10.79 If we define $s = \sqrt{MSE}$, then of which parameter is s an estimate?

10.80 Show that for $k = 2$ populations, $MSE = s_p^2$, where s_p^2 is the pooled variance defined in Section 10.2 on page 402; and hence that $\sqrt{MSE}$ is the pooled sample standard deviation, s_p.

In Exercises 10.81 through 10.84,
a) compute $SSTR$, SSE, and SST using the defining formulas.
b) verify that the one-way ANOVA identity holds.
c) determine $MSTR$ and MSE.
d) obtain the one-way ANOVA table for the data.

10.81 A quality-control engineer recorded the times required by three workers to perform an assembly-line task. Each worker was observed on five randomly selected occasions. The times, to the nearest minute, are as follows:

Hank	Joseph	Susan
8	8	10
10	9	9
9	9	10
11	8	11
10	10	9

10.82 A pig farmer wants to test three different diets designed to maximize weight gain. The farmer randomly selects 12 pigs and divides them randomly into three groups of four pigs each. Each group is given one of the diets. The weight gains, in pounds, after a three-week period are shown below.

Diet A	Diet B	Diet C
10.5	11.3	9.8
10.9	10.9	10.5
10.7	11.8	10.4
10.1	11.8	10.2

10.83 Below we have given the SAT scores for randomly selected students from each of four different high school rank categories.

Top tenth	Second tenth	Second fifth	Third fifth
528	514	649	372
586	457	506	440
680	521	556	495
718	370	413	321
	532	470	424
			332

10.84 The U.S. Bureau of the Census collects data on monthly rents of newly completed apartments by region. Random samples of monthly rents for newly completed apartments in the four U.S. regions gave the following sample data:

Northeast	Midwest	South	West
470	408	428	379
363	386	167	366
413	337	398	444
646	452	573	280
709	359		623
	344		

In Exercises 10.85 and 10.86 we have presented two partially completed ANOVA tables. Fill in the missing entries in each table.

10.85

Source	df	SS	MS = SS/df	F-statistic
Treatment	2		21.652	
Error		84.400		
Total	14			

10.86

Source	df	SS	MS = SS/df	F-statistic
Treatment		2.124	0.708	0.75
Error	20			
Total				

In Exercises 10.87 through 10.92 use Procedure 10.11 to perform the appropriate one-way ANOVA tests.

10.87 A consumer-advocacy group wants to compare four different brands of flashlight batteries. Five randomly selected batteries of each brand are tested. The lifetimes of the batteries to the nearest hour are as follows:

Brand A	Brand B	Brand C	Brand D
42	28	24	20
30	31	36	32
39	31	28	38
28	32	28	28
29	27	33	25

At the 5% significance level, does there appear to be a difference in the mean lifetimes of the four brands of batteries?

10.88 The general manager of a large chain of convenience stores wants to try three different advertising policies. The three policies are
Policy 1: No advertising
Policy 2: Advertise in neighborhood with circulars
Policy 3: Use circulars and advertise in local newspapers

Eighteen stores are randomly selected and divided at random into three groups of six stores. Each group uses one of the three policies. Following the implementation of the policies, sales figures are obtained for each of the stores during a one-month period. The results are displayed, in thousands of dollars, in the following table.

Policy 1	Policy 2	Policy 3
22	21	29
20	25	24
21	25	31
21	20	32
24	22	26
22	26	27

Do the data provide evidence of a difference in mean monthly sales among the three policies? Use $\alpha = 0.01$.

10.89 The U.S. Bureau of Labor Statistics gathers data on hourly earnings of nonsupervisory workers in nonfarm U.S. jobs, by industry. The following data were obtained from random samples of workers in three different industries. Data are in dollars per hour.

Wholesale trade	Finance, Insurance & Real estate	Services
9.78	9.68	9.16
11.86	9.32	9.61
7.60	8.67	3.67
8.05	4.42	
	10.30	

Do the data indicate a difference in mean hourly earnings for nonsupervisory workers in the three industries? Use $\alpha = 0.05$.

10.90 Manufacturers of golf balls seem to always be saying that their ball goes the farthest. A writer for a sports magazine decides to conduct an impartial test. He hires a golf professional to drive five different brands of balls. Each ball is driven four times. The results are as follows. Data are in yards.

Brand 1	Brand 2	Brand 3	Brand 4	Brand 5
249	254	240	251	251
246	247	232	241	263
251	254	247	239	246
244	258	256	245	262

Do the data suggest a difference in the mean distances that the golf pro can drive the five brands of balls? Use $\alpha = 0.05$.

10.91 The U.S. Bureau of Prisons compiles data on the time served by prisoners released from federal institutions for the first time. Independent random samples of released prisoners for five different offense categories yielded the following information on time served (in months):

	Counter-feiting	Drug laws	Firearms	Forgery	Fraud
n_i	15	17	12	10	11
T_i	218	313	218	156	127

$$\Sigma x^2 = 17,769$$

Does there appear to be a difference in the mean times served by prisoners in the five offense groups? Use $\alpha = 0.01$.

10.92 The National Education Association collects data on annual starting salaries of college graduates, by major. Independent random samples of college graduates in marketing, statistics, economics, and computer science provided the following information on annual starting salaries (salary data are given to the nearest thousand dollars):

	Marketing	Statistics	Economics	CS
n_i	35	25	30	34
T_i	683	600	622	870

$$\Sigma x^2 = 64,239$$

Do the data suggest a difference in mean annual starting salaries among the four majors? Take $\alpha = 0.05$.

In Exercises 10.93 through 10.97 we will examine some **confidence interval procedures in one-way ANOVA**.

Suppose independent random samples of sizes n_1, $n_2, \ldots, n_k$ are to be taken from k normally distributed populations with means $\mu_1, \mu_2, \ldots, \mu_k$, respectively.

Further suppose the standard deviations of the k populations are equal. Let $s = \sqrt{MSE}$.

1 A $(1 - \alpha)$-level confidence interval for any particular population mean, say μ_i, is from

$$\bar{x}_i - t_{\alpha/2} \cdot \frac{s}{\sqrt{n_i}} \quad \text{to} \quad \bar{x}_i + t_{\alpha/2} \cdot \frac{s}{\sqrt{n_i}}$$

where $df = n - k$.

2 A $(1 - \alpha)$-level confidence interval for the difference between any two particular population means, say μ_i and μ_j, has endpoints

$$(\bar{x}_i - \bar{x}_j) \pm t_{\alpha/2} \cdot s \sqrt{(1/n_i) + (1/n_j)}$$

where $df = n - k$.

10.93 Refer to Exercise 10.81.
  a) Find a 95% confidence interval for the mean time, μ_2, it takes Joseph to perform the assembly-line task.
  b) Find a 95% confidence interval for the difference, $\mu_1 - \mu_3$, between the mean times it takes Hank and Susan to perform the assembly-line task.

10.94 Refer to Exercise 10.82.
  a) Determine a 95% confidence interval for the mean weight gain, μ_1, after three weeks on Diet A.
  b) Determine a 95% confidence interval for the difference, $\mu_2 - \mu_3$, between the mean weight gains after three weeks on Diet B and three weeks on Diet C.

10.95 Refer to Exercise 10.83.
  a) Obtain a 90% confidence interval for the mean SAT score, μ_3, for students ranked in the second fifth of their high school class.
  b) Obtain a 90% confidence interval for the difference, $\mu_1 - \mu_4$, between mean SAT scores of students ranked in the top tenth and third fifth of their high school class.

10.96 Refer to Exercise 10.84.
  a) Find a 99% confidence interval for the mean monthly rent, μ_2, of newly completed apartments in the Midwest.
  b) Find a 99% confidence interval for the difference, $\mu_1 - \mu_3$, between the mean monthly rents of newly completed apartments in the Northeast and South.

10.97 Refer to Exercise 10.83. Suppose you have obtained a 90% confidence interval for each of the two differences $\mu_1 - \mu_2$ and $\mu_1 - \mu_3$. Can you be 90% confident of both results simultaneously? In other words, can you be 90% confident that both differences are in their corresponding confidence intervals? Explain your answer.

10.98   Recall that $\bar{\bar{x}}$ is the mean of all n pieces of sample data.

a) Show that $\bar{\bar{x}}$ is a weighted average of the k sample means, weighted according to sample size. That is,

$$\bar{\bar{x}} = \frac{n_1\bar{x}_1 + n_2\bar{x}_2 + \cdots + n_k\bar{x}_k}{n_1 + n_2 + \cdots + n_k}$$

b) Prove that if the sample sizes are all equal, then $\bar{\bar{x}}$ is just the mean of the sample means.

10.99   Suppose the sample sizes $n_1, n_2, \ldots, n_k$ are all equal, say to m. Show that, under those circumstances, $MSTR$ is equal to m times the sample variance of the k sample means $\bar{x}_1, \bar{x}_2, \ldots, \bar{x}_k$. Hint: Use part (b) of Exercise 10.98.

10.100   Consider two normally distributed populations with means μ_1 and μ_2. Assume the standard deviations of the two populations are the same. Suppose we wish to perform a hypothesis test to decide whether $\mu_1 \neq \mu_2$. If independent samples are used, describe two hypothesis testing procedures that can be employed to perform the test.

10.101   In this exercise we will derive the shortcut formulas for the sums of squares SST, $SSTR$, and SSE.

a) Show that

$$\Sigma(x - \bar{x})^2 = \Sigma x^2 - \frac{(\Sigma x)^2}{n}$$

where the sums extend over all the data.

b) Use part (a) and the defining formula for SST to deduce the shortcut formula

$$SST = \Sigma x^2 - \frac{(\Sigma x)^2}{n}$$

c) Show that for each j,

$$n_j\left(\frac{T_j}{n_j} - \frac{\Sigma x}{n}\right)^2 = \frac{T_j^2}{n_j} - \frac{2\Sigma x}{n}T_j + \frac{(\Sigma x)^2}{n^2}n_j$$

d) Use part (c) to show that

$$\Sigma n_j\left(\frac{T_j}{n_j} - \frac{\Sigma x}{n}\right)^2 = \Sigma\frac{T_j^2}{n_j} - \frac{(\Sigma x)^2}{n}$$

e) Conclude from part (d) and the defining formula for $SSTR$ that the shortcut formula

$$SSTR = \Sigma\frac{T_j^2}{n_j} - \frac{(\Sigma x)^2}{n}$$

is valid.

f) Why does $SSE = SST - SSTR$?

10.6   Computer packages*

In this section we will illustrate how Minitab can be employed to perform several of the inferential procedures discussed in this chapter. Specifically, we will learn how Minitab can be used to perform the pooled and nonpooled procedures considered in Section 10.2 and the one-way ANOVA procedure considered in Section 10.5.

TWOSAMPLE-T

Procedure 10.5 (page 408) and Procedure 10.6 (page 411) provide step-by-step methods for performing hypothesis tests and obtaining confidence intervals to compare two population means. These procedures are designed to be used with independent samples from two normally distributed populations. Minitab has a program called **TWOSAMPLE-T** that will perform Procedures 10.5 and 10.6 for us. We illustrate the use of this program in the following example.

EXAMPLE 10.17 **Illustrates the use of TWOSAMPLE-T**

A general contractor wants to compare the lifetimes of two major brands of electric water heaters—Eagle and National. The contractor obtained the sample data displayed in Table 10.12. Determine, at the 5% significance level, whether the two brands of water heaters have different mean lifetimes. [Assume that the lifetimes of each brand are approximately normally distributed.]

TABLE 10.12
Lifetimes, in years,
of water heaters
sampled.

Eagle	National
6.9	8.7
7.3	7.0
7.8	8.7
7.4	6.7
7.2	7.8
6.6	8.6
6.2	6.1
8.2	7.5
7.6	7.7
5.7	7.5
5.5	11.2
6.9	6.1
	6.3
	7.0
	10.7

S O L U T I O N
The problem is to perform the two-tailed hypothesis test

$$H_0: \mu_1 = \mu_2 \text{ (mean lifetimes are the same)}$$
$$H_a: \mu_1 \neq \mu_2 \text{ (mean lifetimes are different)}$$

with $\alpha = 0.05$. In Example 10.6 on page 409, we used Procedure 10.5 to perform the hypothesis test. Here we will use TWOSAMPLE-T to do the hypothesis test.

First we enter the sample data in Table 10.12 into the computer. Then we use the command TWOSAMPLE-T followed by the storage locations of the two data sets (C8 and C9). The command and its results are shown in Printout 10.1.

PRINTOUT 10.1
Minitab output for
TWOSAMPLE-T.

```
MTB > TWOSAMPLE-T, C8, C9

TWOSAMPLE T FOR C8 VS C9
        N      MEAN     STDEV    SE MEAN
C8     12     6.942     0.823     0.238
C9     15      7.84      1.53     0.396

95 PCT CI FOR MU C8 - MU C9: (-1.856, 0.05942)
TTEST MU C8 = MU C9 (VS NE): T=-1.95 P=0.065 DF=22.2
```

The first line of the output describes the type of test being performed— TWOSAMPLE T FOR C8 VS C9. The next three lines give the sample size, sample mean, sample standard deviation, and estimated standard error of the mean for each of the two samples. The fifth line displays a 95% confidence interval for the difference, $\mu_1 - \mu_2$, between the two population means. The final line gives a statement of the null and alternative hypotheses, followed by the value of the test statistic ($T = -1.95$), the P-value ($P = 0.065$), and the degrees of freedom ($DF = 22.2$).

Recall that the P-value is the only quantity that we really need to make our decision concerning the hypothesis test. The P-value, which here is 0.065, is the smallest significance level at which we can reject the null hypothesis. In other words, if the P-value is less than the designated significance level, α, then we reject H_0; otherwise, we do not reject H_0. Since the P-value of 0.065 is not less than the designated significance level of $\alpha = 0.05$, we do not reject H_0. That is, the data do not provide sufficient evidence to conclude that there is a difference in mean lifetimes for the two brands of water heaters.   ∎

TWOSAMPLE-T; POOLED

Procedure 10.3 (page 403) and Procedure 10.4 (page 406) provide step-by-step methods for performing hypothesis tests and obtaining confidence intervals to compare two population means. These procedures are designed to be used with independent samples from two normally distributed populations with equal, but unknown, standard deviations.

We sometimes refer to Procedures 10.3 and 10.4 as *pooled-t* procedures. This is because the quantity s_p, which is used in those procedures as an estimate for the common value of the standard deviations of the two populations, is obtained by pooling the sample standard deviations s_1 and s_2. Minitab's version of Procedures 10.3 and 10.4 is **TWOSAMPLE-T** with the subcommand **POOLED**. We will illustrate the use of that program in our next example.

EXAMPLE 10.18   **Illustrates the use of TWOSAMPLE-T; POOLED**

The U.S. National Center for Health Statistics gathers and publishes data on the daily intake of selected nutrients by race and income levels. Suppose we are considering protein intake and want to compare the mean daily intake of people with incomes above the poverty level to the mean daily intake of people with incomes below the poverty level. The data in Table 10.13 give the protein intakes, in grams, over a 24-hour period for random samples of people with incomes above and below the poverty level.

TABLE 10.13

Above poverty level		Below poverty level		
86.0	69.0	51.4	49.7	72.0
59.7	80.2	76.7	65.8	55.0
68.6	78.1	73.7	62.1	79.7
98.6	69.8	66.2	75.8	65.4
87.7	77.2	65.5	62.0	73.3

At the 5% significance level, do the data suggest that people with incomes above the poverty level have a *greater* mean daily intake of protein than those with incomes below the poverty level? [Assume that daily intakes of protein for both people with incomes above and below the poverty level are approximately normally distributed and that the standard deviations for both are equal.]

SOLUTION

We need to perform the hypothesis test

$$H_0: \mu_1 = \mu_2 \text{ (above-poverty mean is not greater)}$$
$$H_a: \mu_1 > \mu_2 \text{ (above-poverty mean is greater)}$$

with $\alpha = 0.05$. Here μ_1 and μ_2 denote the true mean daily intakes of protein for people with incomes above and below the poverty level, respectively. Note that the test is right-tailed.

In Example 10.4 we applied Procedure 10.3 to perform the hypothesis test. Here we will use Minitab. First we enter the two data sets into the computer. Then we use the command TWOSAMPLE-T followed by the storage locations of the two data sets (C10 and C11). Next, we employ the subcommand POOLED to indicate that the pooled procedure is the one we want. Finally, we type the subcommand ALTERNATIVE = 1 to tell the computer to perform a right-tailed test. These commands along with the resulting output are shown in Printout 10.2.

PRINTOUT 10.2

Minitab output for TWOSAMPLE–T; POOLED.

```
MTB >  TWOSAMPLE–T, C10, C11;
SUBC>  POOLED;
SUBC>  ALTERNATIVE=1.

TWOSAMPLE T FOR C10 VS C11
          N       MEAN      STDEV     SE MEAN
C10    10        77.5       11.3       3.59
C11    15       66.29       9.17       2.37

95 PCT CI FOR MU C10  —  MU C11: (2.694, 19.71)
TTEST MU C10  =  MU C11 (VS GT): T=2.72 P=0.0060 DF=23.0
```

The printout displays the same information as TWOSAMPLE-T without the POOLED subcommand. We see that the *P*-value is 0.0060 and since this is less than the designated significance level of $\alpha = 0.05$, we reject H_0. In other words, we have sufficient evidence to claim that people with incomes above the poverty level have a greater mean daily intake of protein than people with incomes below the poverty level. ■

AOVONEWAY

Procedure 10.11 provides a step-by-step method for performing a one-way analysis of variance. A Minitab program called **AOVONEWAY** will perform Procedure 10.11 for us. Let us see how this program works by applying it to the hypothesis test considered in Example 10.16 on page 442.

EXAMPLE 10.19   **Illustrates the use of AOVONEWAY**

The U.S. Energy Information Administration gathers data on residential energy consumption and expenditures. A researcher wants to know if there is a difference in mean annual energy consumption among the four regions of the United States. Independent random samples of households in those four regions yield the following data on last year's energy consumption. The data are given to the nearest 10 million BTU.

TABLE 10.14

Northeast	Midwest	South	West
15	17	11	10
10	12	7	12
13	18	9	8
14	13	13	7
13	15		9
	12		

Do the data provide sufficient evidence to conclude that a difference exists in last year's mean energy consumption among households in the four U.S. regions? Use $\alpha = 0.05$.

SOLUTION

The problem is to perform a one-way ANOVA to test the hypotheses

H_0: $\mu_1 = \mu_2 = \mu_3 = \mu_4$ (means of energy consumption are equal)
H_a: Not all the means are the same

at the 5% significance level. Here μ_1, μ_2, μ_3, and μ_4 are, respectively, last year's mean energy consumption for households in the Northeast, Midwest, South, and West.

To have Minitab perform the hypothesis test, we first enter into the computer the sample data displayed in Table 10.14. Then we type the command AOVONEWAY followed by the storage locations of the four sets of sample data (C12–C15). The command and its results are shown in Printout 10.3.

PRINTOUT 10.3
Minitab output
for AOVONEWAY.

```
MTB > AOVONEWAY, C12-C15

ANALYSIS OF VARIANCE
SOURCE     DF        SS        MS        F
FACTOR      3     97.50     32.50     6.32
ERROR      16     82.30      5.14
TOTAL      19    179.80
                                    INDIVIDUAL 95 PCT CI'S FOR MEAN
                                    BASED ON POOLED STDEV
LEVEL       N      MEAN     STDEV  -------+---------+---------+-----
C12         5    13.000     1.871                 (------*-------)
C13         6    14.500     2.588                    (-----*------)
C14         4    10.000     2.582    (-------*-------)
C15         5     9.200     1.924  (-------*------)
                                   -------+---------+---------+-----
POOLED STDEV =     2.268              9.0       12.0      15.0
```

The first part of the computer printout displays a one-way ANOVA table. This is Minitab's version of the one-way ANOVA table we obtained in Table 10.10 on page 440. Note that Minitab uses the terminology "Factor" instead of "Treatment."

From the ANOVA table we can find almost all the information required to perform the hypothesis test. The first two entries in the DF-column give the degrees of freedom for the F-statistic; thus, df = (3, 16). Consulting Table V we see that for a test at the 5% significance level with df = (3, 16), the critical value is $F_{0.05} = 3.24$. Comparing this to the value of the F-statistic, $F = 6.32$, given in the upper right-hand corner of the ANOVA table, we conclude that the null hypothesis should be rejected. In other words, the data provide sufficient evidence to conclude that a difference exists in last year's mean energy consumption among households in the four U.S. regions. ∎

As you can see, the ANOVA table printed out by Minitab permits us to perform a one-way analysis of variance with absolutely no calculations. This not only saves us a tremendous amount of time, but also eliminates the problem of computational errors.

Printout 10.3 gives us more information than required for a one-way ANOVA. The lower left-hand side of the printout provides a table for the sample sizes, sample means, and sample standard deviations of the four samples. Below that table we find "POOLED STDEV = 2.268." This is the pooled estimate of the common population standard deviation, σ. It can also be obtained by taking the square root of MSE: $\sqrt{MSE} = \sqrt{5.14} = 2.268$. Finally, in the lower right-hand side of the printout, Minitab prints individual 95% confidence intervals for the means μ_1, μ_2, μ_3, and μ_4 of the four populations under consideration.

◆ Chapter Review

KEY TERMS

analysis of variance (ANOVA)*, 433
AOVONEWAY*, 451
df for denominator*, 434
df for numerator*, 434
error mean square (MSE)*, 438
error sum of squares (SSE)*, 438
F-curve*, 434
F-distribution*, 434
independent samples, 391
normal differences, 417
one-way analysis of variance*, 433
one-way ANOVA identity*, 440
one-way ANOVA table*, 440
paired difference, 416

paired samples, 416
pool, 402
POOLED*, 450
pooled sample proportion, 428
pooled sample standard deviation, 402
sampling distribution of the difference
    between two means, 393
sampling distribution of the difference
    between two proportions, 426
total sum of squares (SST)*, 439
treatment mean square (MSTR)*, 437
treatment sum of squares (SSTR)*, 437
TWOSAMPLE-T*, 448

FORMULAS

In the formulas below,

μ_1, μ_2 = population means
$\overline{x}_1, \overline{x}_2$ = sample means
s_1, s_2 = sample standard deviations
n_1, n_2 = sample sizes
p_1, p_2 = population proportions
$\overline{p}_1, \overline{p}_2$ = sample proportions
x_1, x_2 = numbers of "successes"

and

$\overline{d}$ = sample mean of paired differences
s_d = sample standard deviation of paired differences

Test statistic for H_0: $\mu_1 = \mu_2$ (large, independent samples), 395

$$z = \frac{\overline{x}_1 - \overline{x}_2}{\sqrt{(s_1^2/n_1) + (s_2^2/n_2)}}$$

Confidence interval for $\mu_1 - \mu_2$ (large, independent samples), 398

$$(\overline{x}_1 - \overline{x}_2) \pm z_{\alpha/2} \cdot \sqrt{(s_1^2/n_1) + (s_2^2/n_2)}$$

Pooled sample standard deviation, 402

$$s_p = \sqrt{\frac{(n_1 - 1)s_1^2 + (n_2 - 1)s_2^2}{n_1 + n_2 - 2}}$$

Test statistic for H_0: $\mu_1 = \mu_2$ (normal populations, independent samples, and σs unknown but assumed equal), 404

$$t = \frac{\overline{x}_1 - \overline{x}_2}{s_p\sqrt{(1/n_1) + (1/n_2)}}$$

with df = $n_1 + n_2 - 2$.

Confidence interval for $\mu_1 - \mu_2$ (normal populations, independent samples, and σs unknown but assumed equal), 406

$$(\overline{x}_1 - \overline{x}_2) \pm t_{\alpha/2} \cdot s_p\sqrt{(1/n_1) + (1/n_2)}$$

with df = $n_1 + n_2 - 2$.

Test statistic for H_0: $\mu_1 = \mu_2$ (normal populations, independent samples), 408

$$t = \frac{\overline{x}_1 - \overline{x}_2}{\sqrt{(s_1^2/n_1) + (s_2^2/n_2)}}$$

with df = $[(s_1^2/n_1) + (s_2^2/n_2)]^2/[(s_1^2/n_1)^2/(n_1 - 1) + (s_2^2/n_2)^2/(n_2 - 1)]$ rounded down to the nearest integer.

Confidence interval for $\mu_1 - \mu_2$ (normal populations, independent samples), 411

$$(\overline{x}_1 - \overline{x}_2) \pm t_{\alpha/2} \cdot \sqrt{(s_1^2/n_1) + (s_2^2/n_2)}$$

with df = $[(s_1^2/n_1) + (s_2^2/n_2)]^2/[(s_1^2/n_1)^2/(n_1 - 1) + (s_2^2/n_2)^2/(n_2 - 1)]$ rounded down to the nearest integer.

Test statistic for $H_0: \mu_1 = \mu_2$ (paired samples and normal differences), 419

$$t = \frac{\bar{d}}{s_d/\sqrt{n}}$$

with df $= n - 1$.

Confidence interval for $\mu_1 - \mu_2$ (paired samples and normal differences), 421

$$\bar{d} \pm t_{\alpha/2} \cdot \frac{s_d}{\sqrt{n}}$$

with df $= n - 1$.

Pooled sample proportion, 428

$$\bar{p} = \frac{x_1 + x_2}{n_1 + n_2}$$

Test statistic for $H_0: p_1 = p_2$ (large, independent samples), 429

$$z = \frac{\bar{p}_1 - \bar{p}_2}{\sqrt{\bar{p}(1 - \bar{p})}\sqrt{(1/n_1) + (1/n_2)}}$$

Confidence interval for $p_1 - p_2$ (large, independent samples), 431

$$(\bar{p}_1 - \bar{p}_2) \pm z_{\alpha/2} \cdot \sqrt{\bar{p}_1(1 - \bar{p}_1)/n_1 + \bar{p}_2(1 - \bar{p}_2)/n_2}$$

In the formulas below,

$k =$ number of populations	$SSTR =$ treatment sum of squares
$n =$ total number of pieces of data	$SSE =$ error sum of squares
$n_j =$ sample size for Population j	$MSTR =$ treatment mean square
$T_j =$ sum of sample data from Population j	$MSE =$ error mean square
$SST =$ total sum of squares	

One-way ANOVA identity*, 440

$$SST = SSTR + SSE$$

Shortcut formulas in one-way ANOVA*, 441

$$SST = \Sigma x^2 - \frac{(\Sigma x)^2}{n}$$

$$SSTR = \left(\frac{T_1^2}{n_1} + \frac{T_2^2}{n_2} + \cdots + \frac{T_k^2}{n_k}\right) - \frac{(\Sigma x)^2}{n}$$

$$SSE = SST - SSTR$$

$$MSTR = \frac{SSTR}{k - 1}$$

$$MSE = \frac{SSE}{n - k}$$

F-statistic for one-way ANOVA*, 442

$$F = \frac{MSTR}{MSE}$$

with df $= (k - 1, n - k)$.

YOU SHOULD BE ABLE TO . . .

1 use and understand the preceding formulas.

2 perform large-sample inferences to compare the means of two populations using independent samples.

3 perform inferences to compare the means of two normal populations using independent samples when the population standard deviations are unknown and when they are unknown but assumed equal.

4 perform inferences to compare the means of two populations using paired samples, when the population of paired differences is normally distributed.

5 perform large-sample inferences to compare the proportions of two two-category populations using independent samples.

6* use the F-tables, Tables IV and V.

7* explain the essential ideas behind one-way ANOVA.

8* compute the mean squares and sums of squares for a one-way ANOVA using both the defining formulas and the shortcut formulas.

9* construct a one-way ANOVA table.

10* perform a one-way analysis of variance to compare the means of k populations.

REVIEW TEST

1 *Better Homes and Gardens* conducts semi-annual surveys of housing in 100 cities across the United States. One item of interest is the average price for resale homes. Suppose that $n_1 = 50$ randomly selected resale home prices in Phoenix, Arizona have a mean of $\bar{x}_1 = \$81,525$, with a standard deviation of $s_1 = \$29,670$; and that $n_2 = 75$ randomly selected resale home prices in Flint, Michigan have a mean of $\bar{x}_2 = \$47,603$, with a standard deviation of $s_2 = \$12,466$. Do the data suggest that the mean resale price for homes in Phoenix, Arizona *exceeds* that for homes in Flint, Michigan? Perform the test at the 1% significance level.

2 Refer to Problem 1. Determine a 98% confidence interval for the difference, $\mu_1 - \mu_2$, between the mean resale prices of homes in Phoenix, Arizona and Flint, Michigan.

3 Two speed reading programs are to be compared to see whether there is a difference in results. For the test, 10 pairs of people are randomly selected, where each pair consists of people whose present reading speeds are nearly identical. One person from each pair is randomly selected to take Program 1 and the other takes Program 2. After the completion of the speed reading programs, the speeds for the 10 pairs are recorded. Here are the speeds, in words per minute.

Program 1	Program 2
1114	1032
996	1148
979	1074
1125	1076
910	959
1056	1094
1091	1091
1053	1096
996	1032
894	1012

a) At the 10% significance level, do the data provide evidence that there is a difference in mean results for the two speed reading programs?

b) What assumption are you making in performing the hypothesis test in part (a)?

4 Refer to Problem 3. Obtain a 90% confidence interval for the difference, $\mu_1 - \mu_2$, between the mean reading speeds of people using Program 1 and people using Program 2.

5 The U.S. National Science Foundation surveys doctoral scientists for selected characteristics. Suppose that 250 male and 175 female doctoral scientists are selected at random, and that 15 of the males and 19 of the females are unemployed. Does it appear from the data that the percentage of unemployed male doctoral scientists is lower than that for females? Use $\alpha = 0.05$.

6 Refer to Problem 5. Find a 90% confidence interval for the difference, $p_1 - p_2$, between the proportions of unemployed male and female doctoral scientists.

7 A psychology professor teaching at a large university in the Northeast wants to know whether there is a difference in mean IQ between male and female students in attendance. She randomly selects 20 female and 20 male students and has them take IQ tests. The results are as follows:

Female				Male			
130	109	126	116	106	131	109	116
117	100	118	122	114	133	101	120
124	118	131	115	134	134	144	119
120	127	104	129	120	122	122	114
125	130	130	117	107	111	124	110

$(\Sigma x = 2408, \Sigma x^2 = 291{,}396)$ | $(\Sigma x = 2391, \Sigma x^2 = 288{,}239)$

Do the data indicate, at the 5% significance level, that there is a difference in mean IQ between male and female students at the university? [Assume that IQs for both female and male students at the university are approximately normally distributed with equal standard deviations.]

8 Refer to Problem 7. Determine a 95% confidence interval for the difference between the mean IQs of female and male students at the university.

9 Euromonitor Publications Limited provides data on per capita food consumption of major food commodities for various countries. Suppose that random samples of 10 Germans and 15 Russians consumed the following quantities of fish during last year. Data are in kilograms.

Germans		Russians		
17	17	16	21	12
1	9	11	5	23
15	6	19	19	22
10	13	16	23	12
14	11	18	7	17

$(\bar{x}_1 = 11.30, s_1 = 5.06)$ | $(\bar{x}_2 = 16.07, s_2 = 5.61)$

Test whether Germans consume less fish than Russians on the average. Use $\alpha = 0.05$. [Assume that fish consumption is normally distributed.]

10 Refer to Problem 9. Find a 90% confidence interval for the difference, $\mu_1 - \mu_2$, between last year's mean fish consumption by Germans and Russians.

11[*] Consider an F-curve with df $= (24, 5)$.
a) What is the number of degrees of freedom for the numerator?
b) What is the number of degrees of freedom for the denominator?
c) Find the F-value with area 0.01 to its right.
d) Determine $F_{0.05}$.

12[*] The U.S. Federal Bureau of Investigation collects and publishes data on the value of losses due to various types of robbery. Independent random samples of highway-robbery reports, gas-station-robbery reports, and convenience-store-robbery reports gave the following data on value of losses. The data are given to the nearest dollar.

Highway	Gas station	Convenience store
411	314	575
320	356	442
496	379	458
410	424	475
429	365	376
	532	548

a) Compute the sample means and sample standard deviations of the three data sets.
b) Determine $MSTR$ and MSE using the defining formulas.
c) What is $MSTR$ measuring?
d) What is MSE measuring?

13[*] A cereal company wants to test the possible market effect of four different designs for its boxes. Independent random samples of five markets each are selected. Each design is tried in one of the five-market groups. The number of cases sold during the test period are given below.

Design A	Design B	Design C	Design D
41	51	44	58
51	65	58	37
52	35	37	24
50	66	75	54
43	57	54	65

a) Compute SST, $SSTR$, and SSE using the shortcut formulas.
b) Construct the one-way ANOVA table.
c) At the 1% significance level, do the data provide evidence of a difference in mean sales among the four designs?

The statistical inference techniques presented thus far have dealt exclusively with hypothesis tests and confidence intervals for means and proportions. Now we consider some other widely used inferential procedures. These include hypothesis tests concerning the percentage distribution of a population, tests for determining whether two characteristics of a population are statistically dependent, and inferences for the standard deviation of a population. The procedures we examine in this chapter are based on a continuous probability distribution called the *chi-square distribution* and are therefore often referred to as **chi-square procedures.**

Chi-square procedures

CHAPTER OUTLINE

11.1   The chi-square distribution

All of the statistical inference procedures discussed in this chapter rely on a class of continuous probability distributions called **chi-square distributions.** So that in future sections we can concentrate on the development and application of chi-square procedures, we will begin this chapter with a discussion of the chi-square distribution.

As with the normal and t-distributions, probabilities for a random variable having a chi-square distribution are equal to areas under a curve—a χ^2 **(chi-square) curve.** Actually, there are infinitely many χ^2-curves, and we identify the χ^2-curve in question by giving its number of degrees of freedom, just like for t-curves. Figure 11.1 shows three different χ^2-curves.

FIGURE 11.1 χ^2-curves for df $= 5$, 10, and 19.

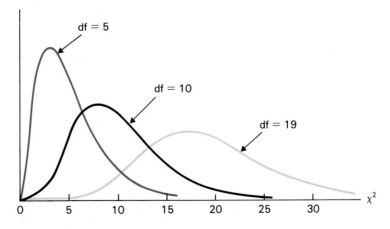

In this figure we can observe some basic properties of χ^2-curves.

KEY FACT Basic properties of χ^2-curves

PROPERTY 1 *The total area under a χ^2-curve is 1.*

PROPERTY 2 *A χ^2-curve starts at 0 on the horizontal axis and extends indefinitely to the right, approaching the horizontal axis as it does so.*

PROPERTY 3 *A χ^2-curve is not symmetrical. It climbs to its high point rapidly and comes back to the axis more slowly—it is **skewed** to the right.*

PROPERTY 4 *As the number of degrees of freedom gets larger, χ^2-curves look increasingly like normal curves.*

To perform a hypothesis test or find a confidence interval that is based on a chi-square distribution, we will need to be able to determine areas under χ^2-curves. A table of areas for χ^2-curves is given in Table VI, which can be found in the left back inside cover of the book. As you can see, this table is similar to the t-distribution table. We will illustrate the use of the χ^2-distribution table in several examples.

EXAMPLE 11.1 **Illustrates how to find the χ^2-value for a specified area**

For a χ^2-curve with 12 degrees of freedom, find the χ^2-value with area 0.025 to its right. See Figure 11.2.

FIGURE 11.2

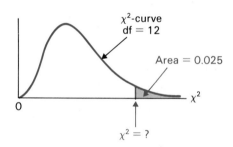

SOLUTION

To determine the χ^2-value in question, we use Table VI. The two outside columns, labelled df, give the number of degrees of freedom. In this case, the number of degrees of freedom is 12, so we concentrate on the row of the table with df = 12. If you go across that row until you are under the column headed $\chi^2_{0.025}$, then you will see the number 23.337. This is the desired χ^2-value. That is, for a χ^2-curve with df = 12, the χ^2-value with area 0.025 to its right is $\chi^2 = 23.337$. See Figure 11.3.

FIGURE 11.3

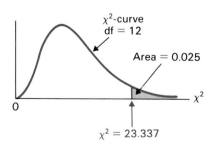

As you might expect, we will use the notation $\chi^2_{0.025}$ to denote the χ^2-value with area 0.025 to its *right*. So, as we have just seen, for a χ^2-curve with df = 12, $\chi^2_{0.025} = 23.337$.   ∎

EXAMPLE 11.2   Illustrates how to find the χ^2-value for a specified area

Find the χ^2-value with area 0.05 to its left, for a χ^2-curve with df = 7. See Figure 11.4.

FIGURE 11.4

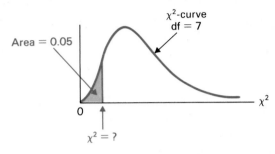

SOLUTION
To determine the χ^2-value in question, we first note that since the total area under the χ^2-curve is 1, the *unshaded* area in Figure 11.4 must equal $1 - 0.05 = 0.95$. Thus, the area to the *right* of the desired χ^2-value is 0.95. This means that the desired χ^2-value is $\chi^2_{0.95}$. From Table VI, we find that for df = 7, $\chi^2_{0.95} = 2.167$. See Figure 11.5.   ∎

FIGURE 11.5

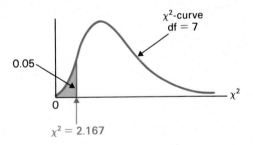

EXAMPLE 11.3   Illustrates how to find the χ^2-values for a specified area

For a χ^2-curve with df = 20, determine the *two* χ^2-values that divide the area under the curve into a middle 0.95 area and two outside 0.025 areas. See Figure 11.6.

FIGURE 11.6

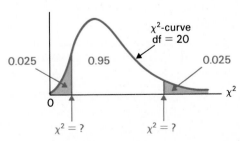

SOLUTION

Since the shaded area on the right is 0.025, we see that the right-hand χ^2-value is $\chi^2_{0.025}$. Using Table VI we find that for df $= 20$, $\chi^2_{0.025} = 34.170$. To obtain the left-hand χ^2-value, we first note that, because the area to the left of this value is 0.025, the area to its right is $1 - 0.025 = 0.975$. Hence the left-hand χ^2-value in Figure 11.6 is $\chi^2_{0.975}$, which by Table VI equals 9.591. Consequently, the two χ^2-values are 9.591 and 34.170, as shown in Figure 11.7.

FIGURE 11.7

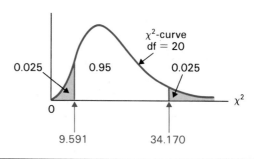

MTB
SPSS

Exercises 11.1

In Exercises 11.1 through 11.8 use Table VI to determine the indicated χ^2-values. Illustrate your work with pictures.

11.1   For a χ^2-curve with 19 degrees of freedom, find the χ^2-value with
a) area 0.025 to its right.
b) area 0.95 to its right.

11.2   For a χ^2-curve with 22 degrees of freedom, find the χ^2-value with
a) area 0.01 to its right.
b) area 0.995 to its right.

11.3   For a χ^2-curve with df $= 10$, determine
a) $\chi^2_{0.05}$
b) $\chi^2_{0.975}$

11.4   For a χ^2-curve with df $= 4$, determine
a) $\chi^2_{0.005}$
b) $\chi^2_{0.99}$

11.5   Consider a χ^2-curve with df $= 8$. Obtain the χ^2-value with
a) area 0.01 to its left.
b) area 0.95 to its left.

11.6   Consider a χ^2-curve with df $= 16$. Obtain the χ^2-value with
a) area 0.025 to its left.
b) area 0.975 to its left.

11.7   Determine the two χ^2-values that divide the area under the curve into a middle 0.95 area and two outside 0.025 areas, for a χ^2-curve with
a) df $= 5$.
b) df $= 20$.

11.8   Determine the two χ^2-values that divide the area under the curve into a middle 0.90 area and two outside 0.05 areas, for a χ^2-curve with
a) df $= 11$.
b) df $= 28$.

11.2   Chi-square goodness-of-fit test

The first statistical inference procedure we will discuss that relies on the chi-square distribution is called the **chi-square goodness-of-fit test.** Among other things, this procedure can be used to test hypotheses concerning the percentage or probability distribution of a population. We begin with an example to illustrate the reasoning behind the chi-square goodness-of-fit test. Following that, we will state and apply the general hypothesis testing procedure used to perform such tests.

EXAMPLE 11.4   **Introduces the chi-square goodness-of-fit test**

The American Medical Association publishes annual information on American physicians in *Physician Characteristics and Distribution in the U.S.* Physicians are broadly classified into four categories—general practice, medical, surgical, and other. Table 11.1 gives a percentage distribution and probability (relative-frequency) distribution for physicians in 1981.

TABLE 11.1
Specialty distribution of
American physicians, 1981.

Specialty	Percent	Probability, p
General practice	14.1%	0.141
Medical	30.6%	0.306
Surgical	26.4%	0.264
Other	28.9%	0.289
Σ	100.0%	1.000

Data from American Medical Association.

The table shows, for instance, that in 1981, 14.1% of all American physicians were in general practice; or, equivalently, that the probability is 0.141 that a randomly selected 1981 American physician was in general practice.

A researcher wants to know whether *this year's* specialty distribution of American physicians has changed from the 1981 distribution. To find out, he first randomly selects 500 American physicians who are currently practicing medicine. The results are displayed in Table 11.2.

TABLE 11.2
Sample results for specialties
of 500 randomly selected
American physicians currently
practicing medicine.

Specialty	Frequency	Percent
General practice	62	12.4%
Medical	172	34.4%
Surgical	130	26.0%
Other	136	27.2%
Σ	500	100.0%

The researcher plans to use the sample data in Table 11.2 to test the hypotheses

H_0: The current specialty distribution of American physicians is the same as the 1981 distribution.
H_a: The current specialty distribution of American physicians is different from the 1981 distribution.

The basic idea behind the hypothesis test is this: Compare the *observed* frequencies in Table 11.2, obtained by sampling, to the frequencies that would be *expected* if the current specialty distribution is the same as the 1981 distribution. If these observed and expected frequencies match up fairly well, then do not reject the null hypothesis; otherwise, reject the null hypothesis.

In order to transform this idea into a procedure for performing the hypothesis test, we need to answer two questions. First, what kind of frequencies should we expect to get from sampling the 500 physicians, if the current specialty distribution is, in fact, the same as the 1981 distribution? Second, how do we decide whether the observed frequencies, obtained from sampling the 500 physicians, match up reasonably well with those that we would expect?

The first question is easy to answer. If this year's specialty distribution is the same as the 1981 distribution, then, for instance, 14.1% of all current American physicians will be in general practice (see Table 11.1). Therefore, in a sample of 500, we would expect about 14.1% of the 500, or 70.5, to be in general practice. Similarly, we would expect about 30.6% of the 500, or 153, to be in the medical category.

The third column of Table 11.3 displays the **expected frequencies** for all four specialties. The expected frequencies are the frequencies we would expect to get by sampling 500 physicians currently practicing in the U.S., if this year's specialty distribution is, in fact, the same as the 1981 distribution given in Table 11.1. Each expected frequency, designated by *E,* is computed using the formula

$$E = np$$

where *n* is the sample size (in this case 500) and *p* is the appropriate probability from the third column of Table 11.1. For instance, the expected frequency for the "surgical" category is

$$E = np = 500 \cdot 0.264 = 132$$

TABLE 11.3   Expected frequencies for the specialties of 500 randomly selected American physicians currently practicing medicine, *if* the current specialty distribution is the same as the 1981 specialty distribution.

Specialty	Probability p	Expected frequency $np = E$
General practice	0.141	$500 \cdot 0.141 = 70.5$
Medical	0.306	$500 \cdot 0.306 = 153.0$
Surgical	0.264	$500 \cdot 0.264 = 132.0$
Other	0.289	$500 \cdot 0.289 = 144.5$

The third column of Table 11.3 provides us with the answer to our first question. It gives the frequencies we would expect to get if the current specialty distribution of physicians in the U.S. is the same as the 1981 distribution.

Our second question, whether the observed frequencies match up reasonably well with the expected frequencies, is harder to answer. We would like to calculate a number that measures how good the fit is. In the second column of Table 11.4, we have repeated from Table 11.2 the **observed frequencies** for all four specialties; that is, the frequencies actually obtained by sampling 500 current U.S. physicians. In the third column of Table 11.4, we have repeated from Table 11.3 the expected frequencies.

TABLE 11.4

Specialty	Observed frequency O	Expected frequency E	Difference $O - E$	Square of difference $(O - E)^2$	$(O - E)^2 / E$
General practice	62	70.5	−8.5	72.25	1.025
Medical	172	153.0	19.0	361.00	2.359
Surgical	130	132.0	−2.0	4.00	0.030
Other	136	144.5	−8.5	72.25	0.500
Σ	500	500.0	0		3.914

To try to measure how well the observed and expected frequencies match up, it is logical to look at their differences, $O - E$. These differences are shown in the fourth column of Table 11.4. As you can see, summing these differences in order to obtain a "total difference" is not very useful—the sum of the $O - E$ values always equals zero. Instead, each difference, $O - E$, is squared (shown in the fifth column) and then divided by its corresponding expected frequency. This gives the values, $(O - E)^2 / E$, shown in the sixth column. The sum of these values,

$$\Sigma (O - E)^2 / E = 3.914$$

is the statistic that is used to measure how well or poorly the observed and expected frequencies match up.

If the null hypothesis is true, then the observed and expected frequencies should be about the same, thus resulting in a small value of the test statistic $\Sigma (O - E)^2 / E$. Consequently, if $\Sigma (O - E)^2 / E$ is too large, then this gives us evidence that the null hypothesis is false.

As we have seen in Table 11.4, $\Sigma (O - E)^2 / E = 3.914$. Can this value be reasonably attributed to chance (sampling error), or is it large enough to indicate that the null hypothesis is false? As usual, to answer this question we need to know the probability distribution of our test statistic. ∎

KEY FACT

Suppose the sample size is large. Then, if the null hypothesis is true, the random variable

$$\Sigma (O - E)^2 / E$$

has approximately a *chi-square distribution*. The number of degrees of freedom is one less than the number of categories.

In Example 11.4, the number of degrees of freedom for the test statistic $\Sigma (O - E)^2 / E$ is df $= 4 - 1 = 3$, since there are four categories (specialties).

We can now state the hypothesis testing procedure to perform a chi-square goodness-of-fit test. For such a test, the null hypothesis is that a population (or random variable) has a specified distribution and the alternative hypothesis is that the population (or random variable) has a distribution different from the one specified in the null hypothesis. Since the null hypothesis is rejected only when the value of the test statistic is too large, the rejection region is always on the right; that is, the test is always *right-tailed*.

PROCEDURE 11.1 To perform a chi-square goodness-of-fit test.

Assumptions
1 All expected frequencies are at least 1.
2 At most 20% of the expected frequencies are less than 5.†

STEP 1 *State the null and alternative hypotheses.*
STEP 2 *Calculate the expected frequencies using the formula*

$$E = np$$

where n denotes the sample size and p denotes the probability for the category given in the null hypothesis.
STEP 3 *Check whether the expected frequencies satisfy Assumptions (1) and (2). [If they do not, this procedure should not be used.]*
STEP 4 *Decide on the significance level, α.*
STEP 5 *The critical value is χ_α^2 with df = k − 1, where k is the number of categories.*

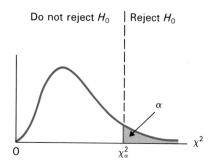

Do not reject H_0 | Reject H_0

α

0   χ_α^2 χ^2

STEP 6 *Compute the value of the test statistic*

$$\chi^2 = \Sigma (O - E)^2 / E$$

STEP 7 *If the value of the test statistic falls in the rejection region, reject H_0; otherwise, do not reject H_0.*
STEP 8 *State the conclusion in words.*

---

† Many texts give the rule that all expected frequencies should be at least 5. Research by W.G. Cochran, a noted statistician, shows that the "rule of 5" is too restrictive.

EXAMPLE 11.5 Illustrates Procedure 11.1

Let us now return to the hypothesis test of Example 11.4 in which a researcher wants to know whether the specialty distribution of American physicians currently in practice is the same as the 1981 distribution. The 1981 distribution is given in Table 11.1 and is repeated here as Table 11.5.

TABLE 11.5 Specialty distribution for American physicians, 1981.

	Specialty	Percent	Probability, p
	General practice	14.1%	0.141
	Medical	30.6%	0.306
	Surgical	26.4%	0.264
	Other	28.9%	0.289
Σ		100.0%	1.000

A random sample of 500 American physicians currently in practice yields the data given in Table 11.6.

TABLE 11.6 Observed frequencies for the specialties of 500 randomly selected American physicians currently practicing medicine.

Specialty	Observed frequency O
General practice	62
Medical	172
Surgical	130
Other	136

At the 5% significance level, do the data suggest that the current specialty distribution of American physicians is different from the 1981 distribution?

SOLUTION

We apply Procedure 11.1.

STEP 1 *State the null and alternative hypotheses.*

The null and alternative hypotheses are

H_0: The current specialty distribution of American physicians is the same as the 1981 distribution given in Table 11.5.

H_a: The current specialty distribution of American physicians is different from the 1981 distribution.

STEP 2 *Calculate the expected frequencies using the formula*

$$E = np$$

where n denotes the sample size and p denotes the probability for the category given in the null hypothesis.

The calculations are summarized in Table 11.7 at the top of the next page.

TABLE 11.7   Expected frequencies for the specialties of 500 randomly selected American physicians currently practicing medicine, *if* the current specialty distribution is the same as the 1981 specialty distribution.

Specialty	Probability p	Expected frequency $np = E$
General practice	0.141	$500 \cdot 0.141 = 70.5$
Medical	0.306	$500 \cdot 0.306 = 153.0$
Surgical	0.264	$500 \cdot 0.264 = 132.0$
Other	0.289	$500 \cdot 0.289 = 144.5$

STEP 3 *Check whether the expected frequencies satisfy Assumptions (1) and (2).*

1 All expected frequencies are at least 1? **Yes** (see Table 11.7).
2 At most 20% of the expected frequencies are less than 5? **Yes**, because none of the expected frequencies are less than 5.

STEP 4 *Decide on the significance level, α.*

We are to perform the test at the 5% significance level, so $\alpha = 0.05$.

STEP 5 *The critical value is χ_α^2 with df = k − 1, where k is the number of categories.*

From Step 4, $\alpha = 0.05$. Also, the number of categories is $k = 4$, since there are four specialties. Using Table VI we find that for df $= k - 1 = 4 - 1 = 3$,

$$\chi_{0.05}^2 = 7.815$$

See Figure 11.8.

FIGURE 11.8

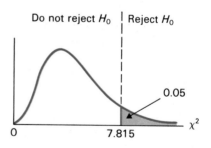

STEP 6 *Compute the value of the test statistic*

$$\chi^2 = \Sigma(O - E)^2 / E$$

Using the observed frequencies from Table 11.6 and the expected frequencies from Table 11.7, we can compute the value of the test statistic. This is done in Table 11.8 on the following page. [Note that we have calculated the sums of the observed frequencies, the expected frequencies, and the differences. Strictly speaking, these sums are not needed. However, they serve as a check for computational errors. The sums, ΣO and ΣE, of the observed and expected frequencies should both equal the sample size n (in this case 500), and the sum of the differences, $\Sigma(O - E)$, should equal zero.]

From the last column of Table 11.8 we find that

$$\chi^2 = \Sigma(O - E)^2/E = 3.914$$

TABLE 11.8

Specialty	Observed frequency O	Expected frequency E	Difference $O - E$	Square of difference $(O - E)^2$	$(O - E)^2/E$
General practice	62	70.5	−8.5	72.25	1.025
Medical	172	153.0	19.0	361.00	2.359
Surgical	130	132.0	−2.0	4.00	0.030
Other	136	144.5	−8.5	72.25	0.500
Σ	500	500.0	0		3.914

STEP 7 *If the value of the test statistic falls in the rejection region, reject H_0; otherwise, do not reject H_0.*

From Step 6 we see that the value of the test statistic is $\chi^2 = 3.914$. Since this does not fall in the rejection region, shown in Figure 11.8, we **do not reject H_0**.

STEP 8 *State the conclusion in words.*

The sample of 500 physicians does *not* provide sufficient evidence to conclude that the current specialty distribution has changed from the 1981 distribution. ∎

MTB
SPSS

The chi-square goodness-of-fit test provides a method for making statistical inferences about the percentage distribution of a population that is divided into k categories. If the number of categories is two (that is, $k = 2$), then the population is a *two-category population*. In this case, it can be shown that the chi-square goodness-of-fit test is equivalent to the z-test for a single population proportion (Procedure 9.4, page 380). See Exercise 11.20.

Exercises 11.2

11.9 Explain why you think the term "goodness-of-fit" is used to describe the type of hypothesis test considered in this section.

11.10 Are the observed frequencies random variables? What about the expected frequencies? Explain your answers.

In Exercises 11.11 through 11.18, apply Procedure 11.1 to perform the appropriate hypothesis test.

11.11 The 1983 distribution of the U.S. resident population by region is as shown below. (Source: U.S. Bureau of the Census.)

Region	Percent
Northeast	21.2%
Midwest	25.2%
South	34.0%
West	19.6%

A random sample of 500 current U.S. residents yields the following data:

Region	Frequency
Northeast	97
Midwest	121
South	176
West	106

At the 5% significance level, do the data suggest that the current distribution of the U.S. resident population by region is different from the 1983 distribution?

11.12 According to the U.S. Bureau of the Census the marital-status distribution of the U.S. population is as follows:

Marital status	Percent
Single	21.5%
Married	63.9%
Widowed	7.7%
Divorced	6.9%

A random sample of 750 males, 25–29 years old, gave the data below.

Marital status	Frequency
Single	289
Married	408
Widowed	0
Divorced	53

At the 1% significance level, does it appear that the marital-status distribution of all 25–29-year-old males is different from that of the population as a whole?

11.13 The table below gives the distribution of the number of years of school completed by U.S. residents 25 years old and over. (Source: U.S. Bureau of the Census.)

Years of school	Relative frequency
8 or less	0.183
9–11	0.153
12	0.346
13–15	0.157
16 or more	0.161

A random sample of 300 residents of Tennessee, 25 years old and over, gave the following statistics:

Years of school	Frequency
8 or less	83
9–11	48
12	95
13–15	36
16 or more	38

Do the data provide evidence that the distribution of the number of years of school completed by Tennessee residents is different from the national distribution? Use $\alpha = 0.01$.

11.14 According to the U.S. Energy Information Administration, the primary-heating-fuel distribution for occupied housing units is as follows:

Primary heating fuel	Percent
Natural gas	56.7%
Fuel oil, kerosene	14.3%
Electricity	16.0%
LPG	4.5%
Wood	6.7%
Other	1.8%

A random sample of 250 occupied housing units built after 1974 yields the frequency distribution below.

Primary heating fuel	Frequency
Natural gas	91
Fuel oil, kerosene	16
Electricity	110
LPG	14
Wood	17
Other	2

At the 5% significance level, does it appear that the primary-heating-fuel distribution for occupied housing units built after 1974 differs from that of all occupied housing units?

11.15 A gambler thinks that a die may be "loaded." That is, he thinks that the probabilities for the six numbers may not be equal. To test his suspicion, he rolls the die 150 times and obtains the following results:

Number	Frequency
1	23
2	26
3	23
4	21
5	31
6	26

Do the data indicate that the die is loaded? Use $\alpha = 0.05$.

11.16 A roulette wheel has 18 red numbers, 18 black numbers, and two green numbers. The table below gives the frequencies with which the ball landed on each color in 200 trials.

Color	Frequency
Red	88
Black	102
Green	10

Do the results suggest that the wheel is out of balance? Use $\alpha = 0.05$.

11.17 The U.S. Federal Bureau of Investigation (FBI) publishes data on the distribution of types of violent crime. Here is the percent distribution for the entire United States.

Violent crime	Percent
Murder	1.6%
Forcible rape	6.4%
Robbery	40.3%
Aggravated assault	51.7%

A random sample of 600 violent-crime reports from the state of New Jersey gave the following statistics.

Violent crime	Frequency
Murder	6
Forcible rape	33
Robbery	292
Aggravated assault	269

Do the data provide evidence that the distribution of types of violent crime in New Jersey is different from the nation as a whole? Use $\alpha = 0.01$.

11.18 The table below gives a percent distribution for the 1983 money income of households in the U.S. (Source: U.S. Bureau of the Census.)

Income level	Percent
Under $5,000	9.2%
$5,000–$9,999	13.7%
$10,000–$14,999	13.0%
$15,000–$19,999	12.0%
$20,000–$24,999	10.8%
$25,000–$34,999	17.0%
$35,000–$49,999	14.0%
$50,000 and over	10.3%

A random sample of 825 household incomes for last year gave the following frequency distribution.

[The income levels are in 1983 constant dollars. That is, they are adjusted for inflation to 1983 levels.]

Income level	Frequency
Under $5,000	70
$5,000–$9,999	117
$10,000–$14,999	92
$15,000–$19,999	111
$20,000–$24,999	101
$25,000–$34,999	140
$35,000–$49,999	111
$50,000 and over	83

Do the data provide evidence that last year's income-level distribution for households has changed from the 1983 distribution? Use $\alpha = 0.05$.

11.19 On page 468 we mentioned that the sums, ΣO and ΣE, of the observed and expected frequencies always equal the sample size n, and that the sum of the differences, $\Sigma(O - E)$, always equals zero.
a) Explain, in words, why $\Sigma O = n$.
b) Show mathematically that $\Sigma E = n$.
c) Prove that $\Sigma(O - E) = 0$.

11.20 Consider a two-category population with population proportion p. Suppose we wish to perform a hypothesis test to decide whether p is different from the value p_0. Discuss the method for performing such a test using
a) the z-test for a single population proportion.
b) the chi-square goodness-of-fit test.
On page 469 we stated that the tests you have discussed in parts (a) and (b) are *equivalent*. What do you think that means?

Testing for normality. Many of the inferential procedures we have studied require that the population(s) being sampled is *normally distributed,* or at least approximately so. For example, the small-sample hypothesis testing procedure for a single population mean (Section 9.5) requires that the population under consideration is approximately normally distributed. Sometimes we are reasonably sure, because of previous experience or theoretical reasons, that the population we are studying is approximately normally distributed. But what if we are not so sure and we want to decide whether or not the population is approximately normal? One way this can be done is to use a chi-square goodness-of-fit test. We illustrate the procedure in the following two exercises.

11.21 Suppose we wish to test whether the ages of U.S. residents are approximately normally distributed. We take a random sample of, say, 500 U.S. residents. The mean and standard deviation turn out to be $\bar{x} = 34.93$ years and $s = 21.82$ years. In addition, we obtain the following frequency distribution:

TABLE 11.9

Age	Observed frequency O
Under 15	91
15–under 25	114
25–under 35	84
35–under 45	59
45–under 65	98
65 or older	54

To use this data to test for normality, we must determine the frequencies that would be expected, if the population of ages is, in fact, normally distributed. We proceed as follows. First, since the population mean, μ, and the population standard deviation, σ, are unknown, we estimate them by the sample mean, $\bar{x} = 34.93$ years, and the sample standard deviation, $s = 21.82$ years:

$$\mu \approx 34.93 \text{ years} \qquad \sigma \approx 21.82 \text{ years}$$

a) Determine the probabilities (percentages), p, associated with each of the age categories in Table 11.9, if the population of ages is normally distributed with mean $\mu = 34.93$ years and standard deviation $\sigma = 21.82$ years. [Fill in the second column of Table 11.10.]

TABLE 11.10

Age	Probability p	Expected frequency E
Under 15		
15–under 25		
25–under 35		
35–under 45		
45–under 65		
65 or older		

b) Use the second column of Table 11.10 and the fact that the sample size is $n = 500$ to fill in the third column of Table 11.10.

Now that we have the observed and expected frequencies, we can perform a chi-square goodness-of-fit test by employing Procedure 11.1. However, there is one modification. For six categories ($k = 6$) we would think that df $= 6 - 1 = 5$. But the degrees of freedom are reduced by one for each of the two parameters, μ and σ, we estimated in order to find the expected frequencies. Thus,

$$\text{df} = 6 - 1 \underset{\substack{\uparrow \\ \text{Usual}}}{} - \underset{\substack{\uparrow \\ \text{For} \\ \text{estimating} \\ \mu \text{ by } \bar{x}}}{1} - \underset{\substack{\uparrow \\ \text{For} \\ \text{estimating} \\ \sigma \text{ by } s}}{1} = 3$$

c) Use Procedure 11.1 to perform the test for normality at the 5% significance level.

11.22 A quality-control engineer wants to know whether the diameters of bolts produced by a machine are approximately normally distributed. From a random sample of 300 bolts, he determines that the sample mean diameter of the bolts is $\bar{x} = 10.00$ mm with a sample standard deviation of $s = 0.10$ mm. Moreover, he obtains the following frequency distribution for the diameters:

Diameter	Observed frequency O
Under 9.8	8
9.8–under 9.9	42
9.9–under 10.0	112
10.0–under 10.1	97
10.1–under 10.2	38
10.2 or over	3

Do the data indicate that the diameters of bolts produced by the machine are not normally distributed? Use $\alpha = 0.05$.

11.3   Chi-square independence test

It is often of interest to know whether there is an association or statistical dependence between two characteristics of a population. For instance, we might want to know whether income level and educational level are statistically dependent; or whether there is an association between success in college and SAT scores.

In trying to answer these kinds of questions, we must generally rely on sample data, since data for the entire population is most often unavailable. Thus, it is usually necessary to employ inferential statistics in order to decide whether two characteristics of a population are dependent. One of the most commonly used methods for making such decisions is called the **chi-square independence test.** The following example introduces and explains the reasoning behind the chi-square test of independence.

EXAMPLE 11.7   **Introduces the chi-square independence test**

A national survey was conducted to obtain information on the alcohol consumption patterns of adults in the U.S. by marital status. A random sample of 1772 residents, 18 years old and over, gave the following data:[†]

TABLE 11.11   Contingency table for alcohol consumption patterns by marital status for 1772 randomly selected U.S. adults.

		Drinks per month			
		Abstain	1–60	Over 60	**Total**
	Single	67	213	74	354
Marital	Married	411	633	129	1173
status	Widowed	85	51	7	143
	Divorced	27	60	15	102
	Total	590	957	225	1772

The table shows, for instance, that 411 of the 1772 people sampled are married and abstain, 1173 are married, and 590 abstain.

Let us use the sample data to see whether there is an association between marital status and alcohol consumption patterns. Specifically, let us perform the hypothesis test

H_0: Marital status and alcohol consumption are independent.
H_a: Marital status and alcohol consumption are dependent.

The idea behind the chi-square independence test is to compare the *observed* frequencies in Table 11.11 with the frequencies that would be expected, if the null hypothesis of independence is true. The test statistic employed to make the comparison is the same as the one used for the goodness-of-fit test, namely,

$$\chi^2 = \Sigma (O - E)^2 / E$$

where O represents observed frequency and E represents expected frequency.

[†] Adapted from: Clark, W. B. and Midanik, L. "Alcohol Use and Alcohol Problems among U.S. Adults: Results of the 1979 National Survey." In: National Institute on Alcohol Abuse and Alcoholism. Alcohol and Health Monograph No 1, Alcohol Consumption and Related Problems. DHHS Pub. No (ADM) 82–1190, 1982.

We will now develop a formula for computing the expected frequencies by determining the E-value for the cell of Table 11.11 corresponding to "married *and* abstain"—the cell in the second row and first column of Table 11.11. To begin, note that the population proportion of all U.S. adults who abstain can be *estimated* by the sample proportion of the 1772 adults sampled who abstain:

$$\text{Number sampled who abstain} \searrow \frac{590}{1772} = 0.333 \ (33.3\%)$$
$$\text{Total number of people sampled} \nearrow$$

If marital status and alcohol consumption are independent (that is, if H_0 is true), then the proportion of married people who abstain will be the same as the proportion of all adults who abstain. Thus, if H_0 is true, the sample proportion 590/1772, or 33.3%, will also be an estimate of the population proportion of married people who abstain.

Now, a total of 1173 of the people sampled are married and, as we have just seen, if H_0 is true, then approximately $\frac{590}{1772}$ ths, or 33.3%, of all married people abstain. Therefore, if H_0 is true, we would *expect* about

$$\frac{590}{1772} \cdot 1173 = 390.6$$

of the people in the survey to be married people who abstain. It is useful to rewrite the left-hand side of the above expected-frequency computation in a slightly different way. By using algebra and referring to Table 11.11, we obtain

$$\text{Expected frequency} = \frac{590}{1772} \cdot 1173 = \frac{1173 \cdot 590}{1772}$$

$$= \frac{(\text{Row total}) \cdot (\text{Column total})}{\text{Sample size}}$$

If we let R denote "Row total" and C denote "Column total," then we can express this compactly as

$$E = \frac{R \cdot C}{n}$$

where, as usual, $E =$ expected frequency and $n =$ sample size.

Using this simple formula, we can obtain the expected frequencies for all 12 cells in Table 11.11. We have already done that for the cell in the second row and first column. For the cell in the upper right-hand corner of the table we get

$$E = \frac{R \cdot C}{n} = \frac{354 \cdot 225}{1772} = 44.9$$

and similar computations give the expected frequencies for the remaining cells. In Table 11.12 we have modified Table 11.11 by placing the expected frequency for each cell beneath the corresponding observed frequency.

TABLE 11.12   Observed and expected frequencies for marital status versus alcohol consumption. Expected frequencies are printed below the observed frequencies.

		Abstain	1–60	Over 60	Total
	Single	67 117.9	213 191.2	74 44.9	354
Marital status	Married	411 390.6	633 633.5	129 148.9	1173
	Widowed	85 47.6	51 77.2	7 18.2	143
	Divorced	27 34.0	60 55.1	15 13.0	102
	Total	590	957	225	1772

Drinks per month

Thus, for instance, the table indicates that the *observed frequency* of adults sampled who are in the "single *and* over 60 drinks per month" category is 74; whereas, if marital status and alcohol consumption are independent, then the *expected frequency* is 44.9.

If the null hypothesis of independence is true, then the observed and expected frequencies should be about the same, resulting in a relatively small value of the test statistic

$$\chi^2 = \Sigma(O - E)^2/E$$

Thus if χ^2 is too large, we will reject the null hypothesis of independence in favor of the alternative hypothesis of dependence.

We now compute the value of χ^2 for the observed and expected frequencies in Table 11.12. The calculations are shown below.

$$\chi^2 = \Sigma(O - E)^2/E$$

$$= (67 - 117.9)^2/117.9 + (213 - 191.2)^2/191.2 + (74 - 44.9)^2/44.9$$

$$+ (411 - 390.6)^2/390.6 + (633 - 633.5)^2/633.5 + (129 - 148.9)^2/148.9$$

$$+ (85 - 47.6)^2/47.6 + (51 - 77.2)^2/77.2 + (7 - 18.2)^2/18.2$$

$$+ (27 - 34.0)^2/34.0 + (60 - 55.1)^2/55.1 + (15 - 13.0)^2/13.0$$

$$= 21.952 + 2.489 + 18.776 + 1.070 + 0.000 + 2.670$$

$$+ 29.358 + 8.908 + 6.856 + 1.427 + 0.438 + 0.324$$

$$= 94.269^†$$

Can this value of the test statistic be reasonably attributed to chance or is it large enough to indicate that the null hypothesis of independence between marital status and alcohol consumption is false? Before we can answer this question, we must know the probability distribution of the χ^2 statistic, $\Sigma(O - E)^2/E$. ∎

---

† Although we have displayed the expected frequencies to one decimal place and the chi-square subtotals to three decimal places, the calculations were done using full calculator accuracy.

KEY FACT

Suppose the sample size is large. Then, if the null hypothesis of independence is true, the random variable

$$\Sigma(O-E)^2/E$$

has approximately a *chi-square distribution*. The number of degrees of freedom is

$$df = (r-1)(c-1)$$

where r and c are the number of rows and columns in the contingency table, respectively.

We will now state the general procedure for performing a chi-square test of independence. For such a test, the null hypothesis is that the two characteristics under consideration are independent, and the alternative hypothesis is that they are dependent. Note that the test is always right-tailed (why?). Also, note carefully the assumptions required for use of this test procedure.

PROCEDURE 11.2 To perform a chi-square independence test.

Assumptions
1 All expected frequencies are at least 1.
2 At most 20% of the expected frequencies are less than 5.

STEP 1 *State the null and alternative hypotheses.*
STEP 2 *Calculate the expected frequencies using the formula*

$$E = \frac{R \cdot C}{n}$$

where $R =$ row total, $C =$ column total, and $n =$ sample size. Place each expected frequency below its corresponding observed frequency.
STEP 3 *Check whether the expected frequencies satisfy Assumptions (1) and (2). [If they do not, this procedure should not be used.]*
STEP 4 *Decide on the significance level, α.*
STEP 5 *The critical value is χ_α^2 with $df = (r-1)(c-1)$, where r and c are the number of rows and columns in the contingency table, respectively.*

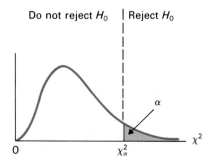

Do not reject H_0 | Reject H_0

STEP 6 *Compute the value of the test statistic*

$$\chi^2 = \Sigma (O - E)^2 / E$$

STEP 7 *If the value of the test statistic falls in the rejection region, reject H_0; otherwise, do not reject H_0.*

STEP 8 *State the conclusion in words.*

EXAMPLE 11.8 ## Illustrates Procedure 11.2

Recall that a random sample of 1772 U.S. adults yielded the data on marital status versus alcohol consumption given in Table 11.11. We repeat that contingency table here as Table 11.13.

TABLE 11.13 Contingency table for alcohol consumption patterns by marital status for 1772 randomly selected U.S. adults.

		Drinks per month			
		Abstain	1–60	Over 60	Total
	Single	67	213	74	354
Marital	Married	411	633	129	1173
status	Widowed	85	51	7	143
	Divorced	27	60	15	102
	Total	590	957	225	1772

Do the data suggest, at the 5% significance level, that marital status and alcohol consumption patterns are statistically dependent?

S O L U T I O N
We employ Procedure 11.2.

STEP 1 *State the null and alternative hypotheses.*

We want to decide whether the characteristics "marital status" and "alcohol consumption" are statistically dependent. Thus, the null and alternative hypotheses are

H_0: **Marital status and alcohol consumption are independent.**
H_a: **Marital status and alcohol consumption are dependent.**

STEP 2 *Calculate the expected frequencies using the formula*

$$E = \frac{R \cdot C}{n}$$

where R = row total, C = column total, and n = sample size. Place each expected frequency below its corresponding observed frequency.

The expected frequency for the "single *and* abstain" cell in the upper left-hand corner of Table 11.13 is

$$E = \frac{R \cdot C}{n} = \frac{354 \cdot 590}{1772} = 117.9$$

Similar computations give the other 11 expected frequencies. See Table 11.14.

TABLE 11.14   Observed and expected frequencies for marital status versus alcohol consumption. Expected frequencies are printed below the observed frequencies.

<table>
<tr><th colspan="5">Drinks per month</th></tr>
<tr><th></th><th>Abstain</th><th>1–60</th><th>Over 60</th><th>Total</th></tr>
<tr><td>Single</td><td>67
117.9</td><td>213
191.2</td><td>74
44.9</td><td>354</td></tr>
<tr><td>Married</td><td>411
390.6</td><td>633
633.5</td><td>129
148.9</td><td>1173</td></tr>
<tr><td>Widowed</td><td>85
47.6</td><td>51
77.2</td><td>7
18.2</td><td>143</td></tr>
<tr><td>Divorced</td><td>27
34.0</td><td>60
55.1</td><td>15
13.0</td><td>102</td></tr>
<tr><td>Total</td><td>590</td><td>957</td><td>225</td><td>1772</td></tr>
</table>

(Marital status labels the rows.)

STEP 3   *Check whether the expected frequencies satisfy Assumptions (1) and (2).*

1   All expected frequencies are at least 1? Yes (see Table 11.14).
2   At most 20% of the expected frequencies are less than 5? Yes, because none of the expected frequencies are less than 5.

STEP 4   *Decide on the significance level, α.*

The test is to be performed at the 5% significance level, so $\alpha = 0.05$.

STEP 5   *The critical value is χ_α^2 with df = (r − 1)(c − 1), where r and c are the number of rows and columns in the contingency table, respectively.*

The number of rows, r, in the contingency table is the number of "marital status" categories, which is four—$r = 4$. The number of columns, c, in the contingency table is the number of "drinks per month" categories, which is three—$c = 3$. Consequently,

$$df = (r − 1)(c − 1) = (4 − 1)(3 − 1) = 3 \cdot 2 = 6$$

Since $\alpha = 0.05$, we see from Table VI that the critical value is

$$\chi_{0.05}^2 = 12.592$$

See Figure 11.9.

FIGURE 11.9

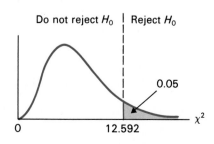

STEP 6 *Compute the value of the test statistic*

$$\chi^2 = \Sigma(O - E)^2/E$$

The observed frequencies (O-values) and expected frequencies (E-values) are displayed in Table 11.14. Using these, we compute the value of the test statistic:

$$\begin{aligned}
\chi^2 &= \Sigma(O - E)^2/E \\
&= (67 - 117.9)^2/117.9 + (213 - 191.2)^2/191.2 \\
&\quad + \cdots + (15 - 13.0)^2/13.0 \\
&= 21.952 + 2.489 + \cdots + 0.324 \\
&= 94.269
\end{aligned}$$

STEP 7 *If the value of the test statistic falls in the rejection region, reject H_0; otherwise, do not reject H_0.*

From Step 6, the value of the test statistic is $\chi^2 = 94.269$, which falls in the rejection region (see Figure 11.9). Thus, we reject H_0.

STEP 8 *State the conclusion in words.*

It appears that marital status and alcohol consumption patterns are statistically dependent. ∎

MTB
SPSS

Concerning Assumptions (1) and (2)

In Procedure 11.2 we made two assumptions about expected frequencies:
1 All expected frequencies are at least 1, and
2 At most 20% of the expected frequencies are less than 5.

What can we do to remedy the situation if these assumptions are violated? Three approaches are possible. We could *combine* rows or columns in order to increase the expected frequencies in those cells where the E-values are too small; we could *eliminate* certain rows or columns where the small expected frequencies occur; or, we could *increase the sample size*. Which, if any, of these modifications is employed depends on the problem under consideration. (See Exercises 11.31 and 11.32.)

Statistical dependence does not imply causation

The chi-square test of independence is used to decide whether there is a statistical dependence between two characteristics of a population. Specifically, the null hypothesis is that the two characteristics are independent and the alternative hypothesis is that they are dependent.

Consequently, if the null hypothesis is rejected, we can conclude that the two characteristics are dependent. This does *not* imply a causal relationship between the two characteristics. For instance, in Example 11.8, we rejected the null hypothesis of independence for marital status and alcohol consumption. This means that there is a statistical dependence between marital status and alcohol consumption. It *does not* mean, for example, that being single *causes* a person to drink more.

Deciding on statistical dependence with population data

It is important to understand that the chi-square test of independence is an inferential procedure, and as such, should be applied only to sample data and not to data for the entire population. How then should we proceed if we have data for the entire population and we want to decide whether there is a statistical dependence between two characteristics of the population?

In that case, we compute the expected frequencies as before, using the formula $E = R \cdot C/N$, where $N =$ population size. If the expected and observed frequencies are *identical* for each cell, then the two characteristics are independent; otherwise they are dependent. It can be shown that this last procedure is equivalent to using the special multiplication rule (Formula 4.6, page 162) to check for independence. See Exercise 11.35.

Exercises 11.3

In Exercises 11.23 through 11.30 employ Procedure 11.2 on page 476 to perform the appropriate hypothesis test, provided the conditions for the use of that procedure are met.

11.23 A study to determine whether annual income is dependent on educational level gave the sample results displayed in the following contingency table.

Years of schooling

Annual income	0–8	9–12	Over 12	Total
Under $10,000	34	36	10	80
$10,000–$24,999	41	72	36	149
$25,000–$39,999	10	78	58	146
$40,000 and over	4	26	45	75
Total	89	212	149	450

At the 1% significance level, do the data provide evidence that annual income and educational level are statistically dependent?

11.24 The Gallup Organization conducts periodic surveys to gauge the support by U.S. adults for regional primary elections. The question asked is: "It has been proposed that four individual primaries be held in different weeks of June during presidential election years. Does this sound like a good idea or a poor idea?" Below is a contingency table for responses by political affiliation, adapted from the results of a May, 1985 Gallup Poll.

Response

Political affiliation	Good idea	Poor idea	No opinion	Total
Republican	266	266	186	
Democrat	308	250	176	
Independent	28	27	21	
Total				1528

a) Fill in the row and column totals.
b) Do the data suggest, at the 5% level of significance, that the feelings of adults on the issue of regional primaries are dependent on political affiliation?

11.25 The U.S. Bureau of Labor Statistics gathers data on occupations of employed workers by sex. Suppose that a random sample of 83 employed workers yields the following data.

Sex

Occupation type	Male	Female	Total
Managerial/Professional	14	10	24
Technical sales Administrative	9	16	25
Service	4	6	10
Other	20	4	24
Total	47	36	83

Do the data indicate that there is an association between "sex" and "occupation type" for employed workers? Use $\alpha = 0.01$.

11.26 In 1981, more than 70 million Americans suffered injuries. More males (40.1 million) were injured than females (30.2 million). Those statistics do not tell us whether males and females tend to be injured in similar circumstances. One set of categories commonly used for the circumstances of accidents is "while at work," "home," "motor vehicle," and "other." In order to study whether accident circumstances differ by sex, a safety official in a large city took a random sample of accident reports. He obtained the following data:

Sex

		Male	Female	Total
Circumstance	While at work	18	4	
	Home	26	28	
	Motor vehicle	4	6	
	Other	36	24	
	Total			

a) Fill in the row and column totals.
b) Determine the sample size.
c) Perform a hypothesis test at the 5% significance level to determine whether "accident circumstance" and "sex" are statistically dependent in the city.

11.27 The FBI keeps data on arrests for violent crimes by the type of crime committed and the age of the person arrested. To test whether there is an association between those two characteristics, 750 arrest records for violent crimes are randomly selected. The sample results are displayed in the contingency table below.

Age

		18–24	25–44	45+	Total
Type of violent crime	Murder	11	16	4	
	Forcible rape	21	26	4	
	Robbery	128	92	6	
	Aggravated assault	162	234	46	
	Total				750

a) Fill in the row and column totals.
b) Is there evidence that an association exists between the type of violent crime committed and the age of the person arrested? Use $\alpha = 0.01$.

11.28 The U.S. National Center for Health Statistics performs national surveys to obtain information on people with acute medical conditions. Acute conditions are counted only if they are medically attended or caused at least one day of restricted activity. Suppose that a random sample of 376 acute conditions revealed the following data. (Here $A =$ infective and parasitic, $B =$ respiratory, $C =$ digestive system, and $D =$ injuries.)

Type of condition

		A	B	C	D	Total
Family income	Under $5,000	4	24	2	8	
	$5,000–$9,999	7	36	4	11	
	$10,000–$19,999	8	35	4	9	
	$20,000–$29,999	12	59	5	17	
	$30,000 and over	19	81	6	25	
	Total					376

a) Fill in the row and column totals.
b) Do the data suggest that there is an association between type of acute condition and family income? Use $\alpha = 0.05$.

11.29 The U.S. Internal Revenue Service (IRS) publishes data on top wealthholders by marital status. A random sample of 487 top wealthholders in 1976 yielded the contingency table below.

Marital status

		Married	Single/ Divorced	Widowed	Total
Net worth	$100,000–$249,999	227	54	63	344
	$250,000–$499,999	60	15	22	97
	$500,000–$999,999	20	4	7	31
	$1,000,000 or more	10	2	3	15
	Total	317	75	95	487

At the 5% significance level, does there appear to be an association between net worth and marital status for top wealthholders?

11.30   The American Bar Foundation publishes statistics on lawyers by selected characteristics. In the contingency table below you will find a cross-classification of 307 randomly selected lawyers by "status in practice" and "size of city practicing in."

Size of city

	Less than 250,000	250,000–499,999	500,000 or more	Total
Government	12	4	14	30
Judicial	8	1	2	11
Private practice	122	31	69	222
Salaried	19	7	18	44
Total	161	43	103	307

(Status in practice)

Do the characteristics "size of city" and "status in practice" for lawyers seem to be dependent? Use $\alpha = 0.05$.

11.31   In Exercise 11.29 it was not possible to perform the chi-square test of independence because the assumptions regarding expected frequencies were not met. As mentioned on page 479, we can use three ways to try to remedy the situation—(1) combine rows or columns, (2) eliminate rows or columns, or (3) increase the sample size.
a) Combine the last two rows of the contingency table in Exercise 11.29 to form a new contingency table.
b) Use the table in part (a) to perform the hypothesis test indicated in Exercise 11.29, if possible.
c) Eliminate the last row of the contingency table in Exercise 11.29 to form a new table.
d) Use the table in part (c) to perform the hypothesis test indicated in Exercise 11.29, if possible.

11.32   In Exercise 11.30 it was not possible to perform the chi-square test of independence because the assumptions regarding expected frequencies were not met. As mentioned on page 479, we can use three ways to try to remedy the situation—(1) combine rows or columns, (2) eliminate rows or columns, or (3) increase the sample size.
a) Combine the first two rows of the contingency table in Exercise 11.30 to form a new contingency table.
b) Use the table in part (a) to perform the hypothesis test described in Exercise 11.30, if possible.
c) Eliminate the second row of the contingency

table in Exercise 11.30 to form a new contingency table.
d) Use the table in part (c) to perform the hypothesis test described in Exercise 11.30, if possible.

11.33   A random sample of 100 members of a union are asked to respond to two questions.

Question 1: Are you happy with your financial situation today?
Question 2: Do you approve of the federal government's economic policies?

The responses are

Question 1

		Yes	No	Total
Question 2	Yes	22	48	70
	No	12	18	30
	Total	34	66	100

a) Calculate χ^2 for this data and test the null hypothesis that response to Question 1 is independent of response to Question 2 for members of this union. Use $\alpha = 0.05$.
b) Suppose that 200 members of the union [twice as many as in part (a)] are sampled and that their responses are in exactly the same proportions as before.

Question 1

		Yes	No	Total
Question 2	Yes	44	96	140
	No	24	36	60
	Total	68	132	200

Calculate χ^2 for this data and compare your answer to the answer from part (a). How would this change affect your hypothesis test?
c) Suppose that 1000 members [10 times as many as in part (a)] are sampled and that their responses are still in exactly the same proportions.

Question 1

		Yes	No	Total
Question 2	Yes	220	480	700
	No	120	180	300
	Total	340	660	1000

Calculate χ^2 for this data and compare your answer to the answer from part (a). How would this change affect your hypothesis test?

d) What happens to χ^2 as sample proportions stay exactly the same, but sample size increases? Does this indicate that you should be cautious about the results of chi-square tests with large sample sizes?

11.34 Suppose that you are given the following two contingency tables whose cells are in the same proportions.

		Total
a	b	$a+b$
c	d	$c+d$
Total $a+c$	$b+d$	$a+b+c+d$

		Total
ma	mb	$ma+mb$
mc	md	$mc+md$
Total $ma+mc$	$mb+md$	$ma+mb+mc+md$

a) Show that if χ_1^2 is the χ^2-statistic for the first table and χ_2^2 is the χ^2-statistic for the second table, then

$$\chi_2^2 = m\chi_1^2$$

b) Prove that a similar result holds for a contingency table with r rows and c columns.

11.35 Consider the *population* data below, which provides a contingency table for males on active military duty by "classification" and "race." Data are in thousands. (Source: U.S. Department of Defense.)

Race

Class	White	Black	Other	Total
Officer	2,401	145	176	2,722
Enlisted	11,085	3,392	1,944	16,421
Total	13,486	3,537	2,120	19,143

a) Suppose a male on active duty is selected at random. Determine the probability of each of the following events:

W = event a white man is selected

A = event an officer is selected

and

$(W \& A)$ = event a white officer is selected

b) Determine whether the events "white" and "officer" are independent by checking whether the *special multiplication rule* holds; that is, by checking whether $P(W \& A) = P(W)P(A)$.

c) Determine whether the observed and expected frequencies for the cell in the upper left-hand corner of the contingency table are equal. [This is equivalent to doing the check in part (b).]

d) Are "classification" and "race" independent characteristics for males on active military duty? Explain.

11.4 Inferences for a population standard deviation

Recall that the *standard deviation, σ,* of a population is a measure of dispersion or variability of the population values. Populations with large standard deviations have a large spread, while those with small standard deviations have a small spread. See Figure 11.10.

FIGURE 11.10

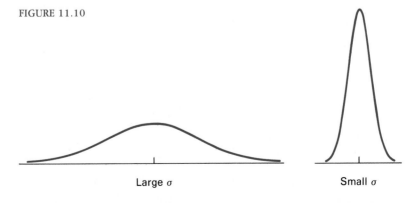

Large σ Small σ

There are many circumstances where the standard deviation of a population is unknown, but where it is nevertheless important to have information about its value. In most cases, it is impossible to determine the value of a population standard deviation exactly. Consequently, we employ inferential statistics to obtain the required information. This section examines procedures for testing hypotheses and finding confidence intervals for a population standard deviation, σ.

EXAMPLE 11.9 **Introduces hypothesis tests for a population standard deviation**

A hardware manufacturer produces 10-millimeter bolts. The manufacturer knows that the diameters of the bolts produced vary somewhat from 10 mm and also from each other. But even if he is willing to accept some variation in bolt diameters, he cannot tolerate too much variation. For, if the variation is too large, then too many of the bolts produced will be unusable.

Thus the manufacturer must make sure that the standard deviation, σ, of the bolt diameters is not unduly large. Since, in this case, it is not possible to find the value of σ exactly, inferential statistics must be employed.

Let us suppose that it has been determined that an acceptable standard deviation, σ, for the bolt diameters is one that is *less than* 0.09 mm.[†] Knowing this, the manufacturer can decide whether or not there is too much variation in the diameters of the bolts being produced by performing the hypothesis test

$$H_0\text{: } \sigma = 0.09 \text{ mm (unacceptable)}$$
$$H_a\text{: } \sigma < 0.09 \text{ mm (acceptable)}$$

If the null hypothesis can be rejected, this will provide evidence that the variation in bolt diameters is appropriate.

The basic idea for carrying out the hypothesis test is

1 Take a random sample of bolts.
2 Compute the *sample* standard deviation, s, of the diameters of the bolts selected.
3 If s is too much smaller than 0.09 mm, reject the null hypothesis in favor of the alternative hypothesis.

The manufacturer takes a random sample of $n = 20$ bolts and carefully measures their diameters. The results are displayed in Table 11.15.

TABLE 11.15
Diameters, in mm, of the 20 randomly selected bolts.

10.03	9.89	9.99	9.96	10.10
10.08	9.95	10.00	9.94	10.01
10.05	9.97	10.03	9.98	10.05
10.03	9.99	10.08	10.02	9.98

The sample standard deviation of the bolt diameters in Table 11.15 is

$$s = \sqrt{\frac{n(\Sigma x^2) - (\Sigma x)^2}{n(n-1)}} = \sqrt{\frac{20(2002.6527) - (200.13)^2}{20 \cdot 19}} = 0.052 \text{ mm}$$

[†] See Exercise 11.52 for an explanation of how that information can be obtained.

Is this value of s too much smaller than 0.09 mm, so that the null hypothesis should be rejected, or can the difference between $s = 0.052$ mm and the null hypothesis value of $\sigma = 0.09$ mm be attributed to sampling error? To answer this question, we must first know the probability distribution of s—**the sampling distribution of the standard deviation.** We therefore turn to a discussion of that sampling distribution, following which we will complete the hypothesis test posed in this example. ∎

The sampling distribution of the standard deviation

Recall that in performing hypothesis tests for a population mean, μ, we found it convenient to use the test statistic

$$\frac{\bar{x} - \mu_0}{s/\sqrt{n}} \cdot$$

instead of $\bar{x}$. Similarly, when testing hypotheses for a population standard deviation, σ, we do not employ s as the test statistic. Rather, we use

$$\frac{n-1}{\sigma_0^2} s^2$$

This test statistic has a probability distribution that will be familiar to you.

KEY FACT The sampling distribution of the standard deviation.[†]

Suppose a random sample of size n is to be taken from a *normally distributed* population with standard deviation σ. Then the random variable

$$\chi^2 = \frac{n-1}{\sigma^2} s^2$$

has the *chi-square distribution with $n - 1$ degrees of freedom.* In other words, probabilities for that random variable are equal to areas under the χ^2-curve with df $= n - 1$.

Hypothesis tests for a population standard deviation

Now that we know the sampling distribution of the standard deviation, we can state a procedure for performing a hypothesis test for σ.

PROCEDURE 11.3 To perform a hypothesis test for a population standard deviation, with null hypothesis H_0: $\sigma = \sigma_0$.

Assumption
Normal population.

STEP 1 *State the null and alternative hypotheses.*

[†] Strictly speaking, the sampling distribution given here is not the sampling distribution of the standard deviation, but is the sampling distribution of a function of the standard deviation.

STEP 2 *Decide on the significance level,* α.

STEP 3 *The critical value(s)*
   a) for a two-tailed test are $\chi^2_{1-\alpha/2}$ *and* $\chi^2_{\alpha/2}$,
   b) for a left-tailed test is $\chi^2_{1-\alpha}$,
   c) for a right-tailed test is χ^2_{α},
with df $= n - 1$. *Use Table VI to find the critical value(s).*

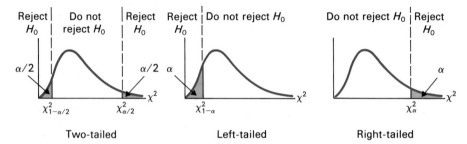

Two-tailed   Left-tailed   Right-tailed

STEP 4 *Compute the value of the test statistic*

$$\chi^2 = \frac{n-1}{\sigma_0^2} s^2$$

STEP 5 *If the value of the test statistic falls in the rejection region, reject* H_0; *otherwise, do not reject* H_0.

STEP 6 *State the conclusion in words.*

EXAMPLE 11.10   **Illustrates Procedure 11.3**

We can now complete the hypothesis test proposed in Example 11.9. Recall that a hardware manufacturer needs to decide whether the standard deviation, σ, of bolt diameters is *less than* 0.09 mm. He randomly samples 20 bolts and obtains their diameters. The results are shown in Table 11.16.

TABLE 11.16
Diameters, in mm,
of the 20 randomly
selected bolts.

10.03	9.89	9.99	9.96	10.10
10.08	9.95	10.00	9.94	10.01
10.05	9.97	10.03	9.98	10.05
10.03	9.99	10.08	10.02	9.98

At the 5% significance level, do the data provide evidence that the standard deviation, σ, of bolt diameters is less than 0.09 mm? [Assume that the population of diameters of all bolts manufactured is normally distributed.]

SOLUTION
We apply Procedure 11.3.

STEP 1 *State the null and alternative hypotheses.*

The null and alternative hypotheses are

$$H_0: \sigma = 0.09 \text{ mm}$$
$$H_a: \sigma < 0.09 \text{ mm}$$

Note that the test is left-tailed since there is a less-than sign ($<$) in the alternative hypothesis.

STEP 2 *Decide on the significance level, α.*

The test is to be performed at the 5% level of significance. Thus, $\alpha = 0.05$.

STEP 3 *The critical value for a left-tailed test is $\chi^2_{1-\alpha}$, with df $= n - 1$.*

We have $\alpha = 0.05$ and $n = 20$. Consequently, df $= 20 - 1 = 19$, and from Table VI we find that the critical value is

$$\chi^2_{1-0.05} = \chi^2_{0.95} = 10.117$$

See Figure 11.11.

FIGURE 11.11

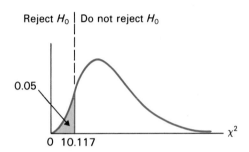

STEP 4 *Compute the value of the test statistic*

$$\chi^2 = \frac{n-1}{\sigma^2_0} s^2$$

First we determine the sample variance s^2. We obtain from the data in Table 11.16 that

$$s^2 = \frac{n(\Sigma x^2) - (\Sigma x)^2}{n(n-1)} = \frac{20(2002.6527) - (200.13)^2}{20 \cdot 19} = 0.0027$$

Since $n = 20$ and $\sigma_0 = 0.09$, the value of the test statistic is

$$\chi^2 = \frac{n-1}{\sigma^2_0} s^2 = \frac{20-1}{(0.09)^2} \cdot 0.0027 = 6.333$$

STEP 5 *If the value of the test statistic falls in the rejection region, reject H_0; otherwise, do not reject H_0.*

From Step 4, the value of the test statistic is $\chi^2 = 6.333$ which falls in the rejection region (see Figure 11.11). Thus, we reject H_0.

STEP 6 *State the conclusion in words.*

It appears that the standard deviation, σ, of bolt diameters is less than 0.09 mm. In other words, the data provide evidence that the variation in bolt diameters is not unduly large. ∎

MTB
SPSS

Confidence intervals for a population standard deviation

Using the Key Fact on page 485, we can also obtain a formula for constructing confidence intervals for a population standard deviation, σ. The derivation of the confidence-interval formula given below is considered in Exercise 11.53.

PROCEDURE 11.4 To find a confidence interval for a population standard deviation, σ.

Assumption
Normal population.

STEP 1 *For a confidence level of $1 - \alpha$, use Table VI to find $\chi^2_{1-\alpha/2}$ and $\chi^2_{\alpha/2}$ for $df = n - 1$.*
STEP 2 *The confidence interval for σ is*

$$\sqrt{\frac{n-1}{\chi^2_{\alpha/2}}} \cdot s \quad to \quad \sqrt{\frac{n-1}{\chi^2_{1-\alpha/2}}} \cdot s$$

where $\chi^2_{1-\alpha/2}$ and $\chi^2_{\alpha/2}$ are found in Step 1, n is the sample size, and s is computed from the actual sample data obtained.

EXAMPLE 11.11 **Illustrates Procedure 11.4**

Refer to Example 11.10. Use the sample data in Table 11.16 to find a 95% confidence interval for the standard deviation, σ, of the diameters of all bolts produced by the manufacturer.

SOLUTION
We employ Procedure 11.4.

STEP 1 *For a confidence level of $1 - \alpha$, use Table VI to find $\chi^2_{1-\alpha/2}$ and $\chi^2_{\alpha/2}$ for $df = n - 1$.*

We want a 95% confidence interval, so the confidence level is $0.95 = 1 - 0.05$. This means that $\alpha = 0.05$. Also $n = 20$, which makes df $= 20 - 1 = 19$. Consulting Table VI we find that

$$\chi^2_{1-\alpha/2} = \chi^2_{1-0.05/2} = \chi^2_{0.975} = 8.907$$

and

$$\chi^2_{\alpha/2} = \chi^2_{0.05/2} = \chi^2_{0.025} = 32.852$$

STEP 2 *The confidence interval for σ is*

$$\sqrt{\frac{n-1}{\chi^2_{\alpha/2}}} \cdot s \quad to \quad \sqrt{\frac{n-1}{\chi^2_{1-\alpha/2}}} \cdot s$$

We have $n = 20$ and from Step 1, $\chi^2_{\alpha/2} = 32.852$ and $\chi^2_{1-\alpha/2} = 8.907$. Also, we found in Example 11.9 that $s = 0.052$ mm. Consequently, a 95% confidence interval for σ is from

$$\sqrt{\frac{20-1}{32.852}} \cdot 0.052 \quad to \quad \sqrt{\frac{20-1}{8.907}} \cdot 0.052$$

or

$$0.040 \quad \text{to} \quad 0.076$$

We can be 95% confident that the standard deviation, σ, of the population of all bolt diameters is somewhere between 0.040 mm and 0.076 mm. ∎

Exercises 11.4

11.36 What does σ measure?

In Exercises 11.37 through 11.42 use Procedure 11.3 on page 485 to perform the appropriate hypothesis test. Assume that the populations are normally distributed.

11.37 Every year thousands of college-bound high-school students take the *Scholastic Aptitude Test* (SAT). This test measures verbal and mathematical abilities of prospective college students. Student scores are reported on a scale that ranges from a low of 200 to a high of 800. This scale was introduced in 1941. At that time, the standard deviation of scores was 100 points. Suppose that a random sample of 25 verbal scores for this year gives the following results:

560	405	367	416	540
347	370	629	570	372
396	439	526	610	339
569	314	228	483	353
379	558	432	648	475

At the 5% significance level, do the data suggest that the standard deviation, σ, for this year's verbal scores is *different* from the 1941 standard deviation of 100? ($\Sigma x = 11{,}325$ and $\Sigma x^2 = 5{,}425{,}835$.)

11.38 A company produces cans of stewed tomatoes with an advertised weight of 14 ounces. Recently the company hired a quality-control engineer. The engineer has performed some statistical analyses and is satisfied that, on the average, the cans do contain 14 ounces of stewed tomatoes. However, she must also be concerned about the *variability* of the weights. Specifically, in order to ensure that the vast majority of the cans have a weight within 10% of the advertised weight of 14 ounces, it is required that the standard deviation, σ, of the weights be *less than* 0.45 oz. A random sample of 10 cans has the weights given below. Data are in ounces.

13.85	13.95	13.90	13.49	14.17
14.33	14.03	13.48	14.27	14.19

Do the data provide evidence that the standard deviation of the weights is less than 0.45 oz? Use $\alpha = 0.05$. ($\Sigma x = 139.66$, $\Sigma x^2 = 1951.2932$.)

11.39 A watch manufacturer claims that the weekly error made by the watches she produces has a standard deviation of about one second. To test this claim, a random sample of 20 watches are set to the correct time. After one week, the error made by each watch is recorded. The results, in seconds, are

0.6	2.3	2.0	−2.1	−1.4
−0.5	1.5	−0.3	0.4	0.6
0.4	−2.2	0.7	0.5	−1.3
−2.0	2.6	−0.8	1.0	−0.6

Does it appear that the standard deviation, σ, of the weekly errors *exceeds* the one-second claim made by the manufacturer? Use $\alpha = 0.01$. ($\Sigma x = 1.4$ and $\Sigma x^2 = 39.32$.)

11.40 The designer of a 100-point aptitude test has attempted to construct the test so that the standard deviation of scores is 10 points. As a preliminary documentation, 30 randomly selected persons are given the test. Their scores are as follows:

83	43	29	37	64	69
79	67	85	61	52	55
70	65	43	64	60	35
76	46	35	43	51	52
61	50	41	87	62	59

Do the data indicate, at the 10% significance level, that the standard deviation of scores for the aptitude test is not 10 points? ($\Sigma x = 1724$ and $\Sigma x^2 = 106{,}172$.)

11.41 A coffee machine is supposed to dispense six fluid ounces of coffee into a styrofoam cup. In reality, the amounts dispensed vary from cup to cup. However, if the machine is working properly, then the vast majority of cups will contain within 10% of the advertised six fluid ounces. This means that the standard deviation of the amounts dispensed should be less than 0.2 fl oz. A random sample of 15 cups gives the following data:

6.2	6.0	5.8	5.9	6.3
6.1	6.2	6.0	6.1	6.0
6.1	6.0	6.1	6.0	5.9

At the 5% significance level, do the data provide evidence that the standard deviation, σ, of the amounts being dispensed is less than 0.2 fl oz?

11.42   The Environmental Protection Agency (EPA) provides mileage estimates for cars and light-duty trucks. According to the EPA, "...the mileages obtained by most drivers will be within plus or minus 15 percent of the [EPA] estimates...." The mileage estimate given for one 1985 model is 23 mpg on the highway. If the claim made in quotes by the EPA is true, then the standard deviation for mileages should be about $0.15 \cdot 23/3 = 1.15$ mpg. Suppose that a random sample of 12 cars of this model yields the following highway mileages:

24.1	23.3	22.5	23.2
22.3	21.1	21.4	23.4
23.5	22.8	24.5	24.3

At the 5% significance level, do these data suggest that the standard deviation, σ, of highway mileages for all 1985 cars of this model is different from 1.15 mpg?

In Exercises 11.43 through 11.48 use Procedure 11.4 on page 488 to find the designated confidence interval.

11.43   Refer to Exercise 11.37. Determine a 95% confidence interval for the standard deviation, σ, of this year's verbal SAT scores.

11.44   Refer to Exercise 11.38. Find a 90% confidence interval for the true standard deviation, σ, of the weights of cans of stewed tomatoes produced by the company.

11.45   Refer to Exercise 11.39. Obtain a 98% confidence interval for the actual standard deviation, σ, of the weekly errors of the watches manufactured.

11.46   Refer to Exercise 11.40. Determine a 90% confidence interval for the actual standard deviation, σ, of scores on the aptitude test.

11.47   Refer to Exercise 11.41. Find a 90% confidence interval for the true standard deviation, σ, of the amounts of coffee being dispensed.

11.48   Refer to Exercise 11.42. Find a 95% confidence interval for the true standard deviation, σ, of highway gas mileages for all 1985 cars of the model in question.

11.49   Refer to Exercise 11.41. Why is it important that the standard deviation, σ, of the amounts of coffee being dispensed not be too large?

11.50   Refer to Exercise 11.42. Why is it useful to know the standard deviation of the gas mileages as well as the mean gas mileage?

11.51   A confidence interval for a population mean, μ, is symmetric about the sample mean, $\bar{x}$. That is, the two endpoints of the confidence interval are equidistant from $\bar{x}$. On the other hand, a confidence interval for a population standard deviation, σ, is not symmetric about the sample standard deviation, s. What is the reason for this difference?

11.52   In the bolt manufacturer problem of Example 11.9, we assumed it had been determined that an acceptable standard deviation, σ, for the bolt diameters is one that is less than 0.09 mm. We will now see how such information might be obtained. Let us suppose the manufacturer has set the tolerance specifications for the 10-millimeter bolts at ± 0.3 mm. In other words, a bolt is considered to be satisfactory if its diameter is between 9.7 and 10.3 mm. Further suppose that the manufacturer has decided that less than 0.1% (one out of a thousand) of the bolts produced should be defective.

a) Let x denote the diameter of a randomly selected bolt. Show that the manufacturer's production criteria can be expressed as

$$P(9.7 \leq x \leq 10.3) > 0.999$$

b) Draw a normal-curve picture that illustrates the equation $P(9.7 \leq x \leq 10.3) = 0.999$. Include an x-axis and a z-axis. [Assume $\mu = 10$ mm.]

c) Deduce from your picture in part (b) that the manufacturer's production criteria are equivalent to the condition

$$\frac{0.3}{\sigma} > z_{0.0005}$$

d) Use your result from part (c) to conclude that the manufacturer's production criteria are equivalent to requiring the standard deviation of bolt diameters to be less than 0.09 mm, that is, $\sigma < 0.09$ mm.

11.53 The purpose of this exercise is to derive the confidence-interval formula given in Step 2 of Procedure 11.4 on page 488.

a) Use the Key Fact on page 485 to show that

$$P\left(\chi^2_{1-\alpha/2} \le \frac{n-1}{\sigma^2}\, s^2 \le \chi^2_{\alpha/2}\right) = 1 - \alpha$$

b) Deduce from part (a) that

$$P\left(\frac{n-1}{\chi^2_{\alpha/2}} \cdot s^2 \le \sigma^2 \le \frac{n-1}{\chi^2_{1-\alpha/2}} \cdot s^2\right) = 1 - \alpha$$

c) Prove that the result in part (b) implies that the interval from

$$\sqrt{\frac{n-1}{\chi^2_{\alpha/2}}} \cdot s \quad \text{to} \quad \sqrt{\frac{n-1}{\chi^2_{1-\alpha/2}}} \cdot s$$

is a $(1 - \alpha)$-level confidence interval for σ.

11.5   Computer packages*

You have probably observed that a large amount of computation is required to perform chi-square tests. Fortunately, statistical computer packages are available that will do virtually all of the work. In this section, we will illustrate the use of computer packages by examining how Minitab can be used to carry out a chi-square independence test.

CHISQUARE

A Minitab program called **CHISQUARE** performs a chi-square independence test on data entered into the computer. To see how this works, we will use CHISQUARE to do the hypothesis test considered in Example 11.8 on page 477.

EXAMPLE 11.12   **Illustrates the use of CHISQUARE**

A random sample of 1772 U.S. adults yielded the data on marital status versus alcohol consumption shown in Table 11.17.

TABLE 11.17 Contingency table for alcohol consumption patterns by marital status for 1772 randomly selected U.S. adults.

		Drinks per month			
		Abstain	1–60	Over 60	Total
	Single	67	213	74	354
Marital	Married	411	633	129	1173
status	Widowed	85	51	7	143
	Divorced	27	60	15	102
	Total	590	957	225	1772

At the 5% significance level, do the data suggest that there is an association between marital status and alcohol consumption patterns?

SOLUTION
We need to perform the hypothesis test

H_0: Marital status and alcohol consumption are independent.
H_a: Marital status and alcohol consumption are dependent.

with $\alpha = 0.05$. In Example 11.8, we applied Procedure 11.2 to accomplish this hypothesis test. Here we will use CHISQUARE.

To begin, we enter the contingency-table data from Table 11.17 into the computer. Then we employ the command CHISQUARE followed by the storage locations of the data (C16–C18). Printout 11.1 displays the command and its output.

PRINTOUT 11.1
Minitab output
for CHISQUARE.

```
MTB > CHISQUARE, C16-C18

Expected counts are printed below observed counts

               C16        C17        C18        Total
    1           67        213         74          354
             117.9      191.2       44.9

    2          411        633        129         1173
             390.6      633.5      148.9

    3           85         51          7          143
              47.6       77.2       18.2

    4           27         60         15          102
              34.0       55.1       13.0

  Total        590        957        225         1772

ChiSq =  21.95 +     2.49 +    18.78 +
          1.07 +     0.00 +     2.67 +
         29.36 +     8.91 +     6.86 +
          1.43 +     0.44 +     0.32 =   94.27
  df = 6
```

The output first displays a table of observed and expected frequencies. This is the Minitab version of Table 11.14 on page 478. After the table, the computer prints the value of the test statistic $\chi^2 = \Sigma(O - E)^2/E$, including the cell by cell subtotals. Thus, we see that the value of the test statistic is $\chi^2 = 94.27$. The final item given is the number of degrees of freedom—df $= 6$.

Now that we have the value of the test statistic and the number of degrees of freedom, it is a simple matter to complete the hypothesis test. For df $= 6$, Table VI shows that the critical value for a test at the 5% significance level is $\chi^2_{0.05} = 12.592$. Since the value of the test statistic is $\chi^2 = 94.27$, which exceeds the critical value of 12.592, we reject H_0. In other words, it appears that marital status and alcohol consumption patterns are statistically dependent. ∎

◆ **Chapter Review**

KEY TERMS

chi-square distribution, 459
chi-square (χ^2) curve, 459

CHISQUARE*, 491
expected frequencies, 464

FORMULAS

In the formulas below,

O = observed frequency
E = expected frequency
k = number of categories in a probability distribution
p = probability or proportion
n = sample size
R = row total in a contingency table
C = column total in a contingency table
r = number of rows in a contingency table
c = number of columns in a contingency table
σ = population standard deviation
s = sample standard deviation

Expected frequencies for a goodness-of-fit test, 466

$$E = np$$

Test statistic for a goodness-of-fit test, 466

$$\chi^2 = \Sigma (O - E)^2 / E$$

with df $= k - 1$.

Expected frequencies for an independence test, 476

$$E = \frac{R \cdot C}{n}$$

Test statistic for an independence test, 477

$$\chi^2 = \Sigma (O - E)^2 / E$$

with df $= (r - 1)(c - 1)$.

Test statistic for H_0: $\sigma = \sigma_0$—normal population, 486

$$\chi^2 = \frac{n - 1}{\sigma_0^2} s^2$$

with df $= n - 1$.

Confidence interval for σ—normal population, 488

$$\sqrt{\frac{n - 1}{\chi_{a/2}^2}} \cdot s \quad \text{to} \quad \sqrt{\frac{n - 1}{\chi_{1 - a/2}^2}} \cdot s$$

(df $= n - 1$).

YOU SHOULD BE ABLE TO . . .

1 use and understand the preceding formulas.

2 use the chi-square distribution table, Table VI.

3 explain the reasoning behind the chi-square goodness-of-fit test and the chi-square independence test.

4 perform a goodness-of-fit test for the percentage distribution of a population.

5 perform an independence test to decide whether there is a statistical dependence between two characteristics of a population.

6 perform a hypothesis test for σ when sampling from a normal population.

7 find a confidence interval for σ when sampling from a normal population.

REVIEW TEST

1 Consider a χ^2-curve with 17 degrees of freedom. Use Table VI to determine
a) $\chi^2_{0.99}$
b) $\chi^2_{0.01}$
c) the χ^2-value with area 0.05 to its right.
d) the χ^2-value with area 0.05 to its left.
e) the two χ^2-values that divide the area under the curve into a middle 0.95 area and two outside 0.025 areas.

2 The U.S. Bureau of the Census and the U.S. Department of Housing and Urban Development publish data on characteristics of new, privately-owned, one-family homes. A percentage distribution for the number of bedrooms in homes completed in 1983 is as follows:

Number of bedrooms	Percent
2 or less	24%
3	59%
4 or more	17%

A recent survey was conducted by the Impulse Research Corporation of Santa Monica, California. The project is called "Homestyle 1988" and is designed to give builders a profile of the next generation of home buyers in three southwestern states—Arizona, California, and Nevada. Researchers conducted telephone interviews with 150 randomly selected potential home buyers between the ages of 24 and 35. They obtained the following data on preferences for the number of bedrooms:

Number of bedrooms	Frequency
2 or less	47
3	93
4 or more	10

Do the data indicate that the actual preference distribution for the number of bedrooms is different from the distribution for the number of bedrooms for homes completed in 1983? Perform the test at the 5% significance level.

3 In a 1985 poll by the Gallup Organization, 1528 randomly selected adults were asked the following question. "The New Jersey Supreme Court recently ruled that all life-sustaining medical treatment may be withheld or withdrawn from terminally ill patients, provided that is what the patients want or would want if they were able to express their wishes. Would you like to see such a ruling in the state in which you live, or not?" Below we have presented a modified version of the sample results which cross-classify response by educational level.

		Response			
		Favor	Oppose	No opinion	Total
Educational level	College grad	264	17	6	287
	Some college	205	26	7	238
	High school grad	461	81	34	576
	Non high school grad	290	81	56	427
	Total	1220	205	103	1528

At the 1% significance level, do the data provide evidence that response and educational level are statistically dependent?

4 IQs measured on the Stanford Revision of the Binet-Simon Intelligence Scale are supposed to have a standard deviation of 16 points. Suppose that 25 randomly selected persons are given an IQ test and that the results are as follows:

91	96	106	116	97
102	96	124	115	121
95	111	105	101	86
88	129	112	82	98
104	118	127	66	102

a) At the 10% significance level, do the data provide evidence that the standard deviation of IQs is not equal to 16? ($\Sigma x = 2588$, $\Sigma x^2 = 273{,}314$)

b) What assumption are you making about IQs in performing the hypothesis test in part (a)?

5 Refer to Problem 4. Determine a 90% confidence interval for the standard deviation of all IQ scores.

It is frequently of interest to know whether two or more variables are related and, if so, how they are related. For instance, is there a relationship between college GPA and SAT scores? If these variables are related, how are they related? The president of a large corporation knows there is a tendency for sales to increase as advertising expenditures increase. But how strong is that tendency, and how can she predict the approximate sales that will result from various advertising expenditures?

In this chapter and the next we will examine some widely employed procedures that are used to analyze the relationship between two variables, such as college GPA and SAT scores. These procedures are part of what is known as *linear regression* and *correlation*. This chapter will be devoted to descriptive methods in regression and correlation. In the next chapter we will study inferential methods in regression and correlation.

Descriptive methods in regression and correlation

CHAPTER OUTLINE

12.1

Linear equations with one independent variable

The methods of simple linear regression require the concept of **linear equations** with one independent variable. The general form of a linear equation with one independent variable is

$$y = b_0 + b_1 x$$

where b_0 and b_1 are fixed numbers, x is the independent variable, and y is the dependent variable.[†] The graph of a linear equation with one independent variable is a **straight line;** moreover, any nonvertical straight line can be described by such an equation.

Three examples of linear equations with one independent variable are

$$y = 4 + 0.2x,$$
$$y = -1.5 - 2x,$$
$$y = -3.4 + 1.8x$$

Figure 12.1, on the next page, shows the straight-line graphs of these three linear equations.

Linear equations with one independent variable occur frequently in applications of mathematics to many different subject areas. These subject areas include the management, life, and social sciences, as well as the physical and mathematical sciences. In Examples 12.1 and 12.2 we will illustrate the use of linear equations in a simple business application.

[†] You may be familiar with the form $y = mx + b$ instead of the form $y = b_0 + b_1 x$. In statistics, the latter form is preferred because it allows a smoother transition to the topic of multiple regression in which there is more than one independent variable.

FIGURE 12.1

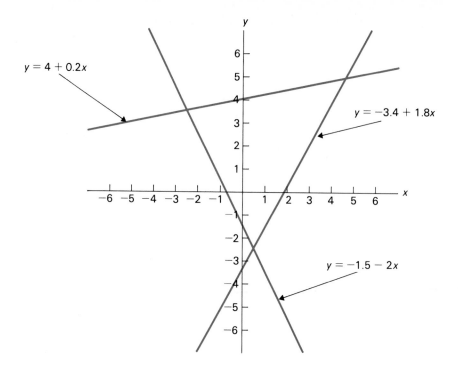

EXAMPLE   12.1   Illustrates linear equations

CJ² Business Services does word processing as one of its basic functions. Its rate is $20/hr plus a $25 disk charge. The total cost to a customer depends, of course, on the number of hours it takes to complete the job. Determine an equation that expresses the total cost of a job in terms of the number of hours required to complete the job.

S O L U T I O N
Let

$$y = \text{total cost}$$

and

$$x = \text{number of hours required}$$

Since the rate for word processing is $20/hr, a job that takes x hours will cost $20x$ plus the $25 disk charge. Thus, the total cost, y, of a job that takes x hours is

$$y = 25 + 20x$$

∎

Note that the equation for the total cost of a word processing job is a *linear equation*. Here $b_0 = 25$ and $b_1 = 20$. Using the equation, we can determine the exact cost for a job once we know the number of hours required. For instance, a job that takes $x = 5$ hours will cost

$$y = 25 + 20 \cdot 5 = \$125$$

and a job that takes $x = 7.5$ hours will cost

$$y = 25 + 20 \cdot 7.5 = \$175$$

In Table 12.1, we have displayed these and a few more cost illustrations.

TABLE 12.1

Time (hrs) x	Cost (\$) y
5.0	125
7.5	175
15.0	325
20.0	425
22.5	475

As we have already mentioned, a linear equation, such as $y = 25 + 20x$, has a straight-line graph. We can obtain the graph for $y = 25 + 20x$ by plotting the points in Table 12.1 and connecting them with a straight line. This is done in Figure 12.2.

FIGURE 12.2
Graph of $y = 25 + 20x$ obtained from points in Table 12.1.

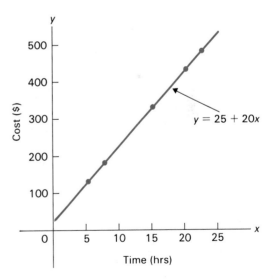

The graph is useful to picture the costs for various times. For instance, a quick glance at the graph shows that a 10-hour job will cost somewhere between $200 and $250. The exact cost is

$$y = 25 + 20 \cdot 10 = \$225$$

Slope and y-intercept

As we said, linear equations and their straight-line graphs are applied in the physical sciences, biology, business, the social sciences, and many other fields. In all these applications, the concepts of **intercept** and **slope** are extremely important.

DEFINITION 12.1 Intercept and slope

For a linear equation $y = b_0 + b_1x$, the **y-intercept** is b_0 and the **slope** is b_1:

$$b_0 = y\text{-intercept}$$
$$b_1 = \text{slope}$$

The y-intercept, b_0, gives the y-value at which the straight line $y = b_0 + b_1x$ intersects the y-axis. The slope, b_1, measures the steepness of the line. More precisely, b_1 indicates how much the y-value on a straight line increases (or decreases) when the x-value increases by one unit. Figure 12.3 summarizes this discussion.

FIGURE 12.3
$y = b_0 + b_1x$

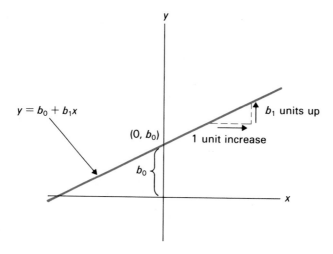

EXAMPLE 12.2   Illustrates the concepts of y-intercept and slope

In Example 12.1, we determined the linear equation that expresses the total cost, y, of a word-processing job in terms of the number of hours, x, to complete the job. The equation is

$$y = 25 + 20x$$

a) Determine the y-intercept and slope of this linear equation.

b) What do the y-intercept and slope represent in terms of the graph of the equation?

c) What do the y-intercept and slope represent in terms of the word process-ing costs?

SOLUTION

a) The y-intercept for the equation is $b_0 = 25$ and the slope is $b_1 = 20$.

b) The y-intercept, $b_0 = 25$, gives the y-value at which the straight line $y = 25 + 20x$ intersects the y-axis. The slope, $b_1 = 20$, indicates that the y-value increases by 20 for every increase in x of one unit. See Figure 12.4.

FIGURE 12.4
Graph of $y = 25 + 20x$.

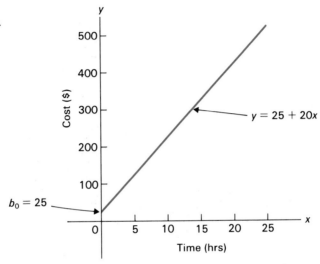

c) In terms of word processing costs, the y-intercept, $b_0 = 25$, represents the total charge for a job that takes zero hours. More to the point, the y-intercept of $25 is a fixed cost that is always there no matter how many hours the job takes. The slope of the equation, $b_1 = 20$, represents the fact that the cost per hour is $20. It is the amount the total charge, y, goes up for every increase of one hour in the time, x, required to complete the job. ∎

A straight line is determined by any two distinct points that lie on the line. This means that the straight-line graph of a linear equation $y = b_0 + b_1 x$ can be ob-tained by simply substituting two different x-values to get two distinct points. For example, to graph the linear equation

$$y = 5 - 3x$$

we can use the x-values $x = 1$ and $x = 3$. The corresponding y-values are $y = 5 - 3 \cdot 1 = 2$ and $y = 5 - 3 \cdot 3 = -4$. Hence, the graph of the equation $y = 5 - 3x$ is the straight line passing through the two points $(1, 2)$ and $(3, -4)$. See Figure 12.5 at the top of the following page.

FIGURE 12.5
Graph of $y = 5 - 3x$.

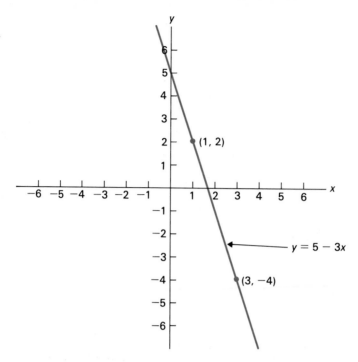

Note that the line in Figure 12.5 slopes downward—the y-values decrease as x increases. This is because the slope of the line is negative: $b_1 = -3 < 0$. Now look back at the line in Figure 12.4. It is the graph of the linear equation $y = 25 + 20x$. The line slopes upward—the y-values increase as x increases. This is because the slope of the line is positive: $b_1 = 20 > 0$. In general, we have the following fact:

KEY FACT

The straight-line graph of the linear equation

$$y = b_0 + b_1 x$$

slopes upward if $b_1 > 0$, slopes downward if $b_1 < 0$, and is horizontal if $b_1 = 0$.

FIGURE 12.6

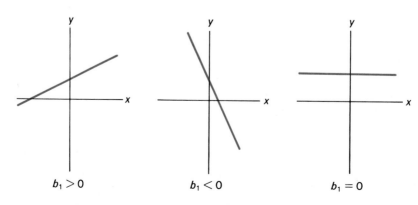

Exercises 12.1

12.1   On July 15, 1985, one of the authors called Avis Rent-A-Car to get the rate for renting a midsize car (a Buick Skylark). The quoted rate on that day was $63.00 per day plus $0.63 per mile. For a *one-day* rental, let x denote the number of miles driven and y denote the total cost.

a) Determine the equation that expresses y in terms of x.

b) Find b_0 and b_1.

c) Construct a table similar to Table 12.1 on page 499 for the x-values 50, 100, and 250 miles.

d) Obtain the graph for the equation in part (a) by plotting the points from part (c) and connecting them with a straight line.

e) Use the graph from part (d) to visually estimate the cost of driving the car 150 miles. Then calculate the cost exactly by employing the equation from part (a).

12.2   Encore Air Conditioning charges $36 per hour plus a $30 service charge. Let x denote the number of hours it takes for a given job and y denote the total cost to the customer.

a) Determine the equation that expresses y in terms of x.

b) Find b_0 and b_1.

c) Construct a table similar to Table 12.1 on page 499 for the x-values 0.5, 1, and 2.25 hours.

d) Obtain the graph for the equation in part (a) by plotting the points from part (c) and connecting them with a straight line.

e) Use the graph from part (d) to visually estimate the cost of a job that takes 1.75 hours. Then calculate the cost exactly by employing the equation from part (a).

12.3   The most commonly used scales for measuring temperature are the Fahrenheit and Celsius scales. If we let y denote Fahrenheit temperature and x denote Celsius temperature, then we can express the relationship between the two scales with the linear equation $y = 32 + 1.8x$.

a) Determine b_0 and b_1.

b) Find the Fahrenheit temperatures corresponding to the Celsius temperatures -40, 0, 20, and 100.

c) Graph the linear equation $y = 32 + 1.8x$ using the four points obtained in part (b).

d) Use the graph from part (c) to visually estimate the Fahrenheit temperature corresponding to 28° Celsius. Then calculate the temperature exactly by employing the linear equation $y = 32 + 1.8x$.

12.4   A ball is thrown straight up with an initial velocity of 64 feet per second. According to the laws of physics, if we let y denote the velocity of the ball after x seconds, then $y = 64 - 32x$.

a) Determine b_0 and b_1 for this linear equation.

b) Find the velocity of the ball after 1, 2, 3, and 4 seconds.

c) Graph the linear equation $y = 64 - 32x$ using the four points obtained in part (b).

d) Use the graph from part (c) to visually estimate the velocity of the ball after 1.5 seconds. Then calculate the velocity exactly by employing the linear equation $y = 64 - 32x$.

In each of Exercises 12.5 through 12.8,

a) determine the y-intercept and slope of the given linear equation.

b) explain what the y-intercept and slope represent in terms of the graph of the equation.

c) explain what the y-intercept and slope represent in terms relating to the given application.

12.5 $y = 63 + 0.63x$ (from Exercise 12.1)

12.6 $y = 30 + 36x$ (from Exercise 12.2)

12.7 $y = 32 + 1.8x$ (from Exercise 12.3)

12.8 $y = 64 - 32x$ (from Exercise 12.4)

In each of Exercises 12.9 through 12.18 you will be given a linear equation. For each exercise,

a) find the y-intercept and slope.

b) determine whether the line slopes upward, slopes downward, or is horizontal, without graphing the equation.

c) graph the equation using two points.

12.9 $y = 3 + 4x$ **12.10** $y = -1 + 2x$

12.11 $y = 6 - 7x$ **12.12** $y = -8 - 4x$

12.13 $y = 0.5x - 2$ **12.14** $y = -0.75x - 5$

12.15 $y = 2$ **12.16** $y = -3x$

12.17 $y = 1.5x$ **12.18** $y = -3$

In each of Exercises 12.19 through 12.26 you will be given the y-intercept, b_0, and slope, b_1, of a straight line. For each exercise,

a) determine whether the line slopes upward, slopes downward, or is horizontal, without graphing the equation.
b) find the equation of the line.
c) graph the equation using two points.

12.19 $b_0 = 5$, $b_1 = 2$

12.20 $b_0 = -3$, $b_1 = 4$

12.21 $b_0 = -2$, $b_1 = -3$

12.22 $b_0 = 0.4$, $b_1 = 1$

12.23 $b_0 = 0$, $b_1 = -0.5$

12.24 $b_0 = -1.5$, $b_1 = 0$

12.25 $b_0 = 3$, $b_1 = 0$

12.26 $b_0 = 0$, $b_1 = 3$

12.27 On page 497 we stated that any *nonvertical* straight-line graph can be described by an equation of the form $y = b_0 + b_1 x$.
a) Why can't a vertical straight line be expressed in that form?
b) What is the form of the equation for a vertical straight line?
c) Does a vertical straight line have a slope? Explain your answer.

12.2   The regression equation

In Examples 12.1 and 12.2, we discussed the linear equation

$$y = 25 + 20x$$

which gives the total cost, y, of a word processing job in terms of the time in hours, x, required to complete the job. Given the amount of time required, x, we can use the equation to determine the *exact* cost of the job, y.

Generally speaking, things are not quite as simple as they are in the word processing situation where one variable (cost) can be predicted *exactly* in terms of another (time required). More often than not, we must be content with rough predictions. In fact, for many circumstances the variable being predicted will vary even for a fixed value of the variable being used to make the prediction. For instance, we cannot predict the exact price, y, of a Datsun Z by just knowing its age, x. Indeed, even for a fixed age, say three years old, the price of a Datsun Z varies from car to car. We must be satisfied with making a rough prediction for the price of a three-year-old Datsun Z, or with an estimate of the average price of three-year-old Datsun Zs.

In Table 12.2 we have presented data on age versus price of Datsun Zs. The data were obtained from the classified ads of the *Arizona Republic*.

TABLE 12.2
Age versus price data for Datsun Zs. Prices are in hundreds of dollars, rounded to the nearest hundred.

Age (yrs) x	Price ($100s) y
5	80
7	57
6	58
6	55
5	70
4	88
7	43
6	60
5	69
5	63
2	118

It is useful to plot the data so that we can visualize any apparent relationships between age and price. Such a plot is called a **scatter diagram.** The scatter diagram for the data in Table 12.2 is given in Figure 12.7.

FIGURE 12.7
Scatter diagram for age versus price data from Table 12.2.

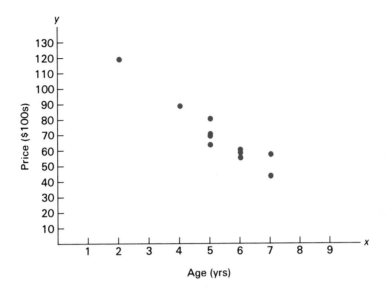

Age (yrs)

Although it is clear from the scatter diagram that the data points do not lie on a straight line, it appears that they are clustered about a straight line. We would like to fit a straight line to the data. Then we could use that line to predict price for Datsun Zs.

Since it is possible to draw many reasonable-looking straight lines through the cluster of points, we need a method to choose the "best" line. The method used is called the **least-squares criterion.** It is based on an analysis of the errors made in using a straight line to fit the data. To introduce the least-squares criterion, we will use a very simple data set. This is done in Example 12.3. We will return to the Datsun Z data shortly.

EXAMPLE 12.3 **Introduces the least-squares criterion**

Let us consider the problem of fitting a straight line to the four data points given in Table 12.3.

TABLE 12.3

x	y
1	1
1	2
2	2
4	6

The scatter diagram for this data is pictured in Figure 12.8 on page 506.

FIGURE 12.8

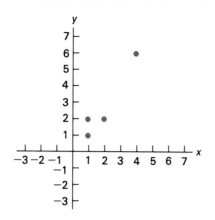

It is possible to fit many straight lines to these data. Figures 12.9 and 12.10 show two such lines.[†]

FIGURE 12.9   FIGURE 12.10

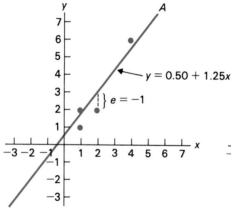  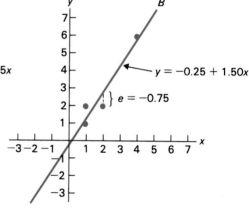

To keep things straight, we use $\hat{y}$ to denote the y-value *predicted* by a straight line for a given value of x. For instance, the y-value predicted by Line A for $x = 2$ is

$$\hat{y} = 0.50 + 1.25 \cdot 2 = 3$$

and the y-value predicted by Line B for $x = 2$ is

$$\hat{y} = -0.25 + 1.50 \cdot 2 = 2.75$$

To obtain a quantitative measure of how well a line fits the data, we first look at the errors, e, made in using the line to predict the y-values of the data points. For instance, Line A predicts a y-value of $\hat{y} = 3$ when $x = 2$. The actual y-value for

---

[†] We should emphasize that these are only two of many straight lines that we could have chosen.

$x = 2$ is $y = 2$ (see Table 12.3). Thus, the error made in using Line A to predict the y-value of the data point $(2, 2)$ is

$$e = y - \hat{y} = 2 - 3 = -1$$

(See Figure 12.9.) The fourth column of Table 12.4 gives the errors made for all four data points by Line A. The fourth column of Table 12.5 gives that for Line B.

TABLE 12.4

Line A: $y = 0.50 + 1.25x$

x	y	$\hat{y}$	e	e^2
1	1	1.75	−0.75	0.5625
1	2	1.75	0.25	0.0625
2	2	3.00	−1.00	1.0000
4	6	5.50	0.50	0.2500
Σ				1.8750

TABLE 12.5

Line B: $y = -0.25 + 1.50x$

x	y	$\hat{y}$	e	e^2
1	1	1.25	−0.25	0.0625
1	2	1.25	0.75	0.5625
2	2	2.75	−0.75	0.5625
4	6	5.75	0.25	0.0625
Σ				1.2500

The rule that decides which line, A or B, fits the data better is as follows: For each line, compute the sum of the squared errors, Σe^2. This is done in the final column of Tables 12.4 and 12.5. The line with the smaller total squared error, in this case Line B, is the one that fits the data better. ∎

With the previous example in mind, we can now state the least-squares criterion for the straight line that best fits a set of data points.

KEY FACT Least-squares criterion

The straight line that best fits a set of data points is the one whose sum of squared errors is smallest.

The straight line that best fits a set of data points according to the least-squares criterion is given a special name.

DEFINITION 12.2 Regression line and regression equation

Regression line: The straight line that fits a set of data points the best according to the least-squares criterion.
Regression equation: The equation of the regression line.

The least-squares criterion tells us what property the best-fitting line to a set of data points must have, but it does not tell us how to find that line. In Formula 12.1 we present the formulas that permit us to actually determine the best-fitting line to a set of data points. The formulas can be derived using elementary calculus (see Exercise 12.46).

FORMULA 12.1 Regression equation

The equation of the best-fitting line (regression line) to a set of n data points is

$$\hat{y} = b_0 + b_1 x$$

where

$$b_1 = \frac{n(\Sigma xy) - (\Sigma x)(\Sigma y)}{n(\Sigma x^2) - (\Sigma x)^2}$$

and

$$b_0 = \frac{1}{n}(\Sigma y - b_1 \Sigma x)$$

EXAMPLE 12.4   Illustrates Formula 12.1

The age versus price data for Datsun Zs are repeated in Table 12.6. Ages are in years. Prices are in hundreds of dollars, rounded to the nearest hundred.

TABLE 12.6

Age (yrs) x	Price ($100s) y
5	80
7	57
6	58
6	55
5	70
4	88
7	43
6	60
5	69
5	63
2	118

a) Determine the regression equation for the data; that is, find the equation of the regression line.

b) Graph the regression equation along with the data points in Table 12.6.

c) Describe the apparent relationship between age and price for Datsun Zs.

d) What does the slope of the regression equation represent in terms of the prices for Datsun Zs?

e) Use the regression equation to predict the price for a three-year-old Z and a four-year-old Z.

SOLUTION

a) To determine the regression equation, we need to compute b_1 and b_0 using Formula 12.1. It is convenient to construct a table of values for x, y, xy, x^2, and their sums. This is presented in Table 12.7.

TABLE 12.7

Table for computing the slope, b_1, and y-intercept, b_0, of the regression equation for the Datsun Z data.

x	y	xy	x^2
5	80	400	25
7	57	399	49
6	58	348	36
6	55	330	36
5	70	350	25
4	88	352	16
7	43	301	49
6	60	360	36
5	69	345	25
5	63	315	25
2	118	236	4
Σ 58	761	3736	326

The slope of the regression equation is therefore

$$b_1 = \frac{n(\Sigma xy) - (\Sigma x)(\Sigma y)}{n(\Sigma x^2) - (\Sigma x)^2}$$

$$= \frac{11(3736) - (58)(761)}{11(326) - (58)^2}$$

$$= \frac{41{,}096 - 44{,}138}{3586 - 3364} = \frac{-3042}{222} = -13.70$$

and the y-intercept is

$$b_0 = \frac{1}{n}(\Sigma y - b_1 \Sigma x)$$

$$= \frac{1}{11}(761 - (-13.70) \cdot 58) = 141.43^\dagger$$

Thus, the regression equation is

$$\hat{y} = 141.43 - 13.70x$$

b) To graph the regression equation, we need only substitute two different x-values to obtain two distinct points. Let us use the x-values $x = 2$ and $x = 8$. The corresponding y-values are

$$\hat{y} = 141.43 - 13.70 \cdot 2 = 114.03$$

and

$$\hat{y} = 141.43 - 13.70 \cdot 8 = 31.83$$

Consequently, the regression line passes through the two points (2, 114.03) and (8, 31.83). In Figure 12.11, at the top of the next page, we have plotted these two points using hollow dots. Drawing a straight line through the two hollow dots yields the regression line, the graph of the regression equation.

---

† In computing b_0 we did not use the rounded-off value of -13.70 for b_1, but kept full calculator accuracy in doing the computation.

FIGURE 12.11
Regression line
and data points for
the Datsun Z data.

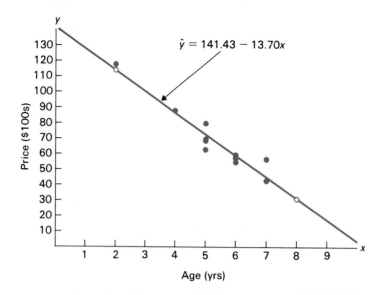

Also included in Figure 12.11 are the data points given in Table 12.6. As we know, the regression line in Figure 12.11 is the straight line that best fits the data points according to the least-squares criterion. That is, it is the straight line whose sum of squared errors is smallest.

c) Here we are to describe the apparent relationship between age and price for Datsun Zs. Since the slope of the regression line is negative, we see that price tends to decrease as age increases—no particular surprise.

d) For this part we are to interpret the slope of the regression equation in terms of the prices for Datsun Zs. To begin, recall that x represents age, in years, and y represents price, in hundreds of dollars. The slope of -13.70, or $-\$1,370$, indicates that Datsun Zs depreciate an estimated \$1,370 per year, at least in the two- to seven-year-old range.

e) Finally, we are to use the regression equation, $\hat{y} = 141.43 - 13.70x$, to predict the price for a three-year-old Z and a four-year-old Z. For a three-year-old Z we have $x = 3$, and so the predicted price is

$$\hat{y} = 141.43 - 13.70 \cdot 3 = 100.33$$

or \$10,033. Similarly, the price the regression equation predicts for a four-year-old Z is

$$\hat{y} = 141.43 - 13.70 \cdot 4 = 86.63$$

or \$8,663. Questions concerning the accuracy and reliability of such predictions will be discussed later. ∎

MTB
SPSS

A warning on the use of linear regression

The idea behind finding a regression line is based on the assumption that the data points are actually scattered about a straight line.[†] In some cases, data points

---

[†] We shall discuss this assumption in detail in Section 13.1, and make it more precise.

FIGURE 12.12

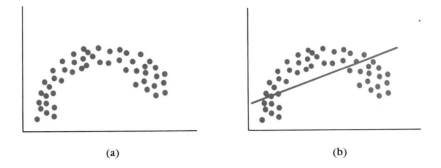

(a)   (b)

may be scattered about a *curve* instead of a straight line, as in Figure 12.12(a). Unfortunately, the formulas for b_0 and b_1 will work for this data set and fit an *inappropriate* straight line to the data. Indeed, the data, which really follows a curve, would be fitted by the straight line shown in Figure 12.12(b).

This procedure is misleading. For instance, it would lead us to predict that *y*-values in Figure 12.12(a) will keep increasing when they have actually begun to decrease. There are techniques that allow us to fit *curves* to data points, as should be done for the data in Figure 12.12(a). We will not cover those methods in this book, but any statistical consultant should be able to advise you on such curve fits. In summary:

KEY FACT

If you plan to find a regression line for a set of data points, first look at a scatter diagram of the data. If the data points do not appear to be scattered about a straight line, do *not* determine a regression line.

Extrapolation

It is important to warn against the danger of *extrapolation* in regression. **Extrapolation** refers to using the regression equation to make predictions for *x*-values that are far removed from the *x*-values in the sample data. Here is why such predictions might be unreliable. Using the regression equation to make predictions is reasonable only when there is a linear relationship between the variables. Although this may be true for values of *x* in the range of the *x*-values in the sample, as evidenced by a scatter diagram, it may not necessarily be true for *x*-values outside that range.

The illustration of age versus price for Datsun Zs provides an excellent example of where extrapolation can be dangerous. The regression equation is

$$\hat{y} = 141.43 - 13.70x$$

The *x*-values of the sample points used in computing this regression equation range from $x = 2$ to $x = 7$; that is, from two to seven years old.

Suppose we *extrapolate* and use the regression equation to predict the price of a 12-year-old Z. The predicted price is

$$\hat{y} = 141.43 - 13.70 \cdot 12 = -22.97$$

or $-\$2,297$. Clearly, this is ridiculous. In fact, if you look in the paper, you will find that a more reasonable prediction for the price of a 12-year-old Z is about $\$3,000$. Thus, although the relationship between age and price of Datsun Zs appears to be linear in the range $x = 2$ to $x = 7$, it is definitely not so in the range $x = 2$ to, say, $x = 12$. Figure 12.13 summarizes the discussion on extrapolation as it applies to age versus price for Datsun Zs.

FIGURE 12.13

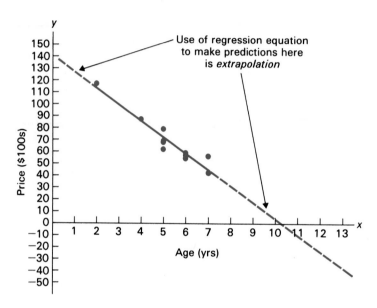

Exercises 12.2

In Exercises 12.28 and 12.29,
a) graph each linear equation together with the given data points.
b) construct tables for x, y, $\hat{y}$, e, and e^2 similar to Tables 12.4 and 12.5 on page 507.
c) determine which line fits the set of data points better according to the least-squares criterion.

12.28 Line A: $y = 1.5 + 0.5x$
Line B: $y = 1.125 + 0.375x$

Data points

x	1	1	5	5
y	1	3	2	4

12.29 Line A: $y = 3 - 0.6x$
Line B: $y = 4 - x$

Data points

x	0	2	2	5	6
y	4	2	0	-2	1

For Exercises 12.30 through 12.37, be sure to save your worksheets. You will need them in later sections. [*Note:* Due to rounding, your answers may deviate somewhat from those given in the back of the book.]

12.30 Refer to Exercise 12.28.
a) Find the regression equation for the data points.
b) Graph the regression equation along with the data points.

12.31 Refer to Exercise 12.29.
a) Find the regression equation for the data points.
b) Graph the regression equation along with the data points.

12.32 Twelve Corvettes, between one and seven years of age, were randomly selected from the classified ads. The following data were obtained on age versus price:

Age (yrs)	Price ($100s)
x	y
7	90
6	90
6	120
4	118
3	175
3	179
7	100
3	128
7	85
6	92
1	173
1	190

a) Determine the regression equation for the data.
b) Graph the regression equation along with the data points.
c) Describe the apparent relationship between age and price for Corvettes.
d) What does the slope of the regression line represent in terms of Corvette prices?
e) Use the regression equation to predict the price of a two-year-old Corvette; a three-year-old Corvette.

12.33 The U.S. National Center for Health Statistics collects and publishes data on heights and weights by age and sex. A random sample of 11 males, aged 18–24 years, gave the following data:

Height (inches)	Weight (lb)
x	y
65	175
67	133
71	185
71	163
66	126
75	198
67	153
70	163
71	159
69	151
69	155

a) Determine the regression equation for the data.

b) Graph the regression equation along with the data points.
c) Describe the apparent relationship between height and weight for 18–24-year-old males.
d) What does the slope of the regression line represent in terms of heights and weights of 18–24-year-old males?
e) Use the regression equation to predict the weight of an 18–24-year-old male who is 67 inches tall; 73 inches tall.

12.34 Hanna Properties, Inc. specializes in custom-home resales in the Equestrian Estates. A random sample of nine currently listed custom homes provided the following information on size and price. The size data are in hundreds of square feet, rounded to the nearest hundred; the price data are in thousands of dollars, rounded to the nearest thousand.

Size	Price
x	y
26	235
27	249
33	267
29	269
29	295
34	345
30	415
40	475
22	195

a) Determine the regression equation for the data.
b) Graph the regression equation along with the data points.
c) Describe the apparent relationship between square-footage and price.
d) What does the slope of the regression line represent in terms of sizes and prices of custom homes in the Equestrian Estates?
e) Use the regression equation to predict the price of a custom home in the Equestrian Estates that has 2600 square feet.

12.35 A report read by a physician indicated that the maximum heart rate an individual can reach during intensive exercise decreases with age. The physician decided to do his own study. Ten randomly selected members of a jogging club performed exercise tests and recorded their peak heart rates.

The results are shown in the following table:

Age	Peak heart rate
x	y
10	210
20	200
20	195
25	195
30	190
30	180
30	185
40	180
45	170
50	165

a) Determine the regression equation for the data.
b) Graph the regression equation along with the data points.
c) Describe the apparent relationship between age and peak heart rate.
d) What does the slope of the regression line represent in terms of age and peak heart rate?
e) Use the regression equation to predict the peak heart rate of a person 22 years old.

12.36 A calculus instructor asked a random sample of eight students to record their study times per lesson in a beginning calculus course. She then made a table for total study times over two weeks and test scores at the end of the two weeks. Here are the results.

Study time (hrs)	Grade (percent)
x	y
10	92
15	81
12	84
20	74
8	85
16	80
14	84
22	80

a) Determine the regression equation for the data.
b) Graph the regression equation along with the data points.
c) Describe the apparent relationship between study time and grade. (Does it surprise you?)
d) What does the slope of the regression line represent in terms of study time and grade?
e) Use the regression equation to predict the grade of a student who studies for 15 hours.

12.37 An economist is interested in the relation between the disposable income of a family and the amount of money spent annually on food. For a preliminary study, he randomly selects eight middle-income families of the same size (father, mother, two children). The results are as follows:

Family disposable income (in thousands of dollars)	Food expenditures (in hundreds of dollars)
x	y
30	55
36	60
27	42
20	40
16	37
24	26
19	39
25	43

a) Determine the regression equation for the data.
b) Graph the regression equation along with the data points.
c) Describe the apparent relationship between disposable income and annual food expenditures.
d) What does the slope of the regression line represent in terms of disposable income and annual food expenditures?
e) Use the regression equation to predict the annual food expenditures for a family with a disposable income of $25,000.

12.38 Which of the following scatter diagrams represent data for which a regression equation should be computed?

12.39 Which of the following scatter diagrams represent data for which a regression equation should be computed?

12.40 The negative relation between study time and grade in Exercise 12.36 has been found by many investigators, and has puzzled them. Can you think of a possible explanation for it?

In Exercises 12.41 through 12.45 we will consider the **sample covariance, s_{xy},** of a set of data points. The sample covariance of n data points is defined by

(1) $$s_{xy} = \frac{\Sigma(x - \bar{x})(y - \bar{y})}{n - 1}$$

12.41 Determine the sample covariance of the data points in Exercise 12.29.

12.42 Determine the sample covariance of the data points in Exercise 12.28.

The sample covariance can be used as an alternate means for computing the slope and y-intercept of the regression equation for a set of data points. The appropriate formulas are

(2) $$b_1 = s_{xy}/s_x^2 \qquad b_0 = \bar{y} - b_1\bar{x}$$

where s_x is the sample standard deviation of the x-values.

12.43 Use the formulas in (2) to find the regression equation for the data points in Exercise 12.29, and compare your answer to the one from Exercise 12.31.

12.44 Apply the formulas in (2) to obtain the regression equation for the data points in Exercise 12.28, and compare your answer to the one from Exercise 12.30.

12.45 a) Prove that the formulas in (2) are equivalent to the ones in Formula 12.1 on page 508.
b) Show that the formula for b_1 given in (2) is equivalent to

$$b_1 = \frac{\Sigma(x - \bar{x})(y - \bar{y})}{\Sigma(x - \bar{x})^2}$$

12.46 In this exercise we will derive the formulas, given in Formula 12.1 on page 508, for the slope and y-intercept of the regression equation. The derivation requires elementary calculus. According to the least-squares criterion, the regression line is the straight line whose sum of squared errors is smallest.
a) Show that the regression line is the straight line, $\hat{y} = b_0 + b_1x$, for which b_0 and b_1 minimize the function

$$f(b_0, b_1) = \Sigma[y - (b_0 + b_1x)]^2$$

b) Use elementary calculus to determine the values of b_0 and b_1 that minimize $f(b_0, b_1)$ and show that these values give the equations in Formula 12.1. *Hint:* Compute the partial derivatives of $f(b_0, b_1)$, set them equal to zero, and solve for b_0 and b_1.

12.3 The coefficient of determination

In Example 12.4, we determined the regression equation

$$\hat{y} = 141.43 - 13.70x$$

for the sample data on age versus price of Datsun Zs. Here x represents age, in years, and $\hat{y}$ predicted price, in hundreds of dollars. We can apply the regression equation to predict the price for a Datsun Z of a given age, x. For instance, we predict that a four-year-old Datsun Z will cost roughly

$$\hat{y} = 141.43 - 13.70 \cdot 4 = 86.63$$

or $8,663.

But how good are such predictions? That is, how useful is the regression equation, $\hat{y} = 141.43 - 13.70x$, for predicting price, y? Are the predictions likely to be close to the actual values or will there tend to be quite a bit of error?

There are several ways that we can attempt to answer questions on how valuable the regression equation is for making predictions. One method is to measure the reduction in the errors made in prediction by using the regression equation as opposed to simply predicting the mean $\bar{y}$. To illustrate the ideas involved, we return to the Datsun Z data.

EXAMPLE   12.5   **Introduces the coefficient of determination**

The age versus price data for Datsun Zs are repeated in Table 12.8. Ages are in years. Prices are in hundreds of dollars, rounded to the nearest hundred.

One way we can employ the information in Table 12.8 to make predictions for the price of a Datsun Z is to ignore age and simply use the mean price, $\bar{y}$, of the Zs sampled. In other words, just use

$$\bar{y} = \frac{\Sigma y}{n} = \frac{761}{11} = 69.18 \ (\$6,918)$$

as our predicted price for a Datsun Z.

Let us see how much (squared) error is made when we predict $\bar{y} = 69.18$ for the observed y-values in the second column of Table 12.8. The required computations are shown in Table 12.9.

TABLE 12.8

Age (yrs)	Price ($100s)
x	y
5	80
7	57
6	58
6	55
5	70
4	88
7	43
6	60
5	69
5	63
2	118

TABLE 12.9[†]

y	$y - \bar{y}$	$(y - \bar{y})^2$
80	10.82	117.03
57	−12.18	148.40
58	−11.18	125.03
55	−14.18	201.12
70	0.82	0.67
88	18.82	354.12
43	−26.18	685.49
60	−9.18	84.31
69	−0.18	0.03
63	−6.18	38.21
118	48.82	2383.21
Σ		4137.64

Thus, the total squared error made when we predict $\bar{y} = 69.18$ for the observed y-values is

$$\Sigma(y - \bar{y})^2 = 4137.64$$

This is called the **total sum of squares, SST.** Thus,

$$SST = \Sigma(y - \bar{y})^2 = 4137.64$$

Now, if the regression equation is useful for predicting price, then we should obtain a reduction in the total squared error by using the regression line values, $\hat{y}$, for our price predictions instead of $\bar{y}$. The total squared error made when we employ the regression equation to predict the observed y-values is computed in Table 12.10. Each $\hat{y}$-value is determined from the regression equation

$$\hat{y} = 141.43 - 13.70x$$

which we obtained on page 509.

---

† Values in this and all other tables in this section are displayed to two decimal places, but computations are done using calculator accuracy.

TABLE 12.10

x	y	$\hat{y}$	$y - \hat{y}$	$(y - \hat{y})^2$
5	80	72.92	7.08	50.14
7	57	45.51	11.49	131.94
6	58	59.22	−1.22	1.48
6	55	59.22	−4.22	17.78
5	70	72.92	−2.92	8.52
4	88	86.62	1.38	1.90
7	43	45.51	−2.51	6.32
6	60	59.22	0.78	0.61
5	69	72.92	−3.92	15.36
5	63	72.92	−9.92	98.38
2	118	114.03	3.97	15.78
Σ				348.22

Hence, the total squared error made when we use the regression line values, $\hat{y}$, for our predictions is

$$\Sigma(y - \hat{y})^2 = 348.22$$

This is called the **error sum of squares, SSE.** Thus,

$$SSE = \Sigma(y - \hat{y})^2 = 348.22$$

   In summary then, if we ignore age and simply use the sample mean of $\bar{y} = 69.18$ for our predictions of the observed y-values, then the total squared error is

$$SST = \Sigma(y - \bar{y})^2 = 4137.64$$

On the other hand, using the regression equation for prediction, the total squared error is

$$SSE = \Sigma(y - \hat{y})^2 = 348.22$$

As you can see, we have obtained a drastic reduction in the total squared error by using the regression equation. The *percentage* reduction is

$$\frac{SST - SSE}{SST} = 1 - \frac{SSE}{SST} = 1 - \frac{348.22}{4137.64} = 0.916$$

or 91.6%. Consequently, by using the regression equation, instead of just $\bar{y}$, we have reduced the total squared error in predicting the observed y-values by 91.6%. This indicates, as we would suspect, that age, x, is extremely useful for predicting price.

   The percentage reduction in the total squared error, $1 - SSE/SST$, obtained by using the regression equation is called the **coefficient of determination** and is denoted by r^2. Thus, as we have just seen, the coefficient of determination for the Datsun Z data is

MTB
SPSS

$$r^2 = 1 - \frac{SSE}{SST} = 1 - \frac{348.22}{4137.64} = 0.916$$

■

The definitions introduced in Example 12.5 are summarized below.

DEFINITION 12.3 Total sum of squares, error sum of squares, and coefficient of determination

Total sum of squares, *SST*:

$$SST = \Sigma(y - \bar{y})^2$$

Error sum of squares, *SSE*:

$$SSE = \Sigma(y - \hat{y})^2$$

Coefficient of determination, r^2:

$$r^2 = 1 - \frac{SSE}{SST}$$

Explained variation

We have introduced the coefficient of determination, r^2, as a descriptive measure of the utility of the regression equation for making predictions. Specifically, r^2 gives the percentage reduction in the total squared error obtained by using the regression equation to predict the observed y-values, instead of just predicting $\bar{y}$.

There is another way to interpret r^2; namely, as the percentage of the variation in the observed y-values that is *explained* by the regression line. To see why, we return to our Datsun Z data.

E X A M P L E   12.6   **Introduces explained variation**

The scatter diagram for the age versus price data of Datsun Zs is reproduced here as Figure 12.14.

FIGURE 12.14
Scatter diagram
for age versus
price of Datsun Zs.

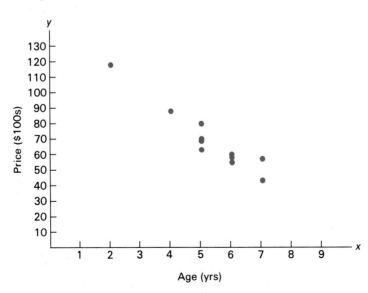

As you can see, there is quite a bit of variation in the observed y-values, ranging from a low of 43 (i.e., $4,300) to a high of 118 (i.e., $11,800). If we now superimpose the regression line, $\hat{y} = 141.43 - 13.70x$, over the data, we obtain the picture in Figure 12.15.

FIGURE 12.15
Regression line and
data points for
Datsun Z data.

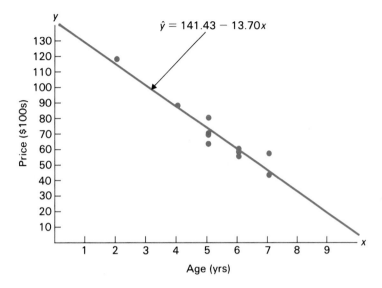

$\hat{y} = 141.43 - 13.70x$

The graph shows that much of the variation in the observed y-values is "explained" by the regression line, in the sense that the regression line predicts a good portion of the type of variability that is perceived.

To describe quantitatively how much of the total variation is explained by the regression line, we proceed as follows. As before, we will use the total sum of squares, *SST,* as our measure of total variation in the observed y-values. We have (see page 516)

$$SST = \Sigma(y - \bar{y})^2 = \Sigma(y - 69.18)^2 = 4137.64$$

Now let us look at a particular observed y-value, say $y = 80$ (corresponding to the data point (5, 80)). In Figure 12.16, at the top of the next page, we have drawn a blow-up of a portion of Figure 12.15 showing the data point (5, 80) and no others. From the figure we see that the deviation of an observed y-value from the mean, $y - \bar{y}$, can be decomposed into two parts—the deviation that is explained by the regression line, $\hat{y} - \bar{y}$, and the remaining unexplained deviation, $y - \hat{y}$. Therefore, the total amount of variation (squared deviation) explained by the regression line is $\Sigma(\hat{y} - \bar{y})^2$. This is called the **regression sum of squares, *SSR*.** To compute *SSR* we need the predicted values, $\hat{y}$, and the mean of the observed y-values, $\bar{y}$. The $\hat{y}$-values are given in the third column of Table 12.10 and are repeated in the first column of Table 12.11. Recalling that $\bar{y} = 69.18$, we obtain the regression sum of squares, *SSR,* from the third column of Table 12.11.

FIGURE 12.16

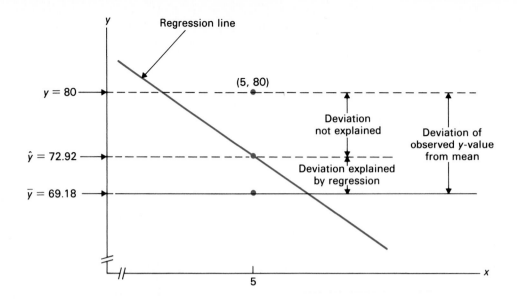

TABLE 12.11

$\hat{y}$	$\hat{y} - \bar{y}$	$(\hat{y} - \bar{y})^2$
72.92	3.74	13.97
45.51	−23.67	560.19
59.22	−9.97	99.31
59.22	−9.97	99.31
72.92	3.74	13.97
86.62	17.44	304.15
45.51	−23.67	560.19
59.22	−9.97	99.31
72.92	3.74	13.97
72.92	3.74	13.97
114.03	44.85	2011.09
Σ		3789.42

Consequently, the regression sum of squares, *SSR,* is equal to

$$SSR = \Sigma(\hat{y} - \bar{y})^2 = 3789.42$$

This is the amount of variation in the observed *y*-values that is explained by the regression. Therefore, the percentage of the total variation in the observed *y*-values that is explained by the regression is

$$\frac{SSR}{SST} = \frac{3789.42}{4137.64} = 0.916 \text{ or } 91.6\%$$

Thus, most of the variation in the observed *y*-values is explained by the regression, and so the regression equation is a good predictor of price. ■

MTB
SPSS

DEFINITION 12.4 Regression sum of squares, *SSR*

$$SSR = \Sigma(\hat{y} - \overline{y})^2$$

You may have noted that, for the Datsun Z data, the percentage of variation in the observed y-values that is explained by the regression, namely,

$$\frac{SSR}{SST} = \frac{3789.42}{4137.64} = 0.916$$

has the same value as the coefficient of determination, r^2:

$$r^2 = 1 - \frac{SSE}{SST} = 1 - \frac{348.22}{4137.64} = 0.916$$

(see page 517). This is no accident! The reason stems from the fact that the total sum of squares, *SST, always* equals the regression sum of squares, *SSR,* plus the error sum of squares, *SSE.* That is,

$$SST = SSR + SSE$$

[For the Datsun Z data we can see that this equation is true since $SST = 4137.64$, $SSR = 3789.42$, and $SSE = 348.22$; and $4137.64 = 3789.42 + 348.22$.] Using the identity $SST = SSR + SSE$, we deduce that

$$\frac{SSR}{SST} = \frac{SST - SSE}{SST} = 1 - \frac{SSE}{SST} = r^2$$

which states that the percentage of variation in the observed y-values that is explained by the regression equals the coefficient of determination.

We now summarize the concepts presented in this section.

KEY FACTS

Total sum of squares, *SST:* The total sum of squares is defined by

$$SST = \Sigma(y - \overline{y})^2$$

and represents the total variation in the observed y-values *or* the total squared error made in using the mean, $\overline{y}$, to predict the observed y-values.

Regression sum of squares, *SSR:* The regression sum of squares is defined by

$$SSR = \Sigma(\hat{y} - \overline{y})^2$$

and represents the amount of variation in the observed y-values that is explained by the regression.

Error sum of squares, *SSE:* The error sum of squares is defined by

$$SSE = \Sigma(y - \hat{y})^2$$

and represents the amount of variation in the observed y-values that is not explained by the regression *or* the total squared error made in using the regression equation to predict the observed y-values.

Regression identity:

$$SST = SSR + SSE$$

Coefficient of determination, r^2: The coefficient of determination is defined by

$$r^2 = 1 - \frac{SSE}{SST}$$

and represents the percentage reduction obtained in the total squared error by using the regression equation to predict the observed y-values, instead of simply the mean $\bar{y}$. The coefficient of determination can also be computed using the formula

$$r^2 = \frac{SSR}{SST}$$

Thus, it also represents the percentage of variation in the observed y-values that is explained by the regression. In any case, r^2 *is a descriptive measure of the utility of the regression equation for making predictions. We have $0 \leq r^2 \leq 1$, and the closer r^2 is to 1, the more useful the regression equation is for making predictions.*

Exercises 12.3

12.47   In this section we introduced a descriptive measure of the utility of the regression equation for making predictions.
  a) What term and symbol are used for that descriptive measure?
  b) Give two interpretations of that descriptive measure.

For Exercises 12.48 through 12.55 be sure to save your worksheets. You will need them in subsequent sections. In each exercise,
a) compute SST, SSR, and SSE.
b) verify the regression identity, $SST = SSR + SSE$.
c) compute the coefficient of determination using both the definition, $r^2 = 1 - SSE/SST$, and the formula $r^2 = SSR/SST$.
d) determine the percentage reduction obtained in the total squared error by using the regression equation to predict the observed y-values, instead of simply using the mean $\bar{y}$.
e) determine the percentage of variation in the observed y-values that is explained by the regression.
f) state how useful the regression equation appears to be for making predictions. (The answer given here is somewhat subjective.)

Note: We have given you the regression equations for the data sets in Exercises 12.48 through 12.55. Those regression equations were obtained in Exercises 12.30 through 12.37, respectively.

Note: Due to rounding, your answers may deviate somewhat from those given in the back of the book.

12.48   The data from Exercise 12.30 of Section 12.2:

x	1	1	5	5
y	1	3	2	4

Regression equation is $\hat{y} = 1.75 + 0.25x$.

12.49   The data from Exercise 12.31 of Section 12.2:

x	0	2	2	5	6
y	4	2	0	−2	1

Regression equation is $\hat{y} = 2.875 - 0.625x$.

12.50   The age versus price data for Corvettes from Exercise 12.32 of Section 12.2:

Age (yrs) x	Price ($100s) y
7	90
6	90
6	120
4	118
3	175
3	179
7	100
3	128
7	85
6	92
1	173
1	190

Regression equation is $\hat{y} = 200.33 - 16x$.

12.51 The height versus weight data for 18–24-year-old males from Exercise 12.33 of Section 12.2:

Height (inches) x	Weight (lb) y
65	175
67	133
71	185
71	163
66	126
75	198
67	153
70	163
71	159
69	151
69	155

Regression equation is $\hat{y} = -174.49 + 4.84x$.

12.52 The size versus price data for custom homes from Exercise 12.34 of Section 12.2:

Size x	Price y
26	235
27	249
33	267
29	269
29	295
34	345
30	415
40	475
22	195

Regression equation is $\hat{y} = -128.333 + 14.444x$.

12.53 The age versus peak-heart-rate data from Exercise 12.35 of Section 12.2:

Age x	Peak heart rate y
10	210
20	200
20	195
25	195
30	190
30	180
30	185
40	180
45	170
50	165

Regression equation is $\hat{y} = 219.78 - 1.09x$.

12.54 The study-time versus grade data from Exercise 12.36 of Section 12.2:

Study time (hrs) x	Grade (percent) y
10	92
15	81
12	84
20	74
8	85
16	80
14	84
22	80

Regression equation is $\hat{y} = 94.87 - 0.85x$.

12.55 The disposable-income versus annual-food-expenditure data from Exercise 12.37 of Section 12.2:

Family disposable income (in thousands of dollars) x	Food expenditures (in hundreds of dollars) y
30	55
36	60
27	42
20	40
16	37
24	26
19	39
25	43

Regression equation is $\hat{y} = 12.86 + 1.21x$.

12.56 Suppose that $r^2 = 0$ for a data set. What can you say about
a) *SSE?*
b) *SSR?*
c) the utility of the regression equation for making predictions?

12.57 Suppose that $r^2 = 1$ for a data set. What can you say about
a) *SSE?*
b) *SSR?*
c) the utility of the regression equation for making predictions?

12.58 In this exercise we will show that the coefficient of determination always lies between 0 and 1, inclusive.
a) Indicate why the quantities *SST, SSR,* and *SSE* are nonnegative.

b) Verify that $SSE \leq SST$. *Hint:* Use part (a) along with the regression identity $SST = SSR + SSE$.

c) Prove that $0 \leq r^2 \leq 1$.

12.59 This exercise provides a proof of the regression identity, $SST = SSR + SSE$. It uses some of the results obtained in the course of deriving the formulas for b_0 and b_1 in Exercise 12.46 of Section 12.2.

a) Show that

$$\Sigma(y - \bar{y})^2 = \Sigma(y - \hat{y})^2 + \Sigma(\hat{y} - \bar{y})^2 + 2\Sigma(y - \hat{y})(\hat{y} - \bar{y})$$

Hint: Write $(y - \bar{y})^2$ as $[(y - \hat{y}) + (\hat{y} - \bar{y})]^2$ and apply the binomial formula.

b) Verify that

$$\Sigma(y - \hat{y})(\hat{y} - \bar{y}) = \Sigma\hat{y}(y - \hat{y}) - \bar{y}\Sigma(y - \hat{y})$$

c) In Exercise 12.46 we found that

$$\Sigma(y - \hat{y}) = 0 \text{ and } \Sigma x(y - \hat{y}) = 0$$

Use these results along with part (b) to show

$$\Sigma(y - \hat{y})(\hat{y} - \bar{y}) = 0$$

d) Deduce the regression identity from parts (a) and (c).

12.4   Linear correlation

We often hear statements pertaining to the correlation or lack of correlation between two variables: "There is a *positive correlation* between advertising expenditures and sales" or "IQ and alcohol consumption are *uncorrelated.*" In this section, we will explain the meaning of such statements.

There are several methods for measuring the correlation between two variables such as advertising expenditures and sales. Probably the most common measure is the **linear correlation coefficient, r.**[†] The linear correlation coefficient is a single number that can be used to describe the strength of the *linear* (straight-line) relationship between two variables. We begin by giving the formula for computing the linear correlation coefficient and by discussing some of its basic properties.

FORMULA 12.3 Linear correlation coefficient

The **linear correlation coefficient, r,** for n data points is given by the formula

$$r = \frac{n(\Sigma xy) - (\Sigma x)(\Sigma y)}{\sqrt{n(\Sigma x^2) - (\Sigma x)^2}\sqrt{n(\Sigma y^2) - (\Sigma y)^2}}[††]$$

The linear correlation coefficient, r, is always between -1 and 1. Values of r close to -1 or $+1$ indicate a strong linear relationship between the variables and that the variable x is a good linear predictor of the variable y—that is, the regression equation is quite useful for making predictions. On the other hand, values of r near 0 indicate a weak linear relationship between the variables and that the variable x is not too useful as a linear predictor of the variable y—that is, the regression equation is not very valuable for making predictions.

[†] Also called the **Pearson product moment correlation coefficient.**
[††] This is actually the shortcut formula for r. The defining formula and its equivalence to Formula 12.3 are discussed in the exercises. See, in particular, Exercise 12.75.

Positive values of r suggest that the variables are **positively linearly corre-lated,** meaning that y tends to increase linearly as x increases, with the tendency being greater the closer that r is to 1. Negative values of r suggest that the variables are **negatively linearly correlated,** meaning that y tends to decrease linearly as x increases, with the tendency being greater the closer that r is to -1. The sign of r is the same as the sign of the slope of the regression line.

Graphically speaking, we can summarize the above discussion as follows (refer to Figure 12.17). If the linear correlation coefficient, r, is close to ± 1, then the data points are clustered closely about the regression line. If the value of r is farther from ± 1, then the data points are more widely scattered about the regression line. And, finally, if the value of r is near 0, then the slope of the regression line is also near 0, thus indicating that there is probably no linear relationship between the variables.

FIGURE 12.17

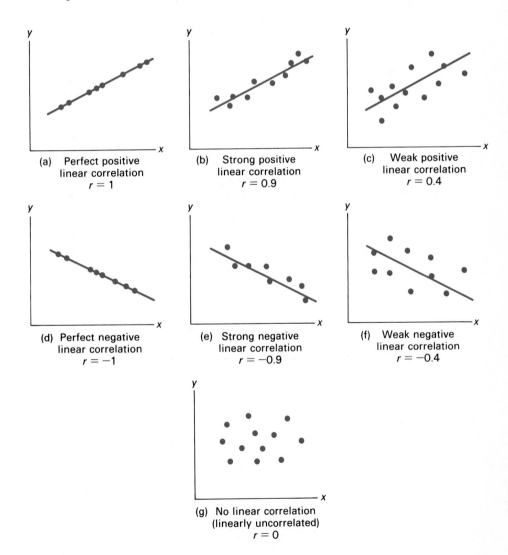

(a) Perfect positive
linear correlation
$r = 1$

(b) Strong positive
linear correlation
$r = 0.9$

(c) Weak positive
linear correlation
$r = 0.4$

(d) Perfect negative
linear correlation
$r = -1$

(e) Strong negative
linear correlation
$r = -0.9$

(f) Weak negative
linear correlation
$r = -0.4$

(g) No linear correlation
(linearly uncorrelated)
$r = 0$

We will now illustrate how to compute and interpret the linear correlation co-efficient of a set of data points. To do this, we return to the data on age and price for a sample of Datsun Zs.

EXAMPLE 12.7   **Illustrates the linear correlation coefficient**

The age versus price data for Datsun Zs are repeated in Table 12.12. Ages are in years. Prices are in hundreds of dollars, rounded to the nearest hundred.

TABLE 12.12

Age (yrs) x	Price ($100s) y
5	80
7	57
6	58
6	55
5	70
4	88
7	43
6	60
5	69
5	63
2	118

a) Compute the linear correlation coefficient, r, of the data.

b) Interpret the value of r obtained in part (a) in terms of the linear relationship between age and price.

c) Discuss the graphical implications of the value of r.

SOLUTION

a) To compute the linear correlation coefficient, r, for the data in Table 12.12, we see from Formula 12.3 that we need a table of values for x, y, xy, x^2, y^2, and their sums. Such a table is given in Table 12.13.

TABLE 12.13

x	y	xy	x^2	y^2
5	80	400	25	6,400
7	57	399	49	3,249
6	58	348	36	3,364
6	55	330	36	3,025
5	70	350	25	4,900
4	88	352	16	7,744
7	43	301	49	1,849
6	60	360	36	3,600
5	69	345	25	4,761
5	63	315	25	3,969
2	118	236	4	13,924
Σ 58	761	3736	326	56,785

Applying Formula 12.3, we obtain

$$r = \frac{n(\Sigma xy) - (\Sigma x)(\Sigma y)}{\sqrt{n(\Sigma x^2) - (\Sigma x)^2}\sqrt{n(\Sigma y^2) - (\Sigma y)^2}}$$

$$= \frac{11(3736) - (58)(761)}{\sqrt{11(326) - (58)^2}\sqrt{11(56,785) - (761)^2}}$$

$$= \frac{41,096 - 44,138}{\sqrt{3586 - 3364}\sqrt{624,635 - 579,121}} = -0.957$$

b) The linear correlation coefficient of $r = -0.957$ suggests that there is a strong negative linear correlation between age and price of Datsun Zs. In particular then, it indicates that as age increases, there is a strong tendency for price to decrease, which is not surprising. It also implies that the regression equation

$$\hat{y} = 141.43 - 13.70x$$

gives reasonably accurate estimates for price based on age.

c) Since the value of the linear correlation coefficient is $r = -0.957$, which is close to -1, the data points should be clustered closely about the regression line. Figure 12.18 shows that this is indeed the case. ∎

MTB
SPSS

FIGURE 12.18
Regression line
and data points
for Datsun Z data.

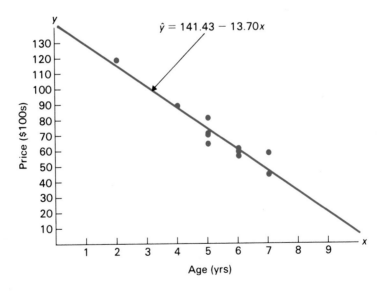

$\hat{y} = 141.43 - 13.70x$

Relationship between the linear correlation coefficient and the coefficient of determination

In Section 12.3 we discussed the *coefficient of determination, r^2*. This descriptive measure has two interpretations: (1) it is the percentage reduction obtained in the total squared error by using the regression equation to predict the observed y-values instead of simply using $\bar{y}$, and (2) it is the percentage of variation in the observed y-values that is explained by the regression. In any case, the coefficient of determination gives us a measure of the utility of the regression equation for making predictions.

Now we have introduced the *linear correlation coefficient, r,* as a descriptive measure of the strength of the linear relationship between the two variables under consideration. We would expect the strength of the linear relationship to also give an indication of the usefulness of the regression equation for making predictions. In other words, we would expect a connection between the linear correlation coefficient and the coefficient of determination. As a matter of fact, the connection is precisely the one suggested by the notation (r = linear correlation coefficient, r^2 = coefficient of determination).

KEY FACT

The coefficient of determination is the square of the linear correlation coefficient.

In Example 12.7 on page 526, we found that the linear correlation coefficient of the age-versus-price data for Datsun Zs is $r = -0.957$. From this and the Key Fact, we can deduce immediately that the coefficient of determination is

$$r^2 = (-0.957)^2 = 0.916$$

This, of course, is the same value we obtained for r^2 in Example 12.5 by using the defining formula for the coefficient of determination, $r^2 = 1 - SSE/SST$.

Generally speaking then, we can compute the coefficient of determination for a data set by using the defining formula, $r^2 = 1 - SSE/SST$, or by first calculating the linear correlation coefficient and then squaring the result. It is almost always easier to employ the latter approach.

A warning on the use of the linear correlation coefficient

We pointed out that the use of linear regression is based on the assumption that the data points are actually scattered about a straight line. The same is true about the use of the linear correlation coefficient. In other words, the linear correlation coefficient, r, is used to describe the strength of the *linear* relationship between two variables. It should be employed as a descriptive measure only when the scatter diagram indicates that the data points are scattered about a straight line. (See Exercise 12.69.)

Correlation is not causation

Two variables may have a high correlation without being causally related. For example, Table 12.14 on the next page displays data on total parimutuel turnover (money wagered) at U.S. race tracks and on college enrollment, for five randomly selected years between 1970 and 1982. (Sources: National Association of State Racing Commissioners and U.S. National Center for Education Statistics.)

TABLE 12.14

Year	Total parimutuel turnover ($millions) x	College enrollment (thousands) y
1970	5,977	8,581
1975	7,862	11,185
1978	10,029	11,260
1981	11,677	12,372
1982	11,888	12,426

The linear correlation coefficient of the data points in Table 12.14 is $r = 0.931$, suggesting that there is a strong positive linear correlation between parimutuel wagering and college enrollment. But this does not mean that there is a causal relationship between the two variables, such as that when people go to race-tracks, they are somehow inspired to go to college. On the contrary, we can only infer that the two variables have a strong tendency to increase (or decrease) simultaneously, and that total parimutuel turnover is a good predictor of college enrollment.

It may happen that two variables are strongly correlated because they are both associated with a third variable. For example, a study was once done to show that teachers' salaries are positively linearly correlated with the dollar amount of liquor sales. This does not imply that the consumption of large amounts of alcohol by society causes an increase in the amount paid to teachers. It *does* imply that there is a positive linear relationship between the variables and that one variable is a useful predictor of the other. A possible explanation for this might be that both of the variables, teachers' salaries and liquor sales, are tied to other variables (such as the rate of inflation) that pull them along together.

Exercises 12.4

In Exercises 12.60 through 12.67, we have repeated the data from Exercises 12.30 through 12.37 of Section 12.2. For each exercise,

a) compute the linear correlation coefficient, r.
b) interpret the value of r in terms of the linear relationship between the two variables under consideration.
c) discuss the graphical interpretation of the value of r and check that it agrees with the graph you obtained in the corresponding exercise in Section 12.2.
d) square the value of r you determined in part (a) and compare your answer to the value of the coefficient of determination you got in the corresponding exercise in Section 12.3.

12.60 The data from Exercise 12.30 of Section 12.2:

x	1	1	5	5
y	1	3	2	4

12.61 The data from Exercise 12.31 of Section 12.2:

x	0	2	2	5	6
y	4	2	0	-2	1

12.62 The age versus price data for Corvettes from Exercise 12.32 of Section 12.2:

Age (yrs) x	Price ($100s) y
7	90
6	90
6	120
4	118
3	175
3	179
7	100
3	128
7	85
6	92
1	173
1	190

12.63   The height versus weight data for 18–24-year-old males from Exercise 12.33 of Section 12.2:

Height (inches) x	Weight (lb) y
65	175
67	133
71	185
71	163
66	126
75	198
67	153
70	163
71	159
69	151
69	155

12.64   The size versus price data for custom homes from Exercise 12.34 of Section 12.2:

Size x	Price y
26	235
27	249
33	267
29	269
29	295
34	345
30	415
40	475
22	195

12.65   The age versus peak-heart-rate data from Exercise 12.35 of Section 12.2:

Age x	Peak heart rate y
10	210
20	200
20	195
25	195
30	190
30	180
30	185
40	180
45	170
50	165

12.66   The study-time versus grade data from Exercise 12.36 of Section 12.2:

Study time (hrs) x	Grade (percent) y
10	92
15	81
12	84
20	74
8	85
16	80
14	84
22	80

12.67   The disposable-income versus annual-food-expenditure data from Exercise 12.37 of Section 12.2:

Family disposable income (in thousands of dollars) x	Food expenditures (in hundreds of dollars) y
30	55
36	60
27	42
20	40
16	37
24	26
19	39
25	43

12.68   A random sample of 10 students in an introductory statistics class gave the following data on heights and final exam scores:

Height (inches) x	Test score y
71	87
68	96
71	66
65	71
66	71
68	55
68	83
64	67
62	86
65	60

a) What sort of value of r would you expect to find for these data? Explain your answer.

b) Compute r.

12.69   Consider the set of data points given below.

x	y
−3	9
−2	4
−1	1
0	0
1	1
2	4
3	9

a) Compute the linear correlation coefficient, r, for this data.

b) Can you conclude from your result in part (a) that the variables x and y are unrelated? Explain.

c) Draw a scatter diagram for the data.

d) Is it appropriate to use the linear correlation coefficient as a descriptive measure for the data? Why?

e) Show that the data are related by the equation $y = x^2$ and graph that equation along with the data points.

12.70   Determine whether r is positive, negative, or zero for each of the data sets pictured in the scatter diagrams below.

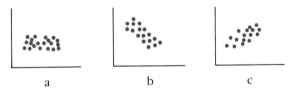

a   b   c

In Exercises 12.41 through 12.45 of Section 12.2, we examined the concept of the *sample covariance* of a set of data points. Recall that the sample covariance, s_{xy}, of n data points is defined by

$$(3) \qquad s_{xy} = \frac{\Sigma(x - \bar{x})(y - \bar{y})}{n - 1}$$

The defining formula for the **linear correlation coefficient** is

$$(4) \qquad r = \frac{s_{xy}}{s_x s_y}$$

where s_x and s_y are the sample standard deviations of the x-values and y-values, respectively.

12.71   Use Formula (4) to compute r for the set of data points in Exercise 12.61, and compare your answer with the value obtained for r in Exercise 12.61.

12.72   Use Formula (4) to compute r for the set of data points in Exercise 12.60, and compare your answer with the value obtained for r in Exercise 12.60.

12.73   In this exercise we will present problems that help us interpret the covariance of a set of data points.

a) Consider the following data set:

x	1	2	3	4	5	6
y	1	3	4	6	6	7

For these data, $\bar{x} = 3.5$ and $\bar{y} = 4.5$. Below we have drawn a coordinate system with a second set of axes passing through the point $(3.5, 4.5)$. Construct a scatter diagram for the above data points on the coordinate system provided.

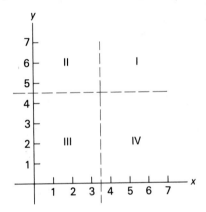

b) The dashed lines above divide the coordinate system into four regions, which we have labelled I, II, III, and IV. For a point (x, y) in region I, $x - \bar{x}$ and $y - \bar{y}$ are both positive, so $(x - \bar{x})(y - \bar{y})$ is positive. Fill in the remainder of the table below using similar reasoning.

Region	Sign of $(x - \bar{x})(y - \bar{y})$
I	+
II	
III	
IV	

c) Without performing any calculations, decide whether the covariance of the data in part (a) will be positive or negative. *Hint:* Use the graph from part (a) and the table from part (b).

d) Use a similar graphing procedure to decide

whether the covariance will be positive or negative for the following data set:

x	1	2	3	4	5	6
y	7	6	6	4	3	1

e) Below are two data sets, one scattered about a straight line of positive slope and the other scattered about a straight line of negative slope.

Complete the statements below.

(i) For data scattered about a straight line of positive slope, the covariance will be _____.

(ii) For data scattered about a straight line of negative slope, the covariance will be _____.

f) Use Formula (4) and the reasoning above, to describe in your own words why r is positive for data scattered around a line of positive slope and negative for data scattered around a line of negative slope.

12.74 In Exercise 12.58, we proved that the coefficient of determination always lies between 0 and 1, inclusive. Use that result and the Key Fact on page 528 to deduce that the linear correlation coefficient always lies between -1 and 1, inclusive.

12.75 Show that, for the linear correlation coefficient, the defining formula, Formula (4), and the shortcut formula, Formula 12.3, are equivalent by establishing the algebraic identity

$$\frac{S_{xy}}{S_x S_y} = \frac{n(\Sigma xy) - (\Sigma x)(\Sigma y)}{\sqrt{n(\Sigma x^2) - (\Sigma x)^2}\sqrt{n(\Sigma y^2) - (\Sigma y)^2}}$$

12.5   Computer packages*

The computations involved in regression and correlation analyses are quite extensive, even for relatively small data sets. Because of this, computer packages play a fundamental role in such analyses.

In this section we will see how Minitab can be used to analyze the age versus price data for Datsun Zs. The data are given in Table 12.2, which is repeated here as Table 12.19. Ages are in years. Prices are in hundreds of dollars, rounded to the nearest hundred.

TABLE 12.19

Age (yrs) x	Price ($100s) y
5	80
7	57
6	58
6	55
5	70
4	88
7	43
6	60
5	69
5	63
2	118

As we noted earlier in the chapter, it is generally a good idea to look at a scatter diagram of the data before proceeding with a regression or correlation analysis. The purpose of this is to check whether the data points appear to be scattered about a straight line, and thereby decide whether it is appropriate to determine a regression equation or compute a correlation coefficient.

To obtain the scatter diagram of the data points in Table 12.19, we first enter the data into the computer. Then we use the command **PLOT** followed by the storage location of the *y*-values ('PRICE') and the storage location of the *x*-values ('AGE'). This command and its results are shown below in Printout 12.1.

PRINTOUT 12.1   Minitab scatter diagram for age versus price data from Table 12.19.

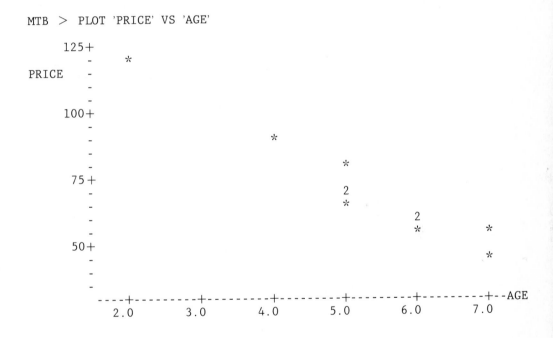

Printout 12.1 is the Minitab analogue of the scatter diagram we drew in Figure 12.7 on page 505. [*Note:* The "2" plotted above the "6.0" indicates there are *two* data points in that vicinity that are either the same or very close together. In this case they are the points (6, 58) and (6, 60).] From the scatter diagram, it does appear that the data points are scattered about a straight line.

Let us now proceed to have Minitab perform a regression analysis on the age versus price data. We type the command **REGRESS** 'PRICE' ON 1 PREDICTOR 'AGE'. This tells Minitab to perform a regression analysis on the data in Table 12.19 with age, *x*, as the predictor variable. Printout 12.2 displays this command and the resulting output.

PRINTOUT 12.2
Minitab output
for REGRESS.

```
MTB > REGRESS 'PRICE' ON 1 PREDICTOR 'AGE'

The regression equation is
PRICE = 141 - 13.7 AGE

Predictor        Coef       Stdev     t-ratio
Constant       141.432      7.538      18.76
AGE            -13.703      1.385      -9.90

s = 6.220      R-sq = 91.6%    R-sq(adj) = 90.6%

Analysis of Variance

SOURCE        DF         SS          MS
Regression     1       3789.4      3789.4
Error          9        348.2        38.7
Total         10       4137.6
```

The first item in Printout 12.2 gives the regression equation—PRICE $= 141 -$
13.7 AGE, which we have previously written as $\hat{y} = 141.43 - 13.70x$. Following
this is a table that gives information about the y-intercept, b_0, and the slope, b_1,
of the regression equation. The row labelled "Constant" provides information
on b_0 and the row labelled "AGE" provides information on b_1. In particular, the
entries 141.432 and -13.703 under the column headed "Coef" are simply the
values of b_0 and b_1, respectively.

Below the table for b_0 and b_1 you will find a line containing three items. The
second item gives the coefficient of determination—R-sq $= 91.6\%$, which we
have expressed as $r^2 = 0.916$ (see page 517).

Next, let us look at the "Analysis of Variance" table. In this table we can find
the three *sums of squares* discussed in Section 12.3. The regression sum of
squares, *SSR,* is the first entry in the "SS" column; thus, $SSR = 3789.4$. The error
sum of squares, *SSE,* is the second entry in the "SS" column; so, $SSE = 348.2$. Finally, the total sum of squares, *SST,* is the third entry in the "SS" column; hence,
$SST = 4137.6$.

As you can see, there are many additional items in the regression output of
Printout 12.2. These items can be used, for example, to perform inferences in regression. We will examine such inferences in Chapter 13.

♦ **Chapter Review**

KEY TERMS

coefficient of determination (r^2), 518
error sum of squares *(SSE),* 518
extrapolation, 511
least-squares criterion, 507
linear correlation coefficient *(r),* 524
linear equation, 497
PLOT*, 533
REGRESS*, 533

regression equation, 507
regression line, 507
regression sum of squares *(SSR),* 521
scatter diagram, 505
slope, 500
straight line, 497
total sum of squares *(SST),* 518
y-intercept, 500

FORMULAS

In the formulas below,

b_0 = y-intercept of regression line
b_1 = slope of regression line
n = sample size (number of data points)
SST = total sum of squares
SSE = error sum of squares
SSR = regression sum of squares
r^2 = coefficient of determination
r = linear correlation coefficient

Regression equation, 508

$$\hat{y} = b_0 + b_1 x$$

where

$$b_1 = \frac{n(\Sigma xy) - (\Sigma x)(\Sigma y)}{n(\Sigma x^2) - (\Sigma x)^2}, \quad b_0 = \frac{1}{n}(\Sigma y - b_1 \Sigma x)$$

Total sum of squares, 518

$$SST = \Sigma(y - \bar{y})^2$$

Error sum of squares, 518

$$SSE = \Sigma(y - \hat{y})^2$$

Regression sum of squares, 521

$$SSR = \Sigma(\hat{y} - \bar{y})^2$$

Regression identity, 521

$$SST = SSR + SSE$$

Coefficient of determination, 518

$$r^2 = 1 - \frac{SSE}{SST} = \frac{SSR}{SST}$$

Linear correlation coefficient, 524

$$r = \frac{n(\Sigma xy) - (\Sigma x)(\Sigma y)}{\sqrt{n(\Sigma x^2) - (\Sigma x)^2}\,\sqrt{n(\Sigma y^2) - (\Sigma y)^2}}$$

YOU SHOULD BE
ABLE TO . . .

1 use and understand the preceding formulas.

2 apply the concepts related to linear equations with one independent variable.

3 explain the idea behind the least-squares criterion.

4 determine and graph the regression equation for a set of data points, interpret the slope of the regres-sion equation, and use the regression equation for making predictions.

5 compute and interpret the coefficient of determi-nation and the three sums of squares (SST, SSE, and SSR).

6 calculate and interpret the linear correlation coeffi-cient, r.

REVIEW TEST

1 A small company has purchased a microcomputer system for $7200 and plans to depreciate the value of the equipment by $1200 per year for six years. Let x denote the age of the equipment, in years, and y denote the value of the equipment, in hundreds of dollars.

   a) Determine the equation that expresses y in terms of x.
   b) Find b_0 and b_1.
   c) Without graphing the equation, decide whether the line slopes upward, slopes downward, or is horizontal. Explain your answer.
   d) Construct a table similar to Table 12.1 on page 499 for the x-values 2 and 5 years.
   e) Obtain the graph for the equation in part (a) by plotting the points from part (d) and connecting them with a straight line.
   f) Use the graph from part (e) to visually estimate the value of the equipment after four years. Then calculate the value exactly by employing the equation from part (a).

2 The director of a large mathematics course hires upper-division science students to grade papers. On each grading day, he records the number of papers graded and the total amount of money paid to the graders. The table below gives the data for 12 randomly selected grading days from last semester.

Number of papers graded (hundreds) x	Cost of grading (dollars) y
16	234
16	220
18	258
22	298
19	273
16	227
18	246
15	210
19	265
17	250
15	223
18	251

   a) Draw a scatter diagram for the data.
   b) Does it appear reasonable to find a regression equation for the data? Explain.
   c) Determine the regression equation for the data, and draw its graph on the scatter diagram from part (a).
   d) Describe the apparent relationship between number of papers graded and cost of grading.
   e) What does the slope of the regression line represent in terms of paper-grading costs?
   f) Use the regression equation to predict the cost of grading 1600 papers.

3 Refer to Problem 2.
   a) Compute SST, SSR, and SSE.
   b) Verify that the regression identity holds.
   c) Calculate the coefficient of determination using the defining formula.
   d) Determine the percentage reduction obtained in the total squared error by using the regression equation to predict the observed y-values, instead of simply using the mean $\bar{y}$.
   e) Determine the percentage of variation in the observed y-values that is explained by the regression.
   f) State how useful the regression equation appears to be for making predictions.

4 Refer to Problem 2.
   a) Compute the linear correlation coefficient using the formula

$$r = \frac{n(\Sigma xy) - (\Sigma x)(\Sigma y)}{\sqrt{n(\Sigma x^2) - (\Sigma x)^2}\sqrt{n(\Sigma y^2) - (\Sigma y)^2}}$$

   b) Interpret the value of r in terms of the linear relationship between the number of papers graded and the cost of grading.
   c) Discuss the graphical implications of the value of r.
   d) Use the value of the linear correlation coefficient computed in part (a) to obtain the coefficient of determination.

In the previous chapter we examined *descriptive methods* in regression and correlation. We discovered how to determine the regression equation for a set of data points and how to use the regression equation to make predictions. We also learned how to compute and interpret the linear correlation coefficient for a set of data points.

Now we will consider some *inferential methods* in regression and correlation. For example, we will see how the regression equation can be used to determine a confidence interval for the mean price of all Datsun Zs of a given age; and how the linear correlation coefficient, *r,* can be used to decide whether there is a negative correlation between the variables age and price of all Datsun Zs.

Inferential methods in regression and correlation

CHAPTER OUTLINE

13.1

The regression model

Up to this point, our study of linear regression and correlation has been purely *descriptive*. For example, we learned how to find the regression line for a sample of data points from two variables without making inferences about the actual line that relates those variables. Additionally, we learned how to use the regression equation to make predictions, without providing explicit statements concerning the accuracy of such predictions.

It is now time to examine *inferential* methods in regression and correlation. For these methods to be applicable, it is necessary that the variables under consideration satisfy certain conditions. In this section, we will discuss the conditions for inferences in regression and correlation. In the next two sections, we will study the inferential procedures themselves.

The regression model

To begin, let us return to the Datsun Z illustration. In Table 13.1 we have reproduced the age versus price data for Datsun Zs. Ages are in years. Prices are in hundreds of dollars, rounded to the nearest hundred.

TABLE 13.1

Age (yrs) x	Price ($100s) y
5	80
7	57
6	58
6	55
5	70
4	88
7	43
6	60
5	69
5	63
2	118

The regression equation for this data is $\hat{y} = 141.43 - 13.70x$ (see page 509). We can use this equation to predict the price of a Datsun Z. But as you can see from the data, we cannot expect such predictions to be completely accurate, since prices vary even for Zs of the same age. For instance, in the sample data there are four five-year-old Zs. Their prices are $8000, $7000, $6900, and $6300. [The *predicted* price for a five-year-old Z is $\hat{y} = 141.43 - 13.70 \cdot 5 = 72.93$, or $7293.]

This variation in price for Zs of the same age should be expected. Cars of the same age have different mileages, interior conditions, paint quality, and so forth. These and a large number of other random factors make prices vary for Zs of the same age. Therefore, we see that for each age, there is an entire population of prices—the prices of all Zs of that age. There is a population of prices for two-year-old Zs, another population of prices for three-year-old Zs, another population of prices for four-year-old Zs, and so on. In other words, to each x-value (age), there corresponds a population of y-values (prices).

With the preceding discussion in mind, we now state the conditions required for using inferential methods in regression analysis. Following that, we will illustrate the meaning of those conditions with reference to the Datsun Z example.

Assumptions for regression inferences

1 *Population regression line:* For each x-value, the *mean* of the corresponding population of y-values lies on a straight line, say, $y = \beta_0 + \beta_1 x$. We refer to this straight line as the **population regression line.**
2 *Equal standard deviations:* The population standard deviation, σ, of the population of y-values corresponding to a given x-value is the *same,* regardless of the x-value.
3 *Normality:* For each x-value, the corresponding population of y-values is *normally distributed.*

In other words, Assumptions (1)–(3) require that there exist constants β_0, β_1, and σ so that for each x-value, the corresponding population of y-values is normally distributed with mean $\beta_0 + \beta_1 x$ and standard deviation σ. These assumptions are often referred to as the **regression model.**

EXAMPLE 13.1 **Illustrates the assumptions for regression inferences**

a) Discuss what it would mean for the assumptions for regression inferences to be satisfied for the variables age, x, and price, y, of Datsun Zs.
b) Display the assumptions graphically.

S O L U T I O N

a) For the assumptions for regression inferences to be satisfied, it would mean there are constants β_0, β_1, and σ such that for each age x, the prices for Zs of that age are normally distributed with mean $\beta_0 + \beta_1 x$ and standard deviation σ. Thus, the prices for two-year-old Zs ($x = 2$) would be normally distributed with mean $\beta_0 + \beta_1 \cdot 2$ and standard deviation σ; the prices for three-year-old Zs ($x = 3$) would be normally distributed with mean $\beta_0 + \beta_1 \cdot 3$ and standard deviation σ;

and so on. [*Note:* The parameters β_0, β_1, and σ are generally unknown and hence must be estimated from sample data.]

b) To display the assumptions graphically, let us first consider Assumption (1). That assumption would require that the mean prices for the various ages of Zs all lie on a straight line $y = \beta_0 + \beta_1 x$. See Figure 13.1.

FIGURE 13.1
Population
regression line.

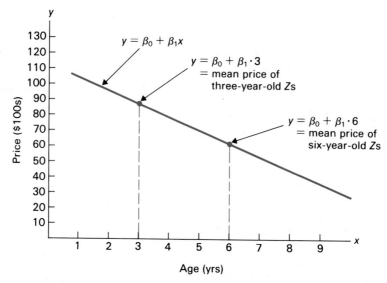

In general, the population regression line is *not* known; in fact, one of the main reasons for determining a sample regression line is to estimate the population regression line. Of course, a sample regression line, in this case $\hat{y} = 141.43 - 13.70x$, ordinarily will not be the same as the population regression line, just as a sample mean, $\bar{x}$, generally will not equal the population mean, μ. We picture the situation in Figure 13.2.

FIGURE 13.2
Population regression
line and sample
regression line.

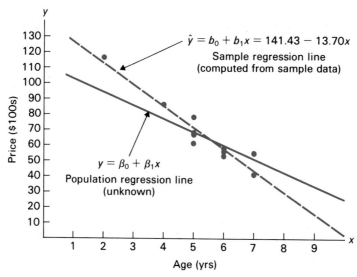

The solid line in Figure 13.2 is the population regression line. The dashed line is a sample regression line, which is the best approximation we can make to the population regression line by using the sample data obtained.[†]

Next we will present a graph that depicts Assumptions (2) and (3). Those assumptions would necessitate that the price distributions for the various ages of Zs are all normally distributed with the same standard deviation, σ. Figure 13.3 shows this for the price distributions of two-year-old, five-year-old, and seven-year-old Zs.

FIGURE 13.3
Price distributions for two-year-old, five-year-old, and seven-year-old Zs under Assumptions (2) and (3).

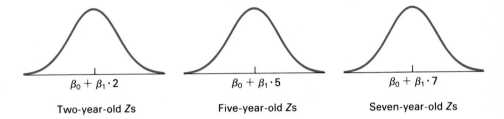

Note that the shapes of the three normal curves in Figure 13.3 are identical. This reflects the fact that the shape of a normal distribution is determined by its standard deviation, and under Assumption (2), the standard deviations of the price distributions are the same.

We can portray all three assumptions for regression inferences, as they pertain to the Datsun Z illustration, by combining Figures 13.1 and 13.3 into a three-dimensional graph. This is shown in Figure 13.4.

FIGURE 13.4
Graphical portrayal of assumptions for regression inferences for Datsun Z illustration.

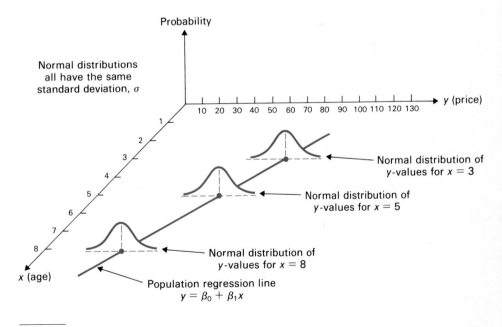

[†] A different sample would yield a different sample regression line.

We should emphasize that Figure 13.4 depicts the various price distributions for Datsun Zs under the presumption that Assumptions (1), (2), and (3) are true. Whether this is actually the case remains to be seen.   ∎

The standard error of the estimate

Let us suppose that we are considering two variables, x and y, for which the assumptions for regression inferences are met. That is, the variables x and y satisfy Assumptions (1)–(3) on page 539. Then, in particular, the population's of y-values corresponding to the various x-values all have the same standard deviation, σ. As we mentioned earlier, the common standard deviation, σ, is usually unknown and must be estimated from the sample data. The statistic used to estimate σ is called the **standard error of the estimate** or the **residual standard deviation** and is defined as follows:

DEFINITION 13.1 Standard error of the estimate

The **standard error of the estimate** is denoted by s_e and is defined by

$$s_e = \sqrt{\frac{SSE}{n-2}}$$

where $SSE = \Sigma (y - \hat{y})^2$.

Recall that SSE is termed the error sum of squares and that it represents the total squared error made in using the regression equation to predict the observed y-values. Thus, roughly speaking, the standard error of the estimate, s_e, is the square root of the average squared difference between the observed and predicted y-values.

To actually compute s_e for a set of sample data, it is generally easier to use the shortcut formula given below, unless SSE has already been computed for some other reason (e.g., to obtain the coefficient of determination).

FORMULA 13.1 Shortcut formula for the standard error of the estimate

The shortcut formula for the standard error of the estimate is

$$s_e = \sqrt{\frac{\Sigma y^2 - b_0(\Sigma y) - b_1(\Sigma xy)}{n-2}}$$

EXAMPLE 13.2   Illustrates the standard error of the estimate

Consider again the age-versus-price data for Datsun Zs given in Table 13.1 on page 538.
   a) Compute the standard error of the estimate, s_e, of the data using the defining formula in Definition 13.1.
   b) Interpret the result from part (a).

SOLUTION

a) We have already computed SSE on page 517 in Example 12.5. We found that SSE = 348.22. Consequently, the standard error of the estimate for the sample data is

$$s_e = \sqrt{\frac{SSE}{n-2}} = \sqrt{\frac{348.22}{11-2}} = 6.22$$

This computation is very easy because the work in calculating SSE had been done previously.

b) Presuming that the variables age, x, and price, y, for Datsun Zs satisfy Assumptions (1)–(3), $s_e = 6.22$ (or $622) is our estimate for the common population standard deviation, σ, of prices for Datsun Zs of any given age. ∎

Exercises 13.1

13.1 State the three assumptions for regression inferences.

In Exercises 13.2 through 13.7 we have repeated the information from Exercises 12.32 through 12.37 of Section 12.2. For each exercise, discuss what it would mean for the assumptions for regression inferences to be satisfied by the variables under consideration.

13.2 Twelve Corvettes, between one and seven years of age, were randomly selected from the classified ads. The following data were obtained on age versus price:

Age (yrs) x	Price ($100s) y
7	90
6	90
6	120
4	118
3	175
3	179
7	100
3	128
7	85
6	92
1	173
1	190

13.3 The U.S. National Center for Health Statistics collects and publishes data on heights and weights by age and sex. A random sample of 11 males, aged 18–24 years, gave the following data:

Height (inches) x	Weight (lb) y
65	175
67	133
71	185
71	163
66	126
75	198
67	153
70	163
71	159
69	151
69	155

13.4 Hanna Properties, Inc. specializes in custom-home resales in the Equestrian Estates. A random sample of nine currently listed custom homes provided the following information on size and price. The size data are in hundreds of square feet, rounded to the nearest hundred; the price data are in thousands of dollars, rounded to the nearest thousand.

Size x	Price y
26	235
27	249
33	267
29	269
29	295
34	345
30	415
40	475
22	195

13.5 A report read by a physician indicated that the maximum heart rate an individual can reach during intensive exercise decreases with age. The physician decided to do his own study. Ten randomly selected members of a jogging club performed exercise tests and recorded their peak heart rates. The results are shown in the following table:

Age x	Peak heart rate y
10	210
20	200
20	195
25	195
30	190
30	180
30	185
40	180
45	170
50	165

13.6 A calculus instructor asked a random sample of eight students to record their study times per lesson in a beginning calculus course. She then made a table for total study times over two weeks and test scores at the end of the two weeks. Here are the results.

Study time (hrs) x	Grade (percent) y
10	92
15	81
12	84
20	74
8	85
16	80
14	84
22	80

13.7 An economist is interested in the relation between the disposable income of a family and the amount of money spent annually on food. For a preliminary study, he randomly selects eight middle-income families of the same size (father, mother, two children). The results are as follows:

Family disposable income (in thousands of dollars) x	Food expenditures (in hundreds of dollars) y
30	55
36	60
27	42
20	40
16	37
24	26
19	39
25	43

For Exercises 13.8 through 13.13,
a) compute the standard error of the estimate, s_e, of the given data using the defining formula in Definition 13.1 on page 542. [*Note:* The error sum of squares, *SSE*, was obtained for each of the data sets in Exercises 12.50 through 12.55 of Section 12.3.]
b) interpret the result from part (a).

13.8 The age versus price data for Corvettes given in Exercise 13.2.

13.9 The height versus weight data for 18–24-year-old males given in Exercise 13.3.

13.10 The size versus price data for custom homes given in Exercise 13.4.

13.11 The age versus peak-heart-rate data given in Exercise 13.5.

13.12 The study-time versus grade data given in Exercise 13.6.

13.13 The disposable-income versus annual-food-expenditure data given in Exercise 13.7.

In Exercises 13.14 and 13.15, compute the standard error of the estimate, s_e,
a) using the defining formula in Definition 13.1 on page 542.
b) using the shortcut formula in Formula 13.1 on page 542.
Perform these computations from scratch; that is, do not employ results obtained from previous exercises.

13.14

x	1	1	5	5
y	1	3	2	4

13.15

x	0	2	2	5	6
y	4	2	0	−2	1

13.16 In this exercise we will derive the shortcut formula, Formula 13.1 on page 542, for the standard error of the estimate.
a) Show that

$$SSE = \Sigma y^2 - b_0 \Sigma(y - \hat{y}) - b_1 \Sigma x(y - \hat{y}) - \Sigma y \hat{y}$$

b) In Exercise 12.46 we found that $\Sigma(y - \hat{y}) = 0$ and $\Sigma x(y - \hat{y}) = 0$. Use those results and part (a) to show that

$$SSE = \Sigma y^2 - b_0(\Sigma y) - b_1(\Sigma xy)$$

c) Deduce Formula 13.1 from part (b).

13.2 Inferences in regression

In this section we will examine several statistical inference procedures for regression. *These inferential techniques presume that Assumptions (1)–(3), given on page 539, are satisfied by the variables under consideration.*

Inferences concerning β_1

The first inferential methods we will study are those concerning the slope, β_1, of the population regression line. Assumptions (1)–(3) imply that for each x-value, the corresponding population of y-values is normally distributed with mean $\beta_0 + \beta_1 x$ and standard deviation σ. If $\beta_1 = 0$, then there is no linear relationship between the variables. Consequently, x will be totally useless as a predictor of y, since, then, the value of x has absolutely nothing to do with the distributions of the y-values. Thus, we can decide whether the variables x and y are linearly related and, consequently, whether x is useful as a predictor of y by performing the hypothesis test

$$H_0: \beta_1 = 0$$
$$H_a: \beta_1 \neq 0$$

Rejection of the null hypothesis indicates that the variables x and y are linearly related, so that the regression is useful for making predictions.

To perform the hypothesis test for β_1, we will employ the statistic b_1, the slope of the sample regression line. From Assumptions (1)–(3), we can deduce the probability distribution of the random variable b_1.

KEY FACT The sampling distribution of the slope of the regression line

Suppose the variables under consideration satisfy Assumptions (1)–(3) and have population regression line $y = \beta_0 + \beta_1 x$. Then the random variable b_1, the slope of the sample regression line, is *normally distributed* and has mean $\mu_{b_1} = \beta_1$ and standard deviation $\sigma_{b_1} = \sigma/\sqrt{\Sigma x^2 - (\Sigma x)^2/n}$.

From the Key Fact, we see that the standardized random variable

$$z = \frac{b_1 - \beta_1}{\sigma/\sqrt{\Sigma x^2 - (\Sigma x)^2/n}}$$

has the standard normal distribution. Since, in general, the common population standard deviation, σ, is unknown, it is necessary to use the standard error of the estimate, s_e, in its place. Replacing σ by s_e in the previous equation, we obtain the random variable

$$\frac{b_1 - \beta_1}{s_e/\sqrt{\Sigma x^2 - (\Sigma x)^2/n}}$$

This random variable does not have the standard normal distribution but, as you might suspect, has a t-distribution.

KEY FACT

Suppose the variables under consideration satisfy Assumptions (1)–(3) and have population regression line $y = \beta_0 + \beta_1 x$. Then the random variable

$$t = \frac{b_1 - \beta_1}{s_e/\sqrt{\Sigma x^2 - (\Sigma x)^2/n}}$$

has the *t-distribution with df* $= n - 2$.

For a null hypothesis of $H_0: \beta_1 = 0$, we can therefore use the random variable

$$t = \frac{b_1}{s_e/\sqrt{\Sigma x^2 - (\Sigma x)^2/n}}$$

as our test statistic and determine the critical values from the t-table, Table III. Specifically, we have the following procedure:

PROCEDURE 13.1 **To perform a hypothesis test to determine whether the slope, β_1, of the population regression line is not zero and, hence, whether the variable x is useful as a predictor of the variable y.**

Assumptions   The Assumptions (1)–(3) for regression inferences.

STEP 1 *State the null and alternative hypotheses.*
STEP 2 *Decide on the significance level, α.*
STEP 3 *The critical values are $\pm t_{\alpha/2}$, with df $= n - 2$. Use Table III to find the critical values.*

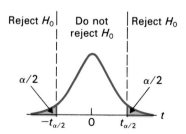

STEP 4 *Compute the value of the test statistic*

$$t = \frac{b_1}{s_e/\sqrt{\Sigma x^2 - (\Sigma x)^2/n}}$$

STEP 5 *If the value of the test statistic falls in the rejection region, reject H_0; otherwise, do not reject H_0.*

STEP 6 *State the conclusion in words.*

EXAMPLE 13.3 **Illustrates Procedure 13.1**

The sample data on age versus price of Datsun Zs are repeated in Table 13.2. Ages are in years. Prices are in hundreds of dollars, rounded to the nearest hundred.

TABLE 13.2

Age (yrs) x	Price ($100s) y
5	80
7	57
6	58
6	55
5	70
4	88
7	43
6	60
5	69
5	63
2	118

Do the data suggest, at the 5% significance level, that the slope, β_1, of the population regression line is not zero and, hence, that age, x, is useful as a predictor of price, y?

SOLUTION

It can be shown that it is reasonable to presume that Assumptions (1)–(3) are valid for the variables age, x, and price, y, of Datsun Zs. Consequently, we can apply Procedure 13.1 to perform the required hypothesis test.

STEP 1 *State the null and alternative hypotheses.*

The null and alternative hypotheses are

$$H_0: \beta_1 = 0$$
$$H_a: \beta_1 \neq 0$$

STEP 2 *Decide on the significance level, α.*

We are to perform the test at the 5% significance level; so, $\alpha = 0.05$.

STEP 3 *The critical values are $\pm t_{\alpha/2}$, with $df = n - 2$.*

From Step 2, $\alpha = 0.05$. Also, from Table 13.2 we see that $n = 11$ and thus, $df = n - 2 = 11 - 2 = 9$. Using Table III, we find that the critical values are

$$\pm t_{\alpha/2} = \pm t_{0.05/2} = \pm t_{0.025} = \pm 2.262$$

See Figure 13.5. at the top of page 548.

FIGURE 13.5

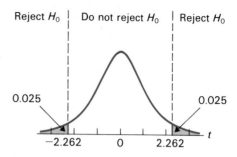

STEP 4 *Compute the value of the test statistic*

$$t = \frac{b_1}{s_e / \sqrt{\Sigma x^2 - (\Sigma x)^2 / n}}$$

In Example 12.4 on page 509, we found that $b_1 = -13.70$, $\Sigma x^2 = 326$, and $\Sigma x = 58$. Also, we determined that $s_e = 6.22$ in Example 13.2 on page 543. Consequently, since $n = 11$, the value of the test statistic is

$$t = \frac{b_1}{s_e / \sqrt{\Sigma x^2 - (\Sigma x)^2 / n}} = \frac{-13.70}{6.22 / \sqrt{326 - (58)^2 / 11}} = -9.895$$

STEP 5 *If the value of the test statistic falls in the rejection region, reject H_0; otherwise, do not reject H_0.*

The value of the test statistic, found in Step 4, is $t = -9.895$. Since this lies in the rejection region, we reject H_0.

STEP 6 *State the conclusion in words.*

Evidently, the slope of the population regression line is not zero and, consequently, the age, x, of a Datsun Z is useful as a predictor of the price, y. ∎

MTB
SPSS

We have seen that by performing the hypothesis test $H_0: \beta_1 = 0$ vs. $H_a: \beta_1 \neq 0$, we can decide whether the variables x and y are linearly related and, consequently, whether the regression is useful for making predictions. In Section 12.3 we introduced the coefficient of determination, r^2, as a descriptive measure of *how* useful the regression equation is for making predictions. This indicates that we should also be able to use r^2 to make inferences concerning the utility of the regression equation and, indeed, this is the case. However, since the resulting test is equivalent to the *t*-test of Procedure 13.1, we will not cover the hypothesis test based on the statistic r^2.

Besides performing hypothesis tests for β_1, we can also determine confidence intervals for that parameter. Recall that the slope of a straight line gives the amount of increase (or decrease) in y resulting from an increase in x by one unit. Also recall that the population regression line, whose slope is β_1, gives the means for the various populations of y-values. Thus, β_1 represents the increase (or decrease) in the mean of the population of y-values due to an increase in x

by one unit. For instance, in the Datsun Z illustration, β_1 is the amount, in hundreds of dollars, that the mean price decreases for every increase in age by one year.

Consequently, it is worthwhile to obtain an estimate of β_1. Using the Key Fact on page 546, we can derive the following confidence interval procedure:

> **PROCEDURE 13.2 To find a confidence interval for the slope, β_1, of the population regression line.**
>
> *Assumptions*   The Assumptions $(1) - (3)$ for regression inferences.
>
> **STEP 1** *For a confidence level of $1 - \alpha$, use Table III to find $t_{\alpha/2}$ for df $= n - 2$.*
> **STEP 2** *The endpoints of the confidence interval are*
>
> $$b_1 \pm t_{\alpha/2} \cdot s_e/\sqrt{\Sigma x^2 - (\Sigma x)^2/n}$$

EXAMPLE 13.4   **Illustrates Procedure 13.2**

Find a 95% confidence interval for the slope, β_1, of the population regression line relating age, x, and price, y, of Datsun Zs.

S O L U T I O N
We apply Procedure 13.2.

STEP 1 *For a confidence level of $1 - \alpha$, use Table III to find $t_{\alpha/2}$ for df $= n - 2$.*

For a 95% confidence interval, $\alpha = 0.05$. Since $n = 11$, df $= 11 - 2 = 9$. Using Table III we find that

$$t_{\alpha/2} = t_{0.025} = 2.262$$

STEP 2 *The endpoints of the confidence interval are*

$$b_1 \pm t_{\alpha/2} \cdot s_e/\sqrt{\Sigma x^2 - (\Sigma x)^2/n}$$

We have from Example 12.4 that $b_1 = -13.70$, $\Sigma x^2 = 326$, and $\Sigma x = 58$. Also, from Example 13.2, $s_e = 6.22$. Therefore, the endpoints of the confidence interval are

$$-13.70 \pm 2.262 \cdot 6.22/\sqrt{326 - (58)^2/11}$$

or

$$-13.70 \pm 3.13$$

Consequently, a 95% confidence interval for β_1 is from

$$-16.83 \quad \text{to} \quad -10.57$$

MTB
SPSS

We can be 95% confident that β_1 is somewhere between -16.83 and -10.57. In other words, we can be 95% confident that the yearly drop in mean price for Datsun Zs is somewhere between $1057 and $1683. ∎

Confidence intervals for means

Recall that the regression equation for the age-price data for Datsun Zs, given in Table 13.1, is $\hat{y} = 141.43 - 13.70x$. We learned that, based on the sample data, this line is our best estimate of the unknown population regression line $y = \beta_0 + \beta_1 x$.

Now, by Assumption (1), the population regression line gives us the means of the price distributions for the various ages of Zs. For instance, the mean price for three-year-old Zs is $\beta_0 + \beta_1 \cdot 3$. Since β_0 and β_1 are unknown, we must be content to estimate the mean, $\beta_0 + \beta_1 \cdot 3$, by the regression equation value

$$\hat{y} = 141.43 - 13.70 \cdot 3 = 100.33$$

or $10,033. This is our point estimate for the mean price of three-year-old Zs. As you know, it would be more informative if we had some idea of the accuracy of this estimate. That is, we would like to determine a confidence interval for the mean price of three-year-old Zs.

To obtain a confidence-interval procedure, we must first know the probability distribution of the random variable $\hat{y}$. This is given in the following Key Fact:

KEY FACT

Suppose the variables under consideration satisfy Assumptions (1)–(3) and have population regression line $y = \beta_0 + \beta_1 x$. Let x_p denote a particular value of the predictor variable x and $\hat{y}_p = b_0 + b_1 x_p$ be the predicted y-value for x_p. Then the random variable $\hat{y}_p$ is *normally distributed* and has mean $\mu_{\hat{y}_p} = \beta_0 + \beta_1 x_p$ and standard deviation

$$\sigma_{\hat{y}_p} = \sigma \sqrt{\frac{1}{n} + \frac{(x_p - \Sigma x/n)^2}{\Sigma x^2 - (\Sigma x)^2/n}}$$

The Key Fact implies that the standardized random variable

$$z = \frac{\hat{y}_p - (\beta_0 + \beta_1 x_p)}{\sigma \sqrt{\dfrac{1}{n} + \dfrac{(x_p - \Sigma x/n)^2}{\Sigma x^2 - (\Sigma x)^2/n}}}$$

has the standard normal distribution. Since σ is usually unknown, we need to replace it by s_e. The resulting random variable has a t-distribution:

KEY FACT

Suppose the variables under consideration satisfy Assumptions (1)–(3) and have population regression line $y = \beta_0 + \beta_1 x$. Let x_p denote a particular value of the predictor variable x and $\hat{y}_p = b_0 + b_1 x_p$ be the predicted y-value for x_p. Then the random variable

$$t = \frac{\hat{y}_p - (\beta_0 + \beta_1 x_p)}{s_e \sqrt{\dfrac{1}{n} + \dfrac{(x_p - \Sigma x/n)^2}{\Sigma x^2 - (\Sigma x)^2/n}}}$$

has the *t-distribution with df = n − 2.*

Recalling that $\beta_0 + \beta_1 x_p$ is the mean of the population of y-values corresponding to x_p, we can use the previous Key Fact to obtain the following confidence-interval procedure for means:

> **PROCEDURE 13.3** To find a confidence interval for the mean of the population of y-values corresponding to a particular x-value, x_p.
>
> **Assumptions** The Assumptions (1)–(3) for regression inferences.
>
> **STEP 1** *For a confidence level of $1 - \alpha$, use Table III to find $t_{\alpha/2}$ for df $= n - 2$.*
> **STEP 2** *Compute the point estimate*
>
> $$\hat{y}_p = b_0 + b_1 x_p$$
>
> *for the mean.*
> **STEP 3** *The endpoints of the confidence interval are*
>
> $$\hat{y}_p \pm t_{\alpha/2} \cdot s_e \sqrt{\frac{1}{n} + \frac{(x_p - \Sigma x/n)^2}{\Sigma x^2 - (\Sigma x)^2/n}}$$

EXAMPLE 13.5 **Illustrates Procedure 13.3**

The sample data on age versus price of Datsun Zs are repeated in Table .13.3. Ages are in years. Prices are in hundreds of dollars, rounded to the nearest hundred.

TABLE 13.3

Age (yrs) x	Price ($\$100$s) y
5	80
7	57
6	58
6	55
5	70
4	88
7	43
6	60
5	69
5	63
2	118

Use the data to find a 95% confidence interval for the mean price of all three-year-old Zs.

SOLUTION
We apply Procedure 13.3.

STEP 1 *For a confidence level of $1 - \alpha$, use Table III to find $t_{\alpha/2}$ for df $= n - 2$.*

We want a 95% confidence interval, which means $\alpha = 0.05$. Since $n = 11$, df $= 11 - 2 = 9$. Consulting Table III, we find that

$$t_{\alpha/2} = t_{0.025} = 2.262$$

STEP 2 *Compute the point estimate*

$$\hat{y}_p = b_0 + b_1 x_p$$

for the mean.

From Example 12.4, the sample regression equation is $\hat{y} = 141.43 - 13.70x$. Here we want $x = x_p = 3$ (three-year-old Zs). Consequently,

$$\hat{y}_p = 141.43 - 13.70 \cdot 3 = 100.33$$

STEP 3 *The endpoints of the confidence interval are*

$$\hat{y}_p \pm t_{\alpha/2} \cdot s_e \sqrt{\frac{1}{n} + \frac{(x_p - \Sigma x/n)^2}{\Sigma x^2 - (\Sigma x)^2/n}}$$

In Example 12.4 we found that $\Sigma x = 58$ and $\Sigma x^2 = 326$; and in Example 13.2 we determined that $s_e = 6.22$. Also, from Step 1, $t_{\alpha/2} = 2.262$, and from Step 2, $\hat{y}_p = 100.33$. Consequently, the endpoints of the confidence interval are

$$100.33 \pm 2.262 \cdot 6.22 \sqrt{\frac{1}{11} + \frac{(3 - 58/11)^2}{326 - (58)^2/11}}$$

or

$$100.33 \pm 8.29$$

Thus, a 95% confidence interval is from

$$92.04 \quad \text{to} \quad 108.62$$

We can be 95% confident that the mean price of all three-year-old Zs is somewhere between \$9,204 and \$10,862. ∎

MTB
SPSS

Prediction intervals

One of the main reasons for determining the regression equation for a sample of data points is to use it for making predictions. The regression equation for the Datsun Z data in Table 13.3 is $\hat{y} = 141.43 - 13.70x$. Thus, for example, our predicted price for a three-year-old Z is

$$\hat{y} = 141.43 - 13.70 \cdot 3 = 100.33$$

or \$10,033. However, since the prices of such cars vary, it makes more sense to find a **prediction interval** for the price of a three-year-old Z, than to give a single predicted value. Prediction intervals are similar to confidence intervals. The term "confidence" is usually reserved for interval estimates of a parameter, such as the *mean price* of all three-year-old Zs. The term "prediction" is used for interval estimates of random variables, such as the *price* of a randomly selected three-year-old Z.

The procedure for finding prediction intervals is quite similar to that for obtaining confidence intervals. The prediction-interval procedure is based on the following fact:

KEY FACT

Suppose the variables under consideration satisfy Assumptions (1)–(3). Let x_p denote a particular value of the predictor variable x, and y_p denote the value of a randomly selected member from the population of all y-values corresponding to x_p. Then the random variable $y_p - \hat{y}_p$ is *normally distributed* and has mean $\mu_{y_p - \hat{y}_p} = 0$ and standard deviation

$$\sigma_{y_p - \hat{y}_p} = \sigma\sqrt{1 + \frac{1}{n} + \frac{(x_p - \Sigma x/n)^2}{\Sigma x^2 - (\Sigma x)^2/n}}$$

If we standardize the random variable $y_p - \hat{y}_p$, then the resulting random variable has the standard normal distribution. But the standardized random variable contains the unknown parameter σ. Thus, we replace σ by s_e. On doing so, we obtain a random variable with a t-distribution. More precisely, we have the following Key Fact:

KEY FACT

Suppose the variables under consideration satisfy Assumptions (1)–(3). Let x_p denote a particular value of the predictor variable x, and y_p denote the value of a randomly selected member from the population of all y-values corresponding to x_p. Then the random variable

$$t = \frac{y_p - \hat{y}_p}{s_e\sqrt{1 + \frac{1}{n} + \frac{(x_p - \Sigma x/n)^2}{\Sigma x^2 - (\Sigma x)^2/n}}}$$

has the *t-distribution with df = n − 2.*

Using this Key Fact, we can develop a prediction-interval procedure. That procedure is given as Procedure 13.4.

PROCEDURE 13.4 To find a prediction interval for a population y-value corresponding to a particular x-value, x_p.

Assumptions The Assumptions (1)–(3) for regression inferences.

STEP 1 *For a prediction level of $1 - \alpha$, use Table III to find $t_{\alpha/2}$ for df = n − 2.*
STEP 2 *Compute the predicted y-value*

$$\hat{y}_p = b_0 + b_1 x_p$$

STEP 3 *The endpoints of the prediction interval are*

$$\hat{y}_p \pm t_{a/2} \cdot s_e \sqrt{1 + \frac{1}{n} + \frac{(x_p - \Sigma x/n)^2}{\Sigma x^2 - (\Sigma x)^2/n}}$$

EXAMPLE 13.6 **Illustrates Procedure 13.4**

Using the sample data in Table 13.3 on page 551, determine a 95% prediction interval for the price of a randomly selected three-year-old Z.

SOLUTION
We employ the step-by-step method given in Procedure 13.4.

STEP 1 *For a prediction level of* $1 - \alpha$, *use Table III to find* $t_{a/2}$ *for df* $= n - 2$.

We want a 95% prediction interval, and so $\alpha = 0.05$. Since $n = 11$, df $= 11 - 2 = 9$. Consulting Table III, we find that

$$t_{a/2} = t_{0.025} = 2.262$$

STEP 2 *Compute the predicted y-value*

$$\hat{y}_p = b_0 + b_1 x_p$$

The sample regression equation is $\hat{y} = 141.43 - 13.70x$. Thus, the predicted y-value for a three-year-old Z is

$$\hat{y}_p = 141.43 - 13.70 \cdot 3 = 100.33$$

STEP 3 *The endpoints of the prediction interval are*

$$\hat{y}_p \pm t_{a/2} \cdot s_e \sqrt{1 + \frac{1}{n} + \frac{(x_p - \Sigma x/n)^2}{\Sigma x^2 - (\Sigma x)^2/n}}$$

From Example 12.4, $\Sigma x = 58$ and $\Sigma x^2 = 326$; and from Example 13.2, $s_e = 6.22$. Also $n = 11$, $t_{a/2} = 2.262$, and $\hat{y}_p = 100.33$. Consequently, the endpoints of the prediction interval are

$$100.33 \pm 2.262 \cdot 6.22 \sqrt{1 + \frac{1}{11} + \frac{(3 - 58/11)^2}{326 - (58)^2/11}}$$

or

$$100.33 \pm 16.33$$

A 95% prediction interval is from

$$84.00 \quad \text{to} \quad 116.66$$

Thus, we can be 95% certain that the price of a randomly selected three-year-old Z will be somewhere between $8,400 and $11,666. ∎

MTB
SPSS

We have just seen that a 95% prediction interval for the price of a randomly selected three-year-old Z is from $8,400 to $11,666. In Example 13.5, we found that a 95% confidence interval for the mean price of all three-year-old Zs is from $9,204 to $10,862. We picture both of these intervals in Figure 13.6.

FIGURE 13.6
Prediction and
confidence intervals
for three-year-old Zs.

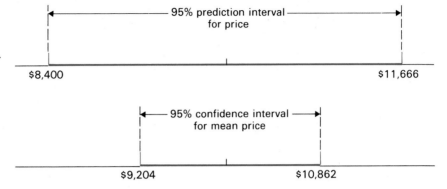

Note that the prediction interval is more extensive than the confidence interval. This is to be expected for the following reason: The error in the estimate of the mean price of three-year-old Zs is due only to the fact that the population regression line is being estimated by a sample regression line, whereas the error in the prediction of the price of a randomly selected three-year-old Z is due to that fact plus the variation in prices for three-year-old Zs.

Exercises 13.2

13.17 What conditions must the variables x and y satisfy in order that it be permissible to apply the inferential methods discussed in this section?

In Exercises 13.18 through 13.23, we have repeated the information from Exercises 12.32 through 12.37 of Section 12.2. For each exercise we also have given the regression equation for the data (found, respectively, in Exercises 12.32 through 12.37) and the standard error of the estimate (found, respectively, in Exercises 13.8 through 13.13). In each exercise,
a) apply Procedure 13.1 on page 546 to decide whether the slope, β_1, of the population regression line is not zero and, hence, whether the variable x is useful as a predictor of the variable y. Perform each hypothesis test at the designated significance level.
b) use Procedure 13.2 on page 549 to find a confidence interval for β_1 at the designated confidence level.
Note: You may presume that Assumptions (1)–(3) for regression inferences are satisfied in each case.
Note: Due to rounding, your answers may deviate somewhat from those given in the back of the book.

13.18 Twelve Corvettes, between one and seven years of age, were randomly selected from the classified ads. The following data were obtained on age versus price:

Age (yrs)	Price ($100s)
x	y
7	90
6	90
6	120
4	118
3	175
3	179
7	100
3	128
7	85
6	92
1	173
1	190

Regression equation is $\hat{y} = 200.33 - 16.00x$, and $s_e = 17.51$. Significance level $= 0.10$. Confidence level $= 0.90$.

13.19 The U.S. National Center for Health Statistics collects and publishes data on heights and weights by age and sex. A random sample of 11 males, aged 18–24 years, gave the following data:

Height (inches) x	Weight (lb) y
65	175
67	133
71	185
71	163
66	126
75	198
67	153
70	163
71	159
69	151
69	155

Regression equation is $\hat{y} = -174.49 + 4.84x$, and $s_e = 16.48$. Significance level $= 0.10$. Confidence level $= 0.90$.

13.20 Hanna Properties, Inc. specializes in custom-home resales in the Equestrian Estates. A random sample of nine currently listed custom homes provided the following information on size and price. The size data are in hundreds of square feet, rounded to the nearest hundred; the price data are in thousands of dollars, rounded to the nearest thousand.

Size x	Price y
26	235
27	249
33	267
29	269
29	295
34	345
30	415
40	475
22	195

Regression equation is $\hat{y} = -128.333 + 14.444x$, $s_e = 54.129$. Significance level $= 0.01$. Confidence level $= 0.99$.

13.21 A report read by a physician indicated that the maximum heart rate an individual can reach during intensive exercise decreases with age. The physi-

cian decided to do his own study. Ten randomly selected members of a jogging club performed exercise tests and recorded their peak heart rates. The results are shown in the following table:

Age x	Peak heart rate y
10	210
20	200
20	195
25	195
30	190
30	180
30	185
40	180
45	170
50	165

Regression equation is $\hat{y} = 219.78 - 1.09x$, and $s_e = 3.51$. Significance level $= 0.05$. Confidence level $= 0.95$.

13.22 A calculus instructor asked a random sample of eight students to record their study times per lesson in a beginning calculus course. She then made a table for total study times over two weeks and test scores at the end of the two weeks. Here are the results.

Study time (hrs) x	Grade (percent) y
10	92
15	81
12	84
20	74
8	85
16	80
14	84
22	80

Regression equation is $\hat{y} = 94.87 - 0.85x$, $s_e = 3.54$. Significance level $= 0.05$. Confidence level $= 0.95$.

13.23 An economist is interested in the relation between the disposable income of a family and the amount of money spent annually on food. For a preliminary study, he randomly selects eight middle-income families of the same size (father, mother, two children). The results are displayed below.

Family disposable income (in thousands of dollars) x	Food expenditures (in hundreds of dollars) y
30	55
36	60
27	42
20	40
16	37
24	26
19	39
25	43

Regression equation is $\hat{y} = 12.86 + 1.21x$, $s_e = 7.68$. Significance level $= 0.01$. Confidence level $= 0.99$.

13.24 Refer to Exercise 13.18.
a) Determine a 90% confidence interval for the mean price of all six-year-old Corvettes.
b) Determine a 90% prediction interval for the price of a randomly selected six-year-old Corvette.
c) Draw graphs similar to the ones in Figure 13.6 on page 555 showing both the 90% confidence interval and the 90% prediction interval from parts (a) and (b).
d) Why is the prediction interval more extensive than the confidence interval?

13.25 Refer to Exercise 13.19.
a) Determine a 90% confidence interval for the mean weight of 18–24-year-old males who are 70 inches tall.
b) Determine a 90% prediction interval for the weight of a randomly selected 18–24-year-old male who is 70 inches tall.

c) Draw graphs similar to the ones in Figure 13.6 on page 555 showing both the 90% confidence interval and the 90% prediction interval from parts (a) and (b).
d) Why is the prediction interval more extensive than the confidence interval?

13.26 Refer to Exercise 13.20.
a) Find a 99% confidence interval for the mean price of 2800-square-foot custom homes in the Equestrian Estates.
b) Find a 99% prediction interval for the price of a randomly selected 2800-square-foot custom home in the Equestrian Estates.

13.27 Refer to Exercise 13.21.
a) Find a 95% confidence interval for the mean peak heart rate of 22-year-olds.
b) Find a 95% prediction interval for the peak heart rate of a randomly selected 22-year-old.

13.28 Refer to Exercise 13.22.
a) Obtain a 95% confidence interval for the mean grade of students who study for 15 hours.
b) Obtain a 95% prediction interval for the grade of a randomly selected student who studies for 15 hours.

13.29 Refer to Exercise 13.23.
a) Obtain a 99% confidence interval for the mean annual food expenditure of families with a disposable income of $25,000.
b) Obtain a 99% prediction interval for the annual food expenditure of a randomly selected family with a disposable income of $25,000.

13.3 Inferences in correlation

Frequently we want to decide whether two variables, x and y, are *linearly* correlated, that is, whether there is a linear relationship between the two variables. As we learned in Section 13.2, this can be accomplished by performing a hypothesis test for the slope, β_1, of the population regression line.

Alternatively, we can perform a hypothesis test for the **population linear correlation coefficient, ρ** (rho). The population linear correlation coefficient, ρ, measures the linear correlation between the population of *all* data points in the same way that the (sample) linear correlation coefficient, r, measures the linear correlation between a sample of data points. Thus, it is ρ that actually describes the linear correlation between two variables.

If $\rho > 0$, the variables are **positively linearly correlated,** meaning that y tends to increase linearly as x increases. If $\rho < 0$, the variables are **negatively linearly correlated,** meaning that y tends to decrease linearly as x increases. If

$\rho = 0$, the variables are **linearly uncorrelated,** meaning that there is no linear relationship between the variables.

As you might suspect, we can use the (sample) linear correlation coefficient, r, as the test statistic for a hypothesis test concerning the population linear correlation coefficient, ρ. For a test with null hypothesis $H_0: \rho = 0$ (i.e., the variables are linearly uncorrelated), the critical values are given in Table VII in the appendix. Thus, we have the following procedure:

PROCEDURE 13.5 To perform a hypothesis test for a population linear correlation coefficient, with null hypothesis $H_0: \rho = 0$.

Assumptions   The Assumptions (1)–(3) for regression inferences.

STEP 1 *State the null and alternative hypotheses.*
STEP 2 *Decide on the significance level, α.*
STEP 3 *The critical value(s)*
  a) *for a two-tailed test are $\pm r_{\alpha/2}$,*
  b) *for a left-tailed test is $-r_\alpha$,*
  c) *for a right-tailed test is r_α,*
with $df = n - 2$. Use Table VII to find the critical value(s).

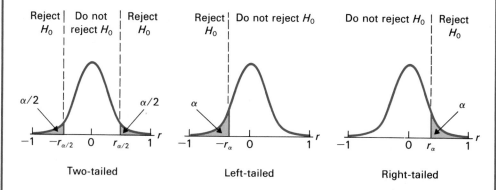

| Two-tailed | Left-tailed | Right-tailed |

STEP 4 *Compute the value of the test statistic*

$$r = \frac{n(\Sigma xy) - (\Sigma x)(\Sigma y)}{\sqrt{n(\Sigma x^2) - (\Sigma x)^2}\sqrt{n(\Sigma y^2) - (\Sigma y)^2}}$$

STEP 5 *If the value of the test statistic falls in the rejection region, reject H_0; otherwise, do not reject H_0.*
STEP 6 *State the conclusion in words.*

Note: The curves in Procedure 13.5 are drawn bell-shaped because when $\rho = 0$ the random variable r is approximately normally distributed.

EXAMPLE   13.7   Illustrates Procedure 13.5

Consider once more the age versus price data for Datsun Zs, which we repeat in Table 13.4. Ages are in years. Prices are in hundreds of dollars, rounded to the nearest hundred.

TABLE 13.4

Age (yrs) x	Price ($100s) y
5	80
7	57
6	58
6	55
5	70
4	88
7	43
6	60
5	69
5	63
2	118

Do the data provide sufficient evidence to conclude that the variables age, x, and price, y, for Datsun Zs are *negatively* linearly correlated? Use $\alpha = 0.05$.

S O L U T I O N

STEP 1 *State the null and alternative hypotheses.*

If we let ρ denote the population linear correlation coefficient for the variables age, x, and price, y, for Datsun Zs, then the null and alternative hypotheses are

$H_0: \rho = 0$ (age and price are linearly uncorrelated)
$H_a: \rho < 0$ (age and price are negatively linearly correlated)

Note that the test is left-tailed since there is a less-than sign ($<$) in the alternative hypothesis.

STEP 2 *Decide on the significance level, α.*

We are to use $\alpha = 0.05$.

STEP 3 *The critical value for a left-tailed test is $-r_\alpha$, with $df = n - 2$.*

We have $n = 11$, so that df $= 9$. Also $\alpha = 0.05$. Consulting Table VII, we find that for df $= 9$, $r_{0.05} = 0.521$. Thus, the critical value is

$$-r_{0.05} = -0.521$$

See Figure 13.7.

FIGURE 13.7

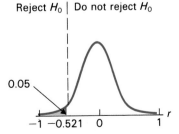

Reject H_0 | Do not reject H_0

0.05

−1 −0.521   0   1 r

STEP 4 *Compute the value of the test statistic*

$$r = \frac{n(\Sigma xy) - (\Sigma x)(\Sigma y)}{\sqrt{n(\Sigma x^2) - (\Sigma x)^2}\sqrt{n(\Sigma y^2) - (\Sigma y)^2}}$$

We have already computed r for the age versus price data in Table 13.4. This was done in Example 12.7 (page 526), where we found that $r = -0.957$.

STEP 5 *If the value of the test statistic falls in the rejection region, reject H_0; otherwise, do not reject H_0.*

The value of the test statistic, found in Step 4, is $r = -0.957$. A glance at Figure 13.7 shows that this falls in the rejection region. Hence, we reject H_0.

STEP 6 *State the conclusion in words.*

Evidently, the variables age and price for Datsun Zs are negatively linearly correlated. In other words, the price of a Datsun Z tends to decrease linearly as its age increases (at least for Zs between two and seven years old). ■

MTB
SPSS

Exercises 13.3

In Exercises 13.30 through 13.35, we have repeated the information from Exercises 12.32 through 12.37 of Section 12.2. For each exercise, we have also given the sample linear correlation coefficient, r, for the data (found, respectively, in Exercises 12.62 through 12.67). In each exercise, apply Procedure 13.5 to perform the indicated hypothesis test. [*Note:* You may presume that Assumptions (1)–(3) for regression inferences are satisfied in each case.]

13.30 Twelve Corvettes, between one and seven years of age, were randomly selected from the classified ads. The following data were obtained on age versus price:

Age (yrs) x	Price ($100s) y
7	90
6	90
6	120
4	118
3	175
3	179
7	100
3	128
7	85
6	92
1	173
1	190

Do the data provide sufficient evidence to conclude that the variables age and price for Corvettes are *negatively* linearly correlated? Use $\alpha = 0.05$. [*Note:* $r = -0.909$]

13.31 The U.S. National Center for Health Statistics collects and publishes data on heights and weights by age and sex. A random sample of 11 males, aged 18–24 years, gave the following data:

Height (inches) x	Weight (lb) y
65	175
67	133
71	185
71	163
66	126
75	198
67	153
70	163
71	159
69	151
69	155

Do the data provide sufficient evidence to conclude that the variables height and weight for 18–24-year-old males are *positively* linearly correlated? Perform the test at the 5% significance level. [*Note:* $r = 0.662$]

13.32 Hanna Properties, Inc. specializes in custom-home resales in the Equestrian Estates. A random sample of nine currently listed custom homes provided the following information on size and price. The size data are in hundreds of square feet, rounded to the nearest hundred; the price data are in thousands of dollars, rounded to the nearest thousand.

Size x	Price y
26	235
27	249
33	267
29	269
29	295
34	345
30	415
40	475
22	195

At the 0.5% significance level, do the variables size and price for resale custom homes in the Equestrian Estates area appear to be *positively* linearly correlated? [*Note:* $r = 0.829$]

13.33 A report read by a physician indicated that the maximum heart rate an individual can reach during intensive exercise decreases with age. The physician decided to do his own study. Ten randomly selected members of a jogging club performed exercise tests and recorded their peak heart rates. The results are shown in the following table:

Age x	Peak heart rate y
10	210
20	200
20	195
25	195
30	190
30	180
30	185
40	180
45	170
50	165

Do the data provide sufficient evidence to conclude that age and peak heart rate are *negatively* linearly correlated? Take $\alpha = 0.025$. [*Note:* $r = -0.971$]

13.34 A calculus instructor asked a random sample of eight students to record their study times per lesson in a beginning calculus course. She then made a table for total study times over two weeks and test scores at the end of the two weeks. Here are the results.

Study time (hrs) x	Grade (percent) y
10	92
15	81
12	84
20	74
8	85
16	80
14	84
22	80

At the 5% significance level, do the data provide sufficient evidence to conclude that study time and grade in beginning calculus courses are linearly correlated? [*Note:* $r = -0.775$]

13.35 An economist is interested in the relation between the disposable income of a family and the amount of money spent annually on food. For a preliminary study, he randomly selects eight middle-income families of the same size (father, mother, two children). The results are displayed below.

Family disposable income (in thousands of dollars) x	Food expenditures (in hundreds of dollars) y
30	55
36	60
27	42
20	40
16	37
24	26
19	39
25	43

Do the data indicate that there is a linear correlation between family disposable income and food expenditures? Use $\alpha = 0.01$. [*Note:* $r = 0.741$]

13.36 A random sample of 10 students in an introductory statistics class gave the following data on heights and final exam scores:

Height (inches) x	Test score y
71	87
68	96
71	66
65	71
66	71
68	55
68	83
64	67
62	86
65	60

Do the data provide sufficient evidence to conclude that the variables height and test score in introductory statistics courses are linearly correlated? Use $\alpha = 0.05$.

13.4 Computer packages*

In Section 12.5 we learned how Minitab can be applied to carry out some of the descriptive methods in regression and correlation. Now we will examine how Minitab can be used to perform several of the inferential procedures discussed in this chapter. As before, we will employ the age-price data for Datsun Zs to illustrate the use of Minitab in performing regression and correlation analyses. We begin by considering inferences about the slope, β_1, of the population regression line.

Inferences concerning β_1

To use Minitab to perform inferences concerning the slope, β_1, of the population regression line, we apply the REGRESS command, as explained in Section 12.5. Printout 12.2 shows the result of applying the REGRESS command to the age-price data for Datsun Zs given in Table 13.1 on page 538. We repeat that printout as Printout 13.1.

PRINTOUT 13.1
Minitab output
for REGRESS.

```
MTB > REGRESS 'PRICE' ON 1 PREDICTOR 'AGE'

The regression equation is
PRICE = 141 - 13.7 AGE

Predictor        Coef       Stdev      t-ratio
Constant       141.432      7.538       18.76
AGE            -13.703      1.385       -9.90

s = 6.220      R-sq = 91.6%     R-sq(adj) = 90.6%

Analysis of Variance

SOURCE          DF         SS          MS
Regression       1       3789.4      3789.4
Error            9        348.2        38.7
Total           10       4137.6
```

In Example 13.3 we applied Procedure 13.1 to determine whether the slope, β_1, of the population regression line is not zero and, hence, whether the age of a Datsun Z is useful as a predictor of price. Specifically, we performed the hypothesis test

$$H_0: \beta_1 = 0$$
$$H_a: \beta_1 \neq 0$$

at the 5% significance level.

Using Printout 13.1 we can perform that hypothesis test quickly and easily. To begin, look at the row labelled AGE. The last entry in that row (under the column headed t-ratio) gives the value of the test statistic

$$t = \frac{b_1}{s_e/\sqrt{\Sigma x^2 - (\Sigma x)^2/n}}$$

Thus, $t = -9.90$. The critical values for the test are $\pm t_{\alpha/2} = \pm t_{0.025}$ for df $= n - 2 = 11 - 2 = 9$. Consulting Table III, we find that the critical values are ± 2.262. Consequently, we see that the value of the test statistic falls in the left-hand portion of the rejection region. This means that we should reject H_0 and conclude that the age of a Datsun Z is useful as a predictor of price.

We can also use Printout 13.1 to obtain a confidence interval for β_1. According to Procedure 13.2 on page 549, the endpoints of a $(1 - \alpha)$-level confidence interval for β_1 are

$$b_1 \pm t_{\alpha/2} \cdot s_e/\sqrt{\Sigma x^2 - (\Sigma x)^2/n}$$

where df $= n - 2$.

Most of the quantities in the above formula can be found in Printout 13.1. The first entry in the row labelled AGE is b_1; consequently, $b_1 = -13.703$. The second entry in the row labelled AGE is the estimated standard deviation of b_1; that is, $s_e/\sqrt{\Sigma x^2 - (\Sigma x)^2/n}$. So,

$$s_e/\sqrt{\Sigma x^2 - (\Sigma x)^2/n} = 1.385$$

For, say, a 95% confidence interval, $\alpha = 0.05$. Then, as we observed earlier, $t_{\alpha/2} = t_{0.025} = 2.262$. Applying the above formula for the endpoints of the confidence interval, we conclude that the endpoints of the desired confidence interval are

$$-13.703 \pm 2.262 \cdot 1.385$$

or

$$-16.84 \quad \text{to} \quad -10.57$$

We can be 95% confident that β_1 is somewhere between -16.84 and -10.57. In other words, we can be 95% confident that the yearly drop in mean price for Datsun Zs is somewhere between $1057 and $1684.

Confidence intervals for means; prediction intervals

Procedure 13.3 on page 551 provides a step-by-step method to find a confidence interval for the mean of the population of y-values corresponding to a particular x-value; and Procedure 13.4 on page 553 provides a step-by-step method to find a prediction interval for a population y-value corresponding to a particular x-value. Alternatively, we can use Minitab to obtain such confidence intervals and prediction intervals. The appropriate command is **REGRESS** with the subcommand **PREDICT.**

We will illustrate the REGRESS;PREDICT command by using it to find a 95% confidence interval for the mean price of all three-year-old Zs and a 95% prediction interval for the price of a randomly selected three-year-old Z. To begin, we enter the sample data on age and price, given in Table 13.1 on page 538, into the computer. Then we apply the REGRESS command as we did in Printout 13.1. On the next line we type the subcommand PREDICT 3. The "3" indicates that we are considering three-year-old Zs. Printout 13.2 shows these commands and the resulting output.

PRINTOUT 13.2
Minitab output for
REGRESS;PREDICT.

```
MTB > REGRESS 'PRICE' ON 1 PREDICTOR 'AGE';
SUBC> PREDICT 3.

The regression equation is
PRICE = 141 − 13.7 AGE

Predictor        Coef        Stdev        t-ratio
Constant      141.432        7.538         18.76
AGE           −13.703        1.385         −9.90

s = 6.220       R-sq = 91.6%     R-sq(adj) = 90.6%

Analysis of Variance

SOURCE         DF          SS           MS
Regression      1       3789.4       3789.4
Error           9        348.2         38.7
Total          10       4137.6

    Fit   Stdev.Fit        95% C.I.            95% P.I.
 100.32        3.66   (  92.04, 108.61) (  83.99, 116.66)
```

The first part of the output in Printout 13.2 is the same as that given by the RE-GRESS command without the PREDICT subcommand. It is only the last two lines of the output that are new. Let us now examine in detail those two lines.

The first item, headed Fit, gives the point estimate, $\hat{y}_p = 100.32$, for the mean price of all three-year-old Zs. This is also the predicted value for the price of a randomly selected three-year-old Z. The second item, headed Stdev.Fit, displays the *estimated* standard deviation of the random variable $\hat{y}_p = b_0 + b_1 x_p$, which is

$$s_e\sqrt{\frac{1}{n} + \frac{(x_p - \Sigma x/n)^2}{\Sigma x^2 - (\Sigma x)^2/n}}$$

Here, of course, $x_p = 3$.

The third item, headed 95% C.I., provides the desired confidence interval. Hence, a 95% confidence interval for the mean price of all three-year-old Datsun Zs is from 92.04 to 108.61. We can be 95% confident that the mean price of all three-year-old Zs is somewhere between $9,204 and $10,861.

Finally, the fourth item, headed 95% P.I., gives the desired prediction interval. Thus, a 95% prediction interval for the price of a randomly selected three-year-old Z is from 83.99 to 116.66. We can be 95% certain that the price of a randomly selected three-year-old Z will be somewhere between $8,399 and $11,666.

You might have noted that we did not specify the confidence level or prediction level in obtaining the confidence and prediction intervals. The reason is that Minitab automatically computes a 95% confidence interval and a 95% prediction interval—and those were the confidence and prediction intervals that we wanted. If we want a confidence level or prediction level other than 95%, then we must proceed somewhat differently. For the details, we refer the reader to the *Minitab Supplement.*

◆ Chapter Review

KEY TERMS

linearly uncorrelated, 558
negatively linearly correlated, 557
population linear correlation
  coefficient (ρ), 557
population regression line, 539
positively linearly correlated, 557

PREDICT*, 564
prediction interval, 552
REGRESS*, 562, 564
regression model, 539
standard error of the estimate (s_e), 542

FORMULAS

In the formulas below,

$b_0 =$ y-intercept of sample regression line
$b_1 =$ slope of sample regression line
$n =$ sample size (number of data points)
$SSE =$ error sum of squares
$r =$ (sample) linear correlation coefficient
$\beta_0 =$ y-intercept of population regression line
$\beta_1 =$ slope of population regression line
$s_e =$ standard error of the estimate
$\rho =$ population linear correlation coefficient

Population regression line, 539

$$y = \beta_0 + \beta_1 x$$

Standard error of the estimate, 542

$$s_e = \sqrt{\frac{SSE}{n-2}} \quad \text{or} \quad s_e = \sqrt{\frac{\Sigma y^2 - b_0(\Sigma y) - b_1(\Sigma xy)}{n-2}}$$

Test statistic for $H_0: \beta_1 = 0$, 546

$$t = \frac{b_1}{s_e / \sqrt{\Sigma x^2 - (\Sigma x)^2 / n}}$$

with df $= n - 2$.

Confidence interval for β_1, 549

$$b_1 \pm t_{a/2} \cdot s_e / \sqrt{\Sigma x^2 - (\Sigma x)^2 / n}$$

(df $= n - 2$).

Confidence interval for the mean of the population of y-values corresponding to x_p, 551

$$\hat{y}_p \pm t_{a/2} \cdot s_e \sqrt{\frac{1}{n} + \frac{(x_p - \Sigma x / n)^2}{\Sigma x^2 - (\Sigma x)^2 / n}}$$

(df $= n - 2$).

Prediction interval for a population y-value corresponding to x_p, 554

$$\hat{y}_p \pm t_{a/2} \cdot s_e \sqrt{1 + \frac{1}{n} + \frac{(x_p - \Sigma x / n)^2}{\Sigma x^2 - (\Sigma x)^2 / n}}$$

Test statistic for H_0: $\rho = 0$, 558

$$r = \frac{n(\Sigma xy) - (\Sigma x)(\Sigma y)}{\sqrt{n(\Sigma x^2) - (\Sigma x)^2} \sqrt{n(\Sigma y^2) - (\Sigma y)^2}}$$

with df $= n - 2$.

YOU SHOULD BE ABLE TO . . .

1 use and understand the preceding formulas.

2 state the assumptions for regression inferences.

3 determine the standard error of the estimate.

4 perform a hypothesis test to determine whether the slope, β_1, of the population regression line is not zero.

5 find a confidence interval for β_1.

6 find a confidence interval for the mean of the population of y-values corresponding to a particular x-value.

7 find a prediction interval for a population y-value corresponding to a particular x-value.

8 perform a hypothesis test for a population linear correlation coefficient, with null hypothesis H_0: $\rho = 0$.

REVIEW TEST

1 A random sample of houses listed for rent in the classified ads yielded the following data on number of bedrooms versus monthly rent:

Number of bedrooms x	Monthly rent ($) y
2	375
4	550
4	625
2	500
3	575
3	465
2	525
3	700
4	750
1	295

Discuss what it would mean for the assumptions for regression inferences to be satisfied by the variables number of bedrooms, x, and monthly rent, y.

2 Refer to Problem 1.
   a) Determine the regression equation for the data.
   b) Compute the standard error of the estimate.
   c) Interpret the result from part (b).

3 Refer to Problems 1 and 2.
   a) Perform a hypothesis test to determine whether the slope, β_1, of the population regression line is not zero and, hence, whether number of bedrooms is useful as a predictor of monthly rent. Use $\alpha = 0.05$.

b) Find a 95% confidence interval for β_1 and interpret your results in words.

c) Determine a 95% confidence interval for the mean monthly rent of three-bedroom homes.

d) Obtain a 95% prediction interval for the monthly rent of a randomly selected three-bedroom home.

e) Why is the prediction interval in (d) more extensive than the confidence interval in (c)?

4 Refer to Problem 1. At the 2.5% significance level, do the data provide sufficient evidence to conclude that the variables "number of bedrooms" and "monthly rent" are positively linearly correlated?

Tables

Table I   Binomial probabilities— $\dbinom{n}{x} p^x (1 - p)^{n-x}$

							p					
n	x	0.1	0.2	0.25	0.3	0.4	0.5	0.6	0.7	0.75	0.8	0.9
1	0	0.900	0.800	0.750	0.700	0.600	0.500	0.400	0.300	0.250	0.200	0.100
	1	0.100	0.200	0.250	0.300	0.400	0.500	0.600	0.700	0.750	0.800	0.900
2	0	0.810	0.640	0.563	0.490	0.360	0.250	0.160	0.090	0.063	0.040	0.010
	1	0.180	0.320	0.375	0.420	0.480	0.500	0.480	0.420	0.375	0.320	0.180
	2	0.010	0.040	0.063	0.090	0.160	0.250	0.360	0.490	0.563	0.640	0.810
3	0	0.729	0.512	0.422	0.343	0.216	0.125	0.064	0.027	0.016	0.008	0.001
	1	0.243	0.384	0.422	0.441	0.432	0.375	0.288	0.189	0.141	0.096	0.027
	2	0.027	0.096	0.141	0.189	0.288	0.375	0.432	0.441	0.422	0.384	0.243
	3	0.001	0.008	0.016	0.027	0.064	0.125	0.216	0.343	0.422	0.512	0.729
4	0	0.656	0.410	0.316	0.240	0.130	0.063	0.026	0.008	0.004	0.002	0.000
	1	0.292	0.410	0.422	0.412	0.346	0.250	0.154	0.076	0.047	0.026	0.004
	2	0.049	0.154	0.211	0.265	0.346	0.375	0.346	0.265	0.211	0.154	0.049
	3	0.004	0.026	0.047	0.076	0.154	0.250	0.346	0.412	0.422	0.410	0.292
	4	0.000	0.002	0.004	0.008	0.026	0.063	0.130	0.240	0.316	0.410	0.656
5	0	0.590	0.328	0.237	0.168	0.078	0.031	0.010	0.002	0.001	0.000	0.000
	1	0.328	0.410	0.396	0.360	0.259	0.156	0.077	0.028	0.015	0.006	0.000
	2	0.073	0.205	0.264	0.309	0.346	0.312	0.230	0.132	0.088	0.051	0.008
	3	0.008	0.051	0.088	0.132	0.230	0.312	0.346	0.309	0.264	0.205	0.073
	4	0.000	0.006	0.015	0.028	0.077	0.156	0.259	0.360	0.396	0.410	0.328
	5	0.000	0.000	0.001	0.002	0.010	0.031	0.078	0.168	0.237	0.328	0.590
6	0	0.531	0.262	0.178	0.118	0.047	0.016	0.004	0.001	0.000	0.000	0.000
	1	0.354	0.393	0.356	0.303	0.187	0.094	0.037	0.010	0.004	0.002	0.000
	2	0.098	0.246	0.297	0.324	0.311	0.234	0.138	0.060	0.033	0.015	0.001
	3	0.015	0.082	0.132	0.185	0.276	0.313	0.276	0.185	0.132	0.082	0.015
	4	0.001	0.015	0.033	0.060	0.138	0.234	0.311	0.324	0.297	0.246	0.098
	5	0.000	0.002	0.004	0.010	0.037	0.094	0.187	0.303	0.356	0.393	0.354
	6	0.000	0.000	0.000	0.001	0.004	0.016	0.047	0.118	0.178	0.262	0.531
7	0	0.478	0.210	0.133	0.082	0.028	0.008	0.002	0.000	0.000	0.000	0.000
	1	0.372	0.367	0.311	0.247	0.131	0.055	0.017	0.004	0.001	0.000	0.000
	2	0.124	0.275	0.311	0.318	0.261	0.164	0.077	0.025	0.012	0.004	0.000
	3	0.023	0.115	0.173	0.227	0.290	0.273	0.194	0.097	0.058	0.029	0.003
	4	0.003	0.029	0.058	0.097	0.194	0.273	0.290	0.227	0.173	0.115	0.023
	5	0.000	0.004	0.012	0.025	0.077	0.164	0.261	0.318	0.311	0.275	0.124
	6	0.000	0.000	0.001	0.004	0.017	0.055	0.131	0.247	0.311	0.367	0.372
	7	0.000	0.000	0.000	0.000	0.002	0.008	0.028	0.082	0.133	0.210	0.478

Table I   (continued)

							p						
n	x	0.1	0.2	0.25	0.3	0.4	0.5	0.6	0.7	0.75	0.8	0.9
8	0	0.430	0.168	0.100	0.058	0.017	0.004	0.001	0.000	0.000	0.000	0.000
	1	0.383	0.336	0.267	0.198	0.090	0.031	0.008	0.001	0.000	0.000	0.000
	2	0.149	0.294	0.311	0.296	0.209	0.109	0.041	0.010	0.004	0.001	0.000
	3	0.033	0.147	0.208	0.254	0.279	0.219	0.124	0.047	0.023	0.009	0.000
	4	0.005	0.046	0.087	0.136	0.232	0.273	0.232	0.136	0.087	0.046	0.005
	5	0.000	0.009	0.023	0.047	0.124	0.219	0.279	0.254	0.208	0.147	0.033
	6	0.000	0.001	0.004	0.010	0.041	0.109	0.209	0.296	0.311	0.294	0.149
	7	0.000	0.000	0.000	0.001	0.008	0.031	0.090	0.198	0.267	0.336	0.383
	8	0.000	0.000	0.000	0.000	0.001	0.004	0.017	0.058	0.100	0.168	0.430
9	0	0.387	0.134	0.075	0.040	0.010	0.002	0.000	0.000	0.000	0.000	0.000
	1	0.387	0.302	0.225	0.156	0.060	0.018	0.004	0.000	0.000	0.000	0.000
	2	0.172	0.302	0.300	0.267	0.161	0.070	0.021	0.004	0.001	0.000	0.000
	3	0.045	0.176	0.234	0.267	0.251	0.164	0.074	0.021	0.009	0.003	0.000
	4	0.007	0.066	0.117	0.172	0.251	0.246	0.167	0.074	0.039	0.017	0.001
	5	0.001	0.017	0.039	0.074	0.167	0.246	0.251	0.172	0.117	0.066	0.007
	6	0.000	0.003	0.009	0.021	0.074	0.164	0.251	0.267	0.234	0.176	0.045
	7	0.000	0.000	0.001	0.004	0.021	0.070	0.161	0.267	0.300	0.302	0.172
	8	0.000	0.000	0.000	0.000	0.004	0.018	0.060	0.156	0.225	0.302	0.387
	9	0.000	0.000	0.000	0.000	0.000	0.002	0.010	0.040	0.075	0.134	0.387
10	0	0.349	0.107	0.056	0.028	0.006	0.001	0.000	0.000	0.000	0.000	0.000
	1	0.387	0.268	0.188	0.121	0.040	0.010	0.002	0.000	0.000	0.000	0.000
	2	0.194	0.302	0.282	0.233	0.121	0.044	0.011	0.001	0.000	0.000	0.000
	3	0.057	0.201	0.250	0.267	0.215	0.117	0.042	0.009	0.003	0.001	0.000
	4	0.011	0.088	0.146	0.200	0.251	0.205	0.111	0.037	0.016	0.006	0.000
	5	0.001	0.026	0.058	0.103	0.201	0.246	0.201	0.103	0.058	0.026	0.001
	6	0.000	0.006	0.016	0.037	0.111	0.205	0.251	0.200	0.146	0.088	0.011
	7	0.000	0.001	0.003	0.009	0.042	0.117	0.215	0.267	0.250	0.201	0.057
	8	0.000	0.000	0.000	0.001	0.011	0.044	0.121	0.233	0.282	0.302	0.194
	9	0.000	0.000	0.000	0.000	0.002	0.010	0.040	0.121	0.188	0.268	0.387
	10	0.000	0.000	0.000	0.000	0.000	0.001	0.006	0.028	0.056	0.107	0.349
11	0	0.314	0.086	0.042	0.020	0.004	0.000	0.000	0.000	0.000	0.000	0.000
	1	0.384	0.236	0.155	0.093	0.027	0.005	0.001	0.000	0.000	0.000	0.000
	2	0.213	0.295	0.258	0.200	0.089	0.027	0.005	0.001	0.000	0.000	0.000
	3	0.071	0.221	0.258	0.257	0.177	0.081	0.023	0.004	0.001	0.000	0.000
	4	0.016	0.111	0.172	0.220	0.236	0.161	0.070	0.017	0.006	0.002	0.000
	5	0.002	0.039	0.080	0.132	0.221	0.226	0.147	0.057	0.027	0.010	0.000
	6	0.000	0.010	0.027	0.057	0.147	0.226	0.221	0.132	0.080	0.039	0.002
	7	0.000	0.002	0.006	0.017	0.070	0.161	0.236	0.220	0.172	0.111	0.016
	8	0.000	0.000	0.001	0.004	0.023	0.081	0.177	0.257	0.258	0.221	0.071
	9	0.000	0.000	0.000	0.001	0.005	0.027	0.089	0.200	0.258	0.295	0.213
	10	0.000	0.000	0.000	0.000	0.001	0.005	0.027	0.093	0.155	0.236	0.384
	11	0.000	0.000	0.000	0.000	0.000	0.000	0.004	0.020	0.042	0.086	0.314

Table I   (continued)

							p					
n	x	0.1	0.2	0.25	0.3	0.4	0.5	0.6	0.7	0.75	0.8	0.9
12	0	0.282	0.069	0.032	0.014	0.002	0.000	0.000	0.000	0.000	0.000	0.000
	1	0.377	0.206	0.127	0.071	0.017	0.003	0.000	0.000	0.000	0.000	0.000
	2	0.230	0.283	0.232	0.168	0.064	0.016	0.002	0.000	0.000	0.000	0.000
	3	0.085	0.236	0.258	0.240	0.142	0.054	0.012	0.001	0.000	0.000	0.000
	4	0.021	0.133	0.194	0.231	0.213	0.121	0.042	0.008	0.002	0.001	0.000
	5	0.004	0.053	0.103	0.158	0.227	0.193	0.101	0.029	0.011	0.003	0.000
	6	0.000	0.016	0.040	0.079	0.177	0.226	0.177	0.079	0.040	0.016	0.000
	7	0.000	0.003	0.011	0.029	0.101	0.193	0.227	0.158	0.103	0.053	0.004
	8	0.000	0.001	0.002	0.008	0.042	0.121	0.213	0.231	0.194	0.133	0.021
	9	0.000	0.000	0.000	0.001	0.012	0.054	0.142	0.240	0.258	0.236	0.085
	10	0.000	0.000	0.000	0.000	0.002	0.016	0.064	0.168	0.232	0.283	0.230
	11	0.000	0.000	0.000	0.000	0.000	0.003	0.017	0.071	0.127	0.206	0.377
	12	0.000	0.000	0.000	0.000	0.000	0.000	0.002	0.014	0.032	0.069	0.282
13	0	0.254	0.055	0.024	0.010	0.001	0.000	0.000	0.000	0.000	0.000	0.000
	1	0.367	0.179	0.103	0.054	0.011	0.002	0.000	0.000	0.000	0.000	0.000
	2	0.245	0.268	0.206	0.139	0.045	0.010	0.001	0.000	0.000	0.000	0.000
	3	0.100	0.246	0.252	0.218	0.111	0.035	0.006	0.001	0.000	0.000	0.000
	4	0.028	0.154	0.210	0.234	0.184	0.087	0.024	0.003	0.001	0.000	0.000
	5	0.006	0.069	0.126	0.180	0.221	0.157	0.066	0.014	0.005	0.001	0.000
	6	0.001	0.023	0.056	0.103	0.197	0.209	0.131	0.044	0.019	0.006	0.000
	7	0.000	0.006	0.019	0.044	0.131	0.209	0.197	0.103	0.056	0.023	0.001
	8	0.000	0.001	0.005	0.014	0.066	0.157	0.221	0.180	0.126	0.069	0.006
	9	0.000	0.000	0.001	0.003	0.024	0.087	0.184	0.234	0.210	0.154	0.028
	10	0.000	0.000	0.000	0.001	0.006	0.035	0.111	0.218	0.252	0.246	0.100
	11	0.000	0.000	0.000	0.000	0.001	0.010	0.045	0.139	0.206	0.268	0.245
	12	0.000	0.000	0.000	0.000	0.000	0.002	0.011	0.054	0.103	0.179	0.367
	13	0.000	0.000	0.000	0.000	0.000	0.000	0.001	0.010	0.024	0.055	0.254
14	0	0.229	0.044	0.018	0.007	0.001	0.000	0.000	0.000	0.000	0.000	0.000
	1	0.356	0.154	0.083	0.041	0.007	0.001	0.000	0.000	0.000	0.000	0.000
	2	0.257	0.250	0.180	0.113	0.032	0.006	0.001	0.000	0.000	0.000	0.000
	3	0.114	0.250	0.240	0.194	0.085	0.022	0.003	0.000	0.000	0.000	0.000
	4	0.035	0.172	0.220	0.229	0.155	0.061	0.014	0.001	0.000	0.000	0.000
	5	0.008	0.086	0.147	0.196	0.207	0.122	0.041	0.007	0.002	0.000	0.000
	6	0.001	0.032	0.073	0.126	0.207	0.183	0.092	0.023	0.008	0.002	0.000
	7	0.000	0.009	0.028	0.062	0.157	0.209	0.157	0.062	0.028	0.009	0.000
	8	0.000	0.002	0.008	0.023	0.092	0.183	0.207	0.126	0.073	0.032	0.001
	9	0.000	0.000	0.002	0.007	0.041	0.122	0.207	0.196	0.147	0.086	0.008
	10	0.000	0.000	0.000	0.001	0.014	0.061	0.155	0.229	0.220	0.172	0.035
	11	0.000	0.000	0.000	0.000	0.003	0.022	0.085	0.194	0.240	0.250	0.114
	12	0.000	0.000	0.000	0.000	0.001	0.006	0.032	0.113	0.180	0.250	0.257
	13	0.000	0.000	0.000	0.000	0.000	0.001	0.007	0.041	0.083	0.154	0.356
	14	0.000	0.000	0.000	0.000	0.000	0.000	0.001	0.007	0.018	0.044	0.229

Table I   (continued)

								p				
n	x	0.1	0.2	0.25	0.3	0.4	0.5	0.6	0.7	0.75	0.8	0.9
15	0	0.206	0.035	0.013	0.005	0.000	0.000	0.000	0.000	0.000	0.000	0.000
	1	0.343	0.132	0.067	0.031	0.005	0.000	0.000	0.000	0.000	0.000	0.000
	2	0.267	0.231	0.156	0.092	0.022	0.003	0.000	0.000	0.000	0.000	0.000
	3	0.129	0.250	0.225	0.170	0.063	0.014	0.002	0.000	0.000	0.000	0.000
	4	0.043	0.188	0.225	0.219	0.127	0.042	0.007	0.001	0.000	0.000	0.000
	5	0.010	0.103	0.165	0.206	0.186	0.092	0.024	0.003	0.001	0.000	0.000
	6	0.002	0.043	0.092	0.147	0.207	0.153	0.061	0.012	0.003	0.001	0.000
	7	0.000	0.014	0.039	0.081	0.177	0.196	0.118	0.035	0.013	0.003	0.000
	8	0.000	0.003	0.013	0.035	0.118	0.196	0.177	0.081	0.039	0.014	0.000
	9	0.000	0.001	0.003	0.012	0.061	0.153	0.207	0.147	0.092	0.043	0.002
	10	0.000	0.000	0.001	0.003	0.024	0.092	0.186	0.206	0.165	0.103	0.010
	11	0.000	0.000	0.000	0.001	0.007	0.042	0.127	0.219	0.225	0.188	0.043
	12	0.000	0.000	0.000	0.000	0.002	0.014	0.063	0.170	0.225	0.250	0.129
	13	0.000	0.000	0.000	0.000	0.000	0.003	0.022	0.092	0.156	0.231	0.267
	14	0.000	0.000	0.000	0.000	0.000	0.000	0.005	0.031	0.067	0.132	0.343
	15	0.000	0.000	0.000	0.000	0.000	0.000	0.000	0.005	0.013	0.035	0.206
20	0	0.122	0.012	0.003	0.001	0.000	0.000	0.000	0.000	0.000	0.000	0.000
	1	0.270	0.058	0.021	0.007	0.000	0.000	0.000	0.000	0.000	0.000	0.000
	2	0.285	0.137	0.067	0.028	0.003	0.000	0.000	0.000	0.000	0.000	0.000
	3	0.190	0.205	0.134	0.072	0.012	0.001	0.000	0.000	0.000	0.000	0.000
	4	0.090	0.218	0.190	0.130	0.035	0.005	0.000	0.000	0.000	0.000	0.000
	5	0.032	0.175	0.202	0.179	0.075	0.015	0.001	0.000	0.000	0.000	0.000
	6	0.009	0.109	0.169	0.192	0.124	0.037	0.005	0.000	0.000	0.000	0.000
	7	0.002	0.055	0.112	0.164	0.166	0.074	0.015	0.001	0.000	0.000	0.000
	8	0.000	0.022	0.061	0.114	0.180	0.120	0.035	0.004	0.001	0.000	0.000
	9	0.000	0.007	0.027	0.065	0.160	0.160	0.071	0.012	0.003	0.000	0.000
	10	0.000	0.002	0.010	0.031	0.117	0.176	0.117	0.031	0.010	0.002	0.000
	11	0.000	0.000	0.003	0.012	0.071	0.160	0.160	0.065	0.027	0.007	0.000
	12	0.000	0.000	0.001	0.004	0.035	0.120	0.180	0.114	0.061	0.022	0.000
	13	0.000	0.000	0.000	0.001	0.015	0.074	0.166	0.164	0.112	0.055	0.002
	14	0.000	0.000	0.000	0.000	0.005	0.037	0.124	0.192	0.169	0.109	0.009
	15	0.000	0.000	0.000	0.000	0.001	0.015	0.075	0.179	0.202	0.175	0.032
	16	0.000	0.000	0.000	0.000	0.000	0.005	0.035	0.130	0.190	0.218	0.090
	17	0.000	0.000	0.000	0.000	0.000	0.001	0.012	0.072	0.134	0.205	0.190
	18	0.000	0.000	0.000	0.000	0.000	0.000	0.003	0.028	0.067	0.137	0.285
	19	0.000	0.000	0.000	0.000	0.000	0.000	0.000	0.007	0.021	0.058	0.270
	20	0.000	0.000	0.000	0.000	0.000	0.000	0.000	0.001	0.003	0.012	0.122

Table II   Areas under the standard normal curve

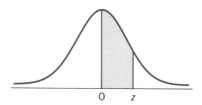

z	0.00	0.01	0.02	0.03	0.04	0.05	0.06	0.07	0.08	0.09
					Second decimal place in z					
0.0	0.0000	0.0040	0.0080	0.0120	0.0160	0.0199	0.0239	0.0279	0.0319	0.0359
0.1	0.0398	0.0438	0.0478	0.0517	0.0557	0.0596	0.0636	0.0675	0.0714	0.0753
0.2	0.0793	0.0832	0.0871	0.0910	0.0948	0.0987	0.1026	0.1064	0.1103	0.1141
0.3	0.1179	0.1217	0.1255	0.1293	0.1331	0.1368	0.1406	0.1443	0.1480	0.1517
0.4	0.1554	0.1591	0.1628	0.1664	0.1700	0.1736	0.1772	0.1808	0.1844	0.1879
0.5	0.1915	0.1950	0.1985	0.2019	0.2054	0.2088	0.2123	0.2157	0.2190	0.2224
0.6	0.2257	0.2291	0.2324	0.2357	0.2389	0.2422	0.2454	0.2486	0.2517	0.2549
0.7	0.2580	0.2611	0.2642	0.2673	0.2704	0.2734	0.2764	0.2794	0.2823	0.2852
0.8	0.2881	0.2910	0.2939	0.2967	0.2995	0.3023	0.3051	0.3078	0.3106	0.3133
0.9	0.3159	0.3186	0.3212	0.3238	0.3264	0.3289	0.3315	0.3340	0.3365	0.3389
1.0	0.3413	0.3438	0.3461	0.3485	0.3508	0.3531	0.3554	0.3577	0.3599	0.3621
1.1	0.3643	0.3665	0.3686	0.3708	0.3729	0.3749	0.3770	0.3790	0.3810	0.3830
1.2	0.3849	0.3869	0.3888	0.3907	0.3925	0.3944	0.3962	0.3980	0.3997	0.4015
1.3	0.4032	0.4049	0.4066	0.4082	0.4099	0.4115	0.4131	0.4147	0.4162	0.4177
1.4	0.4192	0.4207	0.4222	0.4236	0.4251	0.4265	0.4279	0.4292	0.4306	0.4319
1.5	0.4332	0.4345	0.4357	0.4370	0.4382	0.4394	0.4406	0.4418	0.4429	0.4441
1.6	0.4452	0.4463	0.4474	0.4484	0.4495	0.4505	0.4515	0.4525	0.4535	0.4545
1.7	0.4554	0.4564	0.4573	0.4582	0.4591	0.4599	0.4608	0.4616	0.4625	0.4633
1.8	0.4641	0.4649	0.4656	0.4664	0.4671	0.4678	0.4686	0.4693	0.4699	0.4706
1.9	0.4713	0.4719	0.4726	0.4732	0.4738	0.4744	0.4750	0.4756	0.4761	0.4767
2.0	0.4772	0.4778	0.4783	0.4788	0.4793	0.4798	0.4803	0.4808	0.4812	0.4817
2.1	0.4821	0.4826	0.4830	0.4834	0.4838	0.4842	0.4846	0.4850	0.4854	0.4857
2.2	0.4861	0.4864	0.4868	0.4871	0.4875	0.4878	0.4881	0.4884	0.4887	0.4890
2.3	0.4893	0.4896	0.4898	0.4901	0.4904	0.4906	0.4909	0.4911	0.4913	0.4916
2.4	0.4918	0.4920	0.4922	0.4925	0.4927	0.4929	0.4931	0.4932	0.4934	0.4936
2.5	0.4938	0.4940	0.4941	0.4943	0.4945	0.4946	0.4948	0.4949	0.4951	0.4952
2.6	0.4953	0.4955	0.4956	0.4957	0.4959	0.4960	0.4961	0.4962	0.4963	0.4964
2.7	0.4965	0.4966	0.4967	0.4968	0.4969	0.4970	0.4971	0.4972	0.4973	0.4974
2.8	0.4974	0.4975	0.4976	0.4977	0.4977	0.4978	0.4979	0.4979	0.4980	0.4981
2.9	0.4981	0.4982	0.4982	0.4983	0.4984	0.4984	0.4985	0.4985	0.4986	0.4986
3.0	0.4987	0.4987	0.4987	0.4988	0.4988	0.4989	0.4989	0.4989	0.4990	0.4990
3.1	0.4990	0.4991	0.4991	0.4991	0.4992	0.4992	0.4992	0.4992	0.4993	0.4993
3.2	0.4993	0.4993	0.4994	0.4994	0.4994	0.4994	0.4994	0.4995	0.4995	0.4995
3.3	0.4995	0.4995	0.4995	0.4996	0.4996	0.4996	0.4996	0.4996	0.4996	0.4997
3.4	0.4997	0.4997	0.4997	0.4997	0.4997	0.4997	0.4997	0.4997	0.4997	0.4998
3.5	0.4998	0.4998	0.4998	0.4998	0.4998	0.4998	0.4998	0.4998	0.4998	0.4998
3.6	0.4998	0.4998	0.4999	0.4999	0.4999	0.4999	0.4999	0.4999	0.4999	0.4999
3.7	0.4999	0.4999	0.4999	0.4999	0.4999	0.4999	0.4999	0.4999	0.4999	0.4999
3.8	0.4999	0.4999	0.4999	0.4999	0.4999	0.4999	0.4999	0.4999	0.4999	0.4999
3.9	0.5000†									

† For $z \geq 3.90$, the areas are 0.5000 to four decimal places.

Table III   Student's *t*-distribution (Values of t_α)

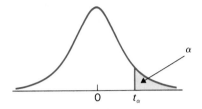

df	$t_{0.10}$	$t_{0.05}$	$t_{0.025}$	$t_{0.01}$	$t_{0.005}$	df
1	3.078	6.314	12.706	31.821	63.657	1
2	1.886	2.920	4.303	6.965	9.925	2
3	1.638	2.353	3.182	4.541	5.841	3
4	1.533	2.132	2.776	3.747	4.604	4
5	1.476	2.015	2.571	3.365	4.032	5
6	1.440	1.943	2.447	3.143	3.707	6
7	1.415	1.895	2.365	2.998	3.499	7
8	1.397	1.860	2.306	2.896	3.355	8
9	1.383	1.833	2.262	2.821	3.250	9
10	1.372	1.812	2.228	2.764	3.169	10
11	1.363	1.796	2.201	2.718	3.106	11
12	1.356	1.782	2.179	2.681	3.055	12
13	1.350	1.771	2.160	2.650	3.012	13
14	1.345	1.761	2.145	2.624	2.977	14
15	1.341	1.753	2.131	2.602	2.947	15
16	1.337	1.746	2.120	2.583	2.921	16
17	1.333	1.740	2.110	2.567	2.898	17
18	1.330	1.734	2.101	2.552	2.878	18
19	1.328	1.729	2.093	2.539	2.861	19
20	1.325	1.725	2.086	2.528	2.845	20
21	1.323	1.721	2.080	2.518	2.831	21
22	1.321	1.717	2.074	2.508	2.819	22
23	1.319	1.714	2.069	2.500	2.807	23
24	1.318	1.711	2.064	2.492	2.797	24
25	1.316	1.708	2.060	2.485	2.787	25
26	1.315	1.706	2.056	2.479	2.779	26
27	1.314	1.703	2.052	2.473	2.771	27
28	1.313	1.701	2.048	2.467	2.763	28
29	1.311	1.699	2.045	2.462	2.756	29
∞	1.282	1.645	1.960	2.326	2.576	∞

Table IV   Values of $F_{0.01}$

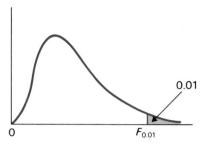

df for numerator

		1	2	3	4	5	6	7	8	9
	1	4052	4999.5	5403	5625	5764	5859	5928	5981	6022
	2	98.50	99.00	99.17	99.25	99.30	99.33	99.36	99.37	99.39
	3	34.12	30.82	29.46	28.71	28.24	27.91	27.67	27.49	27.35
	4	21.20	18.00	16.69	15.98	15.52	15.21	14.98	14.80	14.66
	5	16.26	13.27	12.06	11.39	10.97	10.67	10.46	10.29	10.16
	6	13.75	10.92	9.78	9.15	8.75	8.47	8.26	8.10	7.98
	7	12.25	9.55	8.45	7.85	7.46	7.19	6.99	6.84	6.72
	8	11.26	8.65	7.59	7.01	6.63	6.37	6.18	6.03	5.91
	9	10.56	8.02	6.99	6.42	6.06	5.80	5.61	5.47	5.35
	10	10.04	7.56	6.55	5.99	5.64	5.39	5.20	5.06	4.94
	11	9.65	7.21	6.22	5.67	5.32	5.07	4.89	4.74	4.63
	12	9.33	6.93	5.95	5.41	5.06	4.82	4.64	4.50	4.39
	13	9.07	6.70	5.74	5.21	4.86	4.62	4.44	4.30	4.19
	14	8.86	6.51	5.56	5.04	4.69	4.46	4.28	4.14	4.03
df for denominator	15	8.68	6.36	5.42	4.89	4.56	4.32	4.14	4.00	3.89
	16	8.53	6.23	5.29	4.77	4.44	4.20	4.03	3.89	3.78
	17	8.40	6.11	5.18	4.67	4.34	4.10	3.93	3.79	3.68
	18	8.29	6.01	5.09	4.58	4.25	4.01	3.84	3.71	3.60
	19	8.18	5.93	5.01	4.50	4.17	3.94	3.77	3.63	3.52
	20	8.10	5.85	4.94	4.43	4.10	3.87	3.70	3.56	3.46
	21	8.02	5.78	4.87	4.37	4.04	3.81	3.64	3.51	3.40
	22	7.95	5.72	4.82	4.31	3.99	3.76	3.59	3.45	3.35
	23	7.88	5.66	4.76	4.26	3.94	3.71	3.54	3.41	3.30
	24	7.82	5.61	4.72	4.22	3.90	3.67	3.50	3.36	3.26
	25	7.77	5.57	4.68	4.18	3.85	3.63	3.46	3.32	3.22
	26	7.72	5.53	4.64	4.14	3.82	3.59	3.42	3.29	3.18
	27	7.68	5.49	4.60	4.11	3.78	3.56	3.39	3.26	3.15
	28	7.64	5.45	4.57	4.07	3.75	3.53	3.36	3.23	3.12
	29	7.60	5.42	4.54	4.04	3.73	3.50	3.33	3.20	3.09
	30	7.56	5.39	4.51	4.02	3.70	3.47	3.30	3.17	3.07
	40	7.31	5.18	4.31	3.83	3.51	3.29	3.12	2.99	2.89
	60	7.08	4.98	4.13	3.65	3.34	3.12	2.95	2.82	2.72
	120	6.85	4.79	3.95	3.48	3.17	2.96	2.79	2.66	2.56
	∞	6.63	4.61	3.78	3.32	3.02	2.80	2.64	2.51	2.41

Adapted from D. B. Owen, *Handbook of Statistical Tables.* Courtesy of the Atomic Energy Commission. Reading, MA: Addison-Wesley, 1962.

Table IV   (continued)

			df for numerator								
10	*12*	*15*	*20*	*24*	*30*	*40*	*60*	*120*	*∞*		
6056	6106	6157	6209	6235	6261	6287	6313	6339	6366	*1*	
99.40	99.42	99.43	99.45	99.46	99.47	99.47	99.48	99.49	99.50	*2*	
27.23	27.05	26.87	26.69	26.60	26.50	26.41	26.32	26.22	26.13	*3*	
14.55	14.37	14.20	14.02	13.93	13.84	13.75	13.65	13.56	13.46	*4*	
10.05	9.89	9.72	9.55	9.47	9.38	9.29	9.20	9.11	9.02	*5*	
7.87	7.72	7.56	7.40	7.31	7.23	7.14	7.06	6.97	6.88	*6*	
6.62	6.47	6.31	6.16	6.07	5.99	5.91	5.82	5.74	5.65	*7*	
5.81	5.67	5.52	5.36	5.28	5.20	5.12	5.03	4.95	4.86	*8*	
5.26	5.11	4.96	4.81	4.73	4.65	4.57	4.48	4.40	4.31	*9*	
4.85	4.71	4.56	4.41	4.33	4.25	4.17	4.08	4.00	3.91	*10*	df for denominator
4.54	4.40	4.25	4.10	4.02	3.94	3.86	3.78	3.69	3.60	*11*	
4.30	4.16	4.01	3.86	3.78	3.70	3.62	3.54	3.45	3.36	*12*	
4.10	3.96	3.82	3.66	3.59	3.51	3.43	3.34	3.25	3.17	*13*	
3.94	3.80	3.66	3.51	3.43	3.35	3.27	3.18	3.09	3.00	*14*	
3.80	3.67	3.52	3.37	3.29	3.21	3.13	3.05	2.96	2.87	*15*	
3.69	3.55	3.41	3.26	3.18	3.10	3.02	2.93	2.84	2.75	*16*	
3.59	3.46	3.31	3.16	3.08	3.00	2.92	2.83	2.75	2.65	*17*	
3.51	3.37	3.23	3.08	3.00	2.92	2.84	2.75	2.66	2.57	*18*	
3.43	3.30	3.15	3.00	2.92	2.84	2.76	2.67	2.58	2.49	*19*	
3.37	3.23	3.09	2.94	2.86	2.78	2.69	2.61	2.52	2.42	*20*	
3.31	3.17	3.03	2.88	2.80	2.72	2.64	2.55	2.46	2.36	*21*	
3.26	3.12	2.98	2.83	2.75	2.67	2.58	2.50	2.40	2.31	*22*	
3.21	3.07	2.93	2.78	2.70	2.62	2.54	2.45	2.35	2.26	*23*	
3.17	3.03	2.89	2.74	2.66	2.58	2.49	2.40	2.31	2.21	*24*	
3.13	2.99	2.85	2.70	2.62	2.54	2.45	2.36	2.27	2.17	*25*	
3.09	2.96	2.81	2.66	2.58	2.50	2.42	2.33	2.23	2.13	*26*	
3.06	2.93	2.78	2.63	2.55	2.47	2.38	2.29	2.20	2.10	*27*	
3.03	2.90	2.75	2.60	2.52	2.44	2.35	2.26	2.17	2.06	*28*	
3.00	2.87	2.73	2.57	2.49	2.41	2.33	2.23	2.14	2.03	*29*	
2.98	2.84	2.70	2.55	2.47	2.39	2.30	2.21	2.11	2.01	*30*	
2.80	2.66	2.52	2.37	2.29	2.20	2.11	2.02	1.92	1.80	*40*	
2.63	2.50	2.35	2.20	2.12	2.03	1.94	1.84	1.73	1.60	*60*	
2.47	2.34	2.19	2.03	1.95	1.86	1.76	1.66	1.53	1.38	*120*	
2.32	2.18	2.04	1.88	1.79	1.70	1.59	1.47	1.32	1.00	*∞*	

Table V   Values of $F_{0.05}$

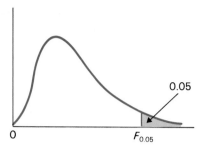

0.05

0          $F_{0.05}$

df for numerator

	1	2	3	4	5	6	7	8	9
1	161.4	199.5	215.7	224.6	230.2	234.0	236.8	238.9	240.5
2	18.51	19.00	19.16	19.25	19.30	19.33	19.35	19.37	19.38
3	10.13	9.55	9.28	9.12	9.01	8.94	8.89	8.85	8.81
4	7.71	6.94	6.59	6.39	6.26	6.16	6.09	6.04	6.00
5	6.61	5.79	5.41	5.19	5.05	4.95	4.88	4.82	4.77
6	5.99	5.14	4.76	4.53	4.39	4.28	4.21	4.15	4.10
7	5.59	4.74	4.35	4.12	3.97	3.87	3.79	3.73	3.68
8	5.32	4.46	4.07	3.84	3.69	3.58	3.50	3.44	3.39
9	5.12	4.26	3.86	3.63	3.48	3.37	3.29	3.23	3.18
10	4.96	4.10	3.71	3.48	3.33	3.22	3.14	3.07	3.02
11	4.84	3.98	3.59	3.36	3.20	3.09	3.01	2.95	2.90
12	4.75	3.89	3.49	3.26	3.11	3.00	2.91	2.85	2.80
13	4.67	3.81	3.41	3.18	3.03	2.92	2.83	2.77	2.71
14	4.60	3.74	3.34	3.11	2.96	2.85	2.76	2.70	2.65
15	4.54	3.68	3.29	3.06	2.90	2.79	2.71	2.64	2.59
16	4.49	3.63	3.24	3.01	2.85	2.74	2.66	2.59	2.54
17	4.45	3.59	3.20	2.96	2.81	2.70	2.61	2.55	2.49
18	4.41	3.55	3.16	2.93	2.77	2.66	2.58	2.51	2.46
19	4.38	3.52	3.13	2.90	2.74	2.63	2.54	2.48	2.42
20	4.35	3.49	3.10	2.87	2.71	2.60	2.51	2.45	2.39
21	4.32	3.47	3.07	2.84	2.68	2.57	2.49	2.42	2.37
22	4.30	3.44	3.05	2.82	2.66	2.55	2.46	2.40	2.34
23	4.28	3.42	3.03	2.80	2.64	2.53	2.44	2.37	2.32
24	4.26	3.40	3.01	2.78	2.62	2.51	2.42	2.36	2.30
25	4.24	3.39	2.99	2.76	2.60	2.49	2.40	2.34	2.28
26	4.23	3.37	2.98	2.74	2.59	2.47	2.39	2.32	2.27
27	4.21	3.35	2.96	2.73	2.57	2.46	2.37	2.31	2.25
28	4.20	3.34	2.95	2.71	2.56	2.45	2.36	2.29	2.24
29	4.18	3.33	2.93	2.70	2.55	2.43	2.35	2.28	2.22
30	4.17	3.32	2.92	2.69	2.53	2.42	2.33	2.27	2.21
40	4.08	3.23	2.84	2.61	2.45	2.34	2.25	2.18	2.12
60	4.00	3.15	2.76	2.53	2.37	2.25	2.17	2.10	2.04
120	3.92	3.07	2.68	2.45	2.29	2.17	2.09	2.02	1.96
∞	3.84	3.00	2.60	2.37	2.21	2.10	2.01	1.94	1.88

df for denominator

Adapted from D. B. Owen, *Handbook of Statistical Tables*. Courtesy of the Atomic Energy Commission. Reading, MA: Addison-Wesley, 1962.

Table V   (continued)

				df for numerator						
10	12	15	20	24	30	40	60	120	∞	
241.9	243.9	245.9	248.0	249.1	250.1	251.1	252.2	253.3	254.3	1
19.40	19.41	19.43	19.45	19.45	19.46	19.47	19.48	19.49	19.50	2
8.79	8.74	8.70	8.66	8.64	8.62	8.59	8.57	8.55	8.53	3
5.96	5.91	5.86	5.80	5.77	5.75	5.72	5.69	5.66	5.63	4
4.74	4.68	4.62	4.56	4.53	4.50	4.46	4.43	4.40	4.36	5
4.06	4.00	3.94	3.87	3.84	3.81	3.77	3.74	3.70	3.67	6
3.64	3.57	3.51	3.44	3.41	3.38	3.34	3.30	3.27	3.23	7
3.35	3.28	3.22	3.15	3.12	3.08	3.04	3.01	2.97	2.93	8
3.14	3.07	3.01	2.94	2.90	2.86	2.83	2.79	2.75	2.71	9
2.98	2.91	2.85	2.77	2.74	2.70	2.66	2.62	2.58	2.54	10
2.85	2.79	2.72	2.65	2.61	2.57	2.53	2.49	2.45	2.40	11
2.75	2.69	2.62	2.54	2.51	2.47	2.43	2.38	2.34	2.30	12
2.67	2.60	2.53	2.46	2.42	2.38	2.34	2.30	2.25	2.21	13
2.60	2.53	2.46	2.39	2.35	2.31	2.27	2.22	2.18	2.13	14
2.54	2.48	2.40	2.33	2.29	2.25	2.20	2.16	2.11	2.07	15
2.49	2.42	2.35	2.28	2.24	2.19	2.15	2.11	2.06	2.01	16
2.45	2.38	2.31	2.23	2.19	2.15	2.10	2.06	2.01	1.96	17
2.41	2.34	2.27	2.19	2.15	2.11	2.06	2.02	1.97	1.92	18
2.38	2.31	2.23	2.16	2.11	2.07	2.03	1.98	1.93	1.88	19
2.35	2.28	2.20	2.12	2.08	2.04	1.99	1.95	1.90	1.84	20
2.32	2.25	2.18	2.10	2.05	2.01	1.96	1.92	1.87	1.81	21
2.30	2.23	2.15	2.07	2.03	1.98	1.94	1.89	1.84	1.78	22
2.27	2.20	2.13	2.05	2.01	1.96	1.91	1.86	1.81	1.76	23
2.25	2.18	2.11	2.03	1.98	1.94	1.89	1.84	1.79	1.73	24
2.24	2.16	2.09	2.01	1.96	1.92	1.87	1.82	1.77	1.71	25
2.22	2.15	2.07	1.99	1.95	1.90	1.85	1.80	1.75	1.69	26
2.20	2.13	2.06	1.97	1.93	1.88	1.84	1.79	1.73	1.67	27
2.19	2.12	2.04	1.96	1.91	1.87	1.82	1.77	1.71	1.65	28
2.18	2.10	2.03	1.94	1.90	1.85	1.81	1.75	1.70	1.64	29
2.16	2.09	2.01	1.93	1.89	1.84	1.79	1.74	1.68	1.62	30
2.08	2.00	1.92	1.84	1.79	1.74	1.69	1.64	1.58	1.51	40
1.99	1.92	1.84	1.75	1.70	1.65	1.59	1.53	1.47	1.39	60
1.91	1.83	1.75	1.66	1.61	1.55	1.50	1.43	1.35	1.25	120
1.83	1.75	1.67	1.57	1.52	1.46	1.39	1.32	1.22	1.00	∞

df for denominator

Table VI   Chi-square distribution (Values of χ_α^2)

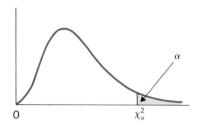

df	$\chi_{0.995}^2$	$\chi_{0.99}^2$	$\chi_{0.975}^2$	$\chi_{0.95}^2$	$\chi_{0.05}^2$	$\chi_{0.025}^2$	$\chi_{0.01}^2$	$\chi_{0.005}^2$	df
1	0.000	0.000	0.001	0.004	3.841	5.024	6.635	7.879	1
2	0.010	0.020	0.051	0.103	5.991	7.378	9.210	10.597	2
3	0.072	0.115	0.216	0.352	7.815	9.348	11.345	12.838	3
4	0.207	0.297	0.484	0.711	9.488	11.143	13.277	14.860	4
5	0.412	0.554	0.831	1.145	11.070	12.832	15.086	16.750	5
6	0.676	0.872	1.237	1.635	12.592	14.449	16.812	18.548	6
7	0.989	1.239	1.690	2.167	14.067	16.013	18.475	20.278	7
8	1.344	1.646	2.180	2.733	15.507	17.535	20.090	21.955	8
9	1.735	2.088	2.700	3.325	16.919	19.023	21.666	23.589	9
10	2.156	2.558	3.247	3.940	18.307	20.483	23.209	25.188	10
11	2.603	3.053	3.816	4.575	19.675	21.920	24.725	26.757	11
12	3.074	3.571	4.404	5.226	21.026	23.337	26.217	28.300	12
13	3.565	4.107	5.009	5.892	22.362	24.736	27.688	29.819	13
14	4.075	4.660	5.629	6.571	23.685	26.119	29.141	31.319	14
15	4.601	5.229	6.262	7.261	24.996	27.488	30.578	32.801	15
16	5.142	5.812	6.908	7.962	26.296	28.845	32.000	34.267	16
17	5.697	6.408	7.564	8.672	27.587	30.191	33.409	35.718	17
18	6.265	7.015	8.231	9.390	28.869	31.526	34.805	37.156	18
19	6.844	7.633	8.907	10.117	30.144	32.852	36.191	38.582	19
20	7.434	8.260	9.591	10.851	31.410	34.170	37.566	39.997	20
21	8.034	8.897	10.283	11.591	32.671	35.479	38.932	41.401	21
22	8.643	9.542	10.982	12.338	33.924	36.781	40.289	42.796	22
23	9.260	10.196	11.689	13.091	35.172	38.076	41.638	44.181	23
24	9.886	10.856	12.401	13.848	36.415	39.364	42.980	45.558	24
25	10.520	11.524	13.120	14.611	37.652	40.646	44.314	46.928	25
26	11.160	12.198	13.844	15.379	38.885	41.923	45.642	48.290	26
27	11.808	12.879	14.573	16.151	40.113	43.194	46.963	49.645	27
28	12.461	13.565	15.308	16.928	41.337	44.461	48.278	50.993	28
29	13.121	14.256	16.047	17.708	42.557	45.722	49.588	52.336	29
30	13.787	14.953	16.791	18.493	43.773	46.979	50.892	53.672	30

Table VII Critical values of r when $\rho = 0$

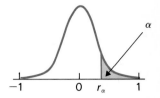

df	$r_{0.05}$	$r_{0.025}$	$r_{0.01}$	$r_{0.005}$	df
1	0.988	0.997	0.9995	0.9999	1
2	0.900	0.950	0.980	0.990	2
3	0.805	0.878	0.934	0.959	3
4	0.729	0.811	0.882	0.917	4
5	0.669	0.754	0.833	0.874	5
6	0.622	0.707	0.789	0.834	6
7	0.582	0.666	0.750	0.798	7
8	0.549	0.632	0.716	0.765	8
9	0.521	0.602	0.685	0.735	9
10	0.497	0.576	0.658	0.708	10
11	0.476	0.553	0.634	0.684	11
12	0.458	0.532	0.612	0.661	12
13	0.441	0.514	0.592	0.641	13
14	0.426	0.497	0.574	0.623	14
15	0.412	0.482	0.558	0.606	15
16	0.400	0.468	0.543	0.590	16
17	0.389	0.456	0.528	0.575	17
18	0.378	0.444	0.516	0.561	18
19	0.369	0.433	0.503	0.549	19
20	0.360	0.423	0.492	0.537	20
21	0.352	0.413	0.482	0.526	21
22	0.344	0.404	0.472	0.515	22
23	0.337	0.396	0.462	0.505	23
24	0.330	0.388	0.453	0.496	24
25	0.323	0.381	0.445	0.487	25
26	0.317	0.374	0.437	0.479	26
27	0.311	0.367	0.430	0.471	27
28	0.306	0.361	0.423	0.463	28
29	0.301	0.355	0.416	0.456	29
30	0.296	0.349	0.409	0.449	30
40	0.257	0.304	0.358	0.393	40
50	0.231	0.273	0.322	0.354	50
60	0.211	0.250	0.295	0.325	60
70	0.195	0.232	0.274	0.302	70
80	0.183	0.217	0.257	0.283	80
90	0.173	0.205	0.242	0.267	90
100	0.164	0.195	0.230	0.254	100

Entries for df = 1 to 23 adapted with permission from D. B. Owen, *Handbook of Statistical Tables*, Addison-Wesley, 1962. Other entries derived using the relation $t_{df} = r\sqrt{df/(1-r^2)}$.

· Answers ·

Chapter 1

Exercises 1.1

1.1 a) The population is the collection of all individuals or items under consideration in a statistical study.
b) A sample is that part of the population from which information is collected.

Exercises 1.2

1.3 Inferential

1.5 Inferential

1.7 Descriptive

1.9 Descriptive

1.11 a) Inferential b) Descriptive c) Descriptive
d) Inferential e) Inferential f) Inferential
g) Descriptive

REVIEW TEST

1 Answers will vary.

2 In conducting an inferential study, information will be obtained from a sample of the population. Generally speaking, the information so obtained will have to be organized and summarized in a clear and effective way prior to the application of inferential methods. Thus, almost any inferential study will involve aspects of descriptive statistics.

3 Descriptive

4 Inferential

5 Inferential

6 Descriptive

7 Inferential. The study makes an inference concerning the percentage of all adult Americans who suffer from at least one psychiatric disorder based on information from a sample of adult Americans.

Chapter 2

Exercises 2.1

2.1 It serves as an aid in the choice of the correct statistical method.

2.3 metric

2.5 ordinal

2.7 a) metric b) ordinal c) frequency
d) qualitative

2.9 a) ordinal
b) metric (could also be considered frequency data)

2.11 qualitative

2.13 height, weight, and age

Exercises 2.2

2.15 No, class limits and class marks do not make sense for qualitative data.

2.17 a) The frequency of a class is the number of data values in the class, whereas the relative frequency of a class is the ratio of the class frequency to the total number of pieces of data.
b) The percentage of a class is 100 times the relative frequency of the class.

2.19

Contents (ml)	Frequency	Relative frequency	Class mark
910–929	1	0.033	919.5
930–949	1	0.033	939.5
950–969	3	0.100	959.5
970–989	9	0.300	979.5
990–1009	7	0.233	999.5
1010–1029	6	0.200	1019.5
1030–1049	2	0.067	1039.5
1050–1069	1	0.033	1059.5
	30	0.999	

2.21

Consumption (mil. BTU)	Frequency	Relative frequency	Class mark
40–49	1	0.02	44.5
50–59	7	0.14	54.5
60–69	7	0.14	64.5
70–79	3	0.06	74.5
80–89	6	0.12	84.5
90–99	10	0.20	94.5
100–109	5	0.10	104.5
110–119	4	0.08	114.5
120–129	2	0.04	124.5
130–139	3	0.06	134.5
140–149	0	0.00	144.5
150–159	2	0.04	154.5
	50	1.00	

2.23

Number of cars sold	Frequency	Relative frequency
0	7	0.135
1	15	0.288
2	12	0.231
3	9	0.173
4	5	0.096
5	3	0.058
6	1	0.019
	52	1.000

2.25

Number of days missed	Frequency	Relative frequency
0	4	0.050
1	2	0.025
2	14	0.175
3	10	0.125
4	16	0.200
5	18	0.225
6	10	0.125
7	6	0.075
	80	1.000

2.27

Starting salary ($thousands)	Frequency	Relative frequency	Class mark
16–under 17	3	0.086	16.5
17–under 18	3	0.086	17.5
18–under 19	5	0.143	18.5
19–under 20	9	0.257	19.5
20–under 21	9	0.257	20.5
21–under 22	4	0.114	21.5
22–under 23	1	0.029	22.5
23–under 24	1	0.029	23.5
	35	1.001	

2.29

Sales (millions)	Frequency	Relative frequency	Class mark
0–under 1	1	0.042	0.5
1–under 2	3	0.125	1.5
2–under 3	10	0.417	2.5
3–under 4	4	0.167	3.5
4–under 5	1	0.042	4.5
5–under 6	1	0.042	5.5
6–under 7	2	0.083	6.5
7–under 8	1	0.042	7.5
8–under 9	1	0.042	8.5
	24	1.002	

2.31

Champion	Frequency	Relative frequency
Oklahoma	2	0.091
Oklahoma State	4	0.182
Iowa State	6	0.273
Michigan State	1	0.045
Iowa	9	0.409
	22	1.000

Exercises 2.3

2.35 A frequency histogram displays the class frequencies on the vertical axis, whereas a relative-frequency histogram displays the class relative frequencies on the vertical axis.

2.37 a)

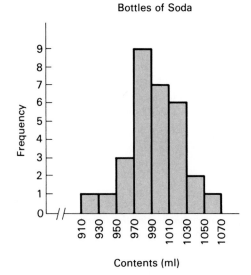

b)

Bottles of Soda

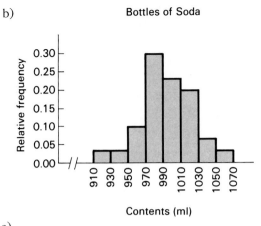

Contents (ml)

2.41   a)

Car Sales

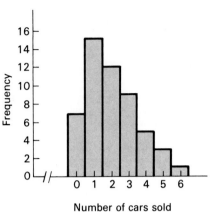

Number of cars sold

2.39   a)

Energy Consumption for Southern Households

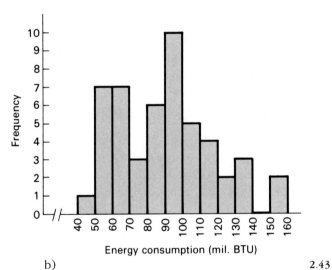

Energy consumption (mil. BTU)

b)

Car Sales

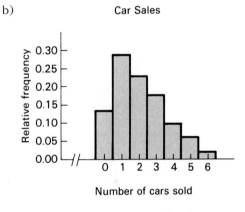

Number of cars sold

b)

Energy Consumption for Southern Households

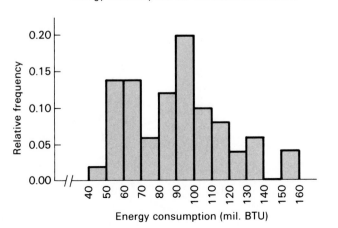

Energy consumption (mil. BTU)

2.43   a)

Absentee Records

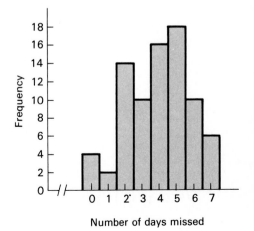

Number of days missed

b)

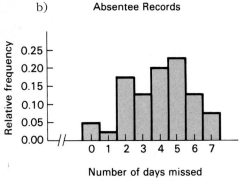

Absentee Records

2.47   Drying Times

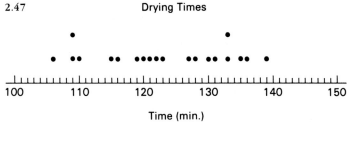

2.45   a)

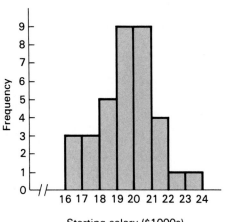

Salaries for Liberal Arts Graduates

2.49   Trucks in Use

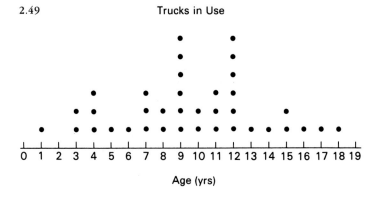

2.51   a)

1963–1984 NCAA Wrestling Championships

b)

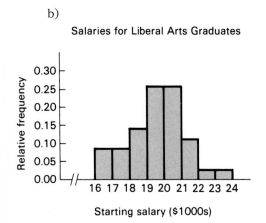

Salaries for Liberal Arts Graduates

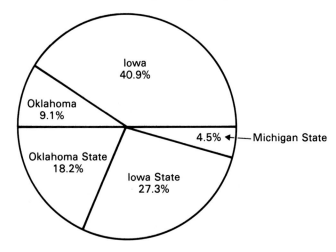

b) 1963–1984 NCAA Wrestling Championships

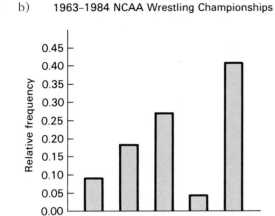

Champion

2.53 a) 20%   b) 25%   c) 7

Exercises 2.4

2.61 a)

91	4
92	
93	
94	6
95	9 7
96	4
97	7 5 4 7
98	6 9 4 8 7
99	0 6 1 9 5 7
100	1
101	4 8 0 7
102	5 8
103	0 1
104	
105	
106	0

b)

91	4
92	
93	
94	6
95	7 9
96	4 (cont.)

97	4 5 7 7
98	4 6 7 8 9
99	0 1 5 6 7 9
100	1
101	0 4 7 8
102	5 8
103	0 1
104	
105	
106	0

2.63 a)

4	5
5	8 4 5 1 0 5 5
6	4 7 9 6 0 6 2
7	7 5 8
8	6 7 1 0 3 3
9	7 6 3 4 9 7 6 0 1 7
10	1 0 9 4 2
11	1 3 1 3
12	9 5
13	0 9 6
14	
15	5 1

b)

4	5
5	0 1 4 5 5 5 8
6	0 2 4 6 6 7 9
7	5 7 8
8	0 1 3 3 6 7
9	0 1 3 4 6 6 7 7 7 9
10	0 1 2 4 9
11	1 1 3 3
12	5 9
13	0 6 9
14	
15	1 5

2.65 a)

2	7 5 4
3	5 9 9 4 7 2 8 4 2
4	1 5 1 5 0 5 6 2 5 9 8 0 1 0 3 0
5	0 5 8 2 0 4 0 2 9 0 9 1
6	0 4 7 6 8 5 7 3 3 1

b)

2	4
2	7 5
3	4 2 4 2
3	5 9 9 7 8
4	1 1 0 2 0 1 0 3 0
4	5 5 5 6 5 9 8
5	0 2 0 4 0 2 0 1
5	5 8 9 9
6	0 4 3 3 1
6	7 6 8 5 7

Exercises 2.5

2.69 c) They give the misleading impression that the district average is much greater relative to the national average than it actually is.

2.71 a) Because it is not in the correct proportion relative to the other bars.

b) Because otherwise the first bar would have to be roughly 30 times the height of the last bar. This would require a lot more space if the other three bars were left as is. On the other hand, if the first bar were kept at about the same height as shown, but without a break, then the last bar would have to be made quite short.

c) No, because the break in the bar aptly warns the reader that it is not in the correct proportion relative to the other bars.

REVIEW TEST

1 a) ordinal   b) metric
    c) qualitative   d) frequency

2 a)

Age at inaug.	Frequency	Relative frequency	Class mark
40–44	2	0.050	42
45–49	5	0.125	47
50–54	12	0.300	52
55–59	12	0.300	57
60–64	6	0.150	62
65–69	3	0.075	67
	40	1.000	

b)

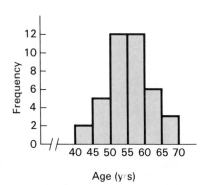

Ages at Inauguration
for First 40 U.S. Presidents

Data from *The World Almanac, 1985*

3

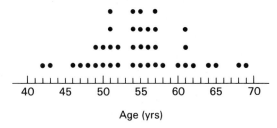

Ages at Inauguration
for First 40 U.S. Presidents

Data from *The World Almanac, 1985*

4 a)
```
4 | 2 3 6 7 8 9 9
5 | 0 0 1 1 1 1 2 2 4 4 4 4 5 5 5 5 6 6 6 7 7 7 7 8
6 | 0 1 1 1 2 4 5 8 9
```

b)
```
4 | 2 3
4 | 6 7 8 9 9
5 | 0 0 1 1 1 1 2 2 4 4 4 4
5 | 5 5 5 5 6 6 6 7 7 7 7 8
6 | 0 1 1 1 2 4
6 | 5 8 9
```

c) The second one (that is, the one with two lines per stem).

5 a)

Number busy	Frequency	Relative frequency
0	1	0.04
1	2	0.08
2	2	0.08
3	4	0.16
4	5	0.20
5	7	0.28
6	4	0.16
	25	1.00

b)

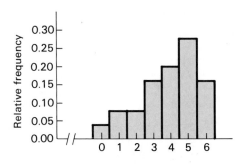

Busy Tellers

6   a)

Accidental Deaths by Type, 1982

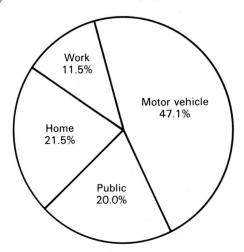

Data from National Safety Council

b)   Accidental Deaths by Type, 1982

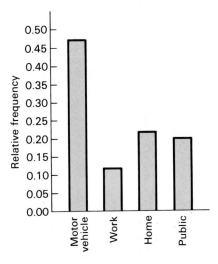

Data from National Safety Council

7   a)

High	Freq.	Relative frequency	Class mark
200–under 400	4	0.114	300
400–under 600	5	0.143	500
600–under 800	5	0.143	700
800–under 1000	13	0.371	900
1000–under 1200	6	0.171	1100
1200–under 1400	2	0.057	1300
	35	0.999	

b)

Dow Jones Highs
1950–1984

Data from *Barron's*

8* b) The percentage of women in the labor force for 1985 is about two and two-thirds times that for 1970.
    c) About one and one-half.
    d) Because it is a truncated graph.
    e) Start the graph at zero instead of 30.

Chapter 3

Exercises   3.1

3.1   To indicate where the center or most typical value of a data set lies.

3.3   mean = 33.5 years, median = 34.5 years

3.5   mean = 40.79¢/lb, median = 41.80¢/lb

3.7   mean = 306.0 ml, median = 306.5 ml

3.9   mean = 193.0 thous. volumes, median = 79.0 thous. volumes

3.11   The *median* because, unlike the mean, it is not affected strongly by the relatively few homes that have an extremely large number of square feet.

3.13   a) mean = 13.9   b) median = 4
       c) The median.

Exercises 3.2

3.19 a) 46   b) 4   c) 11.5

3.21 a) 331,143 mi.   b) 8   c) 41,392.9 mi.

3.23 a) $18,461   b) 18   c) $1025.6

3.25 a) $\bar{x} = \$97.5$

b)

x	x^2	$x - \bar{x}$	$(x - \bar{x})^2$
75	5,625	−22.5	506.25
98	9,604	0.5	0.25
130	16,900	32.5	1056.25
63	3,969	−34.5	1190.25
112	12,544	14.5	210.25
107	11,449	9.5	90.25
Σ 585	60,091	0	3053.50

3.27 a) $\bar{x} = 105.1$

b)

x	x^2	$x - \bar{x}$	$(x - \bar{x})^2$
106	11,236	0.9	0.81
94	8,836	−11.1	123.21
118	13,924	12.9	166.41
109	11,881	3.9	15.21
118	13,924	12.9	166.41
95	9,025	−10.1	102.01
99	9,801	−6.1	37.21
97	9,409	−8.1	65.61
109	11,881	3.9	15.21
106	11,236	0.9	0.81
Σ 1051	111,153	0	692.90

Exercises 3.3

3.33 To indicate the amount of variation in a data set.

3.35 a) 20 years   b) 7.2 years   c) 7.2 years

3.37 a) 10.6¢/lb   b) 3.38¢/lb   c) 3.38¢/lb

3.39 a) 31 ml   b) 8.7 ml   c) 8.7 ml

3.41 a) 501 thousand volumes
b) 205.5 thousand volumes
c) 205.5 thousand volumes

3.43 a) Brand A: $\bar{x} = 9.80$; Brand B: $\bar{x} = 9.80$
b) Brand A: median = 9.7; Brand B: median = 9.7
c) In the amount of variation.
d) The data set for Brand A.
e) Brand A: $s = 0.50$; Brand B: $s = 1.62$
f) Yes, because the data set for Brand A, which has less variation than the data set for Brand B, also has a smaller standard deviation.

3.45 a) 16.1   b) 16.1

Exercises 3.4

3.51 a) Data Set 4   b)

	Data Set 3	Data Set 4
	$\bar{x} = 83$	$\bar{x} = 83$
	$s = 7.8$	$s = 22.3$

c) See Figure A.1 and Figure A.2.

FIGURE A.1   Data Set 3. $\bar{x} = 83$, $s = 7.8$

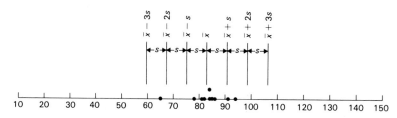

FIGURE A.2   Data Set 4. $\bar{x} = 83$, $s = 22.3$

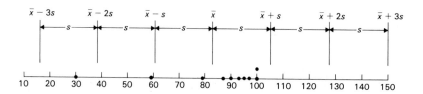

e) Yes. In fact, all of the data in each data set lies within three standard deviations to either side of the mean.

3.53 a) For any data set, at least 36% of the data lies within 1.25 standard deviations to either side of its mean.
b) For any data set, at least 92% of the data lies within 3.5 standard deviations to either side of its mean.
c) For any data set, at least 96% of the data lies within five standard deviations to either side of its mean.

3.55 a) At least 75%; at least 89%.
b) 90%; 100%
c) 90%; 100%
d) Chebychev's theorem does not necessarily give precise estimates for the percentage of data in a data set that lies within k standard deviations to either side of the mean.

3.57 a) See Figure A.3.
b) At least 225 of the 300 lodging establishments have room rates between $23.65 and $58.29.
c) At least 267 of the 300 lodging establishments have room rates between $14.99 and $66.95.

3.59 a) See Figure A.4.
b) At least 188 of the 250 households have vegetable gardens that are between 149 and 1137 sq. ft.
c) At least 223 of the 250 households have vegetable gardens that are at most 1384 sq. ft.

3.61 a) See Figure A.5.
b) At least 23 of the 30 taxable returns have an income tax of at most $6671.0.
c) At least 27 of the 30 taxable returns have an income tax of at most $8886.6.

Exercises 3.5

3.67 a) $\bar{x} = 3.2$ persons   b) $s = 1.3$ persons

3.69 a) $\bar{x} = 21.7$ years   b) $s = 3.7$ years

3.71 a) $\bar{x} = 39.5$ bu/acre, $s = 1.9$ bu/acre

b)

Yield	Frequency
36	1
37	3
38	5
39	8
40	5
41	3
42	1
43	4

c) $\bar{x} = 39.5$ bu/acre, $s = 1.9$ bu/acre

3.73 $\bar{x} = 105.0$, $s = 16.4$

3.75 $\bar{x} = 62.8$ minutes, $s = 10.7$ minutes

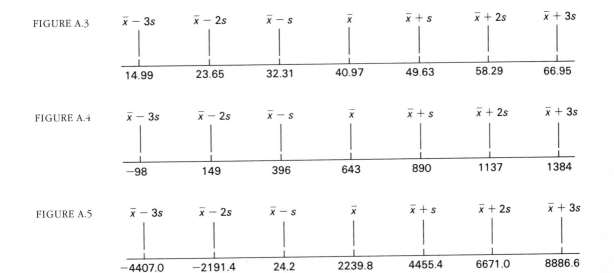

FIGURE A.3

FIGURE A.4

FIGURE A.5

3.77 a) $\bar{x} = 89.7$ mil. BTU, $s = 27.3$ mil. BTU
b) $\bar{x} \approx 89.9$ mil. BTU, $s \approx 27.3$ mil. BTU

3.79 Because the class mark of each class provides only a typical value for the data values in the class and not the data values themselves.

Exercises 3.6

3.83 $Q_1 = 21.0$ hours, $Q_2 = 30.0$ hours, $Q_3 = 35.5$ hours

3.85 $Q_1 = \$1105.5$, $Q_2 = \$1231.0$, $Q_3 = \$1386.0$

3.87 $D_1 = 15.5$ hours, $D_2 = 20.5$ hours, $D_3 = 23.5$ hours, $D_4 = 28.5$ hours, $D_5 = 30.0$ hours, $D_6 = 31.5$ hours, $D_7 = 34.5$ hours, $D_8 = 37.0$ hours, $D_9 = 42.0$ hours

3.89

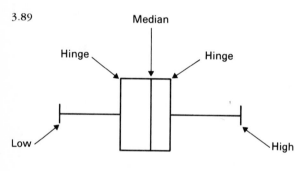

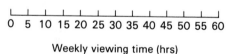

Weekly viewing time (hrs)

3.91

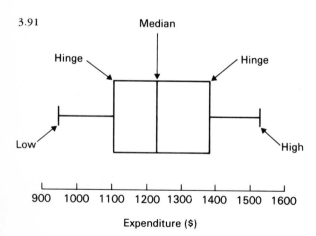

Expenditure ($)

Exercises 3.7

3.95 a) $\bar{x} = 75.0$ inches   b) $s = 2.4$ inches
c) $\mu = 75.0$ inches   d) $\sigma = 2.2$ inches
e) Both means are computed in the same way: Sum the data and then divide by the total number of pieces of data.
f) The sample standard deviation, s, and the population standard deviation, σ, are computed differently. In the defining formula for s we divide by one less than the total number of pieces of data, whereas in the defining formula for σ we divide by the total number of pieces of data.

3.97 a) $\mu = \$92.0$ b) $\sigma = \$28.3$

3.99 a) $\mu = 60.2$ thousand acres
b) $\sigma = 81.2$ thousand acres

3.101 It is a parameter since it is a descriptive measure for a population.

3.103 a) $\mu = 21.7$ years   b) $\sigma = 3.7$ years

3.105 a) $\mu \approx \$500.0$ million
b) $\sigma \approx \$404.0$ million
c) Because the class mark of each class provides only a typical value for the data values in the class and not the data values themselves.

REVIEW TEST

1

	Germans	Russians
Mean	11.3 kg	16.1 kg
Median	12.0 kg	17.0 kg

2 a) The median.   b) The mean.
[*Note:* Answers may vary.]

3 a) $\bar{x} = 4.0$ minutes
b) Range $= 14$ minutes
c) $s = 4.1$ minutes

4 a) & b) See Figure A.6 on the next page.
c) 75%   d) 91.7%
e) Its generality—Chebychev's theorem holds for any data set.

5 a) See Figure A.7 on the next page.
b) At least 27 of the 36 millionaires are between 31.7 and 85.3 years old.
c) At least 32 of the 36 millionaires are between 18.3 and 98.7 years old.

FIGURE A.6

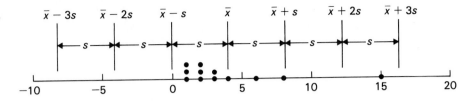

FIGURE A.7

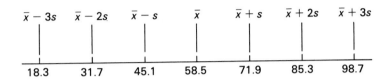

6 a) $Q_1 = 48.0$ years, $Q_2 = 59.5$ years, $Q_3 = 68.5$ years

b)

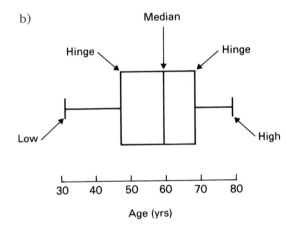

7 a) $\bar{x} = 71.7$ b) $s = 2.5$

8 a) $\mu = 15.58$ thousand   b) $\sigma = 10.23$ thousand

9 a) $\mu \approx \$64.2$ thousand   b) $\sigma \approx \$13.5$ thousand
c) Because the class mark of each class provides only a typical value for the data values in the class and not the data values themselves.

10 a) A sample mean. It is the mean price per gallon for the sample of 10,000 gasoline service stations.
b) $\bar{x}$
c) A sample standard deviation. It is the standard deviation of the prices per gallon for the sample of 10,000 gasoline service stations.
d) s
e) Statistics, since they are descriptive measures for a sample.

Chapter 4

Exercises 4.1

4.1 a) 0.125   b) 0.250   c) 0.750
d) 0   e) 1

4.3 a) 0.283   b) 0.304   c) 0.630
d) 0.210

4.5 a) 0.139   b) 0.500   c) 0.222
d) 0.111

4.7 0.020

4.9 a) 0.740   b) 0.911   c) 0.171

4.11 (b) and (d)

4.13 The event in part (e) is certain; the event in part (d) is impossible.

Exercises 4.2

4.19 $A =$
$B =$
$C =$
$D =$

4.21 $A=$ {HHTT, HTHT, HTTH, THHT, THTH, TTHH}

$B=$ {TTHH, TTHT, TTTH, TTTT}

$C=$ {HHHH, HHHT, HHTH, HHTT, HTHH, HTHT, HTTH, HTTT}

$D=$ {HHHH, TTTT}

4.23   a) (not A) =

The event the die comes up odd.

b) (A & B) =

The event the die comes up four or six.

c) (B or C) =

The event the die does *not* come up three.

4.25   a) (not B) = {HHHH, HHHT, HHTH, HHTT, HTHH, HTHT, HTTH, HTTT, THHH, THHT, THTH, THTT}

The event at least one of the first two tosses is heads.

b) (A & B) = {TTHH}

The event that the first two tosses are tails and the last two are heads.

c) (C or D) = {HHHH, HHHT, HHTH, HHTT, HTHH, HTHT, HTTH, HTTT, TTTT}

The event the first toss is a head or all four tosses are tails.

4.27   a) (not A) is the event that the employee selected missed at least four days. There are 50 employees that missed at least four days.

b) (A & B) is the event that the employee selected missed between one and three days, inclusive. There are 26 employees that missed between one and three days, inclusive.

c) (C or D) is the event that the employee selected missed at least four days. There are 50 employees that missed at least four days. [*Note:* From part (a), we see that (not A) = (C or D).]

4.29   a) (not C) is the event that the person selected is 45 years old or older. There are 30,260 thousand such people.

b) (not B) is the event that the person selected is either under 20 or over 54. There are 20,886 thousand such people.

c) (B & C) is the event that the person selected is between 20 and 44, inclusive. There are 62,718 thousand such people.

d) (A or D) is the event that the person selected is either under 20 or over 54. There are 20,886 thousand such people. [*Note:* From part (b), we see that (not B) = (A or D).]

4.31   a) no   b) yes   c) no

d) yes—events B, C, and D; no

4.33   a) mutually exclusive

b) not mutually exclusive

c) mutually exclusive

d) not mutually exclusive

e) not mutually exclusive

4.35

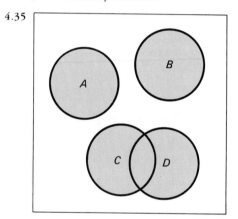

Exercises 4.3

4.37 $\dfrac{f}{N}=\dfrac{2}{10}=0.2$; $P(E)=0.2$

4.39   a) $P(S)=\dfrac{f}{N}=\dfrac{74}{100}=0.74$

b) $S=$ (A or B or C)

c) $P(A)=0.07$, $P(B)=0.28$, $P(C)=0.39$

d) $P(S)=P(A$ or B or $C)=P(A)+P(B)+P(C)$
$=0.07+0.28+0.39=0.74$

4.41   a) 0.816   b) 0.054   c) 0.335

d) 81.6% of U.S. businesses had receipts of under $100,000 in 1980; 5.4% had receipts of at least $500,000; and 33.5% had receipts of between $25,000 and $499,999.

4.43   a) $1-0.07=0.93$ b) $1-0.06=0.94$

4.45   a) $1-0.029=0.971$ b) $1-0.721=0.279$

4.47 a) $P(A) = 0.167$, $P(B) = 0.056$, $P(C) = 0.028$,
$P(D) = 0.056$, $P(E) = 0.028$, $P(F) = 0.139$,
$P(G) = 0.167$
b) 0.223   c) 0.112
d) (i) $10/36 = 0.278$
   (ii) $0.139 + 0.167 - 0.028 = 0.278$

4.49 a) $P(F) = 0.526$, $P(D) = 0.070$, $P(F \& D)$
$= 0.042$
b) 0.554
c) $1 - 0.526 = 0.474$

4.51 $\frac{1}{2} + \frac{1}{10} - \frac{1}{3} = \frac{8}{30} = 0.267$

Exercises 4.4

4.55 a) 8   b) 3274   c) 863
d) 1471   e) 502

4.57 a) 12   b) 108,674.4 thousand
c) 44,670.4 thousand
d) 10,564.6 thousand   e) 686.0 thousand

4.59 a) The missing entries in the second, third, and
fourth rows are, respectively, 157, 247, and 30.
b) 15   c) 690 thousand   d) 313 thousand
e) 128 thousand   f) 1026 thousand

4.61 a) (i) The institution selected is private.
(ii) The institution selected is in the South.
(iii) The institution selected is a public
school in the West.
b) (i) $P(T_2) = 0.551$—55.1% of institutions of
higher education are private.
(ii) $P(R_3) = 0.316$—31.6% of institutions of
higher education are in the South.
(iii) $P(T_1 \& R_4) = 0.096$—9.6% of institutions of higher education are public
schools in the West.

c) **Type**

		Public T_1	Private T_2	$P(R_i)$
Region	Northeast R_1	0.081	0.170	0.251
	Midwest R_2	0.110	0.154	0.264
	South R_3	0.163	0.153	0.316
	West R_4	0.096	0.074	0.170
	$P(T_j)$	0.449	0.551	1.000

4.63 a) (i) The person selected is employed.
(ii) The person selected has completed between 13 and 15 years of school.
(iii) the person selected is an employed person who has completed between 13 and
15 years of school.
b) (i) $P(E_1) = 0.903$ (ii) $P(S_5) = 0.180$
(iii) $P(E_1 \& S_5) = 0.168$
c) (i) $P(E_1 \text{ or } S_5) = 99450.2/108674.4 = 0.915$
(ii) $P(E_1 \text{ or } S_5) = 0.903 + 0.180 - 0.168$
$= 0.915$

d) **Employment status**

		Employed E_1	Unemployed E_2	$P(S_i)$
Years of school completed	Less than 8 S_1	0.033	0.006	0.038
	8 S_2	0.030	0.005	0.035
	9–11 S_3	0.117	0.026	0.143
	12 S_4	0.369	0.042	0.411
	13–15 S_5	0.168	0.012	0.180
	16 or more S_6	0.186	0.006	0.192
	$P(E_j)$	0.903	0.097	1.000

4.65 a) (i) A_3 (ii) T_2 (iii) $(T_1 \& A_5)$
b) (i) $P(A_3) = 0.241$ (ii) $P(T_2) = 0.288$
(iii) $P(T_1 \& A_5) = 0.015$

c) **Tenure of operator**

		Full owner T_1	Part owner T_2	Tenant T_3	Total
Acreage	Under 50 A_1	21.5	3.0	3.4	27.9
	50–179 A_2	22.7	6.3	3.8	32.9
	180–499 A_3	10.6	10.0	3.5	24.1
	500–999 A_4	2.3	5.2	1.2	8.7
	1000+ A_5	1.5	4.3	0.7	6.5
	Total	58.6	28.8	12.6	100.0

Exercises 4.5

4.69 a) $\dfrac{4}{52} = 0.077$ b) $\dfrac{4}{12} = 0.333$

c) $\dfrac{1}{13} = 0.077$ d) $\dfrac{0}{40} = 0$

e) $\dfrac{12}{52} = 0.231$ f) $\dfrac{4}{4} = 1$

g) $\dfrac{3}{13} = 0.231$ h) $\dfrac{8}{48} = 0.167$

4.71 a) $\dfrac{10}{80} = 0.125$ b) $\dfrac{10}{76} = 0.132$

c) $\dfrac{26}{76} = 0.342$

4.73 a) $\dfrac{821}{3274} = 0.251$ b) $\dfrac{555}{1803} = 0.308$

c) $\dfrac{555}{821} = 0.676$

d) 25.1% of all institutions of higher education are in the Northeast; 30.8% of all private institutions of higher education are in the Northeast; 67.6% of all institutions of higher education in the Northeast are private schools.

4.75 a) $P(T_3) = \dfrac{313}{2476} = 0.126$

b) $P(T_3 \,\&\, A_3) = \dfrac{87}{2476} = 0.035$

c) $P(A_3|T_3) = \dfrac{87}{313} = 0.278$

d) $P(A_3|T_3) = \dfrac{P(T_3 \,\&\, A_3)}{P(T_3)} = \dfrac{0.035}{0.126} = 0.278$

4.77 a) 0.187   b) 0.103   c) $\dfrac{0.103}{0.187} = 0.551$

d) $\dfrac{0.103}{0.416} = 0.248$

e) 18.7% of the members of the 98th Congress are senators; 10.3% of the members of the 98th Congress are Republican senators; 55.1% of the senators in the 98th Congress are Republicans; 24.8% of the Republicans in the 98th Congress are senators.

4.79 $\dfrac{0.035}{0.106} = 0.330$ or 33.0%

Exercises 4.6

4.83 $(0.026)(0.292) = 0.008$—0.8% of all families are farm families making at least \$25,000 per year.

4.85 a) $\dfrac{1}{10} = 0.1$ b) $\dfrac{1}{9} = 0.111$

c) $\dfrac{1}{10} \cdot \dfrac{1}{9} = \dfrac{1}{90} = 0.011$ d) $\dfrac{5}{10} \cdot \dfrac{4}{9} = \dfrac{2}{9} = 0.222$

4.87 a) $\dfrac{16}{50} \cdot \dfrac{34}{49} = 0.222$ b) $\dfrac{16}{50} \cdot \dfrac{15}{49} = 0.098$

c) Let $R1$ denote the event that the first governor selected is a Republican, $R2$ denote the event that the second governor selected is a Republican, $D1$ denote the event that the first governor selected is a Democrat, and $D2$ denote the event that the second governor selected is a Democrat. See Figure A.8.

FIGURE A.8

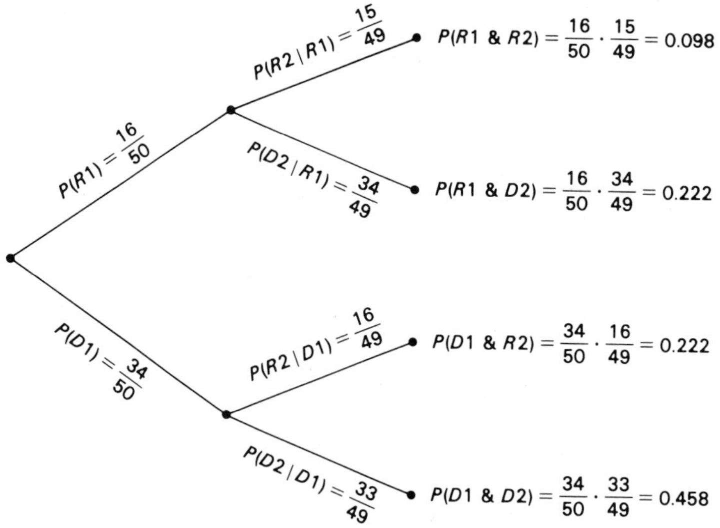

$P(R2|R1) = \dfrac{15}{49}$ $P(R1 \,\&\, R2) = \dfrac{16}{50} \cdot \dfrac{15}{49} = 0.098$

$P(R1) = \dfrac{16}{50}$

$P(D2|R1) = \dfrac{34}{49}$ $P(R1 \,\&\, D2) = \dfrac{16}{50} \cdot \dfrac{34}{49} = 0.222$

$P(D1) = \dfrac{34}{50}$

$P(R2|D1) = \dfrac{16}{49}$ $P(D1 \,\&\, R2) = \dfrac{34}{50} \cdot \dfrac{16}{49} = 0.222$

$P(D2|D1) = \dfrac{33}{49}$ $P(D1 \,\&\, D2) = \dfrac{34}{50} \cdot \dfrac{33}{49} = 0.458$

d) $0.098 + 0.458 = 0.556$

4.89 a) $\frac{9.3}{61.4} = 0.151$ b) $\frac{1.3}{25.8} = 0.050$

c) No, because $P(C_1|S_2) \neq P(C_1)$.
d) No, because $P(S_1) = 0.580$ and $P(S_1|C_2) = 0.458$, so that $P(S_1|C_2) \neq P(S_1)$.

4.91 a) $P(A) = \frac{4}{8} = 0.5$, $P(B) = \frac{4}{8} = 0.5$, $P(C) = \frac{3}{8} =$ 0.375

b) $P(B \mid A) = \frac{2}{4} = 0.5$

c) Yes, since $P(B \mid A) = P(B)$.

d) $P(C \mid A) = \frac{1}{4} = 0.25$

e) No, since $P(C \mid A) \neq P(C)$.

4.93 a) $P(P_1) = 0.584$, $P(C_2) = 0.187$, $P(P_1 \& C_2)$ $= 0.084$
b) $P(P_1)P(C_2) = 0.584 \cdot 0.187 = 0.109$, $P(P_1 \& C_2)$ $= 0.084$. Thus, $P(P_1 \& C_2) \neq P(P_1)P(C_2)$, which means that the events P_1 and C_2 are *not* independent.

4.95 a) $\frac{4}{52} \cdot \frac{4}{52} = 0.006$ b) $\frac{4}{52} \cdot \frac{3}{51} = 0.005$

4.97 a) $\frac{1}{12} = 0.083$ b) 0.5

REVIEW TEST

1 a) $\frac{34,081}{94,427} = 0.361$ b) $\frac{18,339}{94,427} = 0.194$

c)

Adjusted gross income	Number of returns	Event	Probability
Under $10,000	34,081	A	0.361
$10,000–$19,999	24,842	B	0.263
$20,000–$29,999	16,425	C	0.174
$30,000–$39,999	9,863	D	0.104
$40,000–$49,999	4,717	E	0.050
$50,000–$99,999	3,759	F	0.040
$100,000 +	740	G	0.008
Σ	94,427		

2 a) (not J) is the event that the return selected shows an adjusted gross income of at least $100,000. There are 740 thousand such returns.

b) (H & I) is the event that the return selected shows an adjusted gross income of between $20,000 and $49,999. There are 31,005 thousand such returns.
c) (H or K) is the event that the return selected shows an adjusted gross income of at least $20,000. There are 35,504 thousand such returns.
d) (H & K) is the event that the return selected shows an adjusted gross income of between $50,000 and $99,999. There are 3759 thousand such returns.

3 a) not mutually exclusive
b) mutually exclusive
c) mutually exclusive
d) not mutually exclusive

4 a) $P(H) = \frac{34,764}{94,427} = 0.368$, $P(I) = \frac{89,928}{94,427} = 0.952$,
$P(J) = \frac{93,687}{94,427} = 0.992$, $P(K) = \frac{4499}{94,427} = 0.048$

b) $H = (C$ or D or E or $F)$,
$I = (A$ or B or C or D or $E)$,
$J = (A$ or B or C or D or E or $F)$, $K = (F$ or $G)$
c) $P(H) = 0.174 + 0.104 + 0.050 + 0.040 = 0.368$
$P(I) = 0.361 + 0.263 + 0.174 + 0.104 + 0.050 = 0.952$
$P(J) = 0.361 + 0.263 + 0.174 + 0.104 + 0.050 + 0.040 = 0.992$
$P(K) = 0.040 + 0.008 = 0.048$

5 a) $P(\text{not } J) = \frac{740}{94,427} = 0.008$, $P(H \& I) = \frac{31,005}{94,427} = 0.328$,
$P(H$ or $K) = \frac{35,504}{94,427} = 0.376$, $P(H \& K) = \frac{3759}{94,427} = 0.040$
b) $P(J) = 1 - 0.008 = 0.992$

c) $P(H$ or $K) = 0.368 + 0.048 - 0.040 = 0.376$

6 a) 6   b) 13,615 thousand
c) 48,778 thousand   d) 2562 thousand

7 a)   (i) L_3 is the event that the student selected is in college.

(ii) T_1 is the event that the student selected attends a public school.

(iii) (T_1 & L_3) is the event that the student selected attends a public college.

b)   (i) $P(L_3) = \frac{12,174}{56,340} = 0.216$—21.6% of students attend college.

(ii) $P(T_1) = \dfrac{48,778}{56,340} = 0.866$—86.6% of students attend public schools.

(iii) $P(T_1$ & $L_3) = \dfrac{9612}{56,340} = 0.171$—17.1% of students attend public colleges.

c)

	Type		
	Public T_1	Private T_2	$P(L_i)$
Elementary L_1	0.478	0.064	0.542
High school L_2	0.217	0.025	0.242
College L_3	0.171	0.045	0.216
$P(T_j)$	0.866	0.134	1.000

(Level is the row label; Type is the column label.)

d) (i) $P(T_1$ or $L_3) = \dfrac{51,340}{56,340} = 0.911$

(ii) $P(T_1$ or $L_3) = 0.866 + 0.216 - 0.171$
$= 0.911$

8 a) $P(L_3|T_1) = \dfrac{9612}{48,778} = 0.197$—19.7% of students who attend public schools are in college.

b) $P(L_3|T_1) = \dfrac{P(T_1\ \&\ L_3)}{P(T_1)} = \dfrac{0.171}{0.866} = 0.197$

9 a) $P(T_2) = \dfrac{7562}{56,340} = 0.134$,

$P(T_2|L_2) = \dfrac{1400}{13,615} = 0.103$

b) No, because $P(T_2|L_2) \neq P(T_2)$. We see that 10.3% of high school students attend private schools, whereas 13.4% of all students attend private schools.

c) No, because both events can occur; namely, if the student selected is any one of the 1400 thousand students who attend a private high school.

d) We have $P(L_1) = \dfrac{30,551}{56,340} = 0.542$ and $P(L_1|T_1)$
$= \dfrac{26,951}{48,778} = 0.553$. Thus, $P(L_1|T_1) \neq P(L_1)$ and so the event that a student is in elementary school is *not* independent of the event that the student attends public school.

10 a) $\dfrac{3}{50} \cdot \dfrac{19}{49} = 0.023$ b) $\dfrac{28}{50} \cdot \dfrac{27}{49} = 0.309$

c) Let $A1$, $P1$, and $S1$ denote, respectively, the events that the first student selected received a master of arts, a master of public administration, and a

master of science; and let $A2$, $P2$, and $S2$ denote, respectively, the events that the second student selected received a master of arts, a master of public administration, and a master of science. See Figure A.9 on the next page.

d) $0.002 + 0.309 + 0.140 = 0.451$

11   $(0.25)(0.25)(0.25)(0.25) = 0.004$

Chapter 5

Exercises 5.1

5.1 a) 1, 2, 3, 4, 5, 6, and 7   b) $\{x = 5\}$
c) $P(x = 5) = 0.073$—7.3% of U.S. households have exactly five persons.
d) $P(3) = 0.175$

e)

Number of persons x	Probability $P(x)$
1	0.232
2	0.317
3	0.175
4	0.154
5	0.073
6	0.030
7	0.019

5.3 a) 2, 3, 4, 5, 6, 7, 8, 9, 10, 11, and 12
b) $\{y = 7\}$
c) $P(y = 7) = \dfrac{6}{36} = \dfrac{1}{6} = 0.167$
d) $P(11) = \dfrac{2}{36} = \dfrac{1}{18} = 0.056$

e)

Sum of dice y	Probability $P(y)$
2	1/36
3	1/18
4	1/12
5	1/9
6	5/36
7	1/6
8	5/36
9	1/9
10	1/12
11	1/18
12	1/36

FIGURE A.9

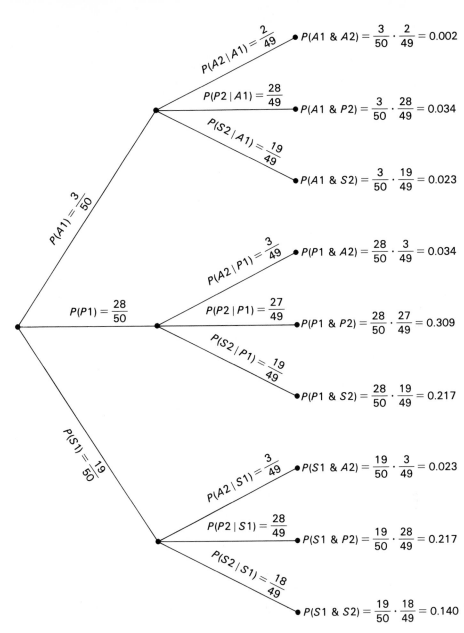

5.5 a) $\{x = 4\}$ b) $\{x \geq 2\}$ c) $\{x < 5\}$
d) $\{2 \leq x < 5\}$ e) 0.21   f) 0.92
g) 0.52   h) 0.44

5.7 a) $\{y = 2\}$ b) $\{y \leq 4\}$ c) $\{y > 1\}$
d) $\{2 \leq y \leq 4\}$ e) 0.134   f) 0.923
g) 0.415   h) 0.338   i) 0.338

5.9 0.975

Exercises 5.2

5.15 a) $\mu_x = 2.685$ b) $\sigma_x = 1.472$

5.17 a) $\mu_y = 7$ b) $\sigma_y = 2.415$

5.19 a) $\mu_x = 4.12$ b) $\sigma_x = 1.59$

5.21 a) $\mu_y = 1.519$ b) $\sigma_y = 1.674$

5.23 a) $P(x=1) = \dfrac{18}{38} = 0.474$,

$P(x=-1) = \dfrac{20}{38} = 0.526$

b) $\mu_x = -0.052$ c) 5.2¢

d) $5.20, $52 e) No

5.25 a) $\mu_w = 0.25$, $\sigma_w = 0.536$ b) 0.25 c) 62.5

Exercises 5.3

5.29 5040, 40,320, 362,880

5.31 a) 10 b) 35 c) 120 d) 792

5.33 a) 10 b) 1 c) 1 d) 126

5.35 a) $p = 0.487$

b)

Outcome	Probability
sss	$(0.487)(0.487)(0.487) = 0.116$
ssf	$(0.487)(0.487)(0.513) = 0.122$
sfs	$(0.487)(0.513)(0.487) = 0.122$
sff	$(0.487)(0.513)(0.513) = 0.128$
fss	$(0.513)(0.487)(0.487) = 0.122$
fsf	$(0.513)(0.487)(0.513) = 0.128$
ffs	$(0.513)(0.513)(0.487) = 0.128$
fff	$(0.513)(0.513)(0.513) = 0.135$

c) See Figure A.10.

d) *ssf, sfs, fss*

e) Each has probability 0.122. Because each probability is obtained by multiplying two success probabilities of 0.487 and one failure probability of 0.513.

5.37 a) $p = 0.2$

FIGURE A.10

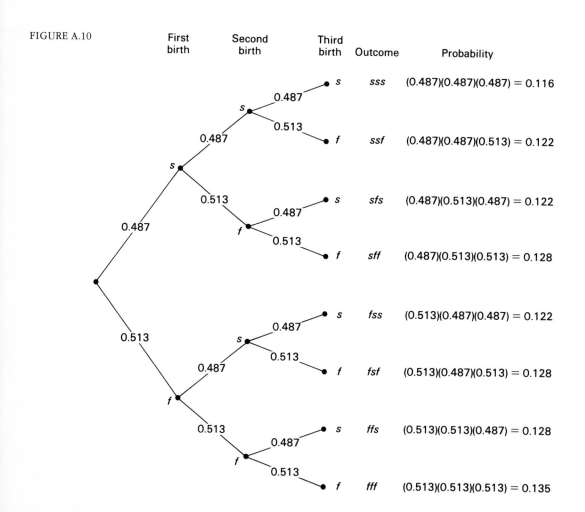

b)

Outcome	Probability
ssss	$(0.2)(0.2)(0.2)(0.2) = 0.0016$
sssf	$(0.2)(0.2)(0.2)(0.8) = 0.0064$
ssfs	$(0.2)(0.2)(0.8)(0.2) = 0.0064$
ssff	$(0.2)(0.2)(0.8)(0.8) = 0.0256$
sfss	$(0.2)(0.8)(0.2)(0.2) = 0.0064$
sfsf	$(0.2)(0.8)(0.2)(0.8) = 0.0256$
sffs	$(0.2)(0.8)(0.8)(0.2) = 0.0256$
sfff	$(0.2)(0.8)(0.8)(0.8) = 0.1024$
fsss	$(0.8)(0.2)(0.2)(0.2) = 0.0064$
fssf	$(0.8)(0.2)(0.2)(0.8) = 0.0256$
fsfs	$(0.8)(0.2)(0.8)(0.2) = 0.0256$
fsff	$(0.8)(0.2)(0.8)(0.8) = 0.1024$
ffss	$(0.8)(0.8)(0.2)(0.2) = 0.0256$
ffsf	$(0.8)(0.8)(0.2)(0.8) = 0.1024$
fffs	$(0.8)(0.8)(0.8)(0.2) = 0.1024$
ffff	$(0.8)(0.8)(0.8)(0.8) = 0.4096$

c) See Figure A.11 on the next page.
d) *sssf, ssfs, sfss, fsss*
e) Each has probability 0.0064. Because each probability is obtained by multiplying three success probabilities of 0.2 and one failure probability of 0.8.

5.39 a) $p = 0.232$

b)

Outcome	Probability
ssss	$(0.232)(0.232)(0.232)(0.232) = 0.003$
sssf	$(0.232)(0.232)(0.232)(0.768) = 0.010$
ssfs	$(0.232)(0.232)(0.768)(0.232) = 0.010$
ssff	$(0.232)(0.232)(0.768)(0.768) = 0.032$
sfss	$(0.232)(0.768)(0.232)(0.232) = 0.010$
sfsf	$(0.232)(0.768)(0.232)(0.768) = 0.032$
sffs	$(0.232)(0.768)(0.768)(0.232) = 0.032$
sfff	$(0.232)(0.768)(0.768)(0.768) = 0.105$
fsss	$(0.768)(0.232)(0.232)(0.232) = 0.010$
fssf	$(0.768)(0.232)(0.232)(0.768) = 0.032$
fsfs	$(0.768)(0.232)(0.768)(0.232) = 0.032$
fsff	$(0.768)(0.232)(0.768)(0.768) = 0.105$
ffss	$(0.768)(0.768)(0.232)(0.232) = 0.032$
ffsf	$(0.768)(0.768)(0.232)(0.768) = 0.105$
fffs	$(0.768)(0.768)(0.768)(0.232) = 0.105$
ffff	$(0.768)(0.768)(0.768)(0.768) = 0.348$

c) *ssff, sfsf, sffs, fssf, fsfs, ffss*
d) Each has probability 0.032. Because each probability is obtained by multiplying two success probabilities of 0.232 and two failure probabilities of 0.768.

Exercises 5.4

5.45 a) $4 \cdot (0.0064) = 0.0256$

b) $\binom{4}{3}(0.2)^3(0.8)^1 = 0.0256$

5.47 a) 0.384   b) 0.519   c) 0.865

d)

Number of girls x	Probability $P(x)$
0	0.135
1	0.384
2	0.365
3	0.116

5.49 a) 0.169   b) 0.958   c) 0.609

d)

Number having a color TV x	Probability $P(x)$
0	0.000
1	0.000
2	0.005
3	0.038
4	0.169
5	0.398
6	0.391

5.51 a) 0.177   b) 0.302   c) 0.996

5.53 a) 0.302   b) 0.738   c) 1

5.55 a) 0.267   b) 0.816   c) 0.451

Exercises 5.5

5.61 a)

x	$P(x)$	$xP(x)$	x^2	$x^2P(x)$
0	0.4096	0.0000	0	0.0000
1	0.4096	0.4096	1	0.4096
2	0.1536	0.3072	4	0.6144
3	0.0256	0.0768	9	0.2304
4	0.0016	0.0064	16	0.0256
Σ		0.8000		1.2800

$\mu_x = \Sigma xP(x) = 0.8$
$\sigma_x = \sqrt{\Sigma x^2 P(x) - \mu_x^2} = \sqrt{1.28 - (0.8)^2} = \sqrt{0.64} = 0.8$
[*Note:* It is only a coincidence that $\mu_x = \sigma_x$.]
b) $\mu_x = np = 4 \cdot (0.2) = 0.8$
$\sigma_x = \sqrt{np(1-p)} = \sqrt{4 \cdot (0.2)(0.8)} = \sqrt{0.64} = 0.8$

5.63 $\mu_x = 5.13$, $\sigma_x = 0.86$

5.65 $\mu_x = 4.404$, $\sigma_x = 1.670$

5.67 $\mu_x = 7.2$, $\sigma_x = 1.2$

5.69 $\mu_x = 1.5$, $\sigma_x = 1.16$

5.71 $\mu_x = 12.5$, $\sigma_x = 3.06$

FIGURE A.11

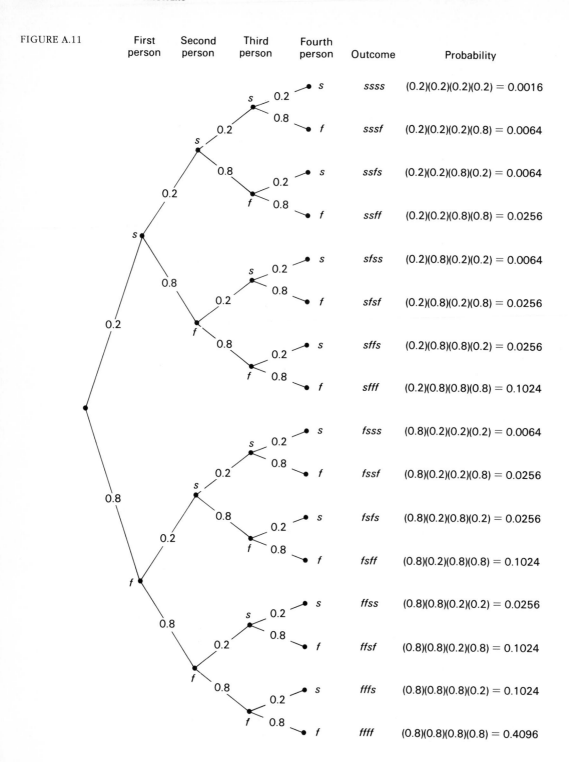

	First person	Second person	Third person	Fourth person	Outcome	Probability
				s	ssss	(0.2)(0.2)(0.2)(0.2) = 0.0016
				f	sssf	(0.2)(0.2)(0.2)(0.8) = 0.0064
				s	ssfs	(0.2)(0.2)(0.8)(0.2) = 0.0064
				f	ssff	(0.2)(0.2)(0.8)(0.8) = 0.0256
				s	sfss	(0.2)(0.8)(0.2)(0.2) = 0.0064
				f	sfsf	(0.2)(0.8)(0.2)(0.8) = 0.0256
				s	sffs	(0.2)(0.8)(0.8)(0.2) = 0.0256
				f	sfff	(0.2)(0.8)(0.8)(0.8) = 0.1024
				s	fsss	(0.8)(0.2)(0.2)(0.2) = 0.0064
				f	fssf	(0.8)(0.2)(0.2)(0.8) = 0.0256
				s	fsfs	(0.8)(0.2)(0.8)(0.2) = 0.0256
				f	fsff	(0.8)(0.2)(0.8)(0.8) = 0.1024
				s	ffss	(0.8)(0.8)(0.2)(0.2) = 0.0256
				f	ffsf	(0.8)(0.8)(0.2)(0.8) = 0.1024
				s	fffs	(0.8)(0.8)(0.8)(0.2) = 0.1024
				f	ffff	(0.8)(0.8)(0.8)(0.8) = 0.4096

REVIEW TEST

1   a) 1, 2, 3, and 4   b) {x = 3}
    c) $P(x = 3) = 0.253$—25.3% of undergraduate students at ASU are juniors.
    d) $P(2) = 0.211$

    e)

Class level x	Probability $P(x)$
1	0.195
2	0.211
3	0.253
4	0.341

2   a) {y = 4}   b) {y ≥ 4}   c) {2 ≤ y ≤ 4}
    d) {y ≥ 1}   e) 0.174   f) 0.322
    g) 0.646   h) 0.948

3   a) $\mu_y = 2.817$ b) 2.817   c) $\sigma_y = 1.504$

4   1, 6, 24, 5040

5   a) 56   b) 56   c) 1
    d) 45   e) 91,390   f) 1

6   a) $p = 0.493$

    b)

Outcome	Probability
sss	$(0.493)(0.493)(0.493) = 0.120$
ssf	$(0.493)(0.493)(0.507) = 0.123$
sfs	$(0.493)(0.507)(0.493) = 0.123$
sff	$(0.493)(0.507)(0.507) = 0.127$
fss	$(0.507)(0.493)(0.493) = 0.123$
fsf	$(0.507)(0.493)(0.507) = 0.127$
ffs	$(0.507)(0.507)(0.493) = 0.127$
fff	$(0.507)(0.507)(0.507) = 0.130$

    c) See Figure A.12.

FIGURE A.12

	First game	Second game	Third game	Outcome	Probability
			s	sss	(0.493)(0.493)(0.493) = 0.120
			f	ssf	(0.493)(0.493)(0.507) = 0.123
			s	sfs	(0.493)(0.507)(0.493) = 0.123
			f	sff	(0.493)(0.507)(0.507) = 0.127
			s	fss	(0.507)(0.493)(0.493) = 0.123
			f	fsf	(0.507)(0.493)(0.507) = 0.127
			s	ffs	(0.507)(0.507)(0.493) = 0.127
			f	fff	(0.507)(0.507)(0.507) = 0.130

d) *ssf, sfs, fss*
e) Each has probability 0.123. Because each probability is obtained by multiplying two success probabilities of 0.493 and one failure probability of 0.507.

7 a) 0.410   b) 0.820

8 a) 0.410   b) 0.820

9 $\mu_x = 3.2$, $\sigma_x = 0.8$

10 Because the trials are not independent and the success probability varies from trial to trial. Because the sample size is small relative to the size of the population.

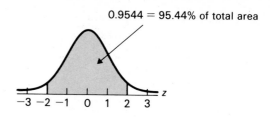

0.9544 = 95.44% of total area

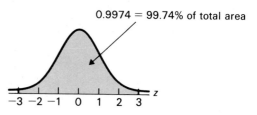

0.9974 = 99.74% of total area

Chapter 6

Exercises **6.1**

6.1 a) 0.3413   b) 0.4772   c) 0.4987
     d) 0.3997   e) 0.4495   f) 0.4750

6.3 a) 0.1772   b) 0.4830   c) 0.3643
     d) 0.4999   e) 0.0199   f) 0.4591

6.5 a) 0.0505   b) 0.0250   c) 0.0099
     d) 0.0049

6.7 a) 0.1359   b) 0.0215   c) 0.1498
     d) 0.0919

6.9 a) 0.9625   b) 0.8264   c) 0.9918
     d) 0.9990

6.11 a) 0.8185   b) 0.9759   c) 0.6789
      d) 0.7638

6.13 a) 0.1587   b) 0.0228   c) 0.1003
      d) 0.0505

6.15 a) 0.9162   b) 0.0645   c) 0.7975

6.17 a) 0.7994   b) 0.8990   c) 0.0500
      d) 0.0198

6.19

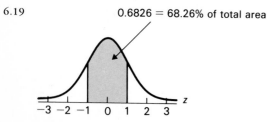

0.6826 = 68.26% of total area

Exercises **6.2**

6.21 The one with parameters $\mu = 1$ and $\sigma = 2$.

6.23 True, because the parameter μ affects only where the normal curve is centered. The shape of the normal curve is determined by the parameter σ.

6.25 a)

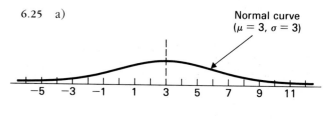

Normal curve
($\mu = 3$, $\sigma = 3$)

b)

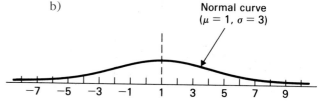

Normal curve
($\mu = 1$, $\sigma = 3$)

c)

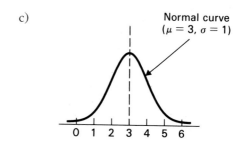

Normal curve
($\mu = 3$, $\sigma = 1$)

6.27   a)

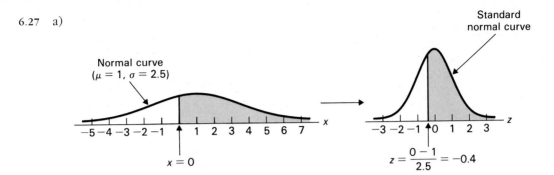

$$\text{Area} = 0.1554 + 0.5000 = 0.6554$$

b)

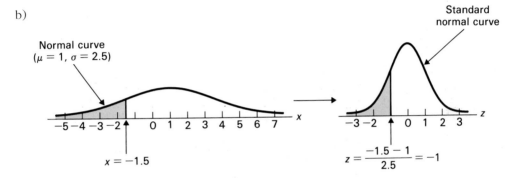

$$\text{Area} = 0.5000 - 0.3413 = 0.1587$$

c)

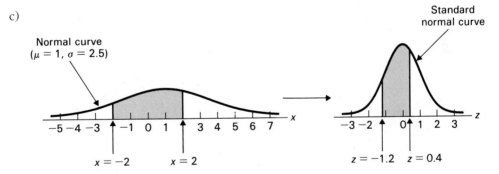

$$\text{Area} = 0.3849 + 0.1554 = 0.5403$$

6.29   a)

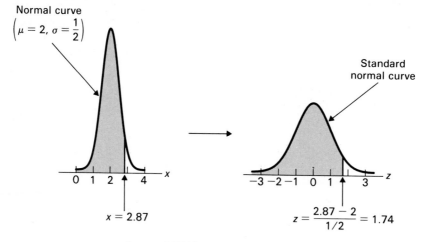

Normal curve
$\left(\mu = 2,\ \sigma = \dfrac{1}{2}\right)$

Standard
normal curve

$x = 2.87$

$z = \dfrac{2.87 - 2}{1/2} = 1.74$

Area $= 0.5000 + 0.4591 = 0.9591$

b)

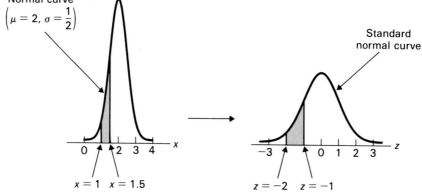

Normal curve
$\left(\mu = 2,\ \sigma = \dfrac{1}{2}\right)$

Standard
normal curve

$x = 1$ $x = 1.5$

$z = -2$ $z = -1$

Area $= 0.4772 - 0.3413 = 0.1359$

c)

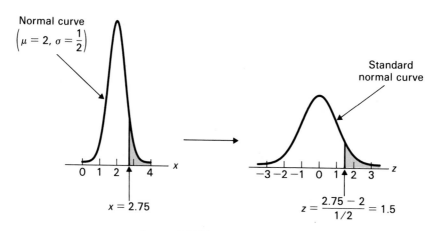

Normal curve
$\left(\mu = 2,\ \sigma = \dfrac{1}{2}\right)$

Standard
normal curve

$x = 2.75$

$z = \dfrac{2.75 - 2}{1/2} = 1.5$

Area $= 0.5000 - 0.4332 = 0.0668$

6.31 a)
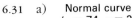
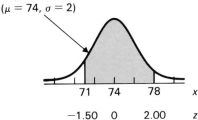

Normal curve
$(\mu = 74, \sigma = 2)$

z-score computations:

Area between
0 and z:

$x = 71 \longrightarrow z = \dfrac{71 - 74}{2} = -1.50$ 0.4332

$x = 78 \longrightarrow z = \dfrac{78 - 74}{2} = 2.00$ 0.4772

Shaded area = 0.4332 + 0.4772 = 0.9104

b)
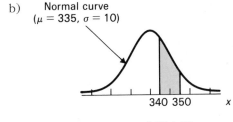

Normal curve
$(\mu = 335, \sigma = 10)$

z-score computations:

Area between
0 and z:

$x = 340 \longrightarrow z = \dfrac{340 - 335}{10} = 0.50$ 0.1915

$x = 350 \longrightarrow z = \dfrac{350 - 335}{10} = 1.50$ 0.4332

Shaded area = 0.4332 − 0.1915 = 0.2417

b)
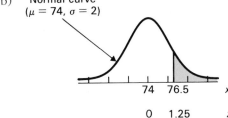

Normal curve
$(\mu = 74, \sigma = 2)$

z-score computation:

Area between
0 and z:

$x = 76.5 \longrightarrow z = \dfrac{76.5 - 74}{2} = 1.25$ 0.3944

Shaded area = 0.5000 − 0.3944 = 0.1056

6.35 a)
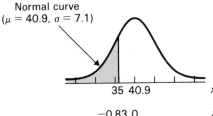

Normal curve
$(\mu = 40.9, \sigma = 7.1)$

z-score computation:

Area between
0 and z:

$x = 35 \longrightarrow z = \dfrac{35 - 40.9}{7.1} = -0.83$ 0.2967

Shaded area = 0.5000 − 0.2967 = 0.2033

6.33 a)
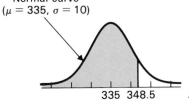

Normal curve
$(\mu = 335, \sigma = 10)$

z-score computation:

Area
between
0 and z:

$x = 348.5 \longrightarrow z = \dfrac{348.5 - 335}{10} = 1.35$ 0.4115

Shaded area = 0.5000 + 0.4115 = 0.9115

b)
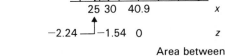

Normal curve
$(\mu = 40.9, \sigma = 7.1)$

z-score computations:

Area between
0 and z:

$x = 25 \longrightarrow z = \dfrac{25 - 40.9}{7.1} = -2.24$ 0.4875

$x = 30 \longrightarrow z = \dfrac{30 - 40.9}{7.1} = -1.54$ 0.4382

Shaded area = 0.4875 − 0.4382 = 0.0493

Exercises **6.3**

6.41   a) Exact percentage = 11.70% (0.1170)
Normal curve area = 12.23% (0.1223)
b) Exact percentage = 39.83% (0.3983)
Normal curve area = 39.14% (0.3914)

6.43   a) 69.43%   b) 97.26%

6.45   a) 13.61%   b) 3.67%

6.47   a) 2   b) −1.75   c) 0   d) See Figure A.13.

6.49   a) 68.26%   b) 95.44%   c) 99.74%
d) 86.64%

6.51   a) $78.20, $112.60   b) $61.00, $129.80
c) $43.80, $147.00   d) See Figure A.14.

6.53   a) 95%   b) 89.9%

Exercises **6.4**

6.55   a) 0.1660   b) 0.1085
c) The probability is 0.1660 that a randomly selected secondary school teacher makes less than $18 thousand per year. The probability is 0.1085 that a randomly selected secondary school teacher makes between $25 thousand and $30 thousand per year.

6.57   a) 0.9633   b) 0.9926
c) 96.33% of the batteries last longer than 20 hours. 99.26% of the batteries last between 15 and 45 hours.

6.59   a) 0.0594   b) 0.2699
c) The probability is 0.0594 that the time of a randomly selected finisher exceeds 75 minutes. The

FIGURE A.13

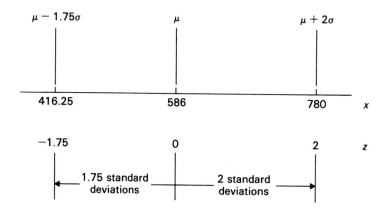

FIGURE A.14

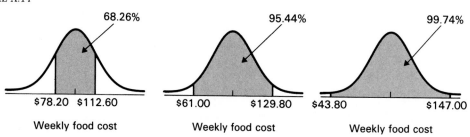

probability is 0.2699 that the time of a randomly selected finisher is either less than 50 minutes or greater than 70 minutes.

6.61 The probability is 0.6826 that x will take a value between 9.9 and 10.1; in other words, about 68.26% of the bolts produced have diameters between 9.9 and 10.1 mm. The probability is 0.9544 that x will take a value between 9.8 and 10.2; in other words, about 95.44% of the bolts produced have diameters between 9.8 and 10.2 mm. The probability is 0.9974 that x will take a value between 9.7 and 10.3; in other words, about 99.74% of the bolts produced have diameters between 9.7 and 10.3 mm.

6.63 The probability is 0.6826 that x will take a value between 24.4 and 35.6; in other words, about 68.26% of the batteries will last between 24.4 and 35.6 hours. The probability is 0.9544 that x will take a value between 18.8 and 41.2; in other words, about 95.44% of the batteries will last between 18.8 and 41.2 hours. The probability is 0.9974 that x will take a value between 13.2 and 46.8; in other words, about 99.74% of the batteries will last between 13.2 and 46.8 hours.

6.65 a) $z = \dfrac{x - 16.3}{17.9}$

b) i) 0.22   ii) 2.70   iii) −0.68

c) No

6.67 a) $z = \dfrac{x - 61}{9}$

b) The number of standard deviations that a randomly selected finisher's time is away from the mean of 61 minutes.

c) The standard normal distribution.

Exercises **6.5**

6.69 a) i) 0.4512   ii) 0.8907
  b) i) 0.4544   ii) 0.8858

6.71 The normal curve with parameters $\mu = 15$ and $\sigma = 2.74$.

6.73 a) 0.0398   b) 0.2835   c) 0.0099

6.75 a) 0.0263   b) 0.8242   c) 0.9991

6.77 a) 0.0233   b) 0.1599   c) 0.0516

REVIEW TEST

1 It is often appropriate to use the normal distribution as the distribution of a population or random variable; and the normal distribution is frequently employed in inferential statistics.

2 a) 0.4932   b) 0.4678   c) 0.2709
  d) 0.0013   e) 0.1305   f) 0.2749
  g) 0.9441   h) 0.9099   i) 0.9104
  j) 0.8426

3 a)

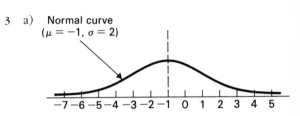

b)

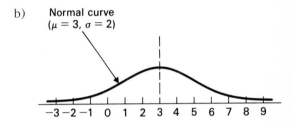

c)

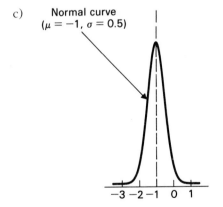

4 a) The second curve.
  b) The first and second curves.
  c) The first and third curves.
  d) The third curve.
  e) The fourth curve.

5 a)   Normal curve
        $(\mu = -1, \sigma = 2.5)$

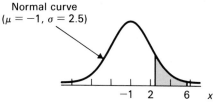

z-score computations: Area between
                             0 and z:

$x = 2 \longrightarrow z = \dfrac{2-(-1)}{2.5} = 1.20$ 0.3849

$x = 6 \longrightarrow z = \dfrac{6-(-1)}{2.5} = 2.80$ 0.4974

Shaded area $= 0.4974 - 0.3849 = 0.1125$

b)   Normal curve
      $(\mu = -1, \sigma = 2.5)$

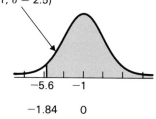

                             Area
                             between
z-score computation: 0 and z:

$x = -5.6 \longrightarrow z = \dfrac{-5.6-(-1)}{2.5} = -1.84$ 0.4671

Shaded area $= 0.4671 + 0.5000 = 0.9671$

c)   Normal curve
      $(\mu = -1, \sigma = 2.5)$

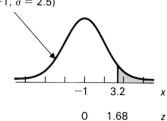

                             Area between
z-score computation: 0 and z:

$x = 3.2 \longrightarrow z = \dfrac{3.2-(-1)}{2.5} = 1.68$ 0.4535

Shaded area $= 0.5000 - 0.4535 = 0.0465$

6 a) 82.76%   b) 89.44%   c) 0.62%

7 a) 1.45   b) -1.8 c) 0

8 a) 400, 600   b) 300, 700   c) 200, 800

9 a) $\mu_x = 339.6$, $\sigma_x = 13.3$
   b) 0.1650   c) 0.0110
   d) The probability is 0.1650 that the gestation period of a randomly selected Morgan horse will be
   between 320 and 330 days. The probability is
   0.0110 that the gestation period of a randomly selected Morgan horse will exceed 370 days.

10 a) 326.3, 352.9   b) 313.0, 366.2
    c) 299.7, 379.5

11 a) $z = \dfrac{x - 339.6}{13.3}$
    b) The number of standard deviations that the ges-
    tation period of a randomly selected Morgan horse
    is away from the mean.
    c) (i) -1.10 (ii) 2.14   (iii) 0
    d) The standard normal distribution.
    e) No. Yes.

12* a) 0.0076   b) 0.9505   c) 0.9988

Chapter 7

Exercises 7.1

7.1 Dentists form a high income group whose incomes are not representative of the incomes of people in general.

7.3 a)

Officials selected	Sample obtained
G, L, S	70, 40, 37
G, L, A	70, 40, 55
G, L, T	70, 40, 50
G, S, A	70, 37, 55
G, S, T	70, 37, 50
G, A, T	70, 55, 50
L, S, A	40, 37, 55
L, S, T	40, 37, 50
L, A, T	40, 55, 50
S, A, T	37, 55, 50

b) $\dfrac{1}{10}, \dfrac{1}{10}, \dfrac{1}{10}$

7.5 a)

Sample
TH, FS, BD, CC
TH, FS, BD, JS
TH, FS, BD, RT
TH, FS, CC, JS
TH, FS, CC, RT
TH, FS, JS, RT
TH, BD, CC, JS
TH, BD, CC, RT
TH, BD, JS, RT
TH, CC, JS, RT
FS, BD, CC, JS
FS, BD, CC, RT
FS, BD, JS, RT
FS, CC, JS, RT
BD, CC, JS, RT

c) $\dfrac{1}{15}, \dfrac{1}{15}$

b) Write the initials of the representatives on six separate pieces of paper, place the six slips of paper into a box, and then, while blindfolded, pick four of the slips of paper.

c) See Figure A.15.

d)

Sample mean $\overline{x}$	Probability $P(\overline{x})$
59.0	0.1
61.0	0.1
62.0	0.1
67.0	0.1
68.0	0.1
70.0	0.2
71.0	0.1
73.0	0.1
79.0	0.1

e) 0.1

f) 0.5. If we take a random sample of two salaries, there is a 50% chance that the mean of the sample selected will be within 4 (that is, $4,000) of the population mean.

Exercises **7.2**

7.7 a) $\mu = 68$ b)

Sample	$\overline{x}$
76, 58	67.0
76, 64	70.0
76, 82	79.0
76, 60	68.0
58, 64	61.0
58, 82	70.0
58, 60	59.0
64, 82	73.0
64, 60	62.0
82, 60	71.0

7.9 b)

Sample	$\overline{x}$
76, 58, 64, 82	70.0
76, 58, 64, 60	64.5
76, 58, 82, 60	69.0
76, 64, 82, 60	70.5
58, 64, 82, 60	66.0

c) See Figure A.16.

d)

Sample mean $\overline{x}$	Probability $P(\overline{x})$
64.5	0.2
66.0	0.2
69.0	0.2
70.0	0.2
70.5	0.2

FIGURE A.15.

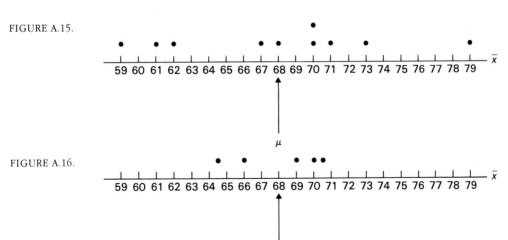

FIGURE A.16.

e) 0

f) 1. If we take a random sample of four salaries, there is a 100% chance (that is, it is certain) that the mean of the sample will be within 4 (that is, $4,000) of the population mean.

7.11 a) $\mu = 15$

b)

Sample	$\overline{x}$
19, 14	16.5
19, 15	17.0
19, 9	14.0
19, 16	17.5
19, 17	18.0
14, 15	14.5
14, 9	11.5
14, 16	15.0
14, 17	15.5
15, 9	12.0
15, 16	15.5
15, 17	16.0
9, 16	12.5
9, 17	13.0
16, 17	16.5

c) See Figure A.17 on the next page.

d)

Sample mean $\overline{x}$	Probability $P(\overline{x})$
11.5	1/15
12.0	1/15
12.5	1/15
13.0	1/15
14.0	1/15
14.5	1/15
15.0	1/15
15.5	2/15
16.0	1/15
16.5	2/15
17.0	1/15
17.5	1/15
18.0	1/15

e) $\frac{1}{15} = 0.067$

f) $\frac{6}{15} = 0.4$. If we take a random sample of two bullfrogs, there is a 40% chance that their mean length will be within 1 cm of the population mean length.

7.13 b)

Sample	$\overline{x}$
19, 14, 15, 9	14.25
19, 14, 15, 16	16.00
19, 14, 15, 17	16.25
19, 14, 9, 16	14.50
19, 14, 9, 17	14.75
19, 14, 16, 17	16.50
19, 15, 9, 16	14.75
19, 15, 9, 17	15.00
19, 15, 16, 17	16.75
19, 9, 16, 17	15.25
14, 15, 9, 16	13.50
14, 15, 9, 17	13.75
14, 15, 16, 17	15.50
14, 9, 16, 17	14.00
15, 9, 16, 17	14.25

c) See Figure A.18 on the next page.

d)

Sample mean $\overline{x}$	Probability $P(\overline{x})$
13.50	1/15
13.75	1/15
14.00	1/15
14.25	2/15
14.50	1/15
14.75	2/15
15.00	1/15
15.25	1/15
15.50	1/15
16.00	1/15
16.25	1/15
16.50	1/15
16.75	1/15

e) $\frac{1}{15} = 0.067$

f) $\frac{10}{15} = 0.667$. If we take a random sample of four bullfrogs, there is a 66.7% chance that their mean length will be within 1 cm of the population mean length.

7.15 b)

Sample	$\overline{x}$
19, 14, 15, 9, 16, 17	15.0

c) See Figure A.19 on the next page.

d)

Sample mean $\overline{x}$	Probability $P(\overline{x})$
15.0	1

FIGURE A.17

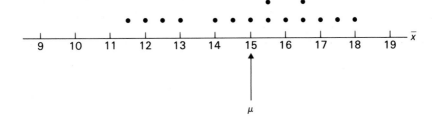

FIGURE A.18

FIGURE A.19

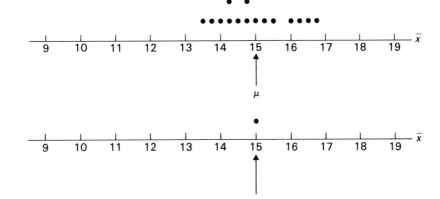

e) 1

f) 1. If we take a random sample of six bullfrogs, there is a 100% chance (that is, it is certain) that their mean length will be within 1 cm of the population mean length.

g) They are the same.

7.17 The larger the sample size, the smaller the sampling error tends to be in estimating μ by $\bar{x}$.

Exercises 7.3

7.21 a) $\mu = 68$

b)

$\bar{x}$	$P(\bar{x})$	$\bar{x}P(\bar{x})$
59.0	0.1	5.9
61.0	0.1	6.1
62.0	0.1	6.2
67.0	0.1	6.7
68.0	0.1	6.8
70.0	0.2	14.0
71.0	0.1	7.1
73.0	0.1	7.3
79.0	0.1	7.9
Σ		68.0

$\mu_{\bar{x}} = \Sigma \bar{x}P(\bar{x}) = 68$

c) $\mu_{\bar{x}} = \mu = 68$

7.23 b)

$\bar{x}$	$P(\bar{x})$	$\bar{x}P(\bar{x})$
64.5	0.2	12.9
66.0	0.2	13.2
69.0	0.2	13.8
70.0	0.2	14.0
70.5	0.2	14.1
Σ		68.0

$\mu_{\bar{x}} = \Sigma \bar{x}P(\bar{x}) = 68$

c) $\mu_{\bar{x}} = \mu = 68$

7.25 a) $\mu_{\bar{x}} = 6.9$, $\sigma_{\bar{x}} = 0.50$

b) $\mu_{\bar{x}} = 6.9$, $\sigma_{\bar{x}} = 0.19$

7.27 a) $\mu_{\bar{x}} = 7.4$, $\sigma_{\bar{x}} = 0.37$

b) $\mu_{\bar{x}} = 7.4$, $\sigma_{\bar{x}} = 0.18$

7.29 Increase the size of the sample.

Exercises 7.4

7.39 a) Normal with parameters μ and $\sigma/\sqrt{n}$.

b) No, since the population being sampled is normally distributed.

c) $\mu_{\bar{x}} = \mu$ and $\sigma_{\bar{x}} = \sigma/\sqrt{n}$

7.41 a) See Figure A.20.

b) Normal with $\mu_{\bar{x}} = 175$ and $\sigma_{\bar{x}} = 9.90$. See Figure A.21.

c) Normal with $\mu_{\bar{x}} = 175$ and $\sigma_{\bar{x}} = 4.67$. See Figure A.22.

7.43 a) 0.7154

b) There is a 71.54% chance that the mean weight, $\bar{x}$, of the nine males obtained will be within five pounds of the true mean weight of $\mu = 175$ lbs.

c) 71.54%   d) 0.95

7.45 a) 0.4972

b) No, because the sample size is at least 30.

c)

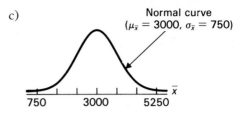

d) 0.9652

e) Because the population standard deviation is so large.

7.47 0.8064

7.49 a) $z = \dfrac{\bar{x} - 100}{5.33}$

FIGURE A.20

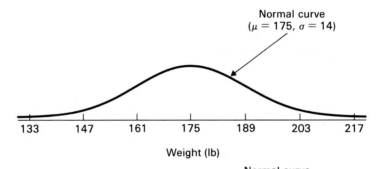

FIGURE A.21

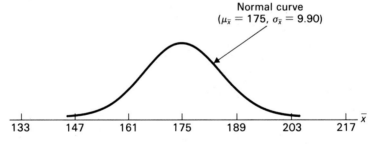

FIGURE A.22

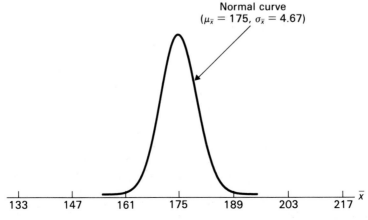

b) standard normal distribution
c) Yes, because the sample size is less than 30.
d) 0.8990

7.51 a) 0.2514   b) 0.0174
c) No, because there is about a 25% chance that a randomly selected bag of water-softener salt will weigh 39 lbs or less, if the company's claim is correct.
d) Yes, because if the company's claim is correct, there is less than a 2% chance that 10 randomly selected bags of water-softener salt will have a mean weight of 39 lbs or less.

REVIEW TEST

1 No, because parents of students at Yale tend to have higher incomes than parents of college students in general.

2 a) not random   b) random

3 a) The error resulting from using the mean income tax, $\bar{x}$, of the 171,700 tax returns sampled as an estimate of the mean income tax, μ, of all 1980 tax returns.
b) $88
c) No, not necessarily. However, increasing the sample size from 171,700 to 250,000 would increase the likelihood for small sampling error.
d) Increase the sample size.

4 a) $\mu_{\bar{x}} = 506$, $\sigma_{\bar{x}} = 47.4$
b) $\mu_{\bar{x}} = 506$, $\sigma_{\bar{x}} = 16.76$
c) Smaller, because $\sigma_{\bar{x}} = \sigma/\sqrt{n}$, and so the larger the sample size, the smaller the value of $\sigma_{\bar{x}}$.

5 a) False. By the central limit theorem, the random variable $\bar{x}$ is approximately normally distributed. Furthermore, $\mu_{\bar{x}} = \mu = 40$ and $\sigma_{\bar{x}} = \sigma/\sqrt{n} = 10/\sqrt{100} = 1$. Thus, $P(30 \le \bar{x} \le 50)$ equals the area under the normal curve with parameters $\mu_{\bar{x}} = 40$ and $\sigma_{\bar{x}} = 1$ that lies between 30 and 50. Applying the usual techniques, we find that area to be 1.0000 to four decimal places. Hence, there is almost a 100% chance that the mean of the sample will be between 30 and 50.
b) Not possible to tell, since we do not know the distribution of the population.
c) True. Referring to part (a), we see that $P(39 \le \bar{x} \le 41)$ equals the area under the normal curve with parameters $\mu_{\bar{x}} = 40$ and $\sigma_{\bar{x}} = 1$ that lies between 39 and 41. Applying the usual techniques, we find that area to be 0.6826. Hence there is about a 68.26% chance that the mean of the sample will be between 39 and 41.

6 a) False. Since the population is normally distributed, so is the random variable $\bar{x}$. Furthermore, $\mu_{\bar{x}} = \mu = 40$ and $\sigma_{\bar{x}} = \sigma/\sqrt{n} = 10/\sqrt{100} = 1$. Hence, as in Problem 5(a), we find that there is almost a 100% chance that the mean of the sample will be between 30 and 50.
b) True. Since the population is normally distributed, percentages for the population are equal to areas under the normal curve with parameters $\mu = 40$ and $\sigma = 10$. Applying the usual techniques, we find that the area under that normal curve between 30 and 50 is 0.6826.
c) True. From part (a), we see that the random variable $\bar{x}$ is normally distributed with $\mu_{\bar{x}} = 40$ and $\sigma_{\bar{x}} = 1$. Hence, as in Problem 5(c), we find that there is about a 68.26% chance that the mean of the sample will be between 39 and 41.

7 a) See Figure A.23.

FIGURE A.23

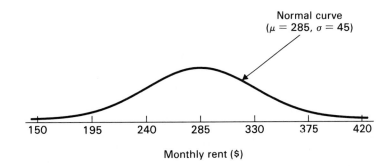

Normal curve
($\mu = 285$, $\sigma = 45$)

150   195   240   285   330   375   420

Monthly rent ($)

b) Normal with $\mu_{\bar{x}} = 285$ and $\sigma_{\bar{x}} = 25.98$. See Figure A.24.

c) Normal with $\mu_{\bar{x}} = 285$ and $\sigma_{\bar{x}} = 15$. See Figure A.25.

8 a) 0.2960

b) There is a 29.6% chance that the mean monthly rent, $\bar{x}$, of the three studio apartments obtained will be within $10 of the population mean monthly rent of $\mu = \$285$.

c) 29.6%   d) 0.9452

9 a) For a normally distributed population, the random variable $\bar{x}$ is normally distributed, regardless of the sample size. Also, we know that $\mu_{\bar{x}} = \mu$. Consequently, since the normal curve for a normally distributed population or random variable is centered at its μ-parameter, all three curves are centered at the same place.

b) Curve B. Since $\sigma_{\bar{x}} = \sigma/\sqrt{n}$, the larger the sample size, the smaller the value of $\sigma_{\bar{x}}$ and, hence, the smaller the spread of the normal curve for $\bar{x}$. Thus, Curve B, which has the smaller spread, corresponds to the larger sample size.

c) Because $\sigma_{\bar{x}} = \sigma/\sqrt{n}$ and the spread of a normal curve is determined by $\sigma_{\bar{x}}$. Thus, different sample sizes result in normal curves with different spreads.

d) Curve B. The smaller the value of $\sigma_{\bar{x}}$, the smaller the sampling error tends to be.

10 a) 0.8530

b) No, because the sample size is large. If the sample size is 20, then we must assume that the population of life insurance amounts is normally distributed in order to answer part (a).

c) 0.9596

11 a) $z = \dfrac{\bar{x} - 419}{6.53}$

b) approximately standard normal

c) 0.0198

d) No, because we do not know the distribution of the population of monthly benefits to retired workers.

12 a) No. Assuming the manufacturer's claim is correct, the probability is 0.1587 that the paint will last 4.5 years or less on a (randomly selected) house

FIGURE A.24

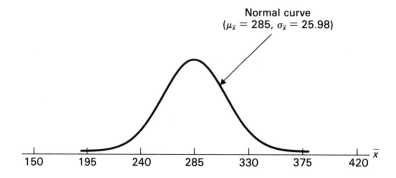

Normal curve
$(\mu_{\bar{x}} = 285,\ \sigma_{\bar{x}} = 25.98)$

150   195   240   285   330   375   420 $\bar{x}$

FIGURE A.25

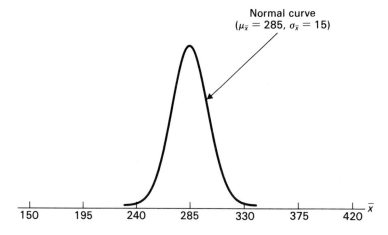

Normal curve
$(\mu_{\bar{x}} = 285,\ \sigma_{\bar{x}} = 15)$

150   195   240   285   330   375   420 $\bar{x}$

painted with the paint. Thus, there is about a 16% chance that the paint would last 4.5 years or less—not substantial evidence against the manufacturer's claim.

b) Yes, because if the manufacturer's claim is correct, there is less than a 0.1% chance that the paint will last an average of 4.5 years or less on 10 (randomly selected) houses painted with the paint.

c) No, because there is about a 26% chance that the paint will last an average of 4.9 years or less on 10 (randomly selected) houses painted with the paint, if the manufacturer's claim is correct.

Chapter 8

Exercises 8.1

8.1 a) 0.8812   b) 6.87   c) 88, 6.37, 7.37
d) We can be 88% confident that the mean birth weight, μ, of all newborns is somewhere between 6.37 lb and 7.37 lb.
e) It may or may not, but we can be 88% confident that it does.

8.3 a) 0.9312   b) 93, 41.84, 51.84
c) We can be 93% confident that the mean retail price, μ, of all science books is somewhere between $41.84 and $51.84.
d) About 93%.

8.5 Because, by the central limit theorem, $\overline{x}$ is approximately normally distributed for large samples (that is, $n \geq 30$), and here $n = 100$.

8.7 A confidence interval for μ provides information on the accuracy of the estimate for μ, whereas a point estimate for μ does not.

Exercises 8.2

8.9 $z_{0.33} = 0.44$

8.11 a) $z_{0.03} = 1.88$ b) $z_{0.005} = 2.575$

8.13 ± 1.645

8.15 a) 617.3 to 668.7 sq. ft.
b) We can be 90% confident that the mean size, μ, of household vegetable gardens in the U.S. is somewhere between 617.3 and 668.7 sq. ft.

8.17 a) $10.85 to $13.17
b) We can be 95% confident that the mean hourly earnings, μ, of all persons employed in the aircraft industry is somewhere between $10.85 and $13.17.

Exercises 8.3

8.19 a) Confidence level = 0.90; $\alpha = 0.10$.
b) Confidence level = 0.99; $\alpha = 0.01$.

8.21 a) 617.3 to 668.7 sq. ft.
b) We can be 90% confident that the mean size, μ, of household vegetable gardens in the U.S. is somewhere between 617.3 and 668.7 sq. ft.

8.23 a) $10.85 to $13.17
b) We can be 95% confident that the mean hourly earnings, μ, of all persons employed in the aircraft industry is somewhere between $10.85 and $13.17.

8.25 a) 133.6 to 140.2 lb
b) We can be 90% confident that the mean weight, μ, of all U.S. females 5 ft 4 inches tall and in the age group 18–24 is somewhere between 133.6 and 140.2 lb.

8.27 a) 131.7 to 142.1 lb
b) The confidence level here is greater than that in Exercise 8.25.
c)

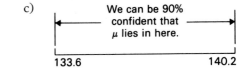

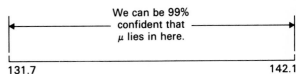

8.29 a) 3.1 to 3.3 persons
b) We can be 95% confident that the mean size, μ, of all U.S. families is somewhere between 3.1 and 3.3 persons.

Exercises 8.4

8.31 a) 25.7 sq. ft.
b) We can be 90% confident that the maximum error made in using $\overline{x}$ to estimate μ is 25.7 sq. ft.
c) 355

8.33 a) $1.16

b) We can be 95% confident that the maximum error made in using $\bar{x}$ to estimate μ is $1.16.

c) 163

d) $12.37 to $13.37

8.35 a) 3.3 lb

b) 271

c) Because σ is unknown.

d) 132.2 to 136.2 lb

Exercises **8.5**

8.39 a) standard normal distribution

b) t-distribution with df $= n - 1$

c) Probabilities for that random variable are equal to areas under the standard normal curve.

d) Probabilities for that random variable are equal to areas under the t-curve with df $= n - 1$.

8.41 a) 1.323

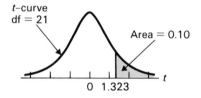

b) 2.518

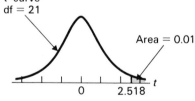

c) -2.080

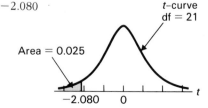

d) ± 1.721

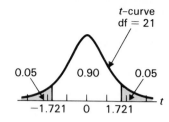

8.43 a) $914.98 to $1136.24

b) Yes, because the confidence interval does not contain, and is to the right of, the 1984 figure of $852.31.

8.45 a) 58.9 to 64.1 bushels/acre

b) The one-acre yields using the new fertilizer are normally distributed.

c) Yes, because the confidence interval does not contain, and is to the right of, the national average of 58.4 bushels per acre.

Exercises **8.6**

8.49 A population proportion, p, is a parameter, since it is a descriptive measure for a population. A sample proportion, $\bar{p}$, is a statistic, since it is a descriptive measure for a sample.

8.51 a) 0.626 to 0.674

b) We can be 95% confident that the percentage of Americans who drink beer, wine, or hard liquor, at least occasionally, is somewhere between 62.6% and 67.4%.

8.53 a) 0.784 to 0.836

b) We can be 99% confident that the percentage of adult Americans who are in favor of "right to die" laws is somewhere between 78.4% and 83.6%.

8.55 a) 0.732 to 0.828

b) We can be 90% confident that the percentage of all real-estate experts who believe next year will be a good time to buy real estate is somewhere between 73.2% and 82.8%.

8.57 Not very well! I applied Procedure 8.3 without checking the assumptions for its use—namely, that the number of successes, x, and the number of failures, $n - x$, are both at least 5. Since the number of successes here is only $(0.008) \cdot 500 = 4$, I should not have used Procedure 8.3.

REVIEW TEST

1   a) 0.9198   b) 92, 16,642, 18,642
    c) We can be 92% confident that the mean household income, μ, of all households of Spanish origin is somewhere between $16,642 and $18,642.
    d) It may or may not, but we can be 92% confident that it does.
    e) About 92%.

2   a) 1.53

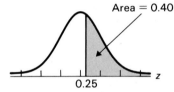

    b) 0.25

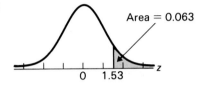

    c) 2.455

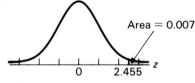

3   54.1 to 62.9 years. We can be 95% confident that the mean age, μ, of all U.S. millionaires is somewhere between 54.1 and 62.9 years old.

4   Only (c) provides a correct interpretation.

5   a) 58.5 to 61.7 hours
    b) We can be 99% confident that the mean battery life, μ, is somewhere between 58.5 and 61.7 hours.

6   a) 1.6 hours
    b) We can be 99% confident that the maximum error made in using $\overline{x}$ to estimate μ is 1.6 hours.
    c) 491
    d) 59.3 to 60.3 hours

7   The one that looks more like the standard normal curve because, as the number of degrees of freedom gets larger, t-curves look increasingly like the standard normal curve.

8   a) Normal distribution with mean $\mu_{\overline{x}} = \mu$ and standard deviation $\sigma_{\overline{x}} = \sigma/\sqrt{n}$.
    b) 0.9544
    c) Probabilities for that random variable are equal to areas under the t-curve with df $= n - 1$.

9   a) 2.101 t-curve

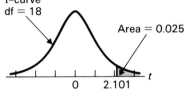

    b) 1.734 t-curve

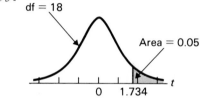

    c) -1.330

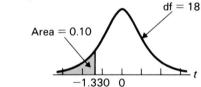

    d) ± 2.878

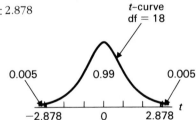

10 a) 6.76 to 8.56 hours
    b) We can be 95% confident that the mean daily viewing time, μ, of all American households is somewhere between 6.76 and 8.56 hours.
    c) No, because the 1983 average of seven hours and two minutes does not lie completely to the left of the confidence interval in part (a).

11 a) 0.491 to 0.591
    b) We can be 95% confident that the percentage of

all registered voters watching the debate who thought Mondale did a better job is somewhere between 49.1% and 59.1%.

c) No, because the confidence interval does not lie completely to the right of 0.5.

Chapter 9

Exercises 9.1

9.1 A hypothesis is a statement that something is true.

9.3 a) 0.2981

b) No, because if $\mu = 1000$ ml, then there is about a 30% chance of obtaining a value for $\bar{x}$ of 996.8 ml or less.

c) 0.0010. Yes, because if $\mu = 1000$ ml, then there is only a 0.1% chance of obtaining a value for $\bar{x}$ of 981.2 ml or less.

9.5 a) 0.0018

b) Yes, because if $\mu = 8.1$ years, then there is only a 0.18% chance of obtaining a value for $\bar{x}$ of 10.6 years or greater.

c) 0.3632. No, because if $\mu = 8.1$ years, then there is about a 36% chance of obtaining a value for $\bar{x}$ of 8.4 years or greater.

9.7 a) 0.3669

b) No, because if $\mu = 106.5$ lb, then there is about a 37% chance of obtaining a value for $\bar{x}$ of 107.3 lb or greater.

c) 0.0099. Yes, because if $\mu = 106.5$ lb, then there is only a 0.99% chance of obtaining a value for $\bar{x}$ of 101.0 lb or less.

9.9 a) $c = 990.0$ ml

b) If $\bar{x} < 990.0$ ml, reject the null hypothesis in favor of the alternative hypothesis. If $\bar{x} \geq 990.0$ ml, do not reject the null hypothesis.

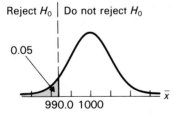

Reject H_0 | Do not reject H_0

0.05

990.0 1000

c) $\bar{x} = 992.6$ ml. Do not reject the null hypothesis.

d) The contents of the 30 bottles selected do *not*

provide sufficient evidence to conclude that the mean content, μ, of all bottles of soda sold by the bottler is less than 1000 ml.

9.11 a) $c = 9.2$ years

b) If $\bar{x} > 9.2$ years, reject the null hypothesis in favor of the alternative hypothesis. If $\bar{x} \leq 9.2$ years, do not reject the null hypothesis.

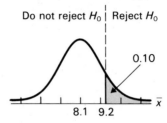

Do not reject H_0 | Reject H_0

0.10

8.1 9.2

c) $\bar{x} = 9.6$ years. Reject the null hypothesis in favor of the alternative hypothesis.

d) The data indicate that this year's mean age for trucks in use exceeds the 1983 mean of 8.1 years.

Exercises 9.2

9.15 STEP 1 $H_0: \mu = \$46.67$, $H_a: \mu > \$46.67$

STEP 2 $\alpha = 0.10$

STEP 3   If $\bar{x} > \$50.24$, reject H_0 in favor of H_a. If $\bar{x} \leq \$50.24$, do not reject H_0.

STEP 4 $\bar{x} = \$58.31$. Reject H_0 in favor of H_a.

STEP 5   The data indicate that this year's mean room rate, μ, for full-service lodging establishments has increased over the 1983 mean of $46.67.

9.17 STEP 1 $H_0: \mu = \$3603$, $H_a: \mu < \$3603$

STEP 2 $\alpha = 0.05$

STEP 3   If $\bar{x} < \$2581.86$, reject H_0 in favor of H_a. If $\bar{x} \geq \$2581.86$, do not reject H_0.

STEP 4 $\bar{x} = \$2239.83$. Reject H_0 in favor of H_a.

STEP 5   The data suggest that the average income tax per taxable return, μ, for last year has decreased from the 1982 mean of $3603.

9.19 STEP 1 $H_0: \mu = 21$ days, $H_a: \mu \neq 21$ days

STEP 2 $\alpha = 0.10$

STEP 3   If $\bar{x} < 16.6$ days or $\bar{x} > 25.4$ days, reject H_0 in favor of H_a. If 16.6 days $\leq \bar{x} \leq 25.4$ days, do not reject H_0.

STEP 4 $\bar{x} = 17.6$ days. Do not reject H_0.

STEP 5   The data do not provide sufficient evidence to indicate that the mean stay, μ, by this year's travelers differs from the 1980 mean of 21 days.

Exercises 9.3

9.21 a) The test statistic is the mean room rate, $\bar{x}$, for the sample of 35 full-service lodging establishments.

b) The rejection region consists of all $\bar{x}$-values to the right of 50.24.

c) The nonrejection region consists of all $\bar{x}$-values to the left of, and including, 50.24.

d) The critical value is 50.24.

e)

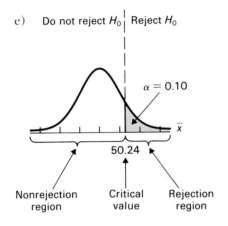

9.23 a) The test statistic is the mean income tax, $\bar{x}$, of the sample of 30 returns.

b) The rejection region consists of all $\bar{x}$-values to the left of 2581.86.

c) The nonrejection region consists of all $\bar{x}$-values to the right of, and including, 2581.86.

d) The critical value is 2581.86.

e)

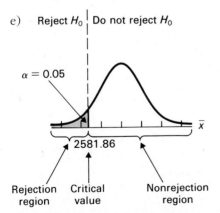

9.25 a) The test statistic is the mean stay, $\bar{x}$, in Europe and the Mediterranean by the sample of 36 U.S. residents who have traveled there this year.

b) The rejection region consists of all $\bar{x}$-values that lie either to the left of 16.6 or to the right of 25.4.

c) The nonrejection region consists of all $\bar{x}$-values between 16.6 and 25.4, inclusive.

d) The critical values are 16.6 and 25.4.

e)

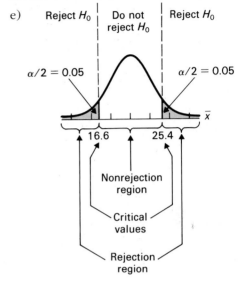

9.27 a) A Type I error would occur if this year's mean room rate for full-service lodging establishments has not increased from the 1983 mean of $46.67, but the results of the sampling lead to the conclusion that it has.

b) A Type II error would occur if this year's mean room rate for full-service lodging establishments has increased over the 1983 mean of $46.67, but the sampling results fail to lead to that conclusion.

c) A correct decision would occur if *either* this year's mean room rate for full-service lodging establishments has not increased from the 1983 mean of $46.67 and the results of the sampling do not lead to the rejection of that fact *or* this year's mean room rate for full-service lodging establishments has increased from the 1983 mean of $46.67 and the results of the sampling lead to rejection of the null hypothesis.

d) Type I error.

e) Correct decision.

9.29 a) A Type I error would occur if this year's mean stay in Europe and the Mediterranean by U.S. travelers is the same as the 1980 mean of 21 days, but the results of the sampling lead to the conclusion that it is not.

b) A Type II error would occur if this year's mean stay in Europe and the Mediterranean by U.S. travelers differs from the 1980 mean of 21 days, but the sampling results fail to lead to that conclusion.

c) A correct decision would occur if *either* this year's mean stay in Europe and the Mediterranean by U.S. travelers is the same as the 1980 mean of 21 days and the results of the sampling do not lead to the rejection of that fact *or* this year's mean stay differs from the 1980 mean of 21 days and the results of the sampling lead to rejection of the null hypothesis.

d) Correct decision.   e) Type II error.

9.31   a) right-tailed   b) left-tailed   c) two-tailed

9.33   True, because the significance level is equal to the probability of making a Type I error.

Exercises **9.4**

9.39   STEP 1 H_0: $\mu = \$280$, H_a: $\mu > \$280$
STEP 2 $\alpha = 0.05$
STEP 3   Critical value $= 1.645$
STEP 4 $z = 0.28$ STEP 5   Do not reject H_0
STEP 6   The data do not provide sufficient evidence to indicate that the mean amount, μ, spent in 1986 for maintenance and repairs has increased over the 1983 mean of $280.

9.41   STEP 1 H_0: $\mu = 18$ mg, H_a: $\mu < 18$ mg
STEP 2 $\alpha = 0.01$
STEP 3   Critical value $= -2.33$
STEP 4 $z = -7.22$ STEP 5   Reject H_0
STEP 6   It appears that adult females under the age of 51 are, on the average, getting less than the RDA of 18 mg of iron.

9.43   a) STEP 1 H_0: $\mu = 50$ lb, H_a: $\mu \neq 50$ lb
STEP 2 $\alpha = 0.05$
STEP 3   Critical values $= \pm 1.96$
STEP 4 $z = 1.13$ STEP 5   Do not reject H_0
STEP 6   The data do not provide sufficient evidence to conclude that the mean weight, μ, of all bags of this dog food differs from the advertised weight of 50 lb.

b) STEPS 1–3 are the same as in part (a)
STEP 4 $z = 2.17$ STEP 5   Reject H_0
STEP 6   The data indicate that the mean weight, μ, of all bags of this dog food differs from the advertised weight of 50 lb.

Exercises **9.5**

9.47   STEP 1 H_0: $\mu = 120$ min., H_a: $\mu > 120$ min.
STEP 2 $\alpha = 0.05$
STEP 3   Critical value $= 1.729$
STEP 4 $t = 1.386$ STEP 5   Do not reject H_0
STEP 6   The data do not provide sufficient evidence to conclude that the mean drying time, μ, is greater than the manufacturer's claim of 120 minutes.

9.49   a) STEP 1 H_0: $\mu = 36$ mo., H_a: $\mu < 36$ mo.
STEP 2 $\alpha = 0.01$
STEP 3   Critical value $= -2.821$
STEP 4 $t = -4.471$ STEP 5   Reject H_0
STEP 6   It appears that the mean life, μ, of the supplier's batteries is less than the claimed 36 months.

b) That battery life is approximately normally distributed.

9.51   STEP 1 H_0: $\mu = 38.5 ¢/\text{lb}$, H_a: $\mu \neq 38.5 ¢/\text{lb}$
STEP 2 $\alpha = 0.05$
STEP 3   Critical values $= \pm 2.145$
STEP 4 $t = 2.622$ STEP 5   Reject H_0
STEP 6   It appears that the current mean retail price, μ, for oranges is different from the 1983 mean of 38.5 cents per pound.

Exercises **9.6**

9.55 $np_0 = 223.3$, $n(1 - p_0) = 126.7$
STEP 1 H_0: $p = 0.638$, H_a: $p \neq 0.638$
STEP 2 $\alpha = 0.05$
STEP 3   Critical values $= \pm 1.96$
STEP 4 $z = -2.15$ STEP 5   Reject H_0
STEP 6   The data indicate that the proportion, p, of voters who think that repair of city streets is an important priority has changed from the 1984 figure of 0.638.

9.57 $np_0 = 15.25$, $n(1 - p_0) = 109.75$
STEP 1 H_0: $p = 0.122$, H_a: $p > 0.122$
STEP 2 $\alpha = 0.10$
STEP 3   Critical value $= 1.28$
STEP 4 $z = 1.02$ STEP 5   Do not reject H_0
STEP 6   The data do not provide sufficient evidence to conclude that the proportion, p, of female physicians is higher now than in 1981.

9.59 $np_0 = 52$, $n(1 - p_0) = 748$
STEP 1 $H_0: p = 0.065$, $H_a: p > 0.065$
STEP 2 $\alpha = 0.01$
STEP 3   Critical value $= 2.33$
STEP 4 $z = 2.58$ STEP 5   Reject H_0
STEP 6   It appears that the actual response rate, p, will exceed 6.5%, so that the decision will be for additional advertising and promotion.

9.61 $np_0 = 30.75$, $n(1 - p_0) = 219.25$
STEP 1 $H_0: p = 0.123$, $H_a: p < 0.123$
STEP 2 $\alpha = 0.05$
STEP 3   Critical value $= -1.645$
STEP 4 $z = -0.53$ STEP 5   Do not reject H_0
STEP 6   The study does not provide sufficient evidence to conclude that in 1983 the proportion, p, of northeastern families below the poverty level was less than the national proportion of 0.123.

REVIEW TEST

1 a) 0.0031
b) Yes, because if $\mu = 20.6$ lb, then there is only a 0.31% chance of obtaining a value for $\bar{x}$ of 23.8 lb or greater.
c) 0.2743. No, because if $\mu = 20.6$ lb, then there is about a 27% chance of obtaining a value for $\bar{x}$ of 21.3 lb or greater.

2 a) $c = 22.5$ lb
b) If $\bar{x} > 22.5$ lb, reject the null hypothesis in favor of the alternative hypothesis. If $\bar{x} \leq 22.5$ lb, do not reject the null hypothesis.

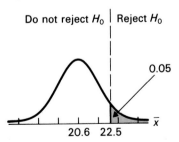

c) $\bar{x} = 21.8$ lb. Do not reject the null hypothesis.
d) The data do not provide sufficient evidence to conclude that last year's mean cheese consumption, μ, is greater than the 1983 mean of 20.6 lb.

3 a) The test statistic is last year's mean cheese consumption, $\bar{x}$, for the 35 randomly selected people.
b) The rejection region consists of all $\bar{x}$-values to the right of 22.5.
c) The nonrejection region consists of all $\bar{x}$-values to the left of, and including, 22.5.
d) The critical value is 22.5.

4 a) A Type I error would occur if last year's mean cheese consumption has not increased over the 1983 mean of 20.6 lb, but the results of the sampling lead to the conclusion that it has.
b) A Type II error would occur if last year's mean cheese consumption has increased over the 1983 mean of 20.6 lb, but the results of the sampling fail to lead to that conclusion.
c) A correct decision would occur if *either* last year's mean cheese consumption has not increased over the 1983 mean of 20.6 lb and the results of the sampling do not lead to the rejection of that fact *or* last year's mean cheese consumption has increased over the 1983 mean of 20.6 lb and the results of the sampling lead to rejection of the null hypothesis.
d) Correct decision
e) Type II error
f) right-tailed

5 $np_0 = 764$, $n(1 - p_0) = 764$
STEP 1 $H_0: p = 0.5$, $H_a: p > 0.5$
STEP 2 $\alpha = 0.01$
STEP 3   Critical value $= 2.33$
STEP 4 $z = 3.89$ STEP 5   Reject H_0
STEP 6   It appears that a majority of all Americans were satisfied with President Reagan's performance.

6 STEP 1 $H_0: \mu = 29$ mpg, $H_a: \mu \neq 29$ mpg
STEP 2 $\alpha = 0.05$
STEP 3   Critical values $= \pm 2.145$
STEP 4 $t = -0.599$ STEP 5   Do not reject H_0
STEP 6   The data do not provide sufficient evidence to indicate that the company's report was incorrect.

7 STEP 1 $H_0: \mu = \$178$, $H_a: \mu < \$178$
STEP 2 $\alpha = 0.05$
STEP 3   Critical value $= -1.645$
STEP 4 $z = -1.43$ STEP 5   Do not reject H_0
STEP 6   The data do not provide sufficient evidence to conclude that the mean value lost due to purse snatching has decreased from the 1983 mean of $178.

Chapter 10

Exercises 10.1

10.1 a) $\mu_1, \sigma_1, \mu_2,$ and σ_2 are parameters; $\bar{x}_1, s_1, \bar{x}_2,$ and s_2 are statistics.
b) $\mu_1, \sigma_1, \mu_2,$ and σ_2 are fixed numbers; $\bar{x}_1, s_1, \bar{x}_2,$ and s_2 are random variables.

10.3 STEP 1 $H_0: \mu_1 = \mu_2, H_a: \mu_1 > \mu_2$
STEP 2 $\alpha = 0.10$
STEP 3   Critical value = 1.28
STEP 4 $z = 1.19$ STEP 5   Do not reject H_0
STEP 6   The data do not provide sufficient evidence to conclude that, on the average, males stay in the hospital longer than females.

10.5 STEP 1 $H_0: \mu_1 = \mu_2, H_a: \mu_1 \neq \mu_2$
STEP 2 $\alpha = 0.05$
STEP 3   Critical values = ± 1.96
STEP 4 $z = 6.32$ STEP 5   Reject H_0
STEP 6   It appears that last year's mean annual fuel expenditure for households using natural gas is different from that for households using only electricity.

10.7 a) STEP 1 $H_0: \mu_1 = \mu_2, H_a: \mu_1 < \mu_2$
STEP 2 $\alpha = 0.05$
STEP 3   Critical value = -1.645
STEP 4 $z = -1.52$ STEP 5   Do not reject H_0
STEP 6   At the 5% significance level, the data do not provide sufficient evidence to conclude that the training program increases sales.

b) STEP 1 $H_0: \mu_1 = \mu_2, H_a: \mu_1 < \mu_2$
STEP 2 $\alpha = 0.10$
STEP 3   Critical value = -1.28
STEP 4 $z = -1.52$ STEP 5   Reject H_0
STEP 6   At the 10% significance level, the data do provide sufficient evidence to conclude that the training program increases sales.

10.9 a) -0.2 to 4.1 days
b) We can be 80% confident that the difference, $\mu_1 - \mu_2$, between the mean lengths of stay in short-term hospitals by males and females is somewhere between -0.2 and 4.1 days.

10.11 a) $175.20 to $332.72
b) We can be 95% confident that the difference, $\mu_1 - \mu_2$, between last year's mean fuel expenditures for households using natural gas and those using

only electricity is somewhere between $175.20 and $332.72.

10.13 a) $-$66.73 to $2.73
b) $-$59.03 to $-$4.97

Exercises 10.2

10.17 STEP 1 $H_0: \mu_1 = \mu_2, H_a: \mu_1 \neq \mu_2$
STEP 2 $\alpha = 0.05$
STEP 3   Critical values = ± 2.101
STEP 4 $t = -3.181$ STEP 5   Reject H_0
STEP 6   There appears to be a difference in mean lasting times between the two paints.

10.19 STEP 1 $H_0: \mu_1 = \mu_2, H_a: \mu_1 < \mu_2$
STEP 2 $\alpha = 0.05$
STEP 3   Critical value = -1.699
STEP 4 $t = -0.587$ STEP 5   Do not reject H_0
STEP 6   The data do not provide sufficient evidence to conclude that mine workers earn less on the average than construction workers.

10.21 a) -2.92 to -0.60 months
b) We can be 95% confident that the difference, $\mu_1 - \mu_2$, between the mean lasting times of the two paints is somewhere between -2.92 and -0.60 months.

10.23 a) $-$1.91 to $0.93
b) We can be 90% confident that the difference, $\mu_1 - \mu_2$, between the mean hourly earnings of non-supervisory mine and construction workers is somewhere between $-$1.91 and $0.93.

10.25 STEP 1 $H_0: \mu_1 = \mu_2, H_a: \mu_1 \neq \mu_2$
STEP 2 $\alpha = 0.05$
STEP 3   Critical values = ± 2.052
STEP 4 $t = -0.955$ STEP 5   Do not reject H_0
STEP 6   The data do not provide sufficient evidence to conclude that there is a difference in mean GPA for sophomores and juniors.

10.27 a) STEP 1 $H_0: \mu_1 = \mu_2, H_a: \mu_1 < \mu_2$
STEP 2 $\alpha = 0.01$
STEP 3   Critical value = -2.518
STEP 4 $t = -4.104$ STEP 5   Reject H_0
STEP 6   The data indicate that Mirror-Sheen has a longer effectiveness time, on the average, than Sureglow.
b) Yes, because the hypothesis test was performed at the 0.01 significance level and the null hypothesis was rejected—there is only a 1% chance that

we would conclude that Mirror-Sheen outlasts Sureglow when in fact it does not.

10.29 a) -0.44 to 0.16
b) We can be 95% confident that the difference, $\mu_1 - \mu_2$, between the mean GPAs of sophomores and juniors at the university is somewhere between -0.44 and 0.16.

10.31 a) -3.8 to -0.9 days
b) We can be 98% confident that the difference, $\mu_1 - \mu_2$, between the mean effectiveness times of Sureglow and Mirror-Sheen is somewhere between -3.8 and -0.9 days.

Exercises **10.3**

10.37 STEP 1 $H_0: \mu_1 = \mu_2, H_a: \mu_1 > \mu_2$
STEP 2 $\alpha = 0.01$
STEP 3 Critical value $= 2.624$
STEP 4 Paired differences, $d = x_1 - x_2$:

1	-3	4	0	7
-1	2	6	3	3
0	2	4	9	-2

STEP 5 $t = 2.696$ STEP 6 Reject H_0
STEP 7 It appears that the running program will, on the average, reduce heart rates.

10.39 STEP 1 $H_0: \mu_1 = \mu_2, H_a: \mu_1 < \mu_2$
STEP 2 $\alpha = 0.05$
STEP 3 Critical value $= -1.761$
STEP 4 Paired differences, $d = x_1 - x_2$, are given.
STEP 5 $t = -1.420$ STEP 6 Do not reject H_0
STEP 7 The data do not provide sufficient evidence to conclude that nonsupervisory mine workers have a smaller average hourly wage than nonsupervisory construction workers.

10.41 STEP 1 $H_0: \mu_1 = \mu_2, H_a: \mu_1 > \mu_2$
STEP 2 $\alpha = 0.05$
STEP 3 Critical value $= 1.833$
STEP 4 Paired differences, $d = x_1 - x_2$:

1	-1	6	1	13
-1	4	-2	4	8

STEP 5 $t = 2.213$ STEP 6 Reject H_0
STEP 7 The data indicate that the mean age of married males is greater than the mean age of married females.

10.43 a) 0.1 to 4.6
b) We can be 98% confident that the difference,

$\mu_1 - \mu_2$, in mean heart rates of people before and after the running program is somewhere between 0.1 and 4.6.

10.45 a) $-\$1.02$ to $\$0.11$
b) We can be 90% confident that the difference, $\mu_1 - \mu_2$, between the mean hourly earnings of nonsupervisory mine and construction workers is somewhere between $-\$1.02$ and $\$0.11$.

10.47 a) 0.6 to 6.0 years
b) We can be 90% confident that the difference, $\mu_1 - \mu_2$, between the mean ages of married males and females is somewhere between 0.6 and 6.0 years.

Exercises **10.4**

10.53 a) p_1 and p_2 are parameters and the other quantities are statistics.
b) p_1 and p_2 are fixed numbers and the other quantities are random variables.

10.55 STEP 1 $H_0: p_1 = p_2, H_a: p_1 \neq p_2$
STEP 2 $\alpha = 0.01$
STEP 3 Critical values $= \pm 2.575$
STEP 4 $z = -2.62$ STEP 5 Reject H_0
STEP 6 It appears that there is a difference in the percentages of American and French households that own washing machines.

10.57 STEP 1 $H_0: p_1 = p_2, H_a: p_1 > p_2$
STEP 2 $\alpha = 0.05$
STEP 3 Critical value $= 1.645$
STEP 4 $z = 4.62$ STEP 5 Reject H_0
STEP 6 We conclude that the percentage of employed workers who have registered to vote exceeds the percentage of unemployed workers who have registered to vote.

10.59 STEP 1 $H_0: p_1 = p_2, H_a: p_1 < p_2$
STEP 2 $\alpha = 0.10$
STEP 3 Critical value $= -1.28$
STEP 4 $z = -0.46$ STEP 5 Do not reject H_0
STEP 6 The data do not provide sufficient evidence to conclude that the percentage of American dentists practicing in the Northeast is smaller than the percentage of American physicians practicing in the Northeast.

10.61 a) -0.140 to -0.002
b) We can be 99% confident that the difference, $p_1 - p_2$, between the proportions of American and

French households that own washing machines is somewhere between -0.140 and -0.002.

10.63 a) 0.102 to 0.212

b) We can be 90% confident that the difference, $p_1 - p_2$, between the proportions of employed and unemployed workers who have registered to vote is somewhere between 0.102 and 0.212.

10.65 a) -0.063 to 0.029

b) We can be 80% confident that the difference, $p_1 - p_2$, between the proportions of American dentists and physicians practicing in the Northeast is somewhere between -0.063 and 0.029.

Exercises **10.5**

10.69 The pooled-t procedure.

10.71 a) 12 b) 7

10.73 a) 1.79 b) 2.29

10.75 a) 2.88 b) 2.10

10.77 a) 2.25 b) 3.21

10.79 The common population standard deviation, σ.

10.81 a) $SSTR = 2.8$, $SSE = 10.8$, $SST = 13.6$
b) $13.6 = 2.8 + 10.8$
c) $MSTR = 1.4$, $MSE = 0.9$
d)

Source	df	SS	MS = SS/df	F-statistic
Treatment	2	2.8	1.4	1.56
Error	12	10.8	0.9	
Total	14	13.6		

10.83 a) $SSTR = 132{,}120.267$, $SSE = 95{,}612.933$, $SST = 227{,}733.200$
b) $227{,}733.200 = 132{,}120.267 + 95{,}612.933$
c) $MSTR = 44{,}040.089$, $MSE = 5{,}975.808$
d)

Source	df	SS	MS = SS/df	F-statistic
Treatment	3	132120.267	44040.089	7.37
Error	16	95612.933	5975.808	
Total	19	227733.200		

10.85

Source	df	SS	MS = SS/df	F-statistic
Treatment	2	43.304	21.652	3.08
Error	12	84.400	7.033	
Total	14	127.704		

10.87 STEP 1 $H_0: \mu_1 = \mu_2 = \mu_3 = \mu_4$
H_a: Not all the means are the same
STEP 2 $\alpha = 0.05$
STEP 3 Critical value $= 3.24$
STEP 4 $SST = 530.95$, $SSTR = 70.95$, $SSE = 460.00$
STEP 5

Source	df	SS	MS = SS/df	F-statistic
Treatment	3	70.95	23.65	0.82
Error	16	460.00	28.75	
Total	19	530.95		

STEP 6 $F = 0.82$; do not reject H_0
STEP 7 The data do not provide sufficient evidence to conclude that there is a difference in mean lifetimes among the four brands of batteries.

10.89 STEP 1 $H_0: \mu_1 = \mu_2 = \mu_3$
H_a: Not all the means are the same
STEP 2 $\alpha = 0.05$
STEP 3 Critical value $= 4.26$
STEP 4 $SST = 60.916$, $SSTR = 5.828$, $SSE = 55.088$
STEP 5

Source	df	SS	MS = SS/df	F-statistic
Treatment	2	5.828	2.914	0.48
Error	9	55.088	6.121	
Total	11	60.916		

STEP 6 $F = 0.48$; do not reject H_0
STEP 7 The data do not provide sufficient evidence to conclude that there is a difference in mean hourly earnings for nonsupervisory workers in the three industries.

10.91 STEP 1 $H_0: \mu_1 = \mu_2 = \mu_3 = \mu_4 = \mu_5$
H_a: Not all the means are the same
STEP 2 $\alpha = 0.01$
STEP 3 Critical value $= 3.65$
STEP 4 $SST = 1384.015$, $SSTR = 406.370$, $SSE = 977.645$
STEP 5

Source	df	SS	MS = SS/df	F-statistic
Treatment	4	406.370	101.593	6.23
Error	60	977.645	16.294	
Total	64	1384.015		

STEP 6 $F = 6.23$; reject H_0
STEP 7 There appears to be a difference in the mean times served by prisoners in the five offense groups.

REVIEW TEST

1 STEP 1 $H_0: \mu_1 = \mu_2$, $H_a: \mu_1 > \mu_2$
  STEP 2 $\alpha = 0.01$
  STEP 3 Critical value = 2.33
  STEP 4 $z = 7.65$ STEP 5 Reject H_0
  STEP 6 It appears that the mean resale price for homes in Phoenix, Arizona, exceeds that for homes in Flint, Michigan.

2 $23,586 to $44,258. We can be 98% confident that the difference, $\mu_1 - \mu_2$, between the mean resale prices of homes in Phoenix, Arizona, and Flint, Michigan, is somewhere between $23,586 and $44,258.

3 a) STEP 1 $H_0: \mu_1 = \mu_2$, $H_a: \mu_1 \neq \mu_2$
  STEP 2 $\alpha = 0.10$
  STEP 3 Critical values = ± 1.833
  STEP 4 Paired differences, $d = x_1 - x_2$:

82	-95	-49	0	-36
-152	49	-38	-43	-118

  STEP 5 $t = -1.766$
  STEP 6 Do not reject H_0
  STEP 7 The data do not provide sufficient evidence to conclude that there is a difference in mean results for the two speed reading programs.
  b) The population of all possible paired differences is normally distributed.

4 -81.5 to 1.5. We can be 90% confident that the difference, $\mu_1 - \mu_2$, between the mean reading speeds of people using Program 1 and people using Program 2 is somewhere between -81.5 and 1.5 words per minute.

5 STEP 1 $H_0: p_1 = p_2$, $H_a: p_1 < p_2$
  STEP 2 $\alpha = 0.05$
  STEP 3 Critical value = -1.645
  STEP 4 $z = -1.82$
  STEP 5 Reject H_0
  STEP 6 It appears that the percentage of unemployed male doctoral scientists is lower than the percentage of unemployed female doctoral scientists.

6 -0.094 to -0.003. We can be 90% confident that the difference, $p_1 - p_2$, between the proportions of unemployed male and female doctoral scientists is somewhere between -0.094 and -0.003.

7 STEP 1 $H_0: \mu_1 = \mu_2$, $H_a: \mu_1 \neq \mu_2$
  STEP 2 $\alpha = 0.05$
  STEP 3 Critical values = ± 1.96
  STEP 4 $t = 0.266$
  STEP 5 Do not reject H_0
  STEP 6 The data do not provide sufficient evidence to conclude that there is a difference in mean IQ between male and female students at the university.

8 -5.4 to 7.1. We can be 95% confident that the difference, $\mu_1 - \mu_2$, between the mean IQs of female and male students at the university is somewhere between -5.4 and 7.1.

9 STEP 1 $H_0: \mu_1 = \mu_2$, $H_a: \mu_1 < \mu_2$
  STEP 2 $\alpha = 0.05$
  STEP 3 Critical value = -1.725
  STEP 4 $t = -2.210$
  STEP 5 Reject H_0
  STEP 6 It appears that Germans consume less fish than Russians, on the average.

10 -8.5 to -1.0. We can be 90% confident that the difference, $\mu_1 - \mu_2$, between last year's mean fish consumption by Germans and Russians is somewhere between -8.5 and -1.0 kilograms.

11* a) 24 b) 5
  c) 9.47 d) 4.53

12* a) $\bar{x}_1 = 413.2$ $\bar{x}_2 = 395.0$ $\bar{x}_3 = 479.0$
    $s_1 = 62.86$ $s_2 = 75.96$ $s_3 = 72.68$
  b) $MSTR = 11583.6$, $MSE = 5076.2$
  c) The variation among the sample means.
  d) The variation within samples.

13* a) $SST = 2936.55$, $SSTR = 228.15$, $SSE = 2708.40$
  b)

Source	df	SS	MS = SS/df	F-statistic
Treatment	3	228.15	76.050	0.45
Error	16	2708.40	169.275	
Total	19	2936.55		

  c) STEP 1 $H_0: \mu_1 = \mu_2 = \mu_3 = \mu_4$
  H_a: Not all the means are the same
  STEP 2 $\alpha = 0.01$
  STEP 3 Critical value = 5.29
  STEPS 4 & 5 (see parts (a) and (b))
  STEP 6 $F = 0.45$; do not reject H_0
  STEP 7 The data do not provide sufficient evidence to conclude that there is a difference in mean sales among the four designs.

Chapter 11

Exercises 11.1

11.1 a) 32.852   b) 10.117

11.3 a) 18.307   b) 3.247

11.5 a) 1.646   b) 15.507

11.7 a) 0.831, 12.832   b) 9.591, 34.170

Exercises 11.2

11.9 Because the hypothesis test is carried out by determining how well the observed frequencies fit the expected frequencies.

11.11 STEP 1 H_0: The current distribution of the U.S. resident population by region is the same as the 1983 distribution.

H_a: The current distribution of the U.S. resident population by region is different from the 1983 distribution.

STEP 2   Expected frequencies:

Region	p	E
Northeast	0.212	106
Midwest	0.252	126
South	0.340	170
West	0.196	98

STEP 3   Assumptions (1) and (2) are satisfied since all expected frequencies are at least 5.
STEP 4 $\alpha = 0.05$
STEP 5   Critical value $= 7.815$
STEP 6 $\chi^2 = 1.827$
STEP 7   Do not reject H_0
STEP 8   The data do not provide sufficient evidence to conclude that the current distribution of the U.S. resident population by region is different from the 1983 distribution.

11.13 STEP 1 H_0: The distribution of the number of years of school completed by Tennessee residents is the same as the national distribution.

H_a: The distribution of the number of years of school completed by Tennessee residents is different from the national distribution.
STEP 2   Expected frequencies:

Years of school	p	E
8 or less	0.183	54.9
9–11	0.153	45.9
12	0.346	103.8
13–15	0.157	47.1
16 or more	0.161	48.3

STEP 3   Assumptions (1) and (2) are satisfied since all expected frequencies are at least 5.
STEP 4 $\alpha = 0.01$
STEP 5   Critical value $= 13.277$
STEP 6 $\chi^2 = 20.037$
STEP 7   Reject H_0
STEP 8   It appears that the distribution of the number of years of school completed by Tennessee residents is different from the national distribution.

11.15 STEP 1 H_0: The die is not loaded.
H_a: The die is loaded.
STEP 2   Expected frequencies:

Number	p	E
1	1/6	25
2	1/6	25
3	1/6	25
4	1/6	25
5	1/6	25
6	1/6	25

STEP 3   Assumptions (1) and (2) are satisfied since all expected frequencies are at least 5.
STEP 4 $\alpha = 0.05$
STEP 5   Critical value $= 11.070$
STEP 6 $\chi^2 = 2.480$
STEP 7   Do not reject H_0
STEP 8   The data do not provide sufficient evidence to conclude that the die is loaded.

11.17 STEP 1 H_0: The distribution of types of violent crime in New Jersey is the same as that for the nation as a whole.
H_a: The distribution of types of violent crime in New Jersey is different from that for the nation as a whole.

STEP 2   Expected frequencies:

Violent crime	p	E
Murder	0.016	9.6
Forcible rape	0.064	38.4
Robbery	0.403	241.8
Aggravated assault	0.517	310.2

STEP 3   Assumptions (1) and (2) are satisfied since all expected frequencies are at least 5.
STEP 4 $\alpha = 0.01$
STEP 5   Critical value = 11.345
STEP 6 $\chi^2 = 18.003$
STEP 7   Reject H_0
STEP 8   The data indicate that the distribution of types of violent crime in New Jersey is different from that for the nation as a whole.

Exercises 11.3

11.23   STEP 1 H_0: Annual income and educational level are independent.
H_a: Annual income and educational level are dependent.
STEP 2   Observed and expected frequencies:

Years of schooling

Annual income	0–8	9–12	Over 12	Total
Under $10,000	34 15.8	36 37.7	10 26.5	80
$10,000–$24,999	41 29.5	72 70.2	36 49.3	149
$25,000–$39,999	10 28.9	78 68.8	58 48.3	146
$40,000 and over	4 14.8	26 35.3	45 24.8	75
Total	89	212	149	450

STEP 3   Assumptions (1) and (2) are satisfied since all expected frequencies are at least 5.
STEP 4 $\alpha = 0.01$
STEP 5   Critical value = 16.812
STEP 6 $\chi^2 = 81.645$
STEP 7   Reject H_0
STEP 8   It appears that annual income and educational level are dependent.

11.25   STEP 1 H_0: Sex and occupation type for employed workers are independent.
H_a: Sex and occupation type for employed workers are dependent.
STEP 2   Observed and expected frequencies:

Sex

Occupation type	Male	Female	Total
Managerial/Professional	14 13.6	10 10.4	24
Technical sales Administrative	9 14.2	16 10.8	25
Service	4 5.7	6 4.3	10
Other	20 13.6	4 10.4	24
Total	47	36	83

STEP 3   Assumptions (1) and (2) are satisfied since all expected frequencies are at least 1, and only 12.5% (1/8) of the expected frequencies are less than 5.
STEP 4 $\alpha = 0.01$
STEP 5   Critical value = 11.345
STEP 6 $\chi^2 = 12.454$
STEP 7   Reject H_0
STEP 8   It appears that sex and occupation type for employed workers are dependent.

11.27   a) See the table below.
b)   STEP 1 H_0: The type of violent crime committed and the age of the person arrested are independent.
H_a: The type of violent crime committed and the age of the person arrested are dependent.
STEP 2 Observed and expected frequencies:

Age

Type of violent crime	18–24	25–44	45+	Total
Murder	11 13.3	16 15.2	4 2.5	31
Forcible rape	21 21.9	26 25.0	4 4.1	51
Robbery	128 97.0	92 110.9	6 18.1	226
Aggravated assault	162 189.8	234 216.9	46 35.4	442
Total	322	368	60	750

STEP 3   Assumptions (1) and (2) are satisfied since all expected frequencies are at least 1, and only 16.7% (2/12) of the expected frequencies are less than 5.
STEP 4 $\alpha = 0.01$
STEP 5   Critical value $= 16.812$
STEP 6 $\chi^2 = 31.241$
STEP 7   Reject H_0
STEP 8   It appears that the type of violent crime committed and the age of the person arrested are dependent.

11.29   STEP 1 H_0: Net worth and marital status for top wealthholders are independent.
H_a: Net worth and marital status for top wealthholders are dependent.
STEP 2   Observed and expected frequencies:

Marital status

Net worth	Married	Single/ Divorced	Widowed	Total
$100,000–$249,999	227 223.9	54 53.0	63 67.1	344
$250,000–$499,999	60 63.1	15 14.9	22 18.9	97
$500,000–$999,999	20 20.2	4 4.8	7 6.0	31
$1,000,000 or more	10 9.8	2 2.3	3 2.9	15
Total	317	75	95	487

STEP 3   Assumption (2) is violated since 25% (3/12) of the expected frequencies are less than 5.

Exercises 11.4

11.37   STEP 1 H_0: $\sigma = 100$, H_a: $\sigma \neq 100$
STEP 2 $\alpha = 0.05$
STEP 3   Critical values $= 12.401, 39.364$
STEP 4 $\chi^2 = 29.561$
STEP 5   Do not reject H_0
STEP 6   The data do not provide sufficient evidence to conclude that the standard deviation, σ, of this year's verbal scores is different from the 1941 standard deviation of 100.

11.39   STEP 1 H_0: $\sigma = 1$ second, H_a: $\sigma > 1$ second
STEP 2 $\alpha = 0.01$
STEP 3   Critical value $= 36.191$
STEP 4 $\chi^2 = 39.222$

STEP 5   Reject H_0
STEP 6   It appears that the standard deviation, σ, of the weekly errors exceeds the one-second claim made by the manufacturer.

11.41   STEP 1 H_0: $\sigma = 0.2$ fl oz, H_a: $\sigma < 0.2$ fl oz
STEP 2 $\alpha = 0.05$
STEP 3   Critical value $= 6.571$
STEP 4 $\chi^2 = 5.933$
STEP 5   Reject H_0
STEP 6   Evidently, the standard deviation, σ, of the amounts being dispensed is less than 0.2 fl oz.

11.43   86.7 to 154.4. We can be 95% confident that the standard deviation of this year's verbal SAT scores is somewhere between 86.7 and 154.4.

11.45   1.04 to 2.27 seconds. We can be 98% confident that the standard deviation, σ, of the weekly errors of the watches manufactured is somewhere between 1.04 and 2.27 seconds.

11.47   0.10 to 0.19 oz. We can be 90% confident that the standard deviation, σ, of the amounts of coffee being dispensed is somewhere between 0.10 and 0.19 oz.

11.49   So as to insure that there will not be a large variation in the amount of coffee dispensed.

REVIEW TEST

1   a) 6.408   b) 33.409
    c) 27.587   d) 8.672
    e) 7.564, 30.191

2   STEP 1 H_0: The actual preference distribution for the number of bedrooms is the same as the distribution by number of bedrooms for homes completed in 1983.
H_a: The actual preference distribution for the number of bedrooms is different from the distribution by number of bedrooms for homes completed in 1983.
STEP 2   Expected frequencies:

Number of bedrooms	p	E
2 or less	0.24	36.0
3	0.59	88.5
4 or more	0.17	25.5

STEP 3 Assumptions (1) and (2) are satisfied since all expected frequencies are at least 5.
STEP 4 $\alpha = 0.05$
STEP 5 Critical value $= 5.991$
STEP 6 $\chi^2 = 13.011$
STEP 7 Reject H_0
STEP 8 It appears that the actual preference distribution for the number of bedrooms is different from the distribution by number of bedrooms for homes completed in 1983.

3 STEP 1 H_0: Response and educational level are independent.
H_a: Response and educational level are dependent.
STEP 2 Observed and expected frequencies:

Response

Educational level	Favor	Oppose	No opinion	Total
College grad	264 / 229.1	17 / 38.5	6 / 19.3	287
Some college	205 / 190.0	26 / 31.9	7 / 16.0	238
High school grad	461 / 459.9	81 / 77.3	34 / 38.8	576
Non high school grad	290 / 340.9	81 / 57.3	56 / 28.8	427
Total	1220	205	103	1528

STEP 3 Assumptions (1) and (2) are satisfied since all expected frequencies are at least 5.
STEP 4 $\alpha = 0.01$
STEP 5 Critical value $= 16.812$
STEP 6 $\chi^2 = 77.837$
STEP 7 Reject H_0
STEP 8 Evidently, response and educational level are dependent.

4 a) STEP 1 H_0: $\sigma = 16$, H_a: $\sigma \neq 16$
STEP 2 $\alpha = 0.10$
STEP 3 Critical values $= 13.848, 36.415$
STEP 4 $\chi^2 = 21.110$
STEP 5 Do not reject H_0
STEP 6 The data do not provide sufficient evidence to conclude that the standard deviation of IQs is not equal to 16.
b) That IQs are normally distributed.

5 12.2 to 19.8. We can be 90% confident that the standard deviation of all IQ scores is somewhere between 12.2 and 19.8.

Chapter 12

Exercises 12.1

12.1 a) $y = 63 + 0.63x$ b) $b_0 = 63$, $b_1 = 0.63$

c)

Miles x	Cost ($) y
50	94.50
100	126.00
250	220.50

d)
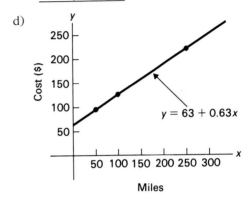

e) About $160; exact cost is $157.50.

12.3 a) $b_0 = 32$, $b_1 = 1.8$
b) $-40, 32, 68, 212$

c)
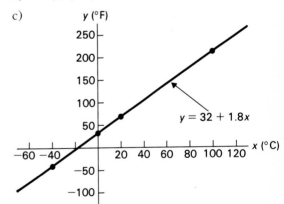

d) About 80°F; exact temperature is 82.4°F

12.5 a) $b_0 = 63$, $b_1 = 0.63$
b) The y-intercept, $b_0 = 63$, gives the y-value at which the straight line $y = 63 + 0.63x$ intersects the y-axis. The slope, $b_1 = 0.63$, indicates that the y-value increases by 0.63 for every increase in x of 1 unit.

c) The y-intercept, $b_0 = 63$, is the cost (in dollars) for driving the car 0 miles. The slope, $b_1 = 0.63$, represents that the cost per mile is $0.63—it is the amount the total cost goes up for each additional mile driven.

12.7 a) $b_0 = 32$, $b_1 = 1.8$

b) The y-intercept, $b_0 = 32$, gives the y-value at which the straight line $y = 32 + 1.8x$ intersects the y-axis. The slope, $b_1 = 1.8$, indicates that the y-value increases by 1.8 for every increase in x of 1 unit.

c) The y-intercept, $b_0 = 32$, gives the Fahrenheit temperature corresponding to $0°C$. The slope, $b_1 = 1.8$, represents that the Fahrenheit temperature increases by $1.8°$ for an increase of the Celsius temperature of $1°$.

12.9 a) $b_0 = 3$, $b_1 = 4$ b) slopes upward

c)

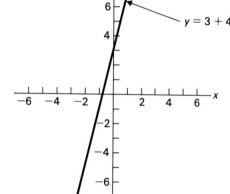

12.11 a) $b_0 = 6$, $b_1 = -7$ b) slopes downward

c)

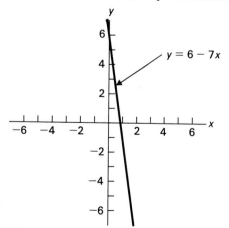

12.13 a) $b_0 = -2$, $b_1 = 0.5$ b) slopes upward

c)

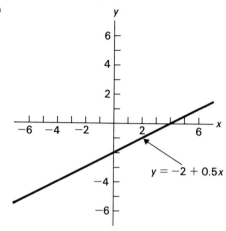

12.15 a) $b_0 = 2$, $b_1 = 0$ b) horizontal

c)

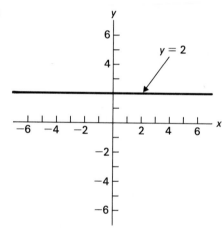

12.17 a) $b_0 = 0$, $b_1 = 1.5$ b) slopes upward

c)

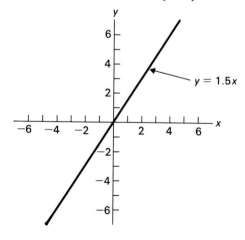

12.19   a) slopes upward   b) $y = 5 + 2x$

c)

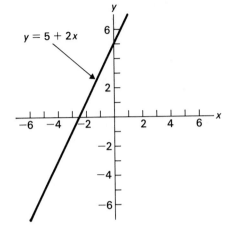

12.25   a) horizontal   b) $y = 3$

c)

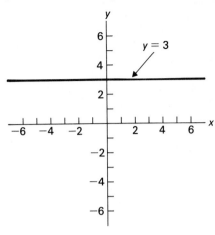

12.21   a) slopes downward   b) $y = -2 - 3x$

c)

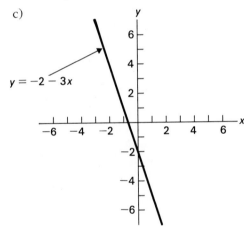

Exercises 12.2

12.29   a)   Line A: $y = 3 - 0.6x$

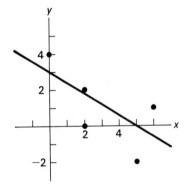

12.23   a) slopes downward   b) $y = -0.5x$

c)

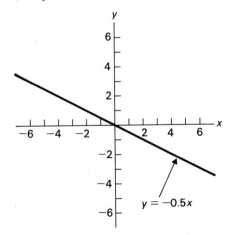

Line B: $y = 4 - x$

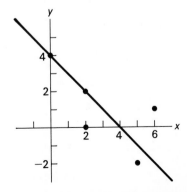

b) Line A: $y = 3 - 0.6x$

x	y	$\hat{y}$	e	e^2
0	4	3.0	1.0	1.00
2	2	1.8	0.2	0.04
2	0	1.8	-1.8	3.24
5	-2	0.0	-2.0	4.00
6	1	-0.6	1.6	2.56
Σ				10.84

Line B: $y = 4 - x$

x	y	$\hat{y}$	e	e^2
0	4	4	0	0
2	2	2	0	0
2	0	2	-2	4
5	-2	-1	-1	1
6	1	-2	3	9
Σ				14

c) Line A

12.31   a) $\hat{y} = 2.875 - 0.625x$

b)

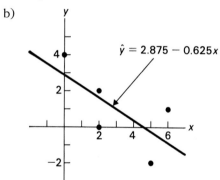

12.33   a) $\hat{y} = -174.49 + 4.84x$

b)

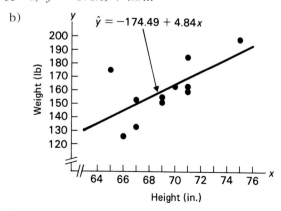

c) Weight tends to increase as height increases.
d) The weights of 18–24-year-old males increase an estimated 4.84 lb for each increase in height of one inch.
e) 149.54 lb; 178.56 lb

12.35   a) $\hat{y} = 219.78 - 1.09x$

b)

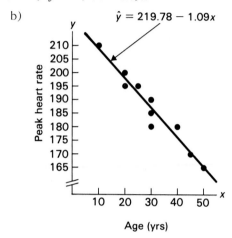

c) Peak heart rate tends to decrease as age increases.
d) The peak heart rate an individual can reach during intensive exercise decreases by an estimated 1.09 for each increase in age of one year.
e) 195.74

12.37   a) $\hat{y} = 12.86 + 1.21x$

b)

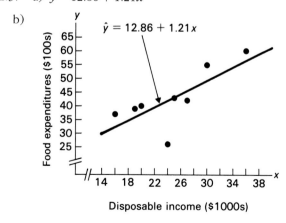

c) Annual food expenditures tend to increase as disposable income increases.

d) Annual food expenditures increase an estimated $121 for each increase in disposable income of $1000.
e) $4321

12.39 Only the second one.

Exercises 12.3

12.47 a) The coefficient of determination, r^2.
b) It represents the percentage reduction obtained in the total squared error by using the regression equation to predict the observed y-values, instead of simply the mean $\bar{y}$. It also represents the percentage of variation in the observed y-values that is explained by the regression.

12.49 a) $SST = 20$, $SSR = 9.375$, $SSE = 10.625$
b) $20 = 9.375 + 10.625$ c) $r^2 = 0.469$
d) 46.9%   e) 46.9%   f) moderately useful

12.51 a) $SST = 4352.91$, $SSR = 1909.46$,
$SSE = 2443.45$
b) $4352.91 = 1909.46 + 2443.45$ c) $r^2 = 0.439$
d) 43.9%   e) 43.9%   f) moderately useful

12.53 a) $SST = 1710$, $SSR = 1611.57$, $SSE = 98.43$
b) $1710 = 1611.57 + 98.43$ c) $r^2 = 0.942$
d) 94.2%   e) 94.2%   f) extremely useful

12.55 a) $SST = 783.50$, $SSR = 429.95$, $SSE = 353.55$
b) $783.50 = 429.95 + 353.55$ c) $r^2 = 0.549$
d) 54.9%   e) 54.9%   f) moderately useful

Exercises 12.4

12.61 a) $r = -0.685$
b) Suggests a moderately strong negative linear correlation.
c) Data points are clustered moderately closely about the regression line.
d) $r^2 = 0.469$

12.63 a) $r = 0.662$
b) Suggests a moderately strong positive linear correlation.
c) Data points are clustered moderately closely about the regression line.
d) $r^2 = 0.439$

12.65 a) $r = -0.971$
b) Suggests an extremely strong negative linear correlation.

c) Data points are clustered extremely closely about the regression line.
d) $r^2 = 0.942$

12.67 a) $r = 0.741$
b) Suggests a moderately strong positive linear correlation.
c) Data points are clustered moderately closely about the regression line.
d) $r^2 = 0.549$

12.69 a) $r = 0$
b) No, only that there is no *linear* relationship between the variables.
c)

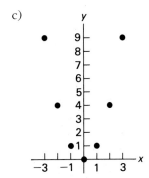

d) No, because the data points are not scattered about a straight line.
e) For each data point (x, y), we have $y = x^2$.

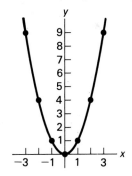

REVIEW TEST

1 a) $y = 72 - 12x$ b) $b_0 = 72$, $b_1 = -12$
c) The line slopes downward since $b_1 < 0$.
d)

Age (yrs)	Value ($100s)
x	y
2	48
5	12

e)

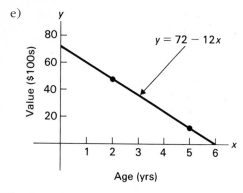

$y = 72 - 12x$

f) About $2500; exact value is $2400.

2 a)

Cost ($) vs **Number graded (hund.)**

b) Yes, because the data points appear to be scattered about a straight line.

c) $\hat{y} = 35.80 + 12.08x$

Cost ($) vs **Number graded (hund.)**

$\hat{y} = 35.80 + 12.08x$

d) Cost of grading tends to increase as the number of papers graded increases.

e) Paper-grading costs increase an estimated $12.08 for each additional 100 papers graded.
f) $229.13

3   a) $SST = 6984.25$, $SSR = 6558.31$, $SSE = 425.94$
   b) $6984.25 = 6558.31 + 425.94$ c) $r^2 = 0.939$
   d) 93.9%   e) 93.9%   f) Extremely useful.

4   a) $r = 0.969$
   b) Suggests an extremely strong positive linear correlation.
   c) Data points are clustered extremely closely about the regression line.
   d) $r^2 = (0.969)^2 = 0.939$

Chapter 13

Exercises 13.1

13.1   (1) For each x-value, the *mean* of the corresponding population of y-values lies on a straight line, $y = \beta_0 + \beta_1 x$. (2) The population standard deviation, σ, of the population of y-values corresponding to a given x-value is the *same*, regardless of the x-value. (3) For each x-value, the corresponding population of y-values is *normally distributed*.

13.3   It would mean there are constants β_0, β_1, and σ such that for each height x, the weights of 18–24-year-old males of that height are normally distributed with mean $\beta_0 + \beta_1 x$ and standard deviation σ.

13.5   It would mean there are constants β_0, β_1, and σ such that for each age x, the peak heart rates that can be reached during intensive exercise by persons of that age are normally distributed with mean $\beta_0 + \beta_1 x$ and standard deviation σ.

13.7   It would mean there are constants β_0, β_1, and σ such that for each disposable-income level x, the annual food expenditures made by middle income families (father, mother, two children) at that level are normally distributed with mean $\beta_0 + \beta_1 x$ and standard deviation σ.

13.9   a) $s_e = 16.48$ lb
   b) Presuming that the variables height, x, and weight, y, for 18–24-year-old males satisfy Assumptions (1) – (3), $s_e = 16.48$ lb is our estimate for the common population standard deviation, σ, for the weights of all 18–24-year-old males of any given height.

13.11 a) $s_e = 3.51$

b) Presuming that the variables age, x, and peak heart rate, y, satisfy Assumptions $(1) - (3)$, $s_e = 3.51$ is our estimate for the common population standard deviation, σ, of peak heart rates for all individuals of any given age.

13.13 a) $s_e = 7.68$

b) Presuming that the variables family disposable income, x, and annual food expenditure, y, satisfy Assumptions $(1) - (3)$, $s_e = 7.68$ ($768) is our estimate for the common population standard deviation, σ, of all annual food expenditures for any given family disposable income.

13.15 a) $s_e = 1.88$ b) $s_e = 1.88$

Exercises 13.2

13.17 The Assumptions $(1) - (3)$ for regression inferences.

13.19 a) STEP 1 $H_0: \beta_1 = 0, H_a: \beta_1 \neq 0$
STEP 2 $\alpha = 0.10$
STEP 3   Critical values $= \pm 1.833$
STEP 4 $t = 2.652$
STEP 5   Reject H_0
STEP 6   It appears that the slope of the population regression line is not zero and, consequently, that the height, x, of an 18–24-year-old male is useful as a predictor of weight, y.

b) 1.49 to 8.18. We can be 90% confident that, for 18–24-year-old males, the increase in mean weight per one inch increase in height is somewhere between 1.49 and 8.18 lb.

13.21 a) STEP 1 $H_0: \beta_1 = 0, H_a: \beta_1 \neq 0$
STEP 2 $\alpha = 0.05$
STEP 3   Critical values $= \pm 2.306$
STEP 4 $t = -11.445$
STEP 5   Reject H_0
STEP 6   It appears that the slope of the population regression line is not zero and, consequently, that age, x, is useful as a predictor of peak heart rate, y.

b) -1.31 to -0.87. We can be 95% confident that the drop in mean peak heart rate per one year increase in age is somewhere between 0.87 and 1.31.

13.23 a) STEP 1 $H_0: \beta_1 = 0, H_a: \beta_1 \neq 0$
STEP 2 $\alpha = 0.01$
STEP 3   Critical values $= \pm 3.707$
STEP 4 $t = 2.701$
STEP 5   Do not reject H_0
STEP 6   The data do *not* provide sufficient evidence to conclude that the slope of the population regression line is not zero (that is, that family disposable income is useful as a predictor of annual food expenditures).

b) -0.45 to 2.88. We can be 99% confident that the change in mean annual food expenditure per $1000 increase in family disposable income is somewhere between $-$45 and $288.

13.25 a) 154.54 to 173.56. We can be 90% confident that the mean weight of 18–24-year-old males who are 70 inches tall is somewhere between 154.54 and 173.56 lb.

b) 132.38 to 195.71. We can be 90% certain that the weight of a randomly selected 18–24-year-old male who is 70 inches tall will be somewhere between 132.38 and 195.71 lb.

c) See Figure A.26.

FIGURE A.26

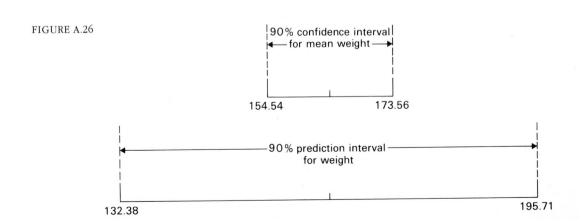

d) The error in the estimate of the mean weight of 18–24-year-old males who are 70 inches tall is due only to the fact that the population regression line is being estimated by a sample regression line; whereas, the error in the prediction of the weight of a randomly selected 18–24-year-old male who is 70 inches tall is due to that fact plus the variation in weights of such males.

13.27 a) 192.64 to 198.85. We can be 95% confident that the mean peak heart rate of 22-year-olds is somewhere between 192.64 and 198.85.
b) 187.08 to 204.40. We can be 95% certain that the peak heart rate of a randomly selected 22-year-old will be somewhere between 187.08 and 204.40.

13.29 a) 33.13 to 53.29. We can be 99% confident that the mean annual food expenditure of families with a disposable income of $25,000 is somewhere between $3313 and $5329.
b) 13.02 to 73.39. We can be 99% certain that the annual food expenditure of a randomly selected family with a disposable income of $25,000 will be somewhere between $1302 and $7339.

Exercises 13.3

13.31 STEP 1 $H_0: \rho = 0$, $H_a: \rho > 0$
STEP 2 $\alpha = 0.05$
STEP 3   Critical value $= 0.521$
STEP 4 $r = 0.662$ STEP 5   Reject H_0
STEP 6   It appears that the variables height and weight for 18–24-year-old males are positively linearly correlated.

13.33 STEP 1 $H_0: \rho = 0$, $H_a: \rho < 0$
STEP 2 $\alpha = 0.025$
STEP 3   Critical value $= -0.632$
STEP 4 $r = -0.971$ STEP 5   Reject H_0
STEP 6   Evidently, the variables age and peak heart rate are negatively linearly correlated.

13.35 STEP 1 $H_0: \rho = 0$, $H_a: \rho \neq 0$
STEP 2 $\alpha = 0.01$
STEP 3   Critical values $= \pm 0.834$
STEP 4 $r = 0.741$ STEP 5   Do not reject H_0
STEP 6   The data do not provide sufficient evidence to conclude that there is a linear correlation between family disposable income and annual food expenditures.

REVIEW TEST

1   It would mean there are constants β_0, β_1, and σ such that for each number of bedrooms, x, the monthly rents for homes with that number of bedrooms are normally distributed with mean $\beta_0 + \beta_1 x$ and standard deviation σ.

2   a) $\hat{y} = 242.29 + 104.90x$
b) $s_e = \$91.62$
c) Presuming the variables number of bedrooms, x, and monthly rent, y, satisfy Assumptions (1) – (3) for regression inferences, $s_e = \$91.62$ is our estimate for the common population standard deviation, σ, of the monthly rents of homes with any given number of bedrooms.

3   a) STEP 1 $H_0: \beta_1 = 0$, $H_a: \beta_1 \neq 0$
STEP 2 $\alpha = 0.05$
STEP 3   Critical values $= \pm 2.306$
STEP 4 $t = 3.547$ STEP 5   Reject H_0
STEP 6   It appears that the slope of the population regression line is not zero and, consequently, that the number of bedrooms, x, is useful as a predictor of monthly rent, y.
b) 36.70 to 173.09. We can be 95% confident that the increase in mean monthly rent per increase in the number of bedrooms by one is somewhere between $36.70 and $173.09.
c) 488.79 to 625.17. We can be 95% confident that the mean monthly rent of three-bedroom homes is somewhere between $488.79 and $625.17.
d) 334.96 to 779.00. We can be 95% certain that the monthly rent of a randomly selected three-bedroom home will be somewhere between $334.96 and $779.00.
e) The error in the estimate of the mean monthly rent of three-bedroom homes is due only to the fact that the population regression line is being estimated by a sample regression line, whereas the error in the prediction of the monthly rent of a randomly selected three-bedroom home is due to that fact plus the variation in monthly rents for three-bedroom homes.

4   STEP 1 $H_0: \rho = 0$, $H_a: \rho > 0$
STEP 2 $\alpha = 0.025$
STEP 3   Critical value $= 0.632$
STEP 4 $r = 0.782$ STEP 5   Reject H_0
STEP 6   Evidently, the variables "number of bedrooms" and "monthly rent" are positively linearly correlated.

· Index ·

TABLE VI   Chi-square distribution (Values of χ^2_α)

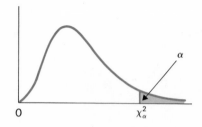

df	$\chi^2_{0.995}$	$\chi^2_{0.99}$	$\chi^2_{0.975}$	$\chi^2_{0.95}$	$\chi^2_{0.05}$	$\chi^2_{0.025}$	$\chi^2_{0.01}$	$\chi^2_{0.005}$	df
1	0.000	0.000	0.001	0.004	3.841	5.024	6.635	7.879	1
2	0.010	0.020	0.051	0.103	5.991	7.378	9.210	10.597	2
3	0.072	0.115	0.216	0.352	7.815	9.348	11.345	12.838	3
4	0.207	0.297	0.484	0.711	9.488	11.143	13.277	14.860	4
5	0.412	0.554	0.831	1.145	11.070	12.832	15.086	16.750	5
6	0.676	0.872	1.237	1.635	12.592	14.449	16.812	18.548	6
7	0.989	1.239	1.690	2.167	14.067	16.013	18.475	20.278	7
8	1.344	1.646	2.180	2.733	15.507	17.535	20.090	21.955	8
9	1.735	2.088	2.700	3.325	16.919	19.023	21.666	23.589	9
10	2.156	2.558	3.247	3.940	18.307	20.483	23.209	25.188	10
11	2.603	3.053	3.816	4.575	19.675	21.920	24.725	26.757	11
12	3.074	3.571	4.404	5.226	21.026	23.337	26.217	28.300	12
13	3.565	4.107	5.009	5.892	22.362	24.736	27.688	29.819	13
14	4.075	4.660	5.629	6.571	23.685	26.119	29.141	31.319	14
15	4.601	5.229	6.262	7.261	24.996	27.488	30.578	32.801	15
16	5.142	5.812	6.908	7.962	26.296	28.845	32.000	34.267	16
17	5.697	6.408	7.564	8.672	27.587	30.191	33.409	35.718	17
18	6.265	7.015	8.231	9.390	28.869	31.526	34.805	37.156	18
19	6.844	7.633	8.907	10.117	30.144	32.852	36.191	38.582	19
20	7.434	8.260	9.591	10.851	31.410	34.170	37.566	39.997	20
21	8.034	8.897	10.283	11.591	32.671	35.479	38.932	41.401	21
22	8.643	9.542	10.982	12.338	33.924	36.781	40.289	42.796	22
23	9.260	10.196	11.689	13.091	35.172	38.076	41.638	44.181	23
24	9.886	10.856	12.401	13.848	36.415	39.364	42.980	45.558	24
25	10.520	11.524	13.120	14.611	37.652	40.646	44.314	46.928	25
26	11.160	12.198	13.844	15.379	38.885	41.923	45.642	48.290	26
27	11.808	12.879	14.573	16.151	40.113	43.194	46.963	49.645	27
28	12.461	13.565	15.308	16.928	41.337	44.461	48.278	50.993	28
29	13.121	14.256	16.047	17.708	42.557	45.722	49.588	52.336	29
30	13.787	14.953	16.791	18.493	43.773	46.979	50.892	53.672	30